# Fisheries and Aquatic Resources of India

# Fisheries and Aquatic Resources of India

*Dr. A. D. Dholakia*
M. Sc. LL.B. (Sp), Ph.D.
Associate Professor,
College of Fisheries, GAU, Veraval, Gujarat

2013
**DAYA PUBLISHING HOUSE**
*A Division of*
**Astral International Pvt. Ltd.**
New Delhi 110 002

First Printed: 2004
Reprint: 2013

ISBN 978-81-7035-339-3



*Published by* : **Daya Publishing House**
A Division of
**Astral International Pvt. Ltd.**
81, Darya Ganj, Near Hindi Park,
Delhi Medical Association Road,
New Delhi - 110 002
Phone: 011-4354 9797, 2327 8134
Fax: +91-11-2324 3060
E-mail: info@astralint.com
Website: www.astralint.com

*Laser Typesetting* : **Classic Computer Services**
Delhi - 110 035

*Printed at* : **Salasar Imaging Systems**
Delhi - 110 035

PRINTED IN INDIA

# ACKNOWLEDGEMENT

I have taken help for numerous sources for the preparation of this book. I am highly thankful to them.

I wish to express my indebtedness to Prof. N.D. Chhaya, Ex. Deputy Commissioner of Fisheries, Government of Gujarat, Ex. Principal, Dean and OSD of College of Fisheries, Faculty of Fisheries, Gujarat Agricultural University, and Ex. M.D. G.F.C.C.A. Ltd. for verifying my manuscripts and for giving useful suggestion, for improving standard of this book and for writing Foreword of this book.

I am highly thankful of Late Mr. R.L. Meena IAS, Commissioner of Fisheries, Government of Gujarat, Gandhinagar for giving permission to use all information published by Department of Fisheries, Government of Gujarat. The Director of C.M.F.R.I., Cochin has also given permission to use their published data. I cannot forget Dr. Sailesh Nayak, Head, Marine Water Resource Deptt., ISRO, Ahmedabad for the help by giving published satellite photographs to be used in this book. I wish to express my gratitude to Madam Nalini Nalak of International Forum for Fish Workers, Trivendram for keen interest and help.

I wish to express my gratitude to Dr. A.Y. Desai, I/C Principal, College of Fisheries, GAU, Veraval and Dean, Faculty of Fisheries, Gujarat Agricultural University and for writing Recommendation for this book.

Loving affection blended with emotions and charm, my son Apar and daughter Ushma has constantly encouraged me to write this book. I received utmost encouragement and appreciation from my wife Sangita. But for the help received from her to keep myself free from any domestic chore whatsoever, the writing of the book would not have been impossible. I wish to record my gratitude to my wife Sangita to whom this book is lovingly dedicated.

**Dr. A.D. Dholakia**

# FOREWORD

Gujarat is blessed with a long coastline, a very broad continental shelf, two Gulfs, considerable estuarine region, together with adequate fresh water resources. As a result, the state has immense potential of Aqua-Bio-Resources.

However, very little authentic reference material is available to students, teachers and research workers on Fisheries and other Aqua-Bio-Resources of the state and country.

Prof A.D. Dholakia has done a yoeman's service by writing a definitive book on Fisheries and other Aqua-Bio-Resources of India in general and Gujarat in particular.

The book deals with many aspects of this vast sector. He has discussed the Fisheries of marine, brackish water and Fresh Water and their different facets. He has also dealt with correlated subjects like craft and gears, fishing grounds, fish diseases, harbours and infrastructure facilities as well as use of modern techniques like remote sensing in use of land development of fisheries and other Aqua-Bio-Resources.

While going through the book, one can realize the amount of painstaking research carried out and tremendous effort put in by the author in creating this book. It can ideally be used as a textbook and mandatory reference source. I have no hesitation in recommending this book to students, teachers and research workers at all levels.

January 30, 2003
Baroda

**N.D. Chhaya**
*Former Dy. Commissioner of Fisheries (GOG)*
*Dean, Faculty of Fisheries (GAU)*
*M.D., G.F.C.C.A. Ltd.*

# PREFACE

My involvement with the Fisheries Department, teaching at various levels and research for over 28 years provided considerable background to push me to give consolidate and proper information. Much information is available on Fisheries and Resources of India. However, they are scattered. Very less information is available and published regarding Fisheries and aquatic resources of Gujarat. As I have done research work and published about 67 papers/articles, I had to refer many references and books. This has provided good collection.

With the help of 233 figures, 65 colour photographs, 128 tables and 373 references, I have tried to explain the topic Fisheries and Aquatic Resources of India with special reference to Gujarat.

Government of Gujarat has established College of Fisheries at Veraval. The details of this book will be able to provide sufficient guidance to the students. It is my belief that University teacher, the researcher, the administrator and the fishery entrepreneur who wish to know about fishery and aquatic resources, this book will provide good reference to them. I have tried to cover many aspects. However, suggestions are always invited for further improvement of the book. I am highly thankful to the publisher for taking interest and extending all co-operation in publishing this book.

**Dr. A.D. Dholakia**

# RECOMMENDATION

Since past 20 years of teaching the subject of Zoology and Fisheries Science I found it very difficult to collect and prepare notes, which really helps the students of Fisheries Science. As we know that the course curricula for Bachelor degree covers so many aspects related to fisheries science like taxonomy, biology, economics, management, instruments used for capture etc. There has been a definite gap in provision of a book, which provides each of these within itself. A suitable manual is needed for these. The author has tried to fill this void in this book with a view that a large number of Indian students and teachers of Fisheries Science will be stimulated.

The book can be used as a text book as well as will provide a good source of reference related to fisheries science and its allied branches.

I congratulate Dr. A.D. Dholakia, College of Fisheries, Gujarat Agricultural University, Veraval (Gujarat) for preparing such a concise edition of resoruce material for fisheries science students and faculties.

**Dr. A.Y. Desai**
*Dean/Principal*
*College of Fisheries*
*Gujarat Agricultural University*
*Veraval (Gujarat)*

# CONTENTS

# 1
# Bombay Duck Fishery

## Introduction

The Bombay Duck, popularly known as "Bombil" is supported by a single species, *Harpodon nehereus* (Ham). It has a wide but discontinuous distribution along the east coast of India. It forms the most important commercial fishery of Gujarat and the peak catch is obtained in the Southern Coast of Saurashtra between Nawabunder and Jafrabad.

Bombay duck fishery forms the important fishery of Gujarat. It forms Av. 16.89% of total marine fish landing of Gujarat.

Bombay duck is the only fish available in Gujarat under family Harpodontiedae. Bombay duck is the predominant fish forming 25.65% in the 'dol' net catch.

## Name in other languages

| | |
|---|---|
| English | Bombay duck |
| Gujarati | Bumla |
| Marathi | Bambil |
| Tamil | Vengarawasi |
| Telugu | Kukahsvahari |
| Orissa | Bombil |
| Bengali | Bomloh, Lutia, Nehri. |

## Classification

| | |
|---|---|
| Order | *Myctoformus* |
| Family | *Harpodon tiedae* |
| Genus | *Harpodon* |
| Species | *Nehereus,* Hamilton Buchanan. |

## Bombay Duck

B. xiii-xxvi, D. 12-13/0, P. 11-12, V. 9, A. 13-15, C. 19, Vert. 38, Caec. Pyl 16(22).

| | |
|---|---|
| Head | Length of head 5.25 to 6, of caudal 6, height of body 6.5 to 7.5 in the total length. |
| Eyes | Diameter 1/16 of the length of the head, 2 diameters from the end of snout and 4 apart. Body compressed, snout short and obtuse. Lower jaw longer, the cleft of the mouth almost reaches the propercular angle. Opercular pieces diaphanous and apparently more or less blended together. |

| | |
|---|---|
| Teeth | Re-curved and erectile. Those of the jaws and some of the palatines have a single barb at the posterior margin of the point, a few of the former are arrow-shaped, in both jaws appear three series of some what distant teeth: the external consists of excessively minute ones, the second of longer, and the international series the longest. Those of the lower jaw are longer than those of the upper, particularly three or four on each side of the symphasis. The pharyngeals and palatals contain two rows. The forth-anterior tooth of each palatal is very long. Fine teeth on the tongue, hyoid bone and the upper margin of the brachial arches. |
| Fins | The first dorsal is situated midway between the snout and the root of the caudal : the ventral opposite the dorsal, midway between the snout and the last ray of the anal : the first anal ray midway between the last ray of the dorsal and the root of the caudal : the adipose dorsal opposite the posterior third of the anal. The point of the ventral reaches to the middle of the anal; a long thin elongated scale is situated as its base. The adipose dorsal is rather large. The caudal divided into three pointed lobes, of which the center is rather the shortest. The pectorals are falciform and in the young reach to the last dorsal ray, but become shorter with age. |
| Scale | Commence opposite the origin of the dorsal fin, all are diaphanous, but difficult to be distinguished in fresh fish. |
| Lateral line | Rather nearer to the back than to the abdomen : it is slightly keeled with about forty fine rectangular membranous pieces, covered with scales like the rest of the body. Intestines the length of the abdominal cavity being destitute of any convolution. |
| Colours | Head, back, and sides semi-transparent like gelatin, light grayish, with minute star like black or brownish dots : anterior part of the abdomen pale, silvery-bluish : rest grayish-white : cheeks and opercles pale silvery bluish, dotted like the body : fins transparent, coloured like the body, but more colsely dotted, so as to appear grayish and black at their extremities, in some specimens the fins are black. Irish golden. |

## Salient Feature for on Field Identification

Light grayish body, elongated and compressed. Eye small with short snout. Mouth wide and armed with slender, recurred and depreciable teeth of unequal size, Dorsal fin followed by an adipose fin. Pelvic fins very long lateral line extended up to the median lobe of caudal fin Max. size 40 cm. (Indian Fishery Atlas MPEDA 1980).

**Fig. 1.1:** ***Harpodon nehereus***

## *Food and Feeding*

Bombay duck is a carnivorous and to some extent cannibalistic in its food habitats. The food consisted of mainly prawns and fishes. Among Crustacea, *Acetes indicus* and *Palaemen tenuipes* and among fish *Coilia dussumieri*, Polynamus, *heptadore typus* formed important constituents of diet.

## *Sex and Fecundity*

Generally when Bombay duck completes two years of its life maturity was observed. i.e. when they reaches about 200 to 210 mm in length. According to Bapat (1970) when *H. nehereus* ranging in size between 229 and 318 mm it was found that number of ova has been found to vary from 14,600 to 1,46,400. A relationship appears to exits between weight of the ovary and number of mature ova. Some worker has observed that the larger females maturing on the second or subsequent occasions produce more ova.

The fish is a continuous breeder with two-peak season in May-July and November-December (Kewalramani, 1972). While according to Chhaya (1990), it is September to January and February to May. The second peak shifts away from Jafrabad and Nawabunder.

## *Maturity*

The maturity of fish is determined between 230 to 240 mm, size. Studies have shown that the fish caught in the Gulf of Cambay have a higher mean length than that of Nawabander and a majority of them have gonads in ripe state. Spent stages were dominant in January, March and May.

## *Growth*

The growth trend of the new recruits at Nawabunder shows an increase of 75 mm in seven months. The mode at 45-60 mm during February shifted to 120-135 mm in September. Similarly at Rajpara the mode at 60-75 mm of March moves to 150-165 mm in October recording a growth of 90 mm in 7 months.

## *Fishing Grounds*

At Jafrabad fishing grounds are hardly at a distance of 12 to 18 km from shore on south coast of Gujarat. The depth of water ranges between 10 to 15 fathoms and sea bottom is soft and muddy.

On south Gujarat coast, fishing is carried out in an area of 1166 sq km and fishing extends to a distant of about 30 km from coast.

## *Fishing Season*

Bombay duck fishery forms quantitatively the most important fishery of Gujarat and the peak catch is obtained in the southern coast of Saurashtra between Nawabander and Jafrabad. Bombay duck is primarily a shoaling fish of open seas. It comes inshore in plentiful in autumn, with fluctuating supplies available during autum and winter.

According to Deshmukh and Kurian (1980) Bombay duck of Gujarat of 1+year and 2+ year of Maharashtra are from different stock.

Along the Saurashtra coast at Jafrabad the fishery is for the short duration from September to January. The recruitment of Juveniles having model size of 45-60 mm in to the fishery was noted in October but in December there was indication of another recruitment into the fishery. The older groups were represented in the fishery during October-November (120-135 mm) and December (225-240 mm).

At Rajpara during October and March juveniles in the size range of 60-75 mm was observed. The older fish 255-270 mm groups enters the fishery in December.

At Nawabander, there was indication of bulk recruitment of young ones of 45-60 mm twice into the fishery during February-March and September. (Deshmukh and Kurian 1980). Landing in Gujarat and its share in Gujarat and India since 1965-66 is given in Table 1.1 while landing pattern of Bombay duck in Gujarat is given in Fig. 1.2.

**Table 1.1: Bombay duck landing in Gujarat and its share in Gujarat and India**

| Year | Total marine Fish landing | Landing of Bombay duck | Per cent of Total Guj. | Marine fish Landing of India | Per cent Share of Guj. in India |
|---|---|---|---|---|---|
| 1965-66 | 109907 | 46752 | 42.54 | 73900 | 63.26 |
| 1970-71 | 151190 | 48411 | 32.02 | 78443 | 61.71 |
| 1975-76 | 208300 | 33402 | 16.03 | 99614 | 33.53 |
| 1979-80 | 206749 | 46371 | 22.43 | 96044 | 48.28 |
| 1980-81 | 218872 | 36270 | 16.57 | 95505 | 37.98 |
| 1981-82 | 220607 | 47025 | 21.32 | 73779 | 63.74 |
| 1982-83 | 192669 | 33259 | 17.26 | 86476 | 38.46 |
| 1983-84 | 223291 | 35120 | 15.73 | 100950 | 34.79 |
| 1984-85 | 290708 | 48581 | 16.71 | 117742 | 41.26 |
| 1985-86 | 306577 | 29031 | 9.47 | 112454 | 25.82 |
| 1986-87 | 315942 | 26733 | 8.46 | 92216 | 28.99 |
| 1987-88 | 327560 | 39804 | 12.15 | 73695 | 54.01 |
| 1988-89 | 414075 | 45243 | 10.92 | 67392 | 67.13 |
| 1989-90 | 432384 | 67841 | 15.69 | 120193 | 56.44 |
| 1990-91 | 500462 | 72712 | 14.53 | 131095 | 55.47 |
| 1991-92 | 530017 | 68387 | 12.9 | 136442 | 50.12 |
| 1992-93 | 609103 | 104039 | 17.08 | 122280 | 85.08 |
| 1993-94 | 619836 | 84093 | 13.57 | 97727 | 86.05 |
| 1994-95 | 645261 | 70890 | 10.99 | 109307 | 64.85 |
| 1995-96 | 598351 | 73194 | 12.23 | 88104 | 83.07 |
| 1996-97 | 660058 | 106055 | 16.07 | 112832 | 93.99 |
| 1997-98 | 702355 | 114707 | 16.33 | 122803 | 93.4 |
| 1998-99 | 551660 | 90493 | 16.4 | 91853 | 98.52 |
| 1999-00 | 670951 | 88166 | 13.14 | 97548 | 90.38 |

*Source* : (1) Commissionerate of Fisheries, Govt. of Gujarat, Gandhinagar (With permission).
(2) CMFRI Published Annual Reports.

## *Length Distribution*

In Gujarat at Nawabander the range observed was 15 to 375 mm. whereas at the other two centers viz. Rajpara and Jafrabad it was from 30 to 360 mm.

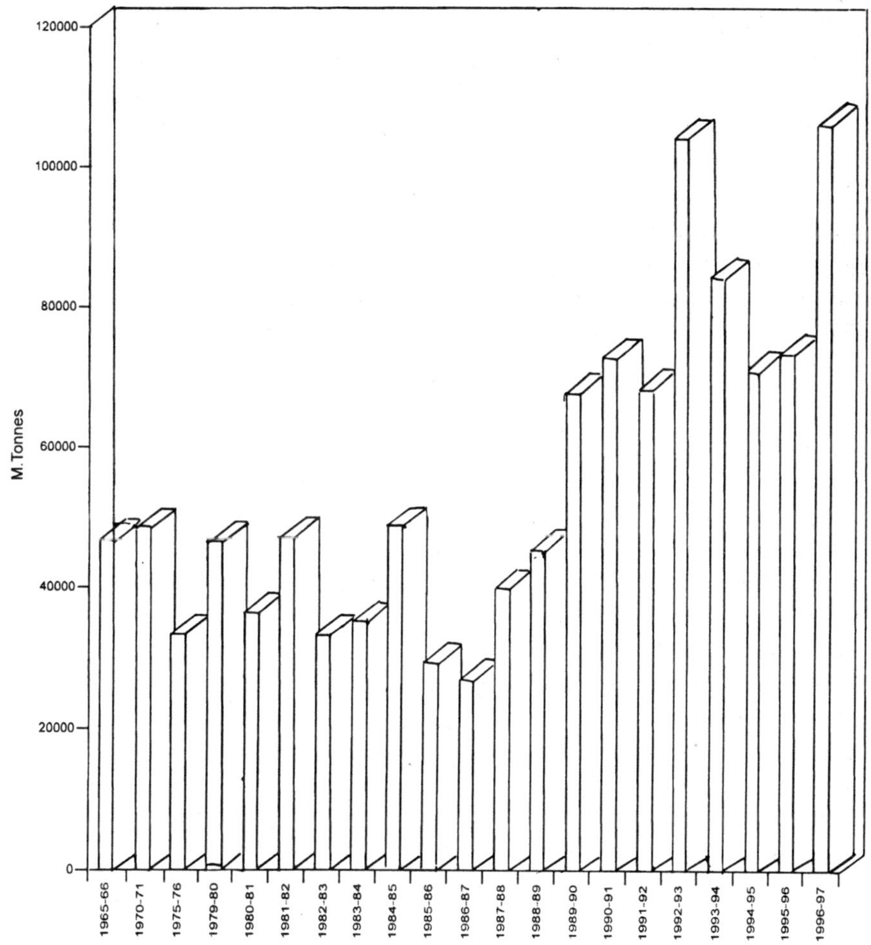


**Fig. 1.2 : Landing Pattern of Bombay duck in Gujarat**

## Craft and Gear

### *Craft*

The boats employed for Bombay duck fishery are locally known as *"Vahan"* ranging from 10 to 20 tones but boats below 10 tones called *"Machwa"* and below 5 tones called *"Lodhiya"* are also operated. Each *Vahan* has a capacity to carry 7 to 9 fishermen and 6 to 9 nets for fishing. The period of fishing extends from 15 to 35 hours. Of this it takes 5 hours to reach the ground and 5 hours to return from fishing grounds. Fishermen prefer to stay at fishing ground, so as to catch fish of 4 tides and land the total catch together. (Vasards and Subedar 1970).

### *Gear*

Generally main gear used for Bombay duck is bag net or "Dol" net. A detail of "Dol" net is given in the craft and gear chapter.

## *Protection of Fishery*

It is found that the spawning of the fish takes place some-where beyond the area of adult abundance, as no egg and larval forms of this fish are obtained in inshore waters of the actual fishing grounds. It is therefore necessary to make selective fishing to give opportunity for fish below 230 size to spawn or else it may result in diminishing returns from the fishery.

It is note worthy that the percentage of Bombay duck in total catch of Gujarat was predominant in sixties (around 1965-66) but has decreased by more than 50% by 1996-97. While Bombay duck represented about 42.59% if total catch, they have come down to 16.07% in 1996-97. This is due to increase in trawlers and gill netters over the years resulting in wider species representation.

## *References*

Anon MPEDA 1977. Indian Fisheries 1947-77. The fifth session of the Indian Ocean Fishery Commission, held at Cochin from 19-26 Oct.-1977.

Alexandar Kurian, Equilibrium state of the fishery for Bombay duck *Harpodon nehereus* (Ham): an outcome of production ecology and exploitation interactions. National Symposium on the utilization of living resources of the Indian seas. Abst: 20.

Bapat, S.V. 1970. *Bull. Cent. Mar. Fish.* Res. Inst., 21: 1-66

Bapat, S.V., S.K. Benerji and D.V. Bal, 1951. *J. Zool. Soc. India,* 3: 341-56.

Chhaya, N.D. 1990. Fishery Science and Management Gujarat Granth Nirman Board, Ahmedabad.

Chopra, B.N., 1939. *J. Bombay. Nat. Hist. Soc.* 4: 221-34.

Deshmukh, V.M. and Alexander Kurian, 1980. The Bombay duck. *Mar. Fish. Infor. Serv.,* T&E Ser., No. 20: 1-11.

Dholakia, A.D. 1980. List of important fishes of Gujarat in Fish processing preservation and quality control, Published by D.P. Dholakia, Ahemdabad, pp. 134-144.

Hora, S.L. 1934 *J. Bombay. Nat. Hist. Soc.* 37: 642-54.

Kewalramani, K.M. 1972. Bombay duck Fishery of Gujarat, Its resources and exploitation-I *Seafood export Journal* IV 10: 1-8

Kewalramani, K.M. 1972. Bombay duck Fishery of Gujarat, Its resources and exploitation-II *Seafood Export Journal,* IV, 15-25.

Sarvaiya, R.T. 1982. "Gujarati Bumla Matsyodyog", Book published by Smt. P.R. Sarvaiya.

Vasavda, G.D., and B.V. Subedar. (1970). Bombay duck fishing Industries at Jafrabad. CIFE Annual Day Souvenir, 39.

# 2
# CAT FISH FISHERY

## Cat Fish

### *Name of Fishery*

Commercially important marine cat fishes belongs to family Plotosidae and Tachysuridae. Cat fish landing in Gujarat accounted for 2.94% in the total marine fish landing of Gujarat.

### *Cat fish species available in Gujarat*

| | Scientific name | English name |
|---|---|---|
| (1) | *Batrachocephalus mino* | Frog headed Cat fish |
| (2) | *Trachysurus caelatus* | Engraved Cat fish |
| (3) | *Tachhysurus Sona* | Dusky Cat fish |
| (4) | *T. jella* | Small eyed Cat fish |
| (5) | *T. dussumieri* | Marine Cat fish |

### *Names in other languages*

| | |
|---|---|
| English | Husky Cat fish |
| Gujarati | Khaga, Khagi, Shigada, Dharwa, Shigati |
| Marathi | Shingala |
| Kannad | Shede |
| Malayalam | Valia etta, Canan etta, Etta |
| Tamil | Keluthu, Thedu |
| Telugu | Jellallu |

### *Classification*

| | |
|---|---|
| Suborder | Siluroidae |
| Family | Trachysuridae |
| Genus | *Tachysurus* |
| Species | *Sona* |

### *Taxonomy*

Synonyms

*Arius sona* (Ham.) Day 457

*Pimelodus sona* Hamilton-Buchanan 1822

*Hexanematichthys sona*, Munro, 1955

*Distinctive characters*

D. 1/7/0; P. 1/12; V. 6; A. 17(6/11) C. 17

| | |
|---|---|
| Head | 4 in the total. |
| Eye | 7 to 9 in the length of head. |
| Barbel | Maxillary barbels reach the end of the head. Three pairs of barbel present. |
| Teeth | Palatine teeth on either side in large triangular patch, emarginate posteriorly, and anteriorly conjoined by small vomarine patch. |
| Spine | Dorsal Spine as long as the head, excluding the snout, Pectoral spine stronger slightly shorter than that of dorsal. |
| Colour | Blueish above, sides leaden grey dull white on belly; all fines blackish terminally. |

## Food and Feeding

**Fig. 2.1:** ***Tachysurus sona*** **(Ham)**

Cat fish are predators and carnivorous. The fish diet consists of eel, Anchoviella, *Acetes.* Crabs are preferred as food especially during the winter months when they constitute 12-20 per cent of the diet. Prawns from the next important item, found during major part of the year. *Acetes indicus* is the predominant species among the prawns. The others include *Peneus monodon, Parapenaeopsis stylifera* etc. *Squilla* is usually found intact in the gut and amounts to 20% in January and form 3.3 per cent of the annual average. (Bal & Rao, 1984).

The intensity of feeding seems to be related to the breeding season. It is highest during the pre spawning period of October-December. During the breeding season the diet is restricted to detritus, Crustaceans, Polychaetes and molluscan. While in the post spawning phase the stomachs are gorged with *Apogen, Leiognathus,* prawns, *Acetes.* There is not much difference between the juveniles and adults in respects of their food habits. *T.Thalassinus,* is also a voracious carnivorous feeder eats mainly Crustaceans, Crabs, Prawns, *Squilla* etc. (Majumdar and Dan 1979, Majumdar, 1969).

## Fecundity

The number of eggs in *T. sona* ranges from 35-258 and in *T.jella* 30-40. The range for *T.sona* is reported comparatively high.

## Maturity

It is reported that in case of *T. sona,* life span has been calculated to be 10 years. The first maturity has been calculated on the bases of K value which drops down first at 247.5 mm length in its second year of life, indicating it to be the length of first maturity.

In marine Cat fish the ovary during its development from the immature stage to maturity undergoes change in its appearance, size and structure as well as in the nature of ova. Though the mature ovary contains ova of all stages of maturity, the ripe ova are well differentiated and distinct from the stock of immature ones. They are large in size and heavy yolked. (Bal and Rao 1984).

## Spawning Season

The Cat fish generally breed once a year, their spawning season appears to be long. They show low fecundity and the number of ova laid varies from species to species and increase with the growth of fish. The males exhibit parental care and carry the fertilised eggs in their buccal cavity until they are hatched. The female dominate Commercial catches almost round the year. Their spawning season varies from species to species. In Gujarat for *T. sona* it is from October to January.

## Fishing Season

In Cat fish fishery three peak periods, of their abundance in a year were observed, February-April, May-July and October-December. The Cat fish show seasonal migration from shallow to deeper grounds during April-June and again in October-December. While from deeper to shallow water during July-September and January-March. (Rao and Dorairaj 1968; Radha Krishnan 1974, Sekharan 1973 a&b).

**Table 2.1 : *Cat fish landing in Gujarat and its share in Gujarat and India***

| Year | Total marine Fish Landing | Landing of Cat fish | Percent of Total Guj. | Marine fish Landing in India | Percent Share of Guj. in India |
|---|---|---|---|---|---|
| 1965-66 | 109907 | 920 | 0.84 | 18900 | 4.87 |
| 1970-71 | 151190 | 2093 | 1.38 | 49743 | 4.21 |
| 1975-76 | 208300 | 6231 | 2.99 | 68689 | 9.07 |
| 1979-80 | 206749 | 5321 | 2.57 | 48817 | 10.90 |
| 1980-81 | 218872 | 6625 | 3.03 | 12805 | 51.74 |
| 1981-82 | 220607 | 6653 | 3.02 | 11794 | 56.41 |
| 1982-83 | 192669 | 6951 | 3.61 | 15222 | 45.66 |
| 1983-84 | 223291 | 3464 | 1.55 | 16768 | 20.66 |
| 1984-85 | 290708 | 9556 | 3.29 | 17125 | 55.80 |
| 1985-86 | 306577 | 10598 | 3.46 | 44277 | 23.94 |
| 1986-87 | 315942 | 11592 | 3.67 | 52750 | 21.98 |
| 1987-88 | 327560 | 13844 | 4.23 | 42696 | 32.42 |
| 1988-89 | 414075 | 14892 | 3.60 | 64055 | 23.25 |
| 1989-90 | 432384 | 13759 | 3.18 | 49294 | 27.91 |
| 1990-91 | 500462 | 12435 | 2.48 | 37605 | 33.07 |
| 1991-92 | 530017 | 17046 | 3.22 | 38999 | 43.71 |
| 1992-93 | 609103 | 17597 | 2.89 | 36156 | 48.67 |
| 1993-94 | 619836 | 20022 | 3.23 | 42736 | 46.85 |
| 1994-95 | 645261 | 24218 | 3.75 | 45161 | 53.63 |
| 1995-96 | 598351 | 23570 | 3.94 | 36090 | 65.31 |
| 1996-97 | 660058 | 17832 | 2.70 | 42762 | 41.70 |
| 1997-98 | 702355 | 21110 | 3.01 | 44684 | 47.24 |
| 1998-99 | 551660 | 21111 | 3.83 | 46093 | 45.80 |
| 1999-00 | 670951 | 24526 | 3.65 | 58332 | 42.05 |

*Source* :- (1) Commissionerate of Fisheries, Govt. of Gujarat, Gandhinagar.
(2) CMFRI Published Annual Reports.

## Craft and Gear

In Gujarat at present Cat fish are caught in Trawl net. However, before Trawlers were in operation they were caught using "machawa" and dol net. In some parts of India Cat fish are caught using hook and line. Bottom set hook and line are preferred, In Kerala, Andhara Pradesh and Orissa boat seines are used.

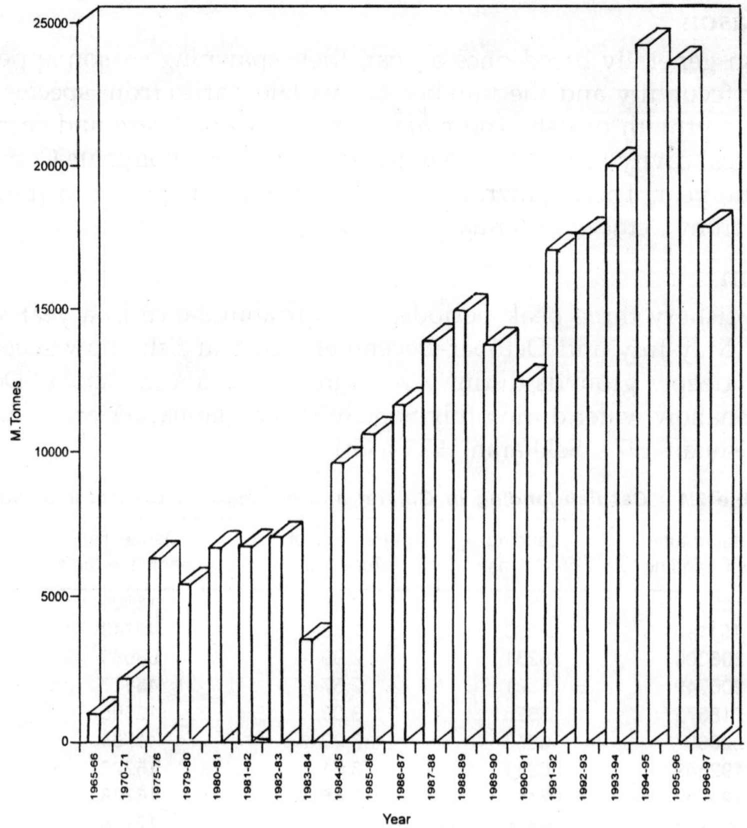


**Fig. 2.2 : Landing Pattern of Cat Fish in Gujarat during last 32 years**

Cat fish landing in Gujarat and its share in Gujarat and India since 1965-66 is given in Table 2.1.

Landing Pattern of Cat fish in Gujarat is given in Fig. 2.2

Analysis of cat fish landing from 1965-66 to 1998-99 i.e. last 33 years shows that Cat fish represents about 3% of total marine fish landing of Gujarat. While in India level Cat fish landing of Gujarat fluctuates between 20 to 55%. During last six years it above 45% of total Cat fish landings of India, while up to 1975-76 it was less than 10%.

## *References*

Bal, D.V. and K.V. Rao, 1984. Marine Fisheries published Tata Graw-Hill Publishing Company Ltd, New Delhi.

Dholakia, A.D. 1986. List of important fishes of Gujarat in Fish processing preservation and quantity control. Published by D.P. Dholakia, Ahemdabad, p. 135.

Karbhai. T.P. 1984. Scientific, common and local Names of Commercially important Marine fishes and shell fishes of Maharashtra and Gujarat. CMFRI Bull. No. 44: 18-23.

Majumdar and S.S. Dan 1979. *Indian J. Fish* 24-(1&2) 90-95.

Majumdar, A.D. 1969. *Indian J. Fish.* 16: 161-69.

Patel M.I. and N.D. Chhaya., 1980. Field key to Fishes of Gujarat, Dept. of Fisheries, Govt. of Gujarat.

Sekharan, K.V. 1973. Proc. Symp. Living resources of the seas around India, CMFRI, 517-36.

# 3
# CEPHALOPOD FISHERY

## Name of Fishery

### *Cephalopods*

Squids, Cuttlefish and Octopuses are the three major groups under Cephalopods. They are soft bodied marine invertibrates grouped under shell fishes.

Over 650 species of Cephalopods occur in all oceans of the world, and are distributed from shallow inshore areas to deep oceanic waters. They widely range in size from tiny seepiolids to giants like *Architeuthis* sp. which grow to a size over 60 feet in total length, out of 650 species of living Cephalopods, about 70 are Commercially important. Out of these 15 species contribute to Cephalopod fishery of India.

### *Classification*

| | | |
|---|---|---|
| Phylum : | Mollusca | |
| Order : | Decapod | |
| Class : | Cephalopoda | |
| Family : | (i) Loignidae | (Squids) |
| | (ii) Sepiidae | (Cuttlefish) |
| | (iii) Octopodidae | (Octopus) |

### *Species Available in Gujarat*

*Neritic squids* (within 200m depth)

1. *Loligo duvaucelii* (d 'orbigny)
2. *Loliolus investigators*
3. *Sepioteuthis lssoniana* (Lesson)
4. *Sepioteuthis arctipinnis* (Gould)

### *Cuttlefishes*

1. *Sepia aculeata* (Ferussac and orbigny)
2. *Sepia pharaonis* (Ehrenberg)
3. *Sepia elliptica*
4. *Sepia prashadi*
5. *Sepiella inermis* (Ferussac and Orbigny)

## Squids

Squids belong to the order Decapod which includes the majority of Cephalopods, possessing a stream-lined elongated torpedo shaped soft body with a pair of fins varing in shape size and disposition. The distinct head in front is with ten circumoral arms provided with toothed suckers or claws or both. An internal shell known as pen or gladius, when present is imbedded in the dorsal mantle skin. The gladius of Squids is almost transparent, thin and chitinous in nature. It varies in shape in different species. Squid exhibit enormous power of swimming and their swift propulsion through water is effected by the combined action of the mantle and the specialized structure called siphon or funnel situated on the ventral side. Such fast moving squids are distributed from shallow waters to varing depths of all oceans.

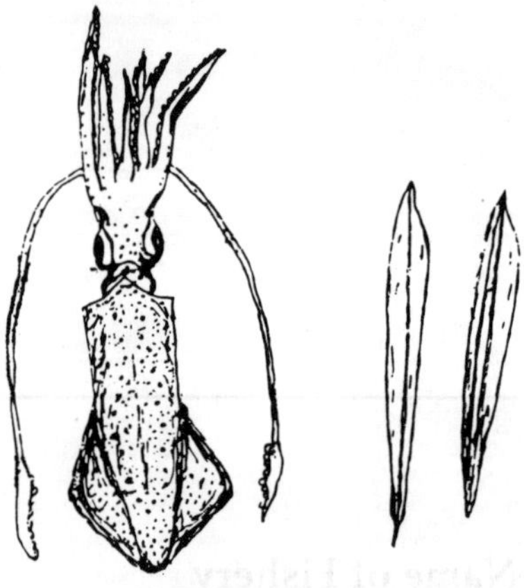

**Fig. 3.1 : *Logio duvaucelii and pen***

## Cuttle Fish

Cuttlefish belong to the order Sepioidae. Like squids, they possess well defined head and ten arms. They have a broad and dorsoventrally flattened body with narrow fins running along the sides to the full length of the body. The fins are lateral and separated at posterior end of the body. The arms are comparatively short and provided with sub-equal suckers mostly arranged in four transverse rows. The two long slender tentacles are retractile into special pockets and used at the time of capturing prey. The characteristic internal calcareous shell known as "Cuttlebone" is calcified and differs in shape and size in different species.

The Cuttlefish, *speciella intermis* (Ferussac and a orbigny) is one of the species of Cephalopods common in trawl catch.

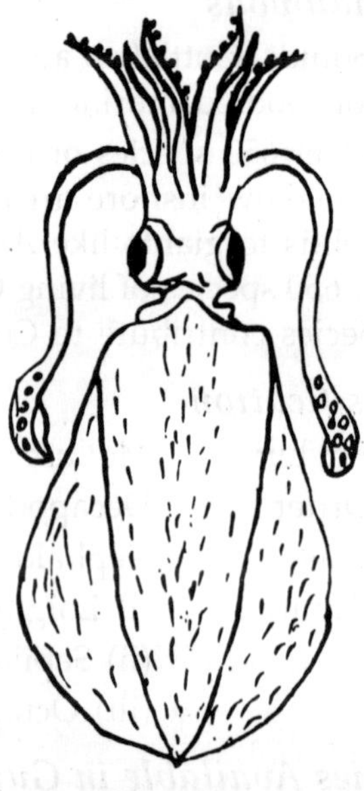

**Fig. 3.2 : *Sepioteuthis arctipinnis***

## Description

*Sepiella intermis* is a small sized Cuttlefish with narrow fins. The mantle has a pore at the posterior extremity and the cuttlebone is without a spine. Shell is moderately small and rather narrowly ovate in outline. The horny margin forms a characteristic thin, broad plate behind the calcareous portion and is highly glossy. The dorsal surface is finely granular and bears two shallow longitudinal grooves which gradually diverge from the posterior to the anterior extremity, and enclose between them a slightly broadly rounded elevation. Shell is chakly white, but the horny margin is pale brown and polished.

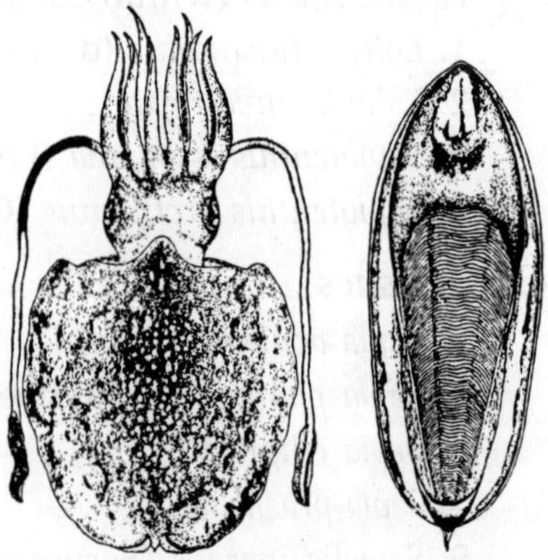

**Fig. 3.3 : *Sepia aculata and cuttle bone***

The eyes are moderate sized. The mantle is Conico-cylindrical and its anterodorsal margin extends forwards over the nuchal region to blunt point. The third and fourth pairs of arms are predominantly keeled. In males, the fourth arm of the left side is hectocotylised and the basal portion of the arm does not have suckers. The total number of suckers on this arm is less (18-39) compared to the number (37-82) of suckers on the other arms. Tentacles are elongated and rounded bearing medium sized tentacular clubs. Funnel is well developed. The jaws are horny and have well developed cutting edges. The dorsal surface is gray brown or purple depending on the chromatophores present. There is a series of white spots on the dorsolateral surface of the mantle. The ventral surface is pale gray or pale brown.

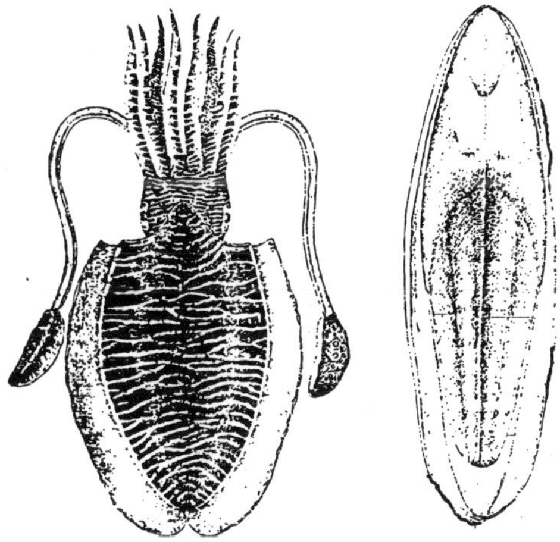

Fig. 3.4 : *Sepia pharaonis and cuttle bone*

## Octopus

*Octopus* is a bottom living form and spends hours lurking in its chosen lair, and often uses the arms in crawling and climbing or in conjunction with the mantle-funnel jet, in a progression over the bottom that can best be described as flowing. It is however a vigorous swimmer, although not so fast as *Logio.* The free part of the funnel is very flexible, and the animal can pounce forward on its prey. The long arms would obviously play an important passive part in locomotion.

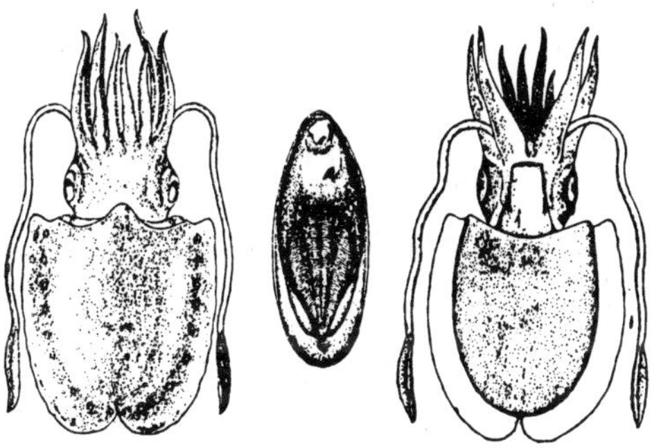

Fig. 3.5 : *Sepia intermis dorsal and ventral view and cuttle bone*

Octopus generally is shy, fascinating and intelligent cephalopod. Their degree of intelligence in often spoken of as shown in their way of escaping a predator by squirting out a dense smoke-screen from their ink glands.

Octopus is declared as protected species in Marine National park.

Fig. 3.6 : *Octopus*

## Taxonomy

Key to the field identification of different genera of Cephalopods to which the common species belong is given below. The external and more familieinternal characters are utilized for distinguishing them.

1. Cephalopods with eight circumoral arms, without tentacels: arm suckers arranged in two rows, without horny rings and stalks. Third arm tip of male spoon-shaped(hectocotylized ............................................................ octopodidae *octopus*).-

Cephalopods with ten arms eight being short and circumoral and two slender and tentacular. Suckers of the arms and tentacles stalked and equipped with armature ........................................................................................................ 2.

2. Shell (cuttlebone) internal calcareous in nature, body ovoid and drosoventrally somewhat flattened. Fins narrow, marginal in position and extending on either side along the entire length of the mantle, not uniting at the end. Arms with mostly quadriserial suckers..........................(Speiidae)........................................................ 3

   Shell (gladius or pen) internal but chitinous in nature. Body cylinderically elongate. Fins either terminal or marginal in position uniting at the apex of the mantle..........4

3. Cuttle bone broadly ova in shape and with a spine at the posterior end. The mantle without a glandular pore at the posterior extremely.............................................*Sepia*

   Cuttlebone smaller in size, ovalin shape and devoid of the spine. The mantle bears a small but distinct glandular pore at the posterior extremely.........................*Sepialla.*

4. Fins triangular or rhoboidal in shape, restricted to the posterior margin of the mantle ........................................................................................................ *Loiogo*

   Fins broad and extending almost to the entire length of the mantle..........*Sepioteuthis.*

## Food and Feeding

The stomach contents consist of prawns (22 to 50%), crabs (1 to 3%), stomatopods (1 to 2.3%), other crustacean (9.3 to 15.2%), fishes (5 to 50%). The species found in the stomach were *Leioganthus* spp., *Upeneus* spp., *Apogon* sp., stomatopods etc. Crabs and stomatopods were evident in the stomach contents form 51 mm onwards.

The intensity of feeding was highest during January to April and May. Compared to males, feeding was generally better in females.

## *Length weight relationship*

Length-weight relationship for male and female of Cuttlefish *Sepiella intermis* are as under :

Males : $\text{Log } W = 1.9601 + (1.9320 \text{ Log - } L)$

$W = 0.01096\ L^{1.9320}$

Females : $\text{Log } W = 2.5493 + (2.3208 \text{ Log. } L)$

$W = 0.002823.\ L^{2.3208}$

## Maturity (Females)

Maturity (Females) stages were denoted as immature, maturing, mature, spent and recovering.

1. In immature females the ovary appears as a thin white patch in the coelomic space in the posterior part of the mantle cavity beneath the dorsal mantle wall. Eggs can be distinguished only by microscopic examination.
2. In maturing females the ovary is bigger in size, the eggs with accumulating yolk are small and are visible to naked eye.
3. Mature ovaries prior to spawning have eggs fully grown.
4. Ripe eggs are translucent with reticulate surface. The mature ovary is compact, occupying a large space in the posterior part of the mantle cavity. The nidamental

glands are full and glossy white and the accessory nidamental glands are orange red.

5. Spent females have only a few number of eggs and the mantle becomes flacid; nidamental glands become thinner and less glossy. Accessory nidamental glands turn pink. Spent females weigh considerably less than the gravid one of corresponding length. Recovery contains developing eggs in addition to a few residual eggs.

   99.9% females of the 71-80 mm. size group were found sexually matured while in the 81-90 mm. size group all were ripe.

## Males

1. Immature males can be identified only by microscopic examination of the testis which appears as a thin and pale white mass in the gonadial part of coelomic space, very close beneath the dorsal mantle wall in the distal part of the mantle cavity.
2. Maturing males can be distinguished by thicker pale white testes which are bigger compared to the previous stage.
3. Ripe testes are thick and turgid containing spermatozoa, the spermatophore sac is bulged with spermatophores.
4. After spawning, the testes becomes flacid and pale and the spermatophore sac shrunken with a few spermatophores only (Unnithan : 1982).

The percentage of ripe male increased gradually from group 51-60 mm and at size range 71-80 mm all males were ripe.

### *Breeding, hatching and parental care in Octopus*

Initially octopus spawns a great number of egg capsules arranged in festoons. It was observed by Sarvesan (1969) that continuously under vigilant care of the female hatching.

Throughout the brooding period, the female never let go the egg festoons which were held close to the body on the dorsal side of the broad extensile interbranhial membrane. The bunch of egg festoons were frequently set in "rocking movement" by the movement of the arms and the body. Several times jets of water were squired over them by the female from the funnel, probably to keep them clean and free from any contamination by otherwise accumulating debris. Towards the close of brooding period the more active and vigilant. Extreme care was taken to remove anything that came near the spawn and food had no appeal to the female.

Just before the commencement of hatching the female occupied one of the corner. The arms were curled backwards and the egg festoons were retained inside the pouch like space formed by the interbranchial membranes. Only the tip of the funnel and eyes were visible between the dorsal pair of arms. At the time of hatching the female frequently directed jets of water from the funnel over the egg festoons and agitated then vigorously with the help of the arms. The egg capsules were then pressed against the "funnel flap" by the inner wall of the mantle opening during respiratory movement. This mechanical stress presumably facilitated the hatching of young. As a result, a few young ones were hatched in the mantle cavity. The young still enclosed in the interbranchial membrane, were "snaked" into the mantle cavity with the respiratory water current and expelled out forcibly through the funnel. 2 to 5 such tiny young one with a size range of 3.3 to 3.8 mm. in dorsal mental length, were pumped off through the funnel at a time . The hatching of the entire brood at varied intervals took about three days. (Servesan, 1969).

## Fishing Season

Peak season for squids is between January and April while lean season was found between June-August.

Similarly peak season for Cuttlefish was found in January, September-October and December. While lean season was reported during June-August.

It is observed that 60% of Cephalopod catch is contributed by cuttlefish and slightly less than 40% is represented by squids. Octopus forms less than 1% to total catch.

The latest squid fishing method developed and used by advance countries are described below in detail.

The commercial fishing of squid in the world are being done by "Jigging" method. In this method "Jigs" used are prepared from 7.5 to 10 cm. long cylindrical lead rod or plastic on the free end one or more crowns of sharp hooks are being fixed. (Fig. 3.8). Previously fishing was done by using hand jig. But now many such jigs are hung in a single line and are operated by special machine just like winch of trawl net. (Dholakia *et al.*, 1980).

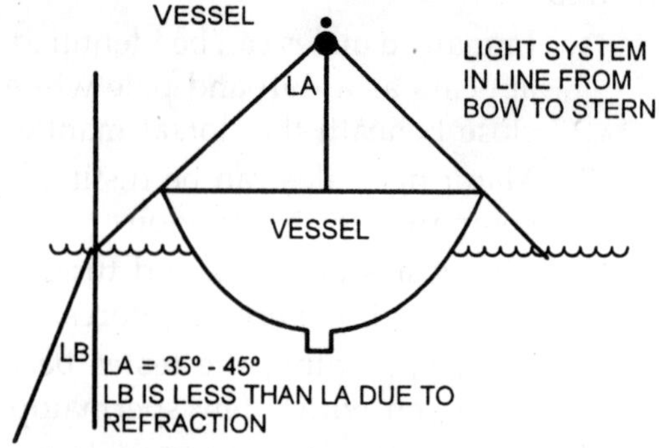


Fig. 3.7 : *Fishing method with light and shade*

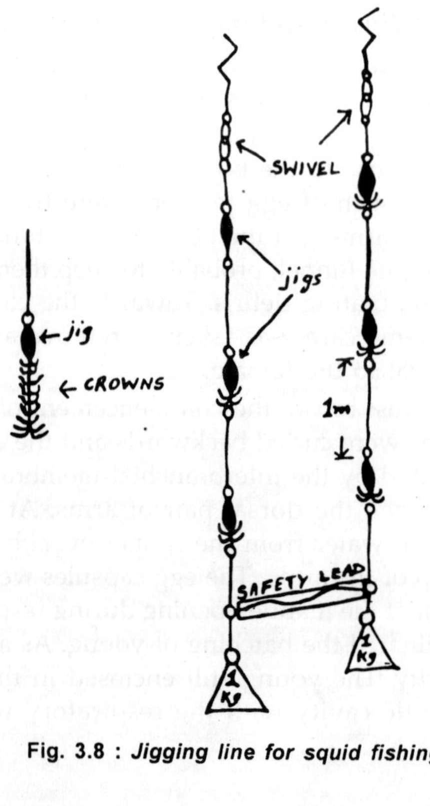


Fig. 3.8 : *Jigging line for squid fishing.*

## Resources

21079 Tones cephalopods are estimated on Indian coasts. While in EEZ., it is estimated at 1,80,000 tones. There is a vast gap between the potential and present landings.

In Gujarat estimated potential for squid is reported as 9000 M.T. while average landing is 3700 T., for Lolio estimated potential is 10,000 T. while landing is 3750 T.

It is estimated that resources potentials that offer the exploitation in the depth beyond the 50 m. for cephalopods in India are 21000 tones. Total marine fish landing of India in 1995 was 2.3 million tones in which cephalopod constitute 5.2%. Hence, still there is a chance of more landing if effort are increased. (CMFRI, 1997).

## Craft and Gear

### *Craft*

In Gujarat cephalopods are caught using trawler and is considered as side catch of prawn fisheries.

### *Fishing method*

As per recent researches the jig lines are tied with one nylon insulated tube. Which allows a branch line to be hauled in so as it is not lost (Fig. 3.8). By using such jig lines 4000 lb./hr of squids were caught in Canada. In different countries these jigs are of different colours. In Norway they are blood red coloured, while in Japan white lead jigs, orange, yellow and red coloured jigs are being used.

As per Jamsson (1979) squid and Cuttlefish are attracted by 0.01 lux intensity of light which would normally occur at a depth of about 40 m. using a vessel with 15 kW of lighting at an angle of between 35 and 45° (Fig. 3.7). Squid tend to concentrate on the boundary between light and shade under the vessel. (Dholakia *et. al.*, 1980)

**Landing pattern of Cephalopod in Gujarat and India and its share**

| Year | Total Maine Fish landing in MT | Landing in Gujarat in MT | Its share in Gujarat | Landing in India in MT | Its share India base |
|---|---|---|---|---|---|
| 76-77 | 126879 | 551 | 0.43 | 10005 | 5.51 |
| 77-78 | 176895 | 1212 | 0.68 | 15931 | 7.61 |
| 78-79 | 229901 | 5077 | 2.21 | 15032 | 33.77 |
| 79-80 | 296749 | 2676 | 1.29 | 11335 | 23.61 |
| 80-81 | 218872 | 5182 | 2.30 | 9548 | 54.30 |
| 81-82 | 220607 | 2516 | 1.14 | 15799 | 15.93 |
| 82-83 | 192669 | 2370 | 1.23 | 19749 | 12.00 |
| 83-84 | 223291 | 4394 | 1.97 | 20964 | 20.96 |
| 84-85 | 290708 | 4376 | 1.56 | 31577 | 13.85 |
| 85-86 | 306577 | 9744 | 3.18 | 42590 | 22.88 |
| 86-87 | 315942 | 10774 | 3.41 | 31875 | 33.80 |
| 87-88 | 327560 | 5206 | 1.59 | 38528 | 13.51 |
| 88-89 | 414075 | 11904 | 2.87 | 54487 | 21.85 |
| 89-90 | 432364 | 13959 | 3.23 | 56287 | 24.80 |
| 90-91 | 500462 | 12380 | 2.47 | 65337 | 18.95 |
| 91-92 | 530017 | 18078 | 3.41 | 90276 | 20.03 |
| 92-93 | 609103 | 21536 | 3.54 | 96889 | 22.23 |
| 93-94 | 619836 | 20142 | 3.25 | 96900 | 20.79 |
| 94-95 | 645261 | 26448 | 4.1 | 116764 | 22.65 |
| 95-96 | 598551 | 24079 | 4.02 | 107071 | 22.49 |
| 96-97 | 660088 | 20285 | 3.07 | 114079 | 17.78 |
| 97-98 | 702355 | 24122 | 3.43 | 107020 | 22.54 |
| 98-99 | 551660 | 18250 | 3.31 | 92377 | 19.75 |

Source : (1) Commissionerate of Fisheries, Govt. of Gujarat (with permission)

(2) C.M.F.R.I. Year Books (Published data)

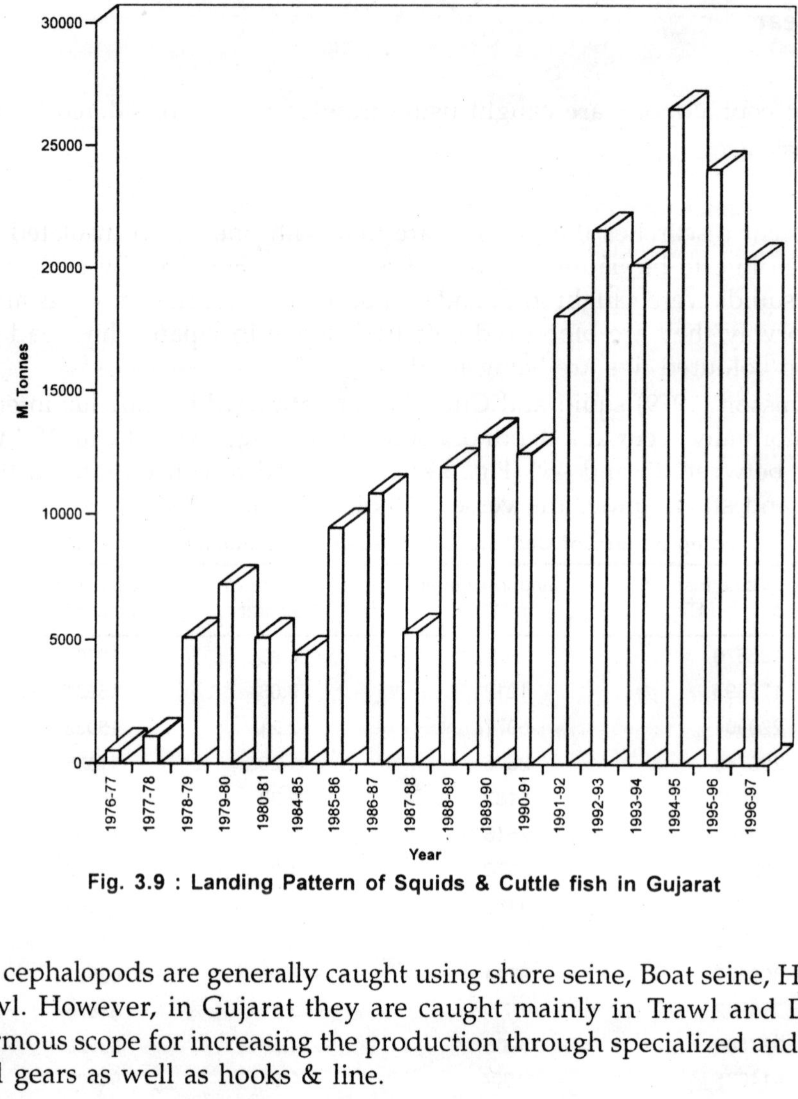


**Fig. 3.9 : Landing Pattern of Squids & Cuttle fish in Gujarat**

## *Gear*

In India cephalopods are generally caught using shore seine, Boat seine, Hooks & Lines, Dol net Trawl. However, in Gujarat they are caught mainly in Trawl and Dol net catch. There is enormous scope for increasing the production through specialized and direct fishing using special gears as well as hooks & line.

## *References*

Annon 1997. Annual report of CMFRI 1996-97 p.p. 144.

Dholakia A.D. 1986. Important marine fishes of Gujarat in "Fish Processing Preservation and quality control" pp.

Dholakia A.D., Y.A. Trivedi and R.G. Dabhi. 1980. Recent trends in Squids and Cuttlefish fishery and Technology. National seminar on recent trends of teaching and research in Aquatic Biology held on 27-29 September 1980 at Bhavnagar.

Sailas E.G., K. Satyanarayan Rao, R. Sarvesan., K. Prabhakaran Nair and M.M. Meiyappan 1982. The exploited squid and cuttlefish resources of India. A review. *Mar. Fish Infor. Serv.*, T&E. Serv, 34: 1-17.

Sarvesan R. 1974. Cephalopods in the Commercial Molluscs of India. CMFRI Bulletin No. 25 : 63-83.

Unnithan, K. Asokakumar 1982. Observation on the biology of Cuttlefish *Sepilla intermis* at Mandapam. *Indian J. Fish.*, 29(1&2) : 101-111.

# 4

# CRAB FISHERY

## Crab Fishery

About 560 species of crabs occur in the Indian region. Of these, only 8 species, namely *Portunus pelagicus, P. sanguinolentus, Scylla serrata, Charybdis (Charybdis) feriatus, C (C) annulata, C. (C) crusiata, Matuta hunaris and Varuna litorata* support localized fishery of some importance in the marine and brackish water regions.

Availability of information on various aspects of the fishery shows that the fishing is restricted to the inshore areas mostly by the operation of indigenous crafts and gears. In most of the places, crabs form an ancillary catch along with other crustacians and fishes.

Crustacean fishery of Gujarat comprises Prawns, Lobsters and Crabs. Among these, crabs form a minor fishery in Jamnagar district of Gujarat.

## Classification

Class *Crustacea*
Order *Decapoda*
Sub-Order *Brachyura*

Author have made an attempt to give a list of edible crabs of Gujarat state. There are 20 species of crabs belonging to five families which are eaten in more or less quantities at various centers along Gujarat coast.

| | Average carapace breadth. |
|---|---|
| I Family : Clalppidae | |
| 1. *Matuta lunaris* (Forskal) | 35 mm |
| 2. *M. planipes* (Fabricue) | 23 mm |
| II Family : Portunidae | |
| 1. *Scyla serrata* (Forskal) | 211 mm |
| 2. *Portunus pelagicus* (Linnaeus) | 167 mm |
| 3. *P. sangunolentus* (Herbst) | 76 mm |
| 4. *Charybdis cruciata* (Herbst) | 133 mm |
| 5. *C. lucifera* (Fabicius) | 82 mm |
| 6. *C. callianassa* (Herbst) | 25 mm |
| 7. *C. orientakis* (Dana) | 46 mm |
| 8. *C. annulata* (Fabricus) | |
| 9. *C. natator* (Herbst) | |

| | |
|---|---|
| 10. *Thalamita prymna* (Herbst) | 65 mm |
| 11. T. crenat (Milne-Edwards) | 51 mm |
| III Family : Xanthida | |
| 1. *Atergatis intergerrius* (Lamarch) | 98 mm |
| 2. *A. fIridus* (Rumph) | 54 mm |
| 3. *Estisus laevimanus* (Randall) | |
| 4. *Xantho scoberriums* Baccolipes (Alcock) | 60 mm |
| 5. *Eripha laviman* (Macleay) | 60 mm |
| IV Family : Ocypodidae | |
| 1. *Macrophtalmus pectimipes* (Cuerin.) | 52 mm |
| V Family : Crapidae | |
| 1. *Veruna literata* (Fabricus) | 41 mm |

**Field Key found useful for species of edible Crabs in Gujarat State**

1. A. Edges of antero lateral border of carapace ridge .......... 2
   B. Edges of antero lateral borders of carapace blunt .......... 12
2. A. Antero lateral borders with only spines .......... 3
   B. Antero lateral borders with only teeth .......... 14
   C. Antero lateral borders with spines and teeth .......... 11
3. A. Carapace white with pinkies dots .......... *Matuta lunaris*
   B. Carapace yellowish with vermicular lines .......... *M. planipes.*
4. A. Antero lateral borders of carapace with four teeth .......... *Etisus laevimanus*
   B. Antero lateral borders of carapace with five teeth .......... 5
   C. Antero lateral border of carapace with six teeth .......... 6
   D. Antero lateral borders of carapace with nine teeth .......... *Scylla serrata.*
5. A. Teeth on antero laterral borders sub equal in size .......... *Thalamita crenata*
   B. Fourth tooth on antero lateral borders rudimentary .......... *T. prymna*
6. A. Last tooth the (6th) on antero lateral borders not longer than rest ..........
   *Charra bdis callianassa.*
   B. Last tooth the (6th) on antero lateral borders not longer than the rest .......... 7
7. A. 2nd tooth of antero lateral borders of carapace rudimentary .......... *C. orientalis*
   B. 2nd tooth of antero lateral borders of carapace not rudimentary .......... 8
8. A. First tooth on the antero lateral borders of carapace acute .......... 9
   B. First tooth on the antero lateral borders of carapace anteriorly truncated and notched .......... C. *cruciata*
9. A. Carapace with four whitish spots .......... *C. lucifora.*
   B. Carapace without whitish spots .......... 10
10. A. Legs with annular bands .......... *C. annulata*
    B. Legs with marginal dotted line .......... *C. matator.*
11. A. Carapace grass coloured with whitish spots .......... *Portunu pelagicus*
    B. Carapace pinkies white with three blood red spots posterierly ..........
    *P. man guinolentus.*

12. A. Antero lateral carapace convex ............ 13
    B. Antero lateral carapace not convex ............ 15
13. A. Carapace surface rough and red coloured ............ *Atergatis integerrinu*
    B. Carapace surface rough ............ 14.
14. A. Carapace surface lumpy with broad shallow depression ........ *Atergatis floridus.*
    B. Caraspace surface with tubercles ............ *Xantho scaberrinus bacalines.*
15. A. Carapace squared with pearly white tubercles .......... *Macrephthalmus pectinipes.*
    B. Carapace not squared (bell shaped) ............ 16.
16. A. Carapace not depressed, distinctly border than long .............. *Eriphia laevimina*
    B. Carapace depressed, nearly equal in length and breadth ............ *Veruna litteata*

## *Charybdis Cruciata* (Herbst)

*Synonymus*

*Portunus crucifere* de Haan, 1980

*Goniosoma cruciferum*, de Man, 1987

*Charybdis curciata*, Chopra and Das, 1937, Chhapgar, 1956.

*llentus Shen*, Bull Fan. Mem. Inst. Biol. (Zool) X P. 81. 1940. *Lupa sanguinolonta*, Barnard, Ann. S. *Afr. Mus* XXXViii, P. 154,

Fig. 4.1 : *Charybdis cruciata* (Herbst)

### Identification

The antero-lateral borders are cut in to nine teeth of which the last is the largest. There is no spine on posterior of the arm of the chelipeds. The anterior male abdominal appendages are quite straight and have no sharp bend near the tip. The margins in the distal part are beset with somewhat stout spines.

Colour dark gray, three blood-red spots ringed with white posteriorly. Average male measurements.

| | |
|---|---|
| Length of carapace | 43 mm |
| Breadth of carapace (Excluding the last lateral spines) | 76 mm |
| Length of cheliped | 94 mm |

(Chapgar 1957)

## *Neptunus (Neptunus) pelagicus* (Linnaeus)

### *Classification*

| | |
|---|---|
| Tribe | Dromiacea |
| Sub tribe | Brachyrhyncha |
| Family | Portunidae |
| Sub family | Lupinae |
| Genus | *Neptunus* |

Sub genus *Neptunus*
Species *Pelagicus* (Linnaeus)

## Synonyms

*Portunus (Portunus) pelagicus,* Rathbun, Dept of Trade and Customs, (Fish) of Australia V.P. 130 (1923)

*Portunus pelagicus,* Suvatti, Dept of Fisheries Bankok, Thailand, p. 171. (1950).

*Neptunus (Neptunus) pelagicus,* Shen, *Bull. Fan. Mem. Inst. Biol. (Zool)* X, p. 81 (1940)

## Identification

This species is distinguished from *Neptunus (Neptunus) sanguinolentus* by the presence of a spine at the far end of the posterior border of the arm of the chelipeds.

The anterior male abdominal appendages are quite straight and have no sharp bend near the tip. The margins in the distal part are beset with a large number of spines than *N. sanguinolentus.*

**Fig. 4.2 : *Neptunus pelagicus* (Sea Crab)**

**Colour:** Colour in the male, pinkies purple, with extensive irregular white spots, the tips of the chelate and the distal segments of the legs purple : Female is sandy coloured.

**Growth :** This species grows four to five inches across the carapace. (Chapgae, 1957).

Average male measures

Length of carapace 57 mm.

Breadth of carapace 167 mm.

**Habitat :** Hydrobiological parameters like fresh water influence during monsoon, Salinity, Temperature, Transparency, turbidity, dissolved oxygen as well as nutrients play an important role in habitate. It was observed that blue crab *P. pelagicus* is able to withstand considerable variations in temperature and salinity.

## *Size frequency*

The size range of the males were up to 165 mm., while the females were 155. Generally, the females were slightly smaller than males and were also fewer in number and constituted 32.46%.

Spawning season : Spawning season of *Portnus* sp. was reported during September to March.

## *Portunus Sanguinolentus* (Herbst)

## *Classification*

Tribe Dromiacea
Sub Tribe Brachyrhyncha
Family Portunidae

Sub family Lupinac
Genus *Neptunus*
Sub genus *Neptunus*
Species *Sanguinolentus* (Herbst)

## Synonyms

*Neptunus sanguinolentus. Portunus (Neptunus) sanguinotentus* (De Hann), *Fauna, Jaonica V P. 38 (1850) Neptunus (Neptunus) sanguib*

Carapace is convex frontal teeth are slightly pointed and anteriorly projected. An "H" shaped deep furrow is noticeable in certain region of carapace. Anterolated teeth are not much anteriorly truncated.

The posterior border of the carapace adjoining the abdominal plate is broad and less convex.

**Fig. 4.3 : *Neptunus sanguinolentus* (Sea Carb)**

Chelate legs are massive; the merus bears three stout spines in the anterior border (inner angle) and two on the posterior border (outer angle) of which one is terminal and the other one is sub-median in position. The carpus has a strong spine at the inner angle; in the outer angle there exists two stout spines. Propodus is with one stout spine at the front of teh apex of the wrist joint and two small ones located side by side behind the finger joint. The colouration of the carapace is predominently green on the dorsal side and cream on the ventral side. Numerous mosic like yellow or white patches are seen on the swimming legs, upper portion of the chelate legs and also on the abdominal flap of the fully mature female.

## Scylla Serrata (Forskal)

Carapace is convex frontal teeth are slightly pointed and anteriorly projected. An "H" shaped deep furrow is noticeable in certain region of carapace. Anterolated teth are not much anteriorly truncated.

**Fig. 4.4 : *Scylla serratla***

The posterior border of the carapace adjoining the abdominal plate is broad, and less convex. Chelate legs are massive; the merus bears three stout spines in the anterior border (inner angle) and two on the posterior border (outer angle) of which one is terminal and the other one is submedian in position. The Carapace has a strong spine at the inner angle; in the outer angle there exists two stout spines. Propodus is with on stout spine at the front of the apex of the wrist joint and two small ones located side by side behind the finger joint. Th colouration of the carapace is redominently green on the dorsal side and cream on the ventral side. Numerous mosic like yellow or white patches are seen on the swimming legs, upper portion of the chelate legs and also on the abdominal flap of the fully mature female.

## Food and Feeding

Mud crabs are carnivorous. They feed on slow moving and bottom dwelling animals such as bivalves/molluscs, small crabs, and dead and decayed animal materials. These crabs are also called "Scavengers".

## Sexual Identity

Sex can be identified in juvenies of S. *Serrata* measuring about 35 mm in carapace width (CW) by shape of the abdominal flap. In male, the abdominal flap is slender and triangular while it is broad and triangular in immature and semi-circular in berried females. In both sexes, the abdominal flap in live crab is folded firmly against ventral side of the body.

## Moulting

The growth in mud cabs in manifested with the shedding of outer shell. Before moulting, new exoskeleton shell is formed below the moulting process, the old shell is cast off. The formation of new shell and casting off of old shell is called as moulting process, which requires energy. The increase in size of crab after moulting takes pace due to absorption of water by the tissue of the body and thus moulted crabs are larger in size. Since the moulted crabs have utilized stored energy for moulting, they weight less and they contain more water. The newly moulted crabs with watery meat and soft exoskeeton are called "Water crabs". Such "Water cabs" remain defenseless and become easy prey to other animals, particularly other hard mud crabs. The newly formed shell of the moulted crabs gets hardened in 3-4 days after moulting. The frequency of moulting is more in juveniles and sub-adults, while it is less in adults.

## Growth

The early juvenile crabs (15 to 60 mm in CW/3 to 20g in total weight) grow at rate of 7 to 12 mm/3 to 13g per month. While juvenile crab 61 to 80 mm/25 to 70g exhibit a monthly growth of 11 to 12 mm/45 to 97g. In the sub-adult and adult stages, the monthly growth worked out to 8 to 10 mm/100 to 130 g. *Scylla serrata* attains maximum size upto 140 mm/ 0.5 kg in the wild. (Kathirvet *et. al.*, 1997)

## Maturity

Attainment of maturity in females can be easily noticed by the change in the shape of abdominal flap, from triangular to half-round/horse-shoe shape. For male there is no external character to identify the maturated ones. The minimum carapace size at maturity observed in *S. serrata* was 140 mm and abdominal width 61 mm. (Radhakrishnan an Samul, 1982) while according to Kathirvel (1997), it was 83 mm. The eggs if matured, are visible by their yellow colour.

## Mating

The copulation takes place between a hard male and freshy moulted soft female. Prior to copulation, a hard male climbs over the back of hard female crab, clasping her by his chelipeds and first two pairs of walking legs. This formation is called as "Pre mating embrace." which lasts for 2-3 days. The pair separates when the female reaches the verge of moulting. The female moults and it is called as "Precopulatory moult." The male assists the female during pre-couplatory mould embraces the soft female again for the actual mating. The male gently turns the female over on her back using their chelipeds. While the female unfolds her abdominal flap and holds the male in position. The copulaton lasts for 6 to 8 hours. During the coupulation act, the male deposits spermatophores in the seminal receptacles of the female. (Kathirvel *et. al.,* 1997).

## Breeding

When the eggs become ripe, they are fertilized by the stored spermatophres. Later, they are extruded and remain as "Sponges," and attached to hair like branches of four pair of appendages of the abdominal flap. The egg mass as whole, attached to the abdominal flap is called "berry". The number of eggs vary from 0.5 to 2.5 million for *S. serrata.* The eggs measure 0.28 to 0.38 mm in diameter. The incubation period is about two weeks, during which time, the colour of the eggs gradually changes from orange to brown and then to black. The stored spermatophores are used for more than two spawning.

## Spawning season

They are continuous breeders as evidenced by the occurrence of ovigerous and berried female throughout the year in the coastal and brackish water regions. However, maximum spawning season of *Portunus* sp. was reported during September to March.

## Fishing season

Though the crabs are caught and eaten throughout the year, the peak fishing season in Gujarat for crab is from July to September. (Radhakrishnan and Samul. 1980.) During these months crab landings are about 300 to 500 kgs per day at Sikka and about 40 to 50 kg at Beyt, Balapur, Salaya, Sachana and Jodia. (Trivedi and Patel 1975.)

## Gears

Following gears and devices are used to catch the crab at various fishing centers. (1) Ad net (Stake net) (2) Sansi (Pair of tongs) (3) Iron rods (4) Cast net (5) Hook and line.

(1) *Ad Net :* Ad net is mad from plastic or cotton No 4. At a time about 40-50 units of about 30 meter length each are used in series. A wall of net supported by about six feet long tree branches is erected during the receding tide. At a time about 20-100 kg of crabs can be caught by this methods.

(2) *Iron Rods :* In Okha, Bet, Balapur and Sikka 2 to 3 feet long iron rods pointed at one end and bent (hooked) at the other end are used to catch crabs. People go to the sea coast during low tide and pierce the carapace of the crabs available in the inter tidal and catch them. The hooked end is used some times to pull crabs out of their burrows.

(3) *Cast Net :* Some time cast net is used to catch *P. pelagicus and P. sanguinolentus* from sandy bottom in the month Jun, July and August. The mesh siz 1.5 to 2.5 cm and it is operated in about 1-2 feet depth at Okha and Bet Balapur.

(4) *Hook and lines : S. serrata* is caught chiefly by hooks and other crabs are mainly collected by commercial trawls operated in inshore and off shore regions. In hook and lines baits like *Octopus, Nehris* or fish are used at hook. Hooks are fixed at the end of about 30 to 40 M long nylone tiwine. They are thrown from shore and whn crabs comes to eat bait they are entangled and caught.

**Table 4.1 : Crabs landing in Gujarat and its share in Gujarat and India**

| Year | Total marine Fish landing | Landing of Crabs | Percent of Total Guj. | Marine fish Landing of India | Percent share of Guj. in India |
|---|---|---|---|---|---|
| 1974-75 | 147133 | 1054 | 0.716 | - | - |
| 1975-76 | 208300 | 38 | 0.018 | 19893 | 0.19 |
| 1976-77 | 125379 | 64 | 0.052 | 19999 | 0.32 |
| 1977-78 | 176895 | 131 | 0.074 | 20068 | 0.65 |
| 1978-79 | 229971 | 192 | 0.083 | 14202 | 1.35 |
| 1979-80 | 206749 | 472 | 0.228 | 20303 | 2.32 |
| 1980-81 | 218872 | 560 | 0.256 | 25386 | 2.21 |
| 1981-82 | 220607 | 599 | 0.272 | 43934 | 1.36 |
| 1982-83 | 192669 | 629 | 0.326 | 52618 | 1.20 |
| 1983-84 | 223291 | 416 | 0.186 | 41130 | 1.01 |
| 1984-85 | 290708 | 1418 | 0.488 | 22232 | 6.38 |
| 1985-86 | 306577 | 1088 | 0.355 | 20755 | 5.24 |
| 1986-87 | 315942 | 2555 | 0.809 | 23098 | 11.06 |
| 1987-88 | 327560 | 2734 | 0.835 | 17422 | 15.69 |
| 1988-89 | 414075 | 978 | 0.236 | 16149 | 6.06 |
| 1989-90 | 432384 | 615 | 0.142 | 23970 | 2.57 |
| 1990-91 | 500462 | 1690 | 0.338 | 28240 | 5.98 |
| 1991-92 | 530017 | 1604 | 0.303 | 26849 | 5.97 |
| 1992-93 | 609103 | 659 | 0.108 | 26938 | 2.45 |
| 1993-94 | 619836 | 1229 | 0.198 | 29631 | 4.15 |
| 1994-95 | 645261 | 1697 | 0.263 | 30610 | 5.54 |
| 1995-96 | 598351 | 2082 | 0.348 | 29049 | 7.17 |
| 1996-97 | 660068 | 1403 | 0.213 | 44963 | 3.12 |
| 1997-98 | 702355 | 1707 | 0.243 | 34293 | 4.98 |
| 1998-99 | 551660 | 1205 | 0.218 | 27689 | 4.35 |
| 1999-00 | 670951 | 1657 | 0.247 | 48384 | 3.42 |

*Source* : 1) Commissionerate of Fisheries, Govt. of Gujarat, Gandhinagar (With permission)
2) CMFRI Published Annual Reports.

## Resources

(1) Estimation of crab resources of India have been made by Rao *et. al.,* (1975.). According to them 43.816 MT was estimated which includes (I) resources from inshore waters of both the East and West Coast up to 15 m (25,348 Mt.). (ii) resources from offshore waters up to 40m (5,259 MT) and brackish water (13,209 MT).

Table 4.1 shows the crab landing in Gujarat, in India and its share since 1974-75. Fig. 4.5 shows the landing pattern and crab in Gujarat since 1974-75.

Prior to 1978-79 crabs wre not considered much important species in Gujarat. After that fishermen started fishing crabs. From the analysis of about 26 years landing it is found that crabs forms less than 0.5% of total marine fish landing in Gujarat and approximate 4 to 7% of total crab landing of India.

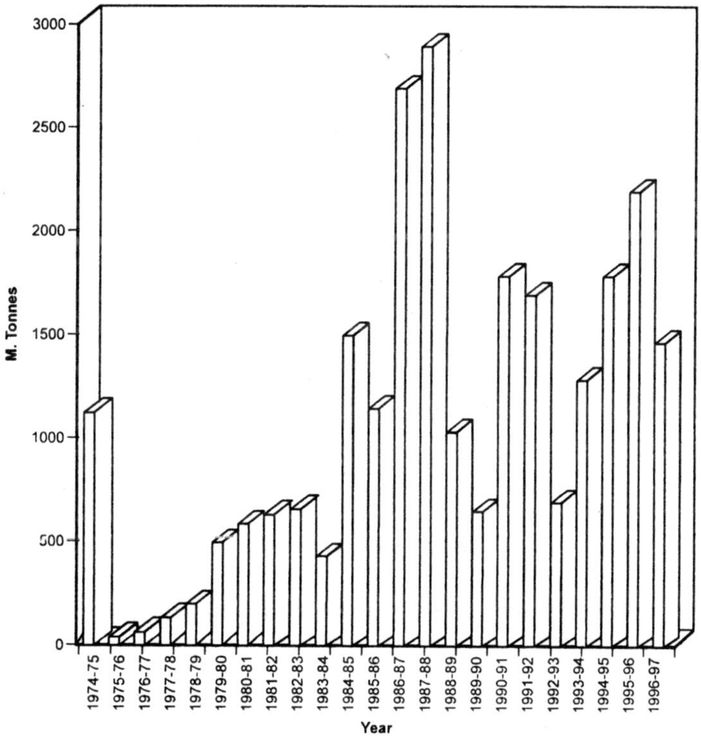


**Fig. 4.5 : Landing Pattern of Crabs in Gujarat**

## *References*

Ansari, Z. A., and S.N. Harkentra. 1975. Crab resources of Goa. *Sea food Export Journal,* 3 : 21-24.

Chhapger, B.F. 1957. On Marine crabs (Decapoda : Brachyura) of Bombay State. *J. Bombay. Nat. Hist. Soc.* 54 : 399 and 503.

Dam Roy, S., 1997. Culture of Mud crab, *Scylla serrata* (Forskal) in Andamans. *Fishing Chimes,* 17 (7) : 26-27.

Dholakia, A.D. 1986. List of important fish and crustacians of Gujarat in "Fish Processing Preservation and quality control." pp. 134-144.

Kathirvel, M.S., S. Srinivasagam, P.K.Ghosh, and C.P. Balasubramanian. 1997. Mud Crab Culture. *Fishing Chimes,* 17 (6); 28-33.

Patel, N.M. and A.D. Dholakia. 1979. Marine Edible Crabs of Gujarat. Dept. of Fisheries. Govt. of Gujarat.

Radhakrishnan, C.K. and C.T. Semul 1980. Prospects of Fishery, Utilization and culture of Crabs in India. *Fish. Technol.,* 17(2) : 67-69.

Radhakrishnan, C.K., and C.T. Semul 1982. Reported on the occurrence of one subspecies of *Scylla serrata* (Forskal) in Cochin Backwaters. *Fish. Technol.,* 19(1) : 5-7.

Rao, P.M., Vedavyasa, M. Thomas, and G. Sudhakar Rao. 1973. *Proc. Symp. on living Resources of Seas around India* CMFRI, 581-591.

Trivedi, Y.A. and N.M. Patel 1975. Crab Fishing gear and Methods and Crab Fishery of Gujarat. *Seafood Export J.* 12 : 17-19.

Vedyasan Rao, P.M., M. Thomas and C. Sudhakaar. 1973. Crab Fishery resources of India. *Proc. Symp. Living Resources of the Sea arround India.* C.M.F.R.I. 582.

# 5

# CHANK FISHERY

## Introduction

Chank is extremely abundant on the East Coast of India being found and fished every where from Cape Comorin to Madrtas although the density of its occurrence appears to thin out north of Point Calimere. The northern limit of this coast may be put as the mouth of Godavari. On the West Coast its geographical distribution is peculiar. Large numbers are fished in the Gulf of Kutch Coast, but Southward of this no traces of the chank is found up to Southern Coast line of Kerala State where this appears and forms a small fishery. (Nayer and Mahadevan, 1974.)

## Species available in India

Hornell (1914, 1915) distinguished 5 well marked subspecies of the central form of *Xancus pyrum* (Linn) in different localities. They are:

(1) *Xancus pyrum Var. Obstusa.*

(2) *Xancus pyrum Var. acuta.*

(3) *Xancus pyrum Var. globosa.*

(4) *Xancus pyrum Var. commoriensis.*

(5) *Xancus pyrum Var. fusus.*

Out of which in Gulf of Kutch *Xancus pyrum Var. acuta* is represented. The breadth in length averages to 1.83 and lower 1.75 in short specimens. The variety *acuta* is called "Jadhi" in commercial parlance.

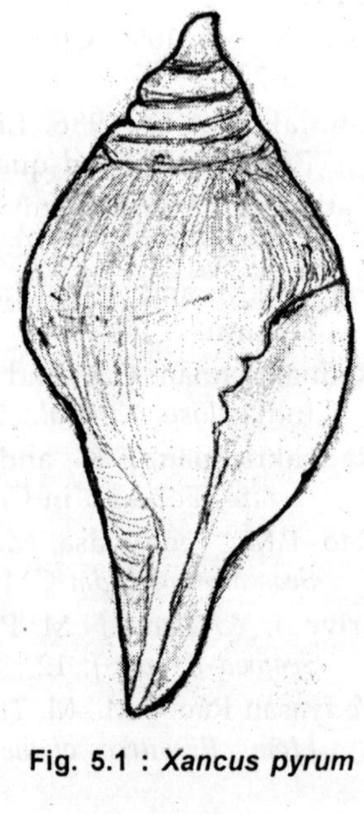

Fig. 5.1 : *Xancus pyrum*

## Identification

The sacred chank *Xancus pyrum Var. acuta* (Bolten) is exploited from the intertidal and reef habitat in the Gulf of Kutch along Jamnagar Dist. They are found on or about the coral reefs where one can wade through the shallow water to pick the chanks (Gokhale, 1963).

Important centres of collection in Gujarat State:

Okha, Salaya, Aramda, Poshitra, Bedia Island, Nora Island, Bharana, Vadinar, Sikka, Pirotan Island, Bedi, Adatra, Beyt

Balapur, etc. On the southern side of the Gulf are important places located approximately over a distance of 130 km.

## Habitat

The ecological studies have revealed that chanks prefer a depth range of 10-27 metres with fairly high content of Calcium in the medium, a sandy bottom, with a mixture of mud in the association of tube dwelling polychaete, worms, echinoderms, filamentous green algae, broken or loose corals. These associations appear to give the animal requisite food material especially the worms (Anon. 1983).

## Reproduction

Normally when males attain 57 mm to 60 mm size, they are mature, while females are bigger than this when mature. Fertilization is internal and the fertilized ova are deposited in spindly twisted chitinous capsule. Such capsules contain about 28-36 eggs. The young ones develop in this enclosed capsule and when they are about 1 cm. in length they are believed to be eating out the portions of the capsules and come out of the capsules. Thus they have a protected life by and large when compared to larvae free swimming forms of other moluscs.

## Protection

The chank is an excellent instance of the acquisition by an animal of characters which appear for all practical purposes absolutely perfect to enable to hold its own with ease in the struggle for existence. Against every one of its known enemies it appears to have suitable means of defence. Several large pits are commonly seen in the ground created by the smaller and larger rays and skates resting on the bottom. Naturally the chanks are exposed to the attack of these predators. But the thick and massive shell of the chank and the animal's semiburrowing habits given it protection against rays and other fishes which have habit of either snapping the protruded feet of this gastropode or wrench the shell. The thick periostracum protects it to a great extent from the insiduous attach of the boring sponges, *Cliona sp.* and its shell-boring cangenors. The strong capsules in constructs for its young gives them adequate protection till they reach a self-supporting stage endowed even at this early period with a resistant shell. The camouflaging colouration of the periostracum with that of the bottom on which the chank lives offers further protection against its presence being discovered by enemies. (Mahadvan and Nayar, 1974)

It is found from the investigations that the major groups of animals which destroy chanks are 1) Sponges, 2) Polychaetes, 3) Molluscus, 4) Sipumculids. Chanks are bored by sponges are more in numbers.

## Fishery

Chank fishery is mostly done by hand picking at low tides in the inter tidal zones in the Gulf of Kutch. The Sindhis, Vadhers, Madhwars and other local communities are involved in this work. About 90% of the chank production of India is obtained by dividing. (Thomas et. al., 1998). Production of chank during 1975 to 1997 from Gujarat are given in Table 5.1 (Dept of Fisheries) chank landing in Kerala from 1964-65 to 1979-80 is given in Table 5.2.

**Table 5.1 : Production of Chank during 1975 to 1997 from Gujarat**

| Year | Large (10 Cm and above) | Medium (8-10 Cm) | Small (6-8 Cm) | Worm eaten | Total |
|---|---|---|---|---|---|
| 1975-76 | 1,165 | 6,644 | 192 | 2,368 | 10,389 |
| 1976-77 | 122 | 776 | 12 | 204 | 1,114 |
| 1977-78 | 196 | 1,223 | 10 | 1,153 | 2,582 |
| 1978-79 | 1,002 | 4,702 | 192 | 1,372 | 7,270 |
| 1979-80 | 465 | 2,126 | 159 | 1,372 | 7,270 |
| 1980-81 | 340 | 2,549 | 1,050 | 2,162 | 6,101 |
| 1981-82 | 262 | 1,372 | 5,134 | 3,561 | 10,329 |
| 1982-83 | 385 | 1,592 | 4,697 | 4,233 | 10,907 |
| 1983-84 | 534 | 2,633 | 6,639 | 4,554 | 14,360 |
| 1984-85 | 271 | 2,063 | 8,667 | 9,898 | 20,899 |
| 1985-86 | 64 | 639 | 3,276 | 3,552 | 7,531 |
| 1986-87 | 117 | 1,045 | 7,874 | 2,031 | 11,067 |
| 1987-88 | 37 | 354 | 2,143 | 408 | 2,942 |
| 1988-89 | 38 | 285 | 2,340 | 593 | 3,257 |
| 1989-90 | 19 | 287 | 1,651 | 835 | 2,792 |
| 1990-91 | 5 | 177 | 2,044 | 1,040 | 3,266 |
| 1991-92 | 5 | 136 | 1,671 | 932 | 2,744 |
| 1992-93 | 9 | 63 | 824 | 1,813 | 2,709 |
| 1993-94 | 13 | 184 | 2,017 | 827 | 3,041 |
| 1994-95 | 7 | 109 | 2,092 | 1,037 | 3,245 |
| 1995-96 | 6 | 79 | 593 | 447 | 1,125 |
| 1996-97 | NIL | 15 | 575 | 208 | 798 |

*Source* : Sujatha Thomas *et al.* 1998.

## Grading and Marketing

The collected chanks are graded according to diameter as large 10 cm and above, medium: 8-10 cm. Small 6-8 cm. Worm eaten and undersized are purchased by Department of Fisheries, Government of Gujarat at fixed price according to size. These chanks are then auctioned mostly to Kolkata based cottage industries. Price during 1980 and 1997 were as under:

| Size in Dia, cm. | Price (Rs)/piece(1980) | Price (Rs)/piece(1997) |
|---|---|---|
| 10 cm and above (Big) | 4.00 | 10.00 |
| 8-10 cm. (Medium) | 3.50 | 07.00 |
| 6-8 cm (Small) | 2.50 | 06.00 |
| Worm eaten and under size | 0.80 | 01.00 to 2.50 |

The sudden decline in production from 1987 onwards may be either due to depletion of resources due to over exploitation or private entrepreneurs are paying high price hence not coming to Department of Fisheries due to nominal rates of purchase.

## *Uses of chanks*

1) At present day the use of chanks for industrial purpose is confined to the chank bangle industry only and for cosmetics.
2) Occupies an important place in the lives of Hindus, being used in worship.
3) They are used in preparing lime for special needs and occasions as the lime made from chank shell is found to be superior quality. But the shells are too valuable to be used for manufacture of lime.
4) In Tamil Nadu, chank powder is used in curing many skin ailments and in the treatment of eye diseases.

**Table 5.2 : Chank landing in Kerala from 1964-65 to 1979-80**

| Year | No of Chanks |
|---|---|
| 1964-65 | 8296 |
| 1965-66 | 11918 |
| 1966-67 | 11500 |
| 1967-68 | 9997 |
| 1968-69 | 14538 |
| 1969-70 | 17814 |
| 1970-71 | 34025 |
| 1971-72 | 41179 |
| 1972-73 | 27850 |
| 1973-74 | N.A. |
| 1974-75 | 17004 |
| 1975-76 | N.A. |
| 1976-77 | 15075 |
| 1977-78 | N.A. |
| 1978-79 | 35117 |
| 1979-80 | 44670 |

*Source* : *Mar. Fish. Infor. Serv.* T&E *Ser.* No 24:1980.

## *References*

Anon. 1983. Wonderful world of Gulf of Kutch. Published by Commissionerate of Fisheries, Govt. of Gujarat, pp. 12-13.

Gokhale, S.V. 1963. Shell fisheries of Saurashtra region, Gujarat State. Department of Fisheries, Gujarat Govt. publication.

Hornell, J. 1914. The sacred chanks of India. A monograph of the Indian Conch. *(Trubinell pyrum) Madras. Fish. Bull.* 7:1-181.

Hornell, J. 1915. The Indian varieties and races of the genus *Turbinella. Mem. Ind. Mus.* 6 : 109-122.

Mahadevan, S. and K. Nagapan Nayer. 1974. Ecology of pearl Oyster and Chank beds. CMFRI Bull No 25: 120

Nayar, Nagapan, K. and S. Mahadevan. 1974. Chank Fisheries and Industries uses of chanks. The commercial Molluscans of India. CMFRI Bull. 25 : 122-140.

Nayar Nagapan K. and K. Satyanararana Rao. 1985. Molluscan Fisheries of India. *Mar. Fish. Infor. Serv.* T. & E *Serv.* No 61. pp. 4-5.

Sujitha Thomas, Joe, K. Kizha Kundam and V.B. Makadia. 1998. A note on Chank Fishery in Gulf of Kutch. Gujarat. *Mar. Fish. Infor. Serv.* T & E *Serv.* No 157 : 20-21.

# 6

# EEL FISHERY

## Introduction

Ele fishery of Gujarat forms (on an average of last 20 years) 1.04% of the Marine Fish landing of Gujarat. From the landing data it can be observed that after 1987-88 landing was reduced and formed about 0.64% of the total Marine fish landing of Gujarat.

## Classification

| | |
|---|---|
| Class | Teleostomi |
| Sub Class | Acanthopterigii |
| Order | Anguilliformes |

In India about 43 species of Eels are recorded out of which following three species are of Commercial Importance.

(1) *Muraenesox talaboides* (Bleekar)
(2) *M. cinereus* (Forskal)
(3) *Angula bengalensis*

## In Gujarat following species of Eels are reported

| Scientific name | English name |
|---|---|
| (1) *Muraenesox talaboides* | Indian Pink Eel |
| (2) *Muraenesox Cinereus* | Silver Conger Eel |
| (3) *Gymnotharax undulatus* | Leopard muray |

In Gujarat, first two species are represented in commercial catch. However, *Gymnotharax undulatus*, locally known as Naro, also contributes to some extent.

## Local name in other languages

| | |
|---|---|
| Gujarati | Wam, Naro |
| Malayalam | Pambu-meen |
| Marathi | Wam |
| Bengali | Bam |
| Tamil | Velangoo |

## Taxonomy

(1) *Muraenesox telabonoides* (Bleekar) Day-658

## Synonyms

*Muraenesox pristis,* Kaup, Apodel Fish. P. 116

*Muraenesox telabonoides,* Bleeker Alt. Ich. Iv, P. 23. T. cliv, f-2;

Gunther et al. Viii, P. 46.

B. xviii-xix, P. 15, D 245-250, C. 10, A. 200-205.

Head — Length of head 0.66 of the distance between end of snout and vent: length of trunk rather less than that of tail. Extent of the cleft of the mouth equals 0.5 of the length of the head.

Eyes — Diameter 3.5 to 3.75 I in the length of the snout and 1 diameter apart.

Teeth — The anterior ones on the vomer are straight, slender, compressed, and elongated, whilst the posterior ones have a small basal lobe in front and behind: the outer mandibular teeth are directed outwards.

Fins — Dorsal commences in advance of the base of the Pectoral, the latter fin equals 3.75 in the length of head.

Colour — Silvery becoming white on a abdomen: Vertical fins yellowish, with a narrow black outer border.

## Field Identification *(M. telabonoides)*

Pectoral fin present. Some free spines absent before dorsal. Extent of cleft of the mouth equals about 0.5 length of the distance between top of snout to gill-openings; dorsal commences well in advance of the gill openings.

*Muraenesox cinereus* (Forskal) Day-662

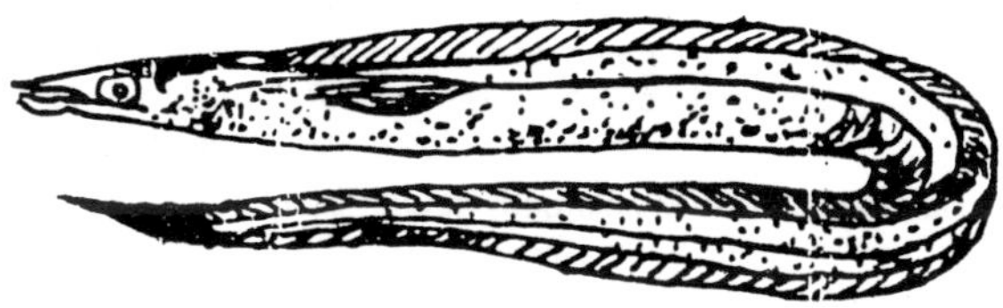

Fig. 6.1 : *Muraenesox talabonoides*

## Synonyms

*Muraena cinerea,* Forsk, Desc. Anim, PP.x and 22

*Muraenesox cinereeus,* Gunther, Catal. Viii, P. 46: Klunz. Verh. Z.b. Gest. Wien. 1871, P. 608.

*Conger lamo,* Schleg. Fauna. Japan. Poiss. P. 262. P1. 114.

## Taxonomy Identification

B. xx-xxii, P. 14-16, C10. A 190-220.

Head — Length of head 2.50 to 2.75 times in the distance between end of snout and vent: length of trunk rather less of than that of the tail. Extent of cleft of mouth equals rather less than 0.5 of the length of the head.

Eyes — Diameter 2.25 to 2.5 in the total length of snout, and 1 diameter apart.

Teeth — The vomerine are straight, compressed, and with a basal lobe anteriorly and posteriorly: The mandibular teeth are much smaller than the vomerine ones, and those in the external row are not bent outwards (as in *M. telabonoides*)

Fins — Dorsal commences slightly in advance of the gill-opening. The Pectoral 3.25 in the length of head.

Colour — Silvery, becoming white on the abdomen: Vertical fins yellowish, with either a narrow or wide outer black edge: pectoral yellow or black.

## Field Identification (*M. Cinereus*)

Extent of cleft of mouth less than 0.5 length of the distance between tip of snout to gill opening; dorsal commences slightly in advance of the gill opening.

## Food and Feeding Habits

Eels are carnivorous fish. The main items of food consists of teleostes (90 per cent) and the rest, of crustaceans and cephalopads (10 per cent) in teleostes Scienids are found to be 63 per cent throughiout the year. (Mohemad, 1955).

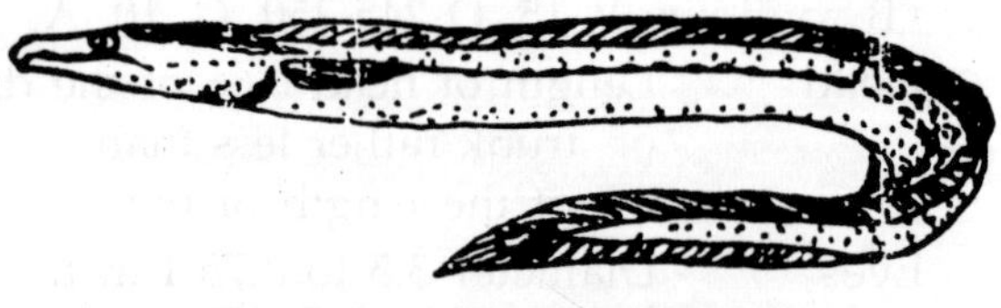

**Fig. 6.2 : *Muraenesox cinereus***

## Sex Ratio

Sex ratio in male-female in *M. talabonoides* as observed on Gujarat, Maharashtra coast was 1:0.62 during January and 1:4.89 in June and around 1:2 during the rest of the year. This indicates that Females are more prominent in June. (Mohemad-1955) Maturity stages of *M. talabonoides* is given in Table 6.1

**Table 6.1 : Maturity Stage of *M. talabonoides***

| State of maturity | Extent of ovary and its appearance | Nature of ova | Ova diametger | Model Size |
|---|---|---|---|---|
| Immature | Ovary thin, ribbon-like transparent. | Not visible to naked eye, Polygonal, Central nucleus. | 0.22 mm. | (Maximum 0.6 mm.) |
| Developing | Slightly enlarged ribbon-like, light-flesh coloured | Not visible to naked eye, yolk formation started, opaque, yellowish in colour | 0.7 mm. | (Maximum 1.1 mm.) |
| Developed | Ovary enlarged, nearly filling the body cavity, yellowish colour. | Ova opaque and yolky; still follicular, ova easily released on teasing | 1.0 mm. | (Maximum 1.4 mm.) |
| Maturing | Enlarged, completely fills the body cavity, yellow coloured | Most ova opaque, amber coloured, cluster of oil globules in transparent ova | 1.2 mm. | (Maximum 1.7 mm.) |
| Mature | Loosely packed ovary completely fills the body cavity. | All ova transparent central nucleus surrounded by translucent yolk; single conspicuous oil globule (0.32 mm.) | 2.2 mm. | — |
| Spent | Ovary thin, blood shot | Most ova small (0.2 mm) Transparent : a few large Transparent ova present | Less than 0.2 mm | — |

*Source* : Mohemad, K.H. 1955. "A study of marine eels of Bombay". *M.Sc. Thesis:* University of Bombay.

## Spawning Season

According to Kagwade (1969), the eel does not shed ova in batches but spawns twice a year, with the primary spawning in April-May and the secondary spawning in September-October.

## Fishing season

Generally in Gujarat fishing is carried out during September to May. However peak period for eel landing is observed during Jan-May and October-November.

Range of length in eel in fish catches was observed between 400 to 2000 mm. Which indicates that in general commercial fishing, juveniles are not found in good quantity. Big eel are known as "Wam while small eels are known as Wamdi". Eel landing in Gujarat and its share in Gujarat and India since 1975-76 is given in Table 6.2.

**Table 6.2 : Eel landing in Gujarat and its share in Gujarat and India**

| Year | Total marine Fish landing | Landing of Eel | Percent of Total Guj. | Marine fish Landing of India | Percent share of Guj. in India |
|---|---|---|---|---|---|
| 1975-76 | 208300 | 2453 | 1.18 | 5710 | 42.96 |
| 1979-80 | 206749 | 4043 | 1.96 | 7155 | 56.51 |
| 1980-81 | 218872 | 5305 | 2.42 | 12082 | 43.91 |
| 1981-82 | 220607 | 3092 | 1.40 | 5027 | 61.51 |
| 1982-83 | 192669 | 3267 | 1.70 | 7479 | 43.68 |
| 1983-84 | 223291 | 3240 | 1.45 | 8302 | 39.03 |
| 1984-85 | 290708 | 2699 | 0.93 | 7112 | 37.95 |
| 1985-86 | 306577 | 4072 | 1.33 | 8302 | 49.05 |
| 1986-87 | 315942 | 3928 | 1.23 | 7887 | 49.80 |
| 1987-88 | 327560 | 2244 | 0.68 | 6202 | 36.18 |
| 1988-89 | 414075 | 3067 | 0.74 | 4309 | 71.18 |
| 1989-90 | 432384 | 2865 | 0.66 | 4888 | 58.61 |
| 1990-91 | 500462 | 3443 | 0.89 | 5090 | 67.64 |
| 1991-92 | 530017 | 2331 | 0.44 | 6873 | 33.92 |
| 1992-93 | 609103 | 3328 | 0.55 | 6121 | 54.37 |
| 1993-94 | 619836 | 3392 | 0.55 | 6725 | 50.44 |
| 1994-95 | 645261 | 3276 | 0.51 | 5904 | 55.49 |
| 1995-96 | 598351 | 4369 | 0.73 | 7030 | 62.15 |
| 1996-97 | 660058 | 3477 | 0.52 | 7792 | 44.62 |
| 1997-98 | 702355 | 4948 | 0.71 | 9594 | 51.57 |
| 1998-99 | 551860 | 3389 | 0.61 | 11287 | 30.03 |
| 1999-00 | 670951 | 3347 | 0.50 | 9187 | 36.43 |

*Source* : (1) Commissionerate of Fisheries, Govt of Gujarat, Gandhinagar (with permission) (2) Annual reports of CMFRI

It is note worthy that the percentage of Eel in total catch of India was between 40 to 60% except in 1988-89. It was 71.18%. The percentage of Eel landing compare to total fish landing of Gujarat is less than one percent after 1986-87. Up to 1986-87 it was more than 1.2% of total fish catch, which may indicate towards over fishing.

## Craft and Gear

### *Craft*

In Gujarat generally Eels are caught in trawlers and some times in non-mechanised boat gillnets.

### Gear

The gear used in Gujarat and Maharashtra coast is the long line, locally known as "Khanda", each unit is made of about 12 sub-units, each consisting 50-60 hooks. This gear is operated at 20-30 m depths.

Appreciable quantities are caught in trawls and dol nets.

Major gear which contribute the eel catch for Gujarat is under:

| | |
|---|---|
| Trawler | 60% |
| Gill net | 06% |
| Non-mechanised | 22% |
| Dol net | 12% |

It is reported that 59-90 m was the most productive area in fishing grounds located near Cambay, Veraval and Kutch. Pelagic trawling yielding a catch rate of 12.8 kg/hr. from 55-99, depth during January and February (Menon et. al., 1998)

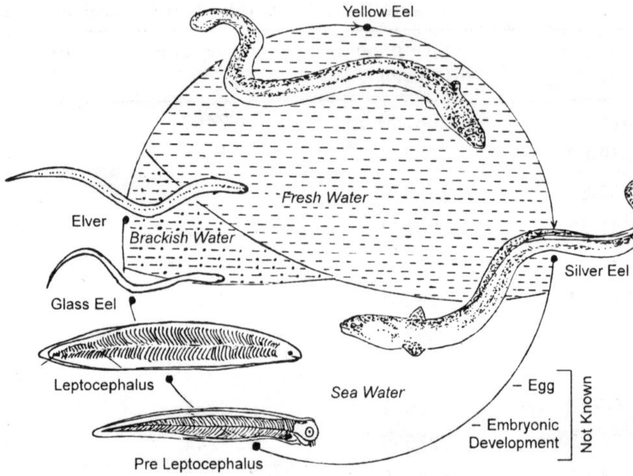


**Fig. 6.3 : Life cycle of eel**

## The eel and its life cycle

Most of the eels spend their life in the sea, but the species of *Anguilla* spend their life both in marine and freshwater environments. The freshwater phase of the eel is taken advantage of and exploited in Eel culture.

As the life cycle of the different species under the genus *Anguilla* are more or less similar a general account of the same is given in Fig. 6.3. *Anguilla* spp bread in the open sea at a depth of 400–500 m. After fertiliztion the egg hatches into a tiny larva, preleptocephalus and later becomes leptophalus. The leptophalus is transparent, leaf like with large number of myotomes and small, pointed head. The leptophali are weak swimmers and their transport/drift from the open sea to the near shore regions is affected by oceanic and coastal currents. The duration of the larval life in different species varies from three months to two and a half years. The leptophali metamorphose into glass Eel, which are thread like and transparent. On getting pigmentation the glass Eels become elvers and approach the coasts. On entering brackish and freshwaters the metamorphosis is completed and the elvers in a few months become young Eels. The latter grow in the brackish water or freshwater for some years and at this phase of the life they are called yellow eels. On nearing sexual

**Fig. 6.4 : *Anguilla* spp.**

maturity the adult yellow eels cease to feed, acquire a silver colouration, leave the fresh and brackish waters and migrate to their spawning ground in the open sea where they breed. After spawning the silver Eels are believed to die.

Thus in the life history of the Eel there are three phases viz.,

(1) Marine larval phase (Leptophalus, glass Eels, elvers.)

(2) Fresh water phase of growth (Yellow Eels)

(3) Adult marine phase of reproduction (Silver Eels).

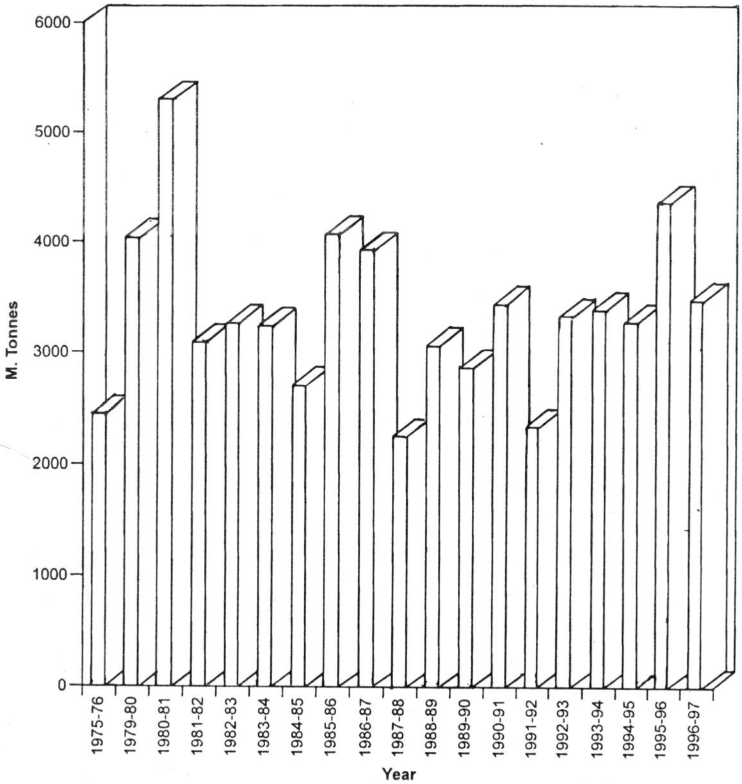

**Landing Pattern of eel in Gujarat from 1975-76, 1996-97**

## Maturity

On Gujarat coast the maturity of *M. talabonoides* was fiound at 1200 mm total length. While on east coast it was 1350 mm.

## *References*


Bal D.V. and K.V. Rao 1984. Fisheries of India, Published by Mcgraw Hill Publishing Company Limited, New Delhi. pp. 283-294.

CMFRI. Fisheries Calendars Bombay fish. 1988. *Mar. fish Infor. Serv,* T. & E., Ser., No. 89.

Dholakia A.D. 1989. Scientific, English and local names of fishes available in Gujarat. In "Fish Processing Preservation and quantity control" Published by D.P. Dholakia, Ahmedabad. pp. 134-144.

Francis Day. 1958. The fishes of India, Published by Mcgraw Hill Publishing Company Limited, New Delhi. pp. 658-662.

Joseph, K.M. 1975. Bull. Expl. Fish Proj. No. 2 pp. 1-53.

Menon, N.G., K. Balachandran and K.S. Scarish 1998. The Eels and Eels fishery of India *Mar. Fish. Infor. Serv.,* T & E., Ser. No. 157: 10-14.

Mohemad, K.H. 1955. "A study on marine Eels of Bombay" M.Sc. *Thesis* University of Bombay.

Patel, M.I. and N.D. Chhaya, 1979. Field key to Fishes of Gujarat. Dept. of Fisheries, Govt. of Gujarat; pp. 16.

# 7

# Elasmobranch Fishery

## Introduction

Fishery are divided into two main groups–one called Elasmobranchs or Cartilaginous fishes characterized by a cartilaginous skeleton and other known as Teleosts or Bony skeleton. The former group belongs all sharks, skates and rays. There are over 37 kinds of sharks and 32 kinds of rays in the Indian seas.

In addition to nature of the skeleton, Elasmobranchs have following peculiar features:

1 The gill-slits 5 to 7 behind the head are exposed. In bony fishes, these are covered by gill cover (Operculum.)

2 The mouth opens downwards whereas in the bony fish it is usually terminal.

3 The body is covered with a tough skin imbedded in which are placoid scales (denticle) which in the mouth become modified into teeth. These placoid scales are very different in structure from the flat overlapping scales of a bony fish.

4 In the tail of sharks, the backbone, instead of being continued straight backwards into the tail-fin, is bent upwards and is found to lie on one side of the upper lobe of the fin while the lower lobe of the fin arises from the lower side of the upturned backbone (heterocereal). In the rays, the tail degenerates into a whip-like structure in which the bcakbone tapers to fine point (leptocercal or leaf tail).

Cartilaginous fishes are divided into two group, viz., (1) Selachoidei comprising sharks, hammerheads, etc, and (2) Batoidei consisting of saw-fishes, etc. Sharks belonging to the first group are distinguished from the other in having the gill-openings at the sides of the neck and is having free pectoral fins not fused with the head whereas in several members of the second group, the gill-opening are situated below the sidcoidal body and the pectoral are joined with the head.

The body in the sharks (Selachoidei) excepting the hammer-heads in cylindrical and tapering at both ends. In the hammer-heads, the heads, this is produced on either side into a flat process carrying the nostrils and eyes outwards the whole body thus resembling a hammer.

The body in the rays (batoidei is not cylindrical; it is either depressed and elongated, e.g., saw-fishes and the plough-fish or it is discoidal with a Whip-like tail, e.g., string rays ad eagle-rays. In saw fishes, the body through some what depressed is still long and shark-like with a well developed tail ending in a heterocercal fin ; the snout (rostrum) is elongated and flattened with a row of large teeth on each side, resembling a saw. In shark the teeth are conical and sharp and in some there are numerous serration; they are adopted for cutting the flesh of the fishes on which they feed. But in rays, which subsist mainly on crabs and molluscs, the teeth are pavements suited for crushing them.

The largest fishes are found among the Cartilaginous Fishes and among Bony-fishes. Some of the sharks grow 6 to a length of 50-70 feet, e.g., the Whale-shark (*Rincodone typpus*) while rays may be broad as 20 feet, e.g., the Eagel-rays (*Dicerobatis spp*). A few of the sharks such as *Carcharias ganeticus and Galeocerdo tigrinus* are notorious as man eaters.

Detailed identification and other aspects of Fishries are mentioned hereunder:

## Pristis perrotteti

### Classification

| | |
|---|---|
| Class | Elasmobranch |
| Sub class | Chondropterygii |
| Order | Plagiostomata |
| Sub order | Batoidei |
| Family | Prostodae |
| Genus | *Pristis* |
| Species | *Perrotteti* |

### Synonyms

*Pristis microdon* (Lathan) Trans, Linn. Soc. li p 280.
*Pristis antiquorum* Costa, Funa. Regn, Pesc. T. viii,

### Names in other languages

| | |
|---|---|
| English | Saw-fish |
| Gujarati | Churiya |
| Tamil | Vala and Ilupa |
| Telgu | Shines and Hachutti meenu. |
| Malayalam | Vala sravu, Vellisravi, and Makara sravu. |
| Kannada | Chakku thatle, Naithatte and Bili sovulu. |

***Pristis perrotteti***

### Identificaton

Head Rostrum of moderate breadth, rather narrow anteriorly, armed with from seventeen to twenty pairs of teeth, mostly of about the same number on either side, and commencing at the base of the snout: the first four pairs being opposite one another, And the breadth of the base of each anteriorly being about equal to 0.33 of the interspace.

Teeth — Larger than in *P. cuspiolatus or P. zysron* oval longer than broad.

Fins — The first dorsal almost entirely in advance of the ventral: second dorsal commences nearly midway between the posterior extremity of the first dorsal and the base of the caudal: these two fins are of equal size: interior margin of caudal with a faint indication of a lobe.

Colours — Reddish-brown superiorly, becoming dull white along the abdominal surface: fish golden with a black edge.

(F. Day, 1958, pp 729)

## *Carcharinus gangeticus* (Mull and Henle)

The Long Tailed Shark

### Classification

| | |
|---|---|
| Order | Plagiostomnta |
| Sub-order | Selachoidei |
| Family | Carchariidae |
| Genus | *Carcharinus* |
| Species | *Gangeticus* |

### Synonyms

*Carcharias gangeticus* Day 415.

*Carcharias japonicus*, Temm. and Schleg. Eaun. Japan. Poiss. P. 302. T. cxxxiii.

*Carcharias gangeticus*, Gunther, Catal vii, p 367.

### Names in other languages

| | |
|---|---|
| English | Long tailed Shark |
| Gujarati | Magra |
| Tamil | Mudan sura |
| Telugu | Sorra |
| Malayalum | Valiya seravu |
| Kannada | Bugg-karuvai, Baliar. |

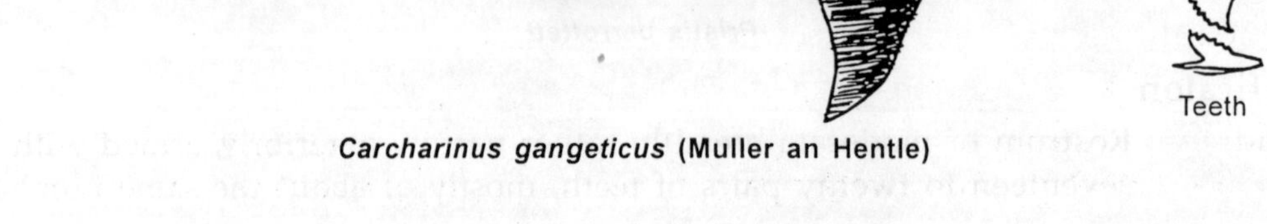


**Carcharinus gangeticus (Muller an Hentle)**

| | |
|---|---|
| Head | Length of the pectoral portion of the snout only 2/3 of the width of the mouth, and equalling half of the interspace between the eye and the middle gill-opening: a very slight grove at the angle of the mouth. Nostrils in the front half of the distance between the end of the snout and the mouth. Gill-opening wider than orbit. |
| Teeth | Twenty-seven to thirty in either jaw, all being serrated: the superior ones almost triangular, their outer edge with a notch, very distinct in the immature, but becoming almost lost with age : those in the lower jaw narrow, erect, and having broad bases. |
| Fins | Pectrol elongated, falciform and extends to beneath the middle of the first dorsal : its internal margin 2/7 of the length of its outer. Origin of the first dorsal slight posterior to the hind edge of the base of the pectrol, its base not extending to above the ventral. The second dorsal of about the same size as the anal, the latter nearer the caudal than the ventral. Length of the caudal 2/7 of the total length. |
| Scales | Small and with rough edges : they are hardly above ½ the size of those in *C. melanopterus.* |
| Colour | Gray superiorly, becoming dull white beneath. Fins gray, the pectrol, ventral, and anal with light edges : Posterior portion of caudal rather dark. (F. Day 1958, p 715) |

Sharks are found all along the coasts of India, but their carrying off human prey is not a common occurrence. The most savage species appear to be the ground sharks of the rivers, as *Carcharias gangeticus,* which seldom looses an opportunity of attacking the bather.

## Carcharinus melanopterus

### Synonyms

*Carcharinus melanopterus* (Q. G.)

### Names in other languages

| | |
|---|---|
| English | Black-finned shark |
| Gujarati | Magru |
| Tamil | Koppulisure, Katta sura, Kala Kumattai sura. |
| Telugu | Guna sura, Caval sura, Mukkan sura. |
| Malaylam | Bokka sura, Mookan sravu |
| Kanada | Neikaplethantte |

| | |
|---|---|
| Head | Length of the pectoral portion of the snout two-third the width of the mouth and considerably less than the distance between the eye and the first gill-opening: a very short groove at the angle of the jaws. Nostrils nearer to the extremity of the snout than to the end of the mouth. Snout rounded and very obtuse. |
| Teeth | 24-25(31)/24-25(31), the upper broad, flattened, oblique, with a very sight notch or rather concavity on the outer side and serrated: the lower much narrower, pointed, on a broad base and also serrated, these serration being |

generally minute, sometimes only on one side of the tooth and are occasionally wanting, especially in the lower jaw.

| | |
|---|---|
| Fins | Pectrol falciform, extending to beneath the middle or even hind edge of the base of the dorsal, its outer edge three times as long as its internal one. Base of first dorsal a little nearer the pectoral than its posterior end is to the ventral. Second dorsal opposite and similar to the anal. Length of cauldal 3.75 to 4.25 in the total length. |
| Scales | Comparatively large, lineated, but with almost smooth edges. |
| Colour | Brown or bluish-gray superiorly, fading to dull white beneath ends of all the fins deep black, iris gray, stained darker superiorly. (F. Day. 1958, p 715) |

## Classification

| | |
|---|---|
| *Class* | *Elasmobranch* |
| Sub class | Chondropterygii |
| Order | Plagiostomata |
| Sub order | Selachoidei |
| Family | Carchariidae |
| Genus | *Zygaena* |
| Species | *blochii* (Cuvier) |

## Synonyms

## Names in other languages

| | |
|---|---|
| English | Hammer headed shark, Aro headed hammer head shark |
| Gujarati | Kannar, Kanmushi |
| Tamil | Komban Sorrah, Nettaikomban sorrah, Kuttaikomban sorrah. |
| Telugu | Sapa sora and Koma surrah |
| Malayalam | Kannankodi |
| Kannada | Kebichatte |

***Sphyrna blochii* (Cuvier)**

Head Each lateral expansion of the side of the head from twice to thrice as broad, and with a deep groove along the anterior edge, reaching externally nearly as far as the orbit. The posterior edge of the lateral expansion about as long as its anterior edge, while the external edge is nearly straight. Nostril much nearer the mouth than the eye.

Eyes Situated near the upper angle of the external edge of the lobe of the head.

Teeth Oblique, externally notched, and smooth in their entire extent.

Fins The dorsal arises sllightly behind the base of the pectrol, extending nearly half way to above the ventral: second dorsal arising above the posterior half of the anal, than which it is much smaller.

Colour A deep gray or brownish-gray, becoming lighter beneath: the fins are usually of a slightly deeper colour than the body.

(F. Day, 1958, p 719)

## *Trygon Sephen*

Family Trygonidae.

Genus *Trygon*

Species *sephen*

## Synonyms

*Dasyatis sephan* (Forsk)

*Raja sephan* (Forsk), Deser, An p. 17

*Raja sancur* Ham. Buch. Fish. Ganger, p.2.

*Hypolophus sephan* Mull. and Henle. p. 170.

## Names in other Languages

English Cow tailed Ray

Gujarati Bpvar, Boor

Tamil Advalan

Telugu Volugiri, Tenkee

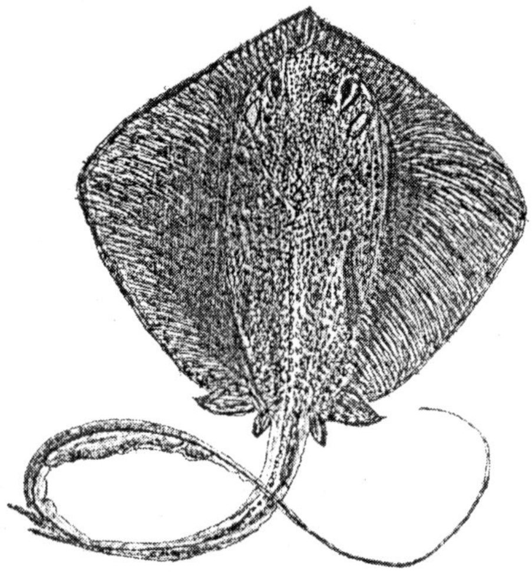

***Trygon sephan*** **(Day)**

## Identification

Body Disk rather broader than long; the length of tail three to four times that of the body. Snout more obstuse in adults.

Teeth Lower dental laminae somewhat pointed, the angularly bent for its reception. 33/25 rows transversely and 20/25 vertically opposite the symphysis, they are all flattened, those in the center of the upper jaw being the smallest.

Fins Tail with broad lower cutaneous fold: There may be one or two serrated spines situated rather behind its anterior third.

Scales Upper surface of the head and body and base of tail covered by thick, concave, or flat-headed, several sided, tubercles. In the scapular region and central line of the back are two in the young, and three in the adult, of large, smooth, approximating tubercles, the middle and largest of which is usually oval, the second in size posterior and heart shaped, the anterior or smallest round or heart shaped.

Colour — The upper surface of the immature is reddish-brown: but in the adult lead-coloured, becoming black in the posterior two third of the tail. some examples have a raw of black blotches near their outer margin.

Tail — Tail long, armed with a serrated spine.
(F. Day 1958, pp 749)

## *Narcine timlei* (Day)

## (Electric Ray)

An electric organ situated between the pectoral fin and the head. In India they do not appear to be used as food. It attains at least eighteen inches in length.

### Classification

| | |
|---|---|
| Order | Plagiostomata |
| Sub order | Selachoidaei |
| Family | Torpedinidae |
| Genus | *Narcine* |
| Species | *timlei* |

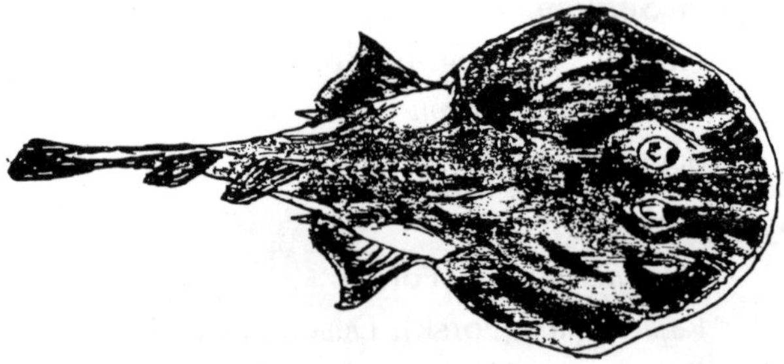

***Narcine timlei*** **(Day)**

### Synonyms

*Raja timlei,* Bl. Schn., p. 359
*Narcin indica,* Henle, Narcin. P. 35.

### Identification

Head — Outline of disk somewhat rounded, while it is broader than long: along the side of the tail is broad skinny keel reaching the base of the caudal fin. Caudal portion of fish rather longer than the body. Nasal valves confluent, forming a quadrangular skinny flap which is rather elongated in the middle. Spiracle just behind the orbit and not tuberculated on the edge.

Teeth — Flattened anteriorly, the internal ones with a small median cusp: the dental plate only embraces the central half of either jaw, the lips being thickened and continuous at the angles. The form of the dental plate varies, being either angular pointing downwards or similar to what obtains in the upper jaw.

Fins — The anterior dorsal usually commences just behind the ventral but occassionally over their posterior extremity, it is somewhat smaller than the posterior: in the young its apex forms an angle, which in adult becomes rounded. The distance between the two dorsal equal that between the posterior dorsal and the caudal. The size and shape of the two dorsal is similar: Caudal with its hinder edge rounded and confluent with the inferior.

Colour Body and fins reddish-brown above with numerous irregularly sized chocolate coloured spots: lower surface white. Pupil triangular, apex below iris golden. In the immature the spots have a white margin. Some examples are marked all over with large brown blotches much wider than the ground colours. In others no spots at all exist. These colour does not depend on sex, age or locality.

(F. Day. 1968, 733)

## *Rhynchobatus djiddnesis (Day)*

These fishes are destructive amongst marine crustacea and molluscs: and are said to live in large droves. Owing to the great injury done by them to the beds of pearl oysters at ceylon, it was proposed to enclose such with stakes to prevent their ingress. This sp. grows to at least six feet in length.

## Classification

| | |
|---|---|
| Family | Rhinobatidae |
| Genus | *Rhynchobatus* |
| Species | *djeddensis* |

## Synonyms

*Rhynchobatus djiddnesis* (Forsk), Raja *djeddensis,* Forsk, Descr, Amin p. 18, *Rhinobatus djisensis,* Bl. Sch. P.356, *Rhynchobatus laevis,* Mul and Henle, p.111.

## Names in other languages

| | |
|---|---|
| English | White spotted shovel nose ray. |
| Gujarati | Bhuther, Dhose |
| Tamil | Kachu ulvuai |
| Telugu | Walawah tenkee, Nul utavi. |
| Malayalam | Varithalai |
| Kannada | Etti baliar. |

***Rhynchobatus djiddnesis* (Forskal)**

## *Identification*

| | |
|---|---|
| Head | Snout elongated, the distance between the mouth and end of the snout equals 1/4 to 1/5 of the entire length excluding the caudal fin, being shortest in adults. |
| Eye | Rather large; the spiracle close behind the orbit. |
| Teeth | Oval, wider than broad with a horizontal cusp. Across the center of each 40-42/40-42. Twenty to twenty five verticle rows across the middle of both jawa. Dental plate with a central and smaller lateral elevation corresponding emarginations exists in the upper jaw. |
| Fins | The first dorsal commences opposite the center of the base of the ventral. The second dorsal is half near to the base of the caudal than to the posterior extremity of the first dorsal; it is smaller than lateral fin; but its shape is the same. |
| Scale | Minute, of irregular shapes and size keeled; a number of tubecles, directed backwards, exists its upper edge to above the spircle. |
| Colour | The immature are dull-gray above, whitish sometimes tinged with red beneath. A dark or black band on the upper eyelid and dark spot beneath on either side of the snout: also there is usally, but not invariably, a black spot at the root of the pectoral, which may have several small white ones round it. The body and some times the pectrol fin is spotted with whitish, or light gray. Iris golden. The adult is of a dull-gray above and lighter on the abdomen. |
| | (F. Day, 1958, pp 731). |

## Food and Feeding

Fishes of this group are carnivorous fishes except *Rincodon Sp. Rincodon sp.* feed only phytoplantons. Generally sharks move in shoals. They eat every fish coming on way. Eyes of these fishes are weak, however smelling capacity is powerful. They can smell blood and baits from very far distance. Sharks and Skates prey their feed from any height. While Rays are a bottom feeder hence they prey from bottoms only. Main food for Rays are prawns, molluscan detritus etc. These fishes deposits fat either in liver and/or in body under skin. Hence, many times they withstand without food for a long period.

## Habitat

Elasmobranch fishes habitat in shallow waters near shore as well as in deep sea. From the fish landing in Gujarat it was observed that sharks generally available in deeper water 50 to 80 m depth water while Skates and Rays which are bottom fishes were found in 20 to 40 m depth near shore.

## Reproduction

In this class of fish male is having special organ known as 'Clasper'. With the help of clasper fertilization is done in female body. For breeding fishes are coming to shallow water. Different species in this group of fishes gives birth to babies, while some lays fertilized eggs, using any of the following three methods.

### *(a) Oviparous method*

In this method eggs are kept in deep sea. At the time of laying eggs special type of thick gum like secretion is released which will keep all eggs combined with its filamental fiber. Such group of eggs are known as "Sailor's purse" or "Marmaid Purse". Biggest size of such purse is between 14 to 18 cm while smallest size is between 3.3 to 6 cm.

### *(b) Viviparous Method*

In this method female gives birth to babies. However, they are connected by one tube with mother. And through the tube they are getting nourishment as well as breathing from mother. When babies are capable to take their food by themselves, connecting tube with mother is broken and small babies are independent.

### *(c) Ova-viviparous Method*

In this method embrio are well developed in the body of the mother. Only when embrio are ripe female gives direct birth to babies.

Generally female fish release 5 to 30 either eggs or babies. Maturation time of embrio varies between 4.5 months to 15 months.

**Table 7.1 : Elasmobranch landing in Gujarat and its share in Gujarat and India**

| Year | Total Marine Fish landing | Landing of Elasmobranch | Percent of Total Gujarat | Marine fish Landing of India | Per cent share of Guj. in India |
|---|---|---|---|---|---|
| 1965-66 | 109907 | 2522 | 2.29 | 32100 | 7.86 |
| 1970-71 | 151190 | 5891 | 3.90 | 43294 | 13.61 |
| 1975-76 | 208300 | 10017 | 4.81 | 65230 | 15.36 |
| 1979-80 | 206749 | 4794 | 2.32 | 52843 | 9.07 |
| 1980-81 | 218872 | 6306 | 2.88 | 57862 | 10.90 |
| 1981-82 | 220607 | 6801 | 3.08 | 56009 | 12.14 |
| 1982-83 | 192669 | 7270 | 3.77 | 64316 | 11.30 |
| 1983-84 | 223291 | 7352 | 3.29 | 70046 | 10.50 |
| 1984-85 | 290708 | 1769 | 0.61 | 57757 | 3.06 |
| 1985-86 | 306577 | 10290 | 3.36 | 52805 | 19.49 |
| 1986-87 | 315942 | 12240 | 3.87 | 51898 | 23.58 |
| 1987-88 | 327560 | 11326 | 3.46 | 56585 | 20.02 |
| 1988-89 | 414075 | 11551 | 2.79 | 56847 | 20.32 |
| 1989-90 | 432384 | 12785 | 2.96 | 49979 | 25.58 |
| 1990-91 | 500462 | 9580 | 1.91 | 72748 | 13.17 |
| 1991-92 | 530017 | 15707 | 2.96 | 50420 | 31.15 |
| 1992-93 | 609103 | 16301 | 2.68 | 63132 | 25.82 |
| 1993-94 | 619836 | 17645 | 2.85 | 66423 | 26.56 |
| 1994-95 | 645261 | 15395 | 2.39 | 69274 | 22.22 |
| 1995-96 | 598351 | 19422 | 3.25 | 58634 | 33.12 |
| 1996-97 | 660058 | 15256 | 2.31 | 75304 | 20.26 |
| 1997-98 | 702355 | 21387 | 3.05 | 75301 | 28.40 |
| 1998-99 | 551860 | 14702 | 2.66 | 65902 | 22.31 |
| 1999-00 | 670951 | 16588 | 2.47 | 72963 | 22.73 |

Source: (1) Commissionerate of Fisheries, Govt. of Gujarat, Gandhinagar (with permission)
(2) CMFRI Published Annual Reports.

## Fishing Season

Fishing on west coast of India is between October and March. While on east coast of India it is between May and September.

## Craft and Gear

### *Craft*

In Gujarat for Shark fishing wooden cannoes were used. Now a days cannoes type fiber glass boats are being used. Similarly Gill netter boats are also being used for fishing. They are also caught in trawler.

### *Gear*

Generally they are caught in Gill nets. In Gujarat special bottom set net (Jada jal) are used for shark and skate fishing. However, being a bigger fishes they are caught using hook and line using different type of bait on single hooks in long line. Long line are popular in West coast. The larger sharks and rays are harpooned. When during monsoon other fishing is closed or stopped fishermen of Gujarat specially near Veraval from Jaleshwar Bandar going for bigger spp specially *Rhincodon sp.* using hooks.

### *References*

Anon C.M.F.R.I. Cochin, News letter No 32.

Chhaya N.D. 1990. Fisheries management and Science. Published by Gujarat Granth Nirman Board, Govt. of Gujarat.

Devansan, D.W., and K. Chidambaram. 1948. The common food fishes of the Madras presidency.

Dholakia, A.D. 1986. Scientific, English and local names of fishes available in Gujarat in Fish processing preservation and quality control. Published by D.P.Dholakia, pp 134-144.

Dholakia, A.D. and G.D.Vasavda, 1982. Commercial production of shark liver oil in Gujarat. Presented at International Seminar on Harvest and Post Harvest Technology of Fish at Cochin Book published in 1985, pp 634-637.

Dholakia, A.D., N.D.Chhaya and Y.A.Trivedi. 1980. Shark-Harmful marine davil or a useful commercial possiblities? "Matsyavikas" (I); 60:9-13 and (II); 61:12-13. Published by Commissioner of Fisheries, Govt. of Gujarat, Ahmedabad.

Dholakia, A.D., H.K. Gadhia and S.V.Vyas. 1982. Need of Veterinary grade shark liver oil in poultry culture. "Matsya Vikas", 75:23:26 Published by Commissioner of Fisheries, Govt. of Gujarat, Ahmedabad.

Fransis, Day. 1958. The Fishes of India. Published by William Dawson & Sons Ltd., London. pp. 715, 719, 729, 740.

Sukumaran, T. and A.D. Dholakia 1977. Shark Liver Oil-A good medicine. "Matsya Vikas", 41-42:32:34. Published by Director of Fisheries, Govt. of Gujarat, Ahmedabad.

# 8
# Edible Oyster and Mollusc Fishery

## Edible Oyster Fishery

It is said that "Oysters" is scientifically the best known marine animal. The living species of oyster are grouped into two families *Ostreidae* and *Gryphoidae*. Under the family *Ostreidae* six genera are recorded and under *Gryphoidae* there are two genera.

Of the six species of oyster found in Indian waters the first four mentioned below are known to be commercially viable species.

| | Scientific name | Commercial English name |
|---|---|---|
| 1) | *Crassostrea madrasensis* (Preston) | Indian back water Oyster |
| 2) | *C. rivularis* (Gold) | Chinese Oyster |
| 3) | *C. grypoides* (Schlotheim) | West coast Oyster |
| 4) | *Saccostrea cuculata* (Born) | Indian Rock Oyster |
| 5) | *Saxostrea cuculata* (Aqati and Rai) | Bombay Oyster |
| 6) | Hystissa hyotis (Linnaeus) | Giant Oyster |

Out of these no 2, 3, 4, and 5 are available on Saurashtra (Gujarat) coast and Gulf of Kutch. In Gujarat they are known as "Edible" or "Tar".

Oysters occur in shallow waters and form subsistence fisheries in several coastal regions along the Indian coast. Of the six species *Cressostrea madrasensis* is the dominant species having wide distribution. The annual catch which was estimated at less than 4000 tones was substantially increased in the recent years due to farming practices in the estuarine systems. Characters like high tolerance to environmental variation and rapid growth have made oysters a candidate species for commercial farming in some parts of India. Being a filter feeder, it gives high production per unit area and eliminates the cause for additional expenditure on supplementary feed. (Pillai, 2000).

## Morphology

The edible oyster is sedentary animal. The soft body of the animal is encased by two valve-lower cupped valve (left valve) and upper flat right valve. While the lower valve is commented to the substratum of the upper valve acts as a lid to open and close by hinge mechanism connecting both the valves. The movement of the valves is accomplished by the contraction and relaxation of thick abductor muscle.The shell consists of three layers. The outer most layer is known as periostracum. The inner layer is the nacreous layer which is thin, hard and usually shiny. The middle layer is thick and chalky. The shell is primarily composed of calcium carbonate. The periostracum is almost all protein. The texture size,

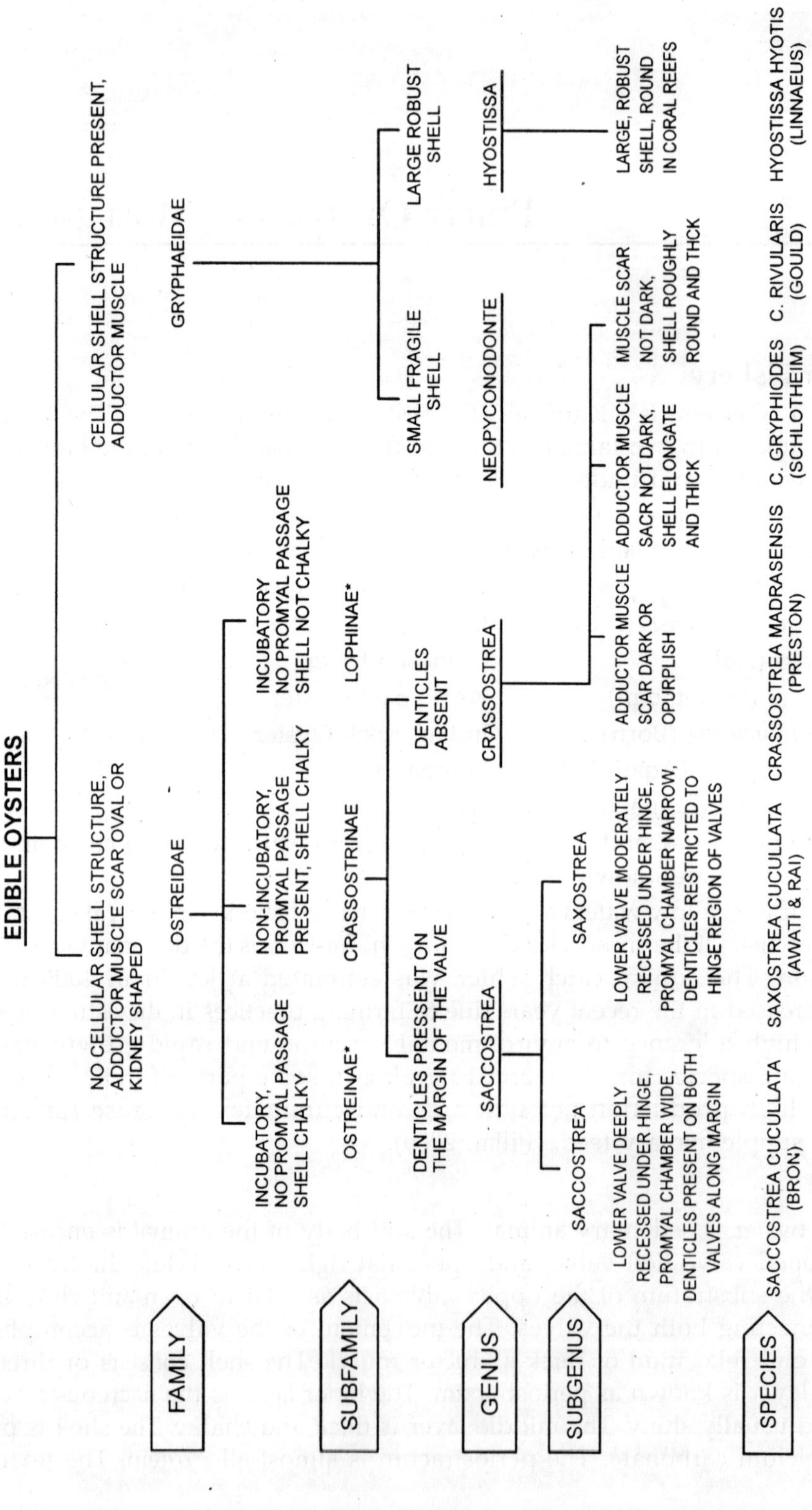


**Fig. 8.1 : Chart to identify the Edible oysters.**

shape, marking and colour of the shell very between species and from place to place. The same species grown in different environment exhibit different in shell characters.

Fig 8.2 shows the difference between *Crassostrea gryphoides* and *Saccostrea cuculata* which are main available species of Gujarat (Dholakia, 1996).

## Anatomy

With upper valve removed and the oyster lying on the left valve, the body of the oyster lies with the mouth near the hinge or the beak and the posterior side is the rounded end or the bill. The long axis of the oyster is actually the height, but common usage indicates this as the length. The dimension perpendicular to the height in the place of the shell is the length (dorso-ventral axis), but referred to as width in usage. A mantle covers the soft body except the adductor muscle area. (Fig. 8.3).

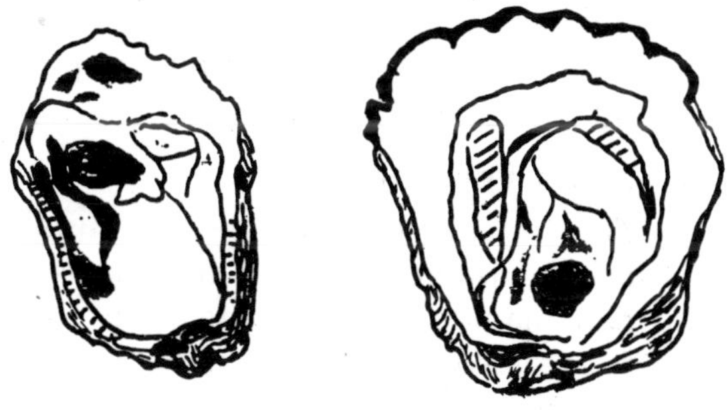

**Crassostrea gryphoides** **Saccostrea cuculata**

**Fig. 8.2 : Species of Edible Oyster**

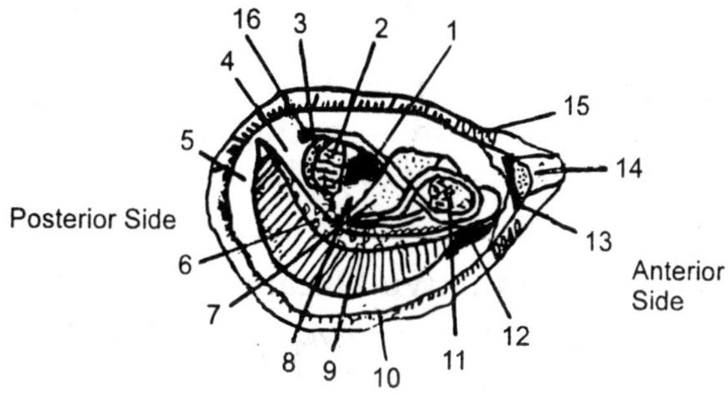


**Fig. 8.3 : Anatomy of an Edible Oyster**

## Mantle

A mantle fold spread as a pair of two thick shirts from the hinge line down towards the bill, covering the viscera. The edge of the mantle have three lobs or folds which are mainly sensory and assist in controlling the flow of water to the body. The mantle edge lies close to the edge if the shell where most of the shell is deposited. By secretion of the mantle edge the shell increases in the length and width.

**Table 8.1. Malacological difference between the oyster *Crassostrea gryphoides and C. madrasensis***

| | *C. gryphoides* | *C. madrasensis* |
|---|---|---|
| 1 | Adductor muscle roundest, bean-shaped rarely ablong and elliptical | Adductor muscle elliptical or oblong |
| 2 | Pigmentation of the mantle pronounced on the edge of the outer and inner lobes. The pigmentation also extends on the surface of the inner lobe. | Pigmentation if present, is mainly on the middlle and inner lobs. It does not or rarely extends on the surface of the inner lobe. |
| 3 | Mantle pigmented black | Mantle without or with pigmentation. |
| 4 | Anal portion of the rectum 1 to 2 mm in length and markedly directed outwards. | Anal portion of the rectum ½ to 1 mm in length sand slightly directed out words. |
| 5 | The anal tip is simple, at times slightly funnel-shaped. | The anal tip with variously folded collar. |
| 6 | Anal opening situated at the corner of the posterior and ventral margins of the adductor muscle. | Anal opening situated at the middle of the ventral margin of the adductor muscle. |

*Source* : Durve, V. S., 1973.

## Identification

Chart-I shows the identification of different species of Edible Oysters of Indian waters. (CMFRI-1992). Malacological difference between the oyster *Crassostrea gryphoides and C. madrasesnsis* was done by Durva 1993 and is shown in Table 8.1 colour photograph & edible oyster is given as Fig 8.4.

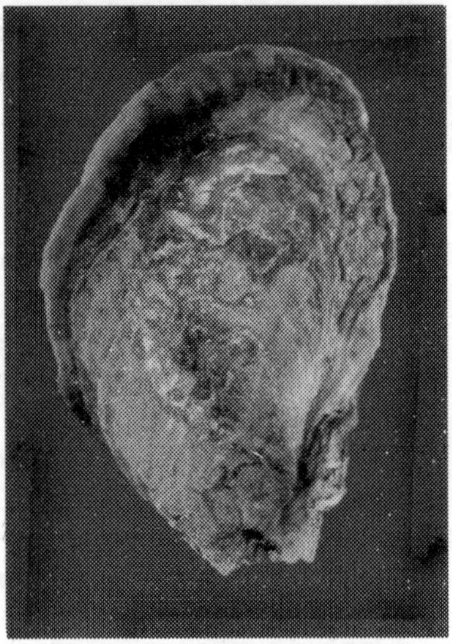

**Fig. 8.4 :** ***Edible Oyster Crassostrea Grophoides***

## Food and Feeding

The food of oysters mainly consists of organic detritus and photoplanktonic organisms such as diatoms and nanoplanktons spores and particulate matter of sea weeds are also found in stomach. Further, oysters are capable of absorbing dissolved organic matter in the water through the surfaces of the fills, palps and mantle.

The uptake of food largely depends on the pumping rate of oysters. In sheltered areas under normal temperature and salinity an oyster can pump 29 to 30 liters of water/hour. Under tropical conditions in relatively sheltered areas on an average an oyster will pumping nearly 20 hours in a day. This feature facilitates the oyster to grow fast.

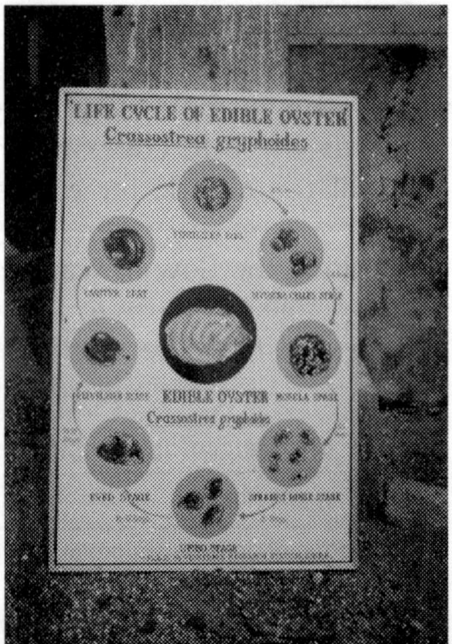


**Fig. 8.5 Life cycle of Edible Oyster**

## The Reproductive Organs

In cupped oyster (*Crassostrea),* the sexes are separate. the ovary and the testis consists of a series of branching tubules or follicles on each side of the body covering the visceral organs. The ripe eggs and sperms pass along a series of tubules by the ciliary's action in these tubules which finally merge in a tube along the dorsal side of the body. Two separate stystems of genital canals are found one on each side of the oyster which open into the epibranchial chamber and form there gametes are discharged to the exterior.

## Growth

Growth of oyster is expressed in terms of increment in length (length of the shell) and weight. The shelll growth is correlated with the growth of the living tissues. However, the growth of oyster varies from place to place and is elated to food availability and environmental conditions, particularly temperature and salinity. Oysters of the length range 60 mm to 90 mm are ideal and 30% of them should be of age group (60-75 mm) in order to be asssured of the availability of males in the brood stock. Life cyce of C. *grypoides* is given in Fig 8.5 (Dholakia. 1996).

## Spawning Season

Biannual spawning habit is observed with peak spawning in April-May. Second season is September-October. For natural spat collection September-October period was not so effective.

## Breeding

Oysters generally are dioecious, but hermaphrodites are not uncommon. Young oysters (*C. madrasensis*) function primarily as male (60-75%) and later becomes female. The eggs of *C. madrasensis* are 48 to 90 U in diameter. (CMFRI - 1992). At a time a single female *C. madrasensis* of 80 to 90 mm length will spawn 10-15 million eggs. Eggs lose their viability totally at the end of 24 hours in temperate regions and under tropical condition it is lost within 4 hours. In fertilized eggs of *C. madrasensis,* the first cleavage occurs immediately after the formation of two polar bodies within 45 minutes.

## Larval rearing

The sequence of development of the *C. madrasensis* larvae from straight hinge to pediveliger stage is as follows.

| Stage | Size in u | Hours/Days |
|---|---|---|
| 1) Straight-hinge | 60-70 | 20 hours |
| 2) Early umbo | 100 | 3rd day |
| 3) Mid umbo | 150 | 7th day |
| 4) Advanced umbo | 260-270 | 12th to 15th day |
| 5) Eyed larvae | 280-290 | 13th to 17th day |
| 6) Pediveliger | 330 to 250 | 14th to 18th day |

## Resource and availability

The oyster resources of the country are restricted but, the oyster beds are distributed in several centers along the Indian coast. Details is given in Table 8.2.

## Spat Collection from Natural Breeding Ground

Of the several types of spat collectors like bamboo mats, strings of coconut shells, run of green mussel and oyster shell valves and country tiles (with and without lime coating) tried in different locations around the bed. The most satisfactory results were obtained from twice lime coated semi cylindrical tiles lid completely submerged on wooden platform (Mahadevan S. *et. al.* 1980)

## Spat removal

Spat settled on collectors were allowed to grow on them up to a size of 30 mm-35 mm. After which removal of individual spat is easily done without injuring the fleshy interior. Pressure slightly exerted dislodges the spat without damage.

## Marketable Size

General marketable size of edible oyster is 85-90 mm reaches in 12 months with meat weight of 8-10%, each oyster weighing 120 g (shell on) with a meat weight of 10 g.

**Table 8.2 : Availability of Edible Oyster**

| Species | State | Area of availability |
|---|---|---|
| 1) *Creassostrea madrasensis* | Orissa | Bhauda Estuary near Sonapur, Miouth of Chilka lake. |
| | Andhra Pradesh | Sarada Estuary near Waltair, Bhimunipatnam Back water, Banks of Upputer Canal, Kadinada, Krishnapatnam, Goplaapalli Back water. |
| | Andaman and Nicobar | Island of Port Blair, Havelock island, Mayahender. |
| | Tamil Nadu | Near Madras, Pulicat Back water, Ennore Back waters, Vaigai Estuary near Athankarai, Tuticorin. |
| | Kerala | Ashtamudi and Vembanad lake, Mahe Estuary, Creeks of Darnadam. Valapatinam, Nileshwar. |
| | Karnataka | Nethravathi, Mulki, Udyavora, Venkatapur, Coondapoor and Kali Estuary. |
| 2) *Crassostrea gryphoides* | Karnataka | All Creeks |
| | Goa | All Creeks |
| | Maharashtra | All Creeks |
| | Gujarat | Mundra, Joraoara, Harshad, Miyani. Poshitra, Balapur. |
| 3) *C. rivularis* | Gujarat | Coastal waters and creeks |
| | Maharashtra | Mahim, Ratnagiri and Jaytapur. |
| 4) *Saccostrea cuculata* | | This species is found on rocky substrata in marine environment in shallow coastal and intertida coast of India and Andaman and Lakshadweep Island. |

## Production

Production of an average of 0.48 tones of oyster's from a single rack has been demonstrated by CMFRI. In one hectare it is possible to put up at least 280 racks and stock nearly one million oyster resulting in an yield of a minimum of 135 m.t. oyster (with shell) with a total meat yield of 13,5m.t.

## Nutritive value

The meat of the oyster is low calorie and highly palatable if grown in ideal conditions. It consists of 8-11% proteins, various carbohydrates and minerals. The fat content is less than 2.5%. This is because; the absence of fat makes them more nutritious and easily digestible. The oysters are rich in copper and Iron as an equivalent portion of livers. The successive accumulation of copper will give green colour to the meat and will diminish its texture and flavour. Further more, they contain much manganese, calcium, phosphorus and many times more iodine than beef. (CMFRI 1977).

Dholakia (1998) has collected and stationwise identified marine water molluscans of Gujarat is shown in Table 8.5.

## Utilization of Oyster shell

The shell constitute about 85 % of the total weight of oyster and contains 52-55% calcium oxide and is used in the manufacture of calcium, carbide, lime, fertilizers and cement. Further the oyster shells are useful as spat collectors of oyster spats. The shell are broken to suitable size and used as poultry grit. Recently fancy buttons are prepared from oyster shells.

## Predators and Foulers

Crabs, fishes, starfishes, polychaetes and gastropods are the predators that affect the oyster population. Excessive growth of algae on oyster bed may create anoxic condition leading to their mortality. Fouling organisms such as barnacles, ascidians, other mollusks and algae are considered mainly nuisance.

## Diseases and Parasites

Large scale mortalities of oysters were reported from time to time from temperate waters. Recently histopanthological studies have thrown some light one the possible factors responsible for may instances of large-scale mortalities. Some epizootic viruses are suspected to be reponsible for the cause of diseases and mortalities.

*"Dermo"* a dreaded disease caused by fungus *Perkinsus marinus* infected heavy mortality among East American Oysters. Gill disease and shell disease attributed to pathogenic fungus, caused heavy mortalities among Japanese and European oysters. A disease caused by the haplosporidian parasite *Minchinia nelsoni* was responsible for heavy mortality in American oysters.

Oyster beds affected by the above disease have revived after a long time. Large and thick oyster beds often are prone to be affected by diseases. Super intensive farming of oysters has to be avoided in order to avoid these problems.

## Pollution

Pollution plays an important role in the ecology of oyster beds. Considerable quantity of industrial wastewaters, agricultural run-off, dust from thermal power station and oil enters the sea. Oysters accumulate certain heavy metals and pose health hazards to those who consume them. Hence assessment of the level of pollution has to be made before starting culture work. Further, continuous pollution monitoring is to be carried out in oyster growing areas in order to ensure the quality of the meat. (CMRFI, 1992) .

## Mussel Culture

Coastal aquaculture and mariculture in India occupy an area of about 1,20,000 ha. Providing employment to more than 2,00,000 people. However, as the present region under production forms only 10% of the identified potential area in the coastal belt, there is great scope for increased thrust for the development of mariculture.

The national annual muscle production which stood at less than 10,000 tonnes till the beginning of this decade, has more than double due to development of muscle farms in the coastal areas. Internal market development for bivalve meat through consumer awareness programme supported the growth of this new industry. (Pillai, 2000) He has also given hatchery protocol for bivalve, edible oyster and pearl oyster. (Fig. 8.5) Culture flow chart for mussles is given in Fig. 8.7.

## Methods for mussel farming in India

### *Rock Method*

This method is especially suited for estuaries and shallow seas. Bamboo or Casurina poles are driven into the bottom spaced 1-2 m apart. These stakes are connected horizontally with poles. The horizontal poles should be above the level of water at high tide. Seeded rope can be suspended into the water for farming from these poles. Three seed rope can be suspended from one square meter area of the rack.

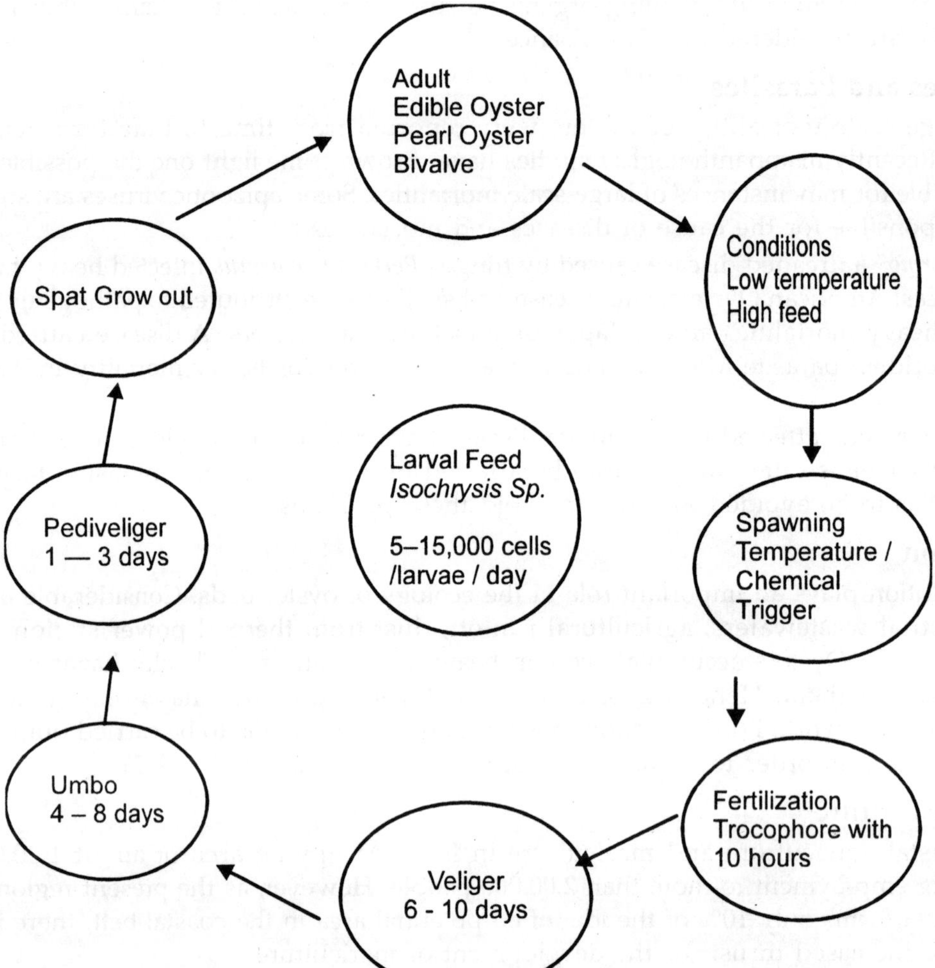


**Fig. 8.6 : Hatchery protocol for bivalve, edible oyster and pearl oyster**

## *Long line method*

This method is considered ideal for unprotected open sea conditions. Synthetic rope of 16-20 mm diameter is used for the long like (Main line). The main line is suspended with 200 liter barrel tied to it; spaced at 5 m. The long line and barrels are anchored position at either ends using concrete blocks and nylon ropes. Seeded ropes are suspended from the long line.

## *Raft Culture*

Ideal for open sea conditions which are not rough. Square or rectangular rafts are made with sturdy Bamboo or Casurina poles. Buoyancy for the raft is given by tying 5 barrels of

**MUSSELS (Green and Brown) Culture flow chart**

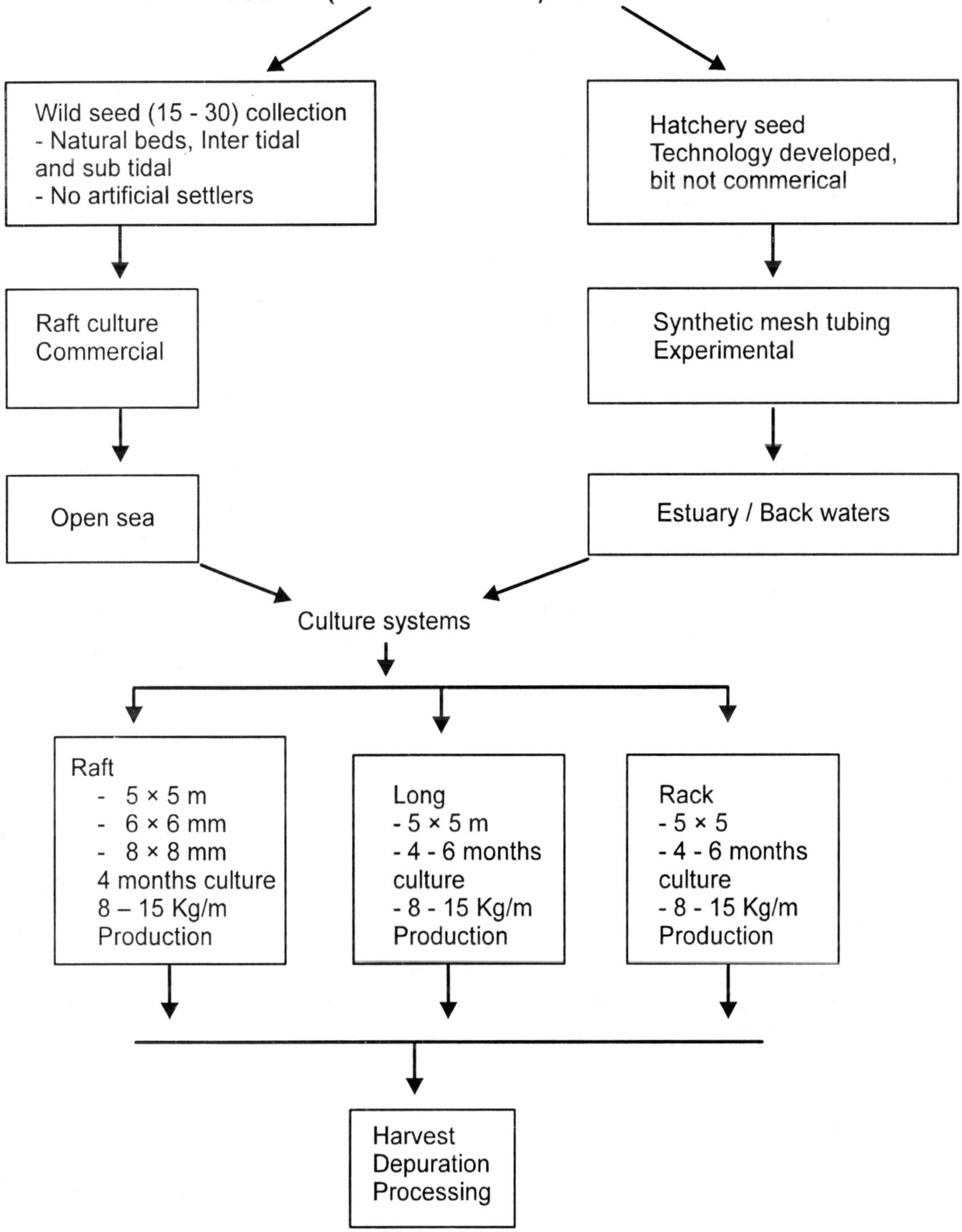


**Fig. 8.7 : Mussel culture flow chart.**

200 liters capacity (metal oil painted with anticorrosive paint or synthetic barrel) ideal size of the raft is 5 X 5 m. The rafts are to be positioned at suitable site in the sea using anchors. (grapnel, granite, concrete). Three seeded ropes can be suspended from one square meter area of the raft.

## References


Anon. 1977. Edible Oysters. C.M.F.R.I. News letter No 5, Nov. -76-March 77.

Anon. 1992. The Indian Edible Oyster-Technology for seed production and farming. C.M.F.R.I. Cochin.

Dholakia, A.D.1996. Edible oyster of Gujarat and its Culture, Gujarat Agricultural University, Sikka. Annual Report Fisheries Research Station, Guj. Agri. Uni. Sikka.

Dholakia A.D. 1998.

Druve, V. S. 1973. Malacological difference between the oyster *Crassostrea graphoides and C. madrasensis.*

Kasinathan, R. and M. Marimuthu 1978. Studies on the Biology of *Crassostrea madrasensis (Reston)* from Cuddaloree Coast. National Seminar of Edible Mollusca, Ab. No 37, p.27

Mahadevan, S.K. Ramadoes, M.E. Rajapandian, D.C.V. Easterson. Edible Oyster farming.

Pillai, V. N. 2000. Bivalve Mari culture in India (Pearl Oyster, Edible Oyster and Mussel and oyster). *Asia Pecific Asso. of Agri. Research Institutions* F.A.O. Bangkok.

Sarvaiya, R.T. and N.D. Chhaya, 1983. Saurashtra Molluscus. Smt. P. R. Sarvaiya. Ahmedabad.

Thangavelu, R. 1978. Ecology of settlement of spat of the Backwater oyster, *Crassostrea madrasensis* in Pulicat Lake. National seminar on Edible Molluscan Ab. No 7, p. 38.

Naskar, K. R. 1991. Plant succession of the mangrove Ecosystem of Sundarbans in West Bengal India, Abstract 15. *National Symposium on Conservation and Management of Living Resources.*

Pandya, J. P. 1993. Some aspects of the distribution of Macro fauna in relation to the size of Mangrove plants in the international zone at Rozy Port of Jamnagar. Department of Fisheries, Govt. of Gujarat.

Ramachandran, K. K. and C. N. Mahanan, 1987. Perspectives in Management of Mangroves in Kerala with special reference to Kumakon. *Proc. National Seminar on Estuarine Management* Edt. N. B. Nair : 252-257.

Singh, H. S. 1994. Status report of Mangroves in Gujarat State Gujarat Forest Department, Government of Gujarat.

Singh, H. S. 1994. Management plan for Marine National Park. Conservator of Forest, Marine National Park. Jamnagar.

Singh, H. S. 2000. Mangroves in Gujarat-Current status and conservation strategy. Gujarat Ecological Education and Research (GEER) Foundation, Gandhinagar, p. 128.

Singh, H. S. 2002. Marine Protected Areas in India. Status of Coastal Wetlands and their conservation in India, pp. 9-10.

# 9

# GHOL - DARA FISHERY

## Introduction

The north-west region, consisting of Bombay, Cambay, Veraval. Porbander, Dwarka and Kutch areas, has the most extensive shelf area (about 200,000 km$^2$) and an abundance of quality fish such as Ghol. (*Pseudosciaena diacanthus* (Blekar), Dara (*Polynemus indicus*) (show), Koth (*Otolithoides brunneus*) (Day) and fair abundance of prawns in some areas. Large quantity landing was observed from Kutch, near Porbander, Jamnagar and Navabandar Rajpara area.

This group of fishes are belonging to class Pisces, subclass-Teleostei and order: Acanthoterygii Family Genus and species of Ghol-Dara, Koth etc., are given in details as under.

### *Classification-Dara*

| | |
|---|---|
| Family | Polynemidae |
| Genus | Polynemus |
| Species | Indicus |

### *Names in other languages*

| | |
|---|---|
| English | Giant thread fin |
| Gujarati | Dara |
| Marathi | Lukwah, Bhat |
| Tamil | Yeta |
| Burma | Kwey-yeng |

### Polynemus indicus *show.* *Day-179*

Synonyms *Polynemus indicus* show, zool, v.p.155; swainson, Fishes, ii, p. 334; Mcclell. Cal. J.N.H., P. 179 pl vi;

*Polynemus uronemus* cuv. and val, iii, p. 385.

*Polydactylus indicus* show.

B. vii, D.8 $\frac{1}{13\text{-}14}$, P. 15 + v, V. 1/5, A. $\frac{2\text{-}3}{11\text{-}12}$,

C. 17, L.I. 70-75, L. tr. 7/13, Vert 5/19, Cae. Ply. many.

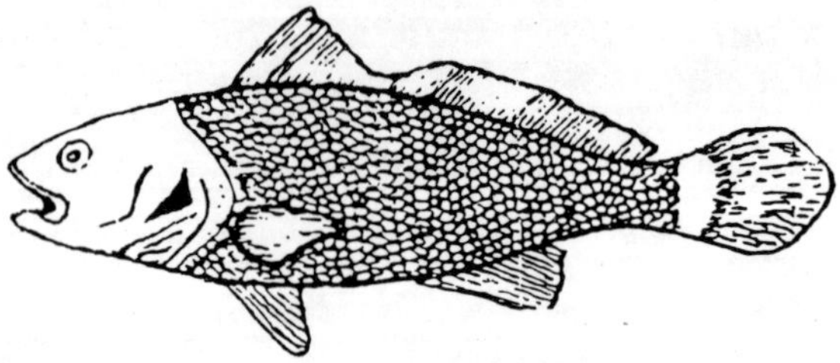

Fig. 9.1 : *Polynemus Indicus*

| | |
|---|---|
| Head | Length of head 4.25 to 4.50, of Pectoral 0.17 to 0.15 of caudal 0.28, height of body 0.17 of the total length including the filamentous prolongation of the tail. |
| Eyes | Diameter 0.14 of length of head, 1 to 1.25 diameter from end of snout and 2 apart. Height of head equals 0.66 and its width 0.20 of its length. The maxilla reaches to far behind the orbit, Preopercle with a rounded and produced angle, its posterior margin moderately serrated, and having a strong tooth above its angle Interorbital space rather flat. |
| Teeth | Villiform in the jaws, in a somewhat semilunar band on the vomer, and in a wide cuneiform band broadest anteriorly on the palatines, the last being emerginate internally about its center. |
| Fins | Spines of first dorsal weak, the third having rather a filamentous prolongation. The fin one third higher than long, second dorsal highest in front where it equals the anal and is one third higher than its base is long: upper edge of the fin concave. Pectoral with its rays branched having five articulated free rays, the upper of which is longest reaching nearly to the anal, which letter fin extends posteriorly some distance behind the vertical from the end of the second dorsal, its lower edge is concave. The distance between the bases of the ventral and anal fins is longer than the head. Caudal deeply lunated having pointed lobes which usually have filamentous termination, the lower being mostly the longer. Free portion of tail in its least depth equal to 0.40 of the length of the head. |
| Air-bladder | Oval and thick, occupying the entire length of the abdomen and posteriorly prolonged amongst the caudal muscles. It adheres to the vertebrae from the third to seventh, whilst from either side towards the ventral surface it has from 28 to 35 appendages. |
| Lateral-line | Continued along the lower caudal lobe almost to its end. Vertical fins rather densely scaled. |
| Colour | Back purplish-black, abdomen silvery-white, dashed with gold. First and second dorsal also anal stained with black, as is like wise the lower half of the opercle. Caudal with many black points. |
| Size. | It attains an average of 4 feet in length, but is rarely above 10 kgs weight. |

### *Classification–Ghol*

| | |
|---|---|
| Family | Sciaenidae |
| Genus | Sciaena |
| Species | Diacanthus |

## *Names in other languages*

| | |
|---|---|
| English | Two spinned jew fish, spotted crocker |
| Gujarati | Ghol |
| Marathi | Ghol |
| Telugu | Nella, Kachili |
| Malayalam | Cora |
| Kannada | Badi, Dholi |
| Tamil | Kurab |

## *Sciaena diacanthus* Day-189

## *Synonyms*

*Pseudosciaena dicanthus,* Bleeker, Mem. Scien. 1874, P. 27; *Johnius diacanthus,* Cantal, P.67; Bleeker, Java, P. 326; Kner, Novara Fische P.133.

B. vii, D. $10/\frac{1}{23\text{-}24}$, P. 18-19, V, 1/5. A. $\frac{3}{7}$, C. 17, L1. 72/52/160. Ltr. 7/18, Caec Pyl.8.

**Head** Length of head 0.25 to 4.25 of caudal, 0.17 height of body, 4.5 to 0.20 of the total length.

**Eyes** Diameter 4.50 to 0.17 of length of head, 1.25 to 1.50 diameter from end of snout and also apart.

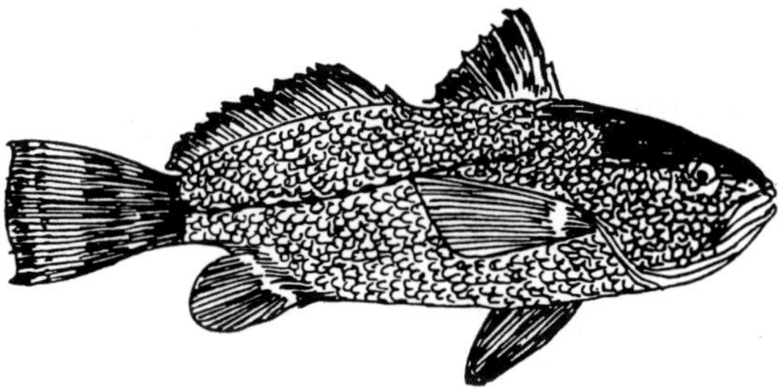

Fig. 9.2 : *Sciaena diacanthus*

**Snout** Slightly inflated, upper jaw a little longer, cleft of mouth oblique, the maxilla reaches to below the last third of the orbital. The distance from eye to upper jaw equals 2/3 of diameter of the orbit. Margin of preopercle crenulated in adult in the immature the angle is distinculated; two opercular points snout with three open pores across its base, and three opening along the free edge of the skin, but no lateral lobe. Five open pores under the symphysis of the lower jaw. Shoulder flap finally serrated.

**Teeth** Villiform in either jaw, with an outer raw of curved, conical and distantly placed ones in the premaxillaries, and an inner enlarged raw in the mandibles.

**Fins** Dorsal spines weak, increasing in length of the third and fourth which are 0.25 higher of than the rays, and 2 to 2.4 in height of the body. Pectoral equals the length of the head behind the middle of the eyes. Ventral reaches nearly half way to the base of the anal. Second anal spine moderately strong, at least half as long as the rays and equals to 2.75 in the length of head: the length of the base of the fin equals 0.33 to 0.28 of base of the soft dorsal. Caudal wedge shaped.

**Scales** Ctenoid except on the snout and below eyes.

**Lateral-line** Becomes straight above the end of the anal fin, the tubes are arborescent posteriorly.

Colour — Brownish-gray shot with silver along the back, which below the lateral line gradually fades to dull silvery-gray, head of the same colour glossed with purple. Fins yellowish, with black dots. Eyes golden. In the immature as up to foot and a half in length or even more, the fins are grayish with dark edges and the dorsal has two raws of dark spots : the caudal also has black spots and black edge. In still younger specimens the back and upper half of the body has many black spots, and the young are as a rule vertically banded.

Size — Attains at least 5 feet in length, normally.

## *Classification-koth*

| | |
|---|---|
| Family | Sciaenidae |
| Genus | Otolithus |
| Species | Brunneus |

## *Names in other languages*

| | |
|---|---|
| English | Rosy Jew fish, crocker. |
| Gujarati | Koth |
| Marathi | Koth |

## **Otolithus brunneus** ***Day-195***

## *Synonyms*

*Sciaenoides brunneus* Day

*Otolithus brunneus,* Journal Linn. Soc. 1873, P.524.

B. vii, D.9./ $\frac{1}{27\text{-}28}$ , p.18, V.1/5, A, $\frac{3}{7}$ , C. 17

L. I. $\frac{130}{\frac{102}{130}}$ ,L. tr. 21/34

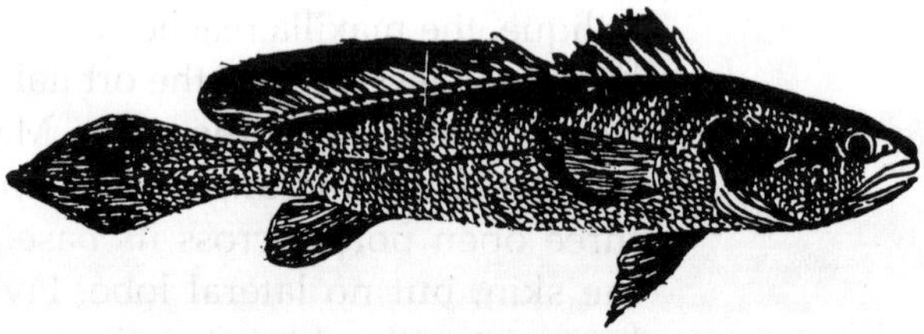

Fig. 9.3 : *Otolithus brunneus*

Head — Length of head, 0.25 to 0.22 of caudal 0.18 to 0.17 height of body 0.18 to 0.17 of the total length. Width of the head equals half its length, its height equal 0.6 of its length.

Eyes — Diameter 0.14 to 0.12 of length of head, 1.50 diameters from end of snout, and 2 apart.

Jaw — The jaws of about equal length, or the lower slightly the shorter. Cleft of mouth some what oblique, the maxilla reaches to below the hind edge of the eye. Preopercle with a few widely separated serration's, most apparent at its rounded angle. Opercle with two obtuse points. No open glands on snout: two badly marked pores on the front of the lower jaw below the symphsis. Shoulder-lobe finely fimbriated along its free edge.

| | |
|---|---|
| Teeth | An outer row of curved conical teeth in the upper jaw, having about two villiform rows internally, a small curved carine on either side of the certre of the upper jaw : an outer villiform row and an inner single row of irregularly sized and rather distantly placed conical teeth internally in the lower jaw, with a small canine-like one on either side of the symphysis, having a few vlliform ones posteriorly, few small teeth are present external to the enlarged row. |
| Fins | Dorsal spines weak and having filamentous termination, the third to sixth of some what the same length and equal to about 2.25 in the height of the body. Second dorsal highest in its last thrid, where it equals or exceeds that of the spinous dorsal. Pectoral as long as the head posterior to the middle of the eye: Ventral only reahces one third of the distance to the vent. Anal spines weak and short, 0.33 of length of rays, the length of base of the fin equal to 5.50 or 5.17 of that of the soft dorsal. Caudal pointed. |
| Scale | Rather oval, cycloid, and transverse by elongated on the head, ctenoid and vertically elongated on the body: a few very fine ones over the base of the soft dorasl and anal fins. |
| Lateral-line | Gently curves to below the first forth of the second dorsal, where it becomes straight. |
| Colour | Brownish, becoming golden below: fins darkest externally gray to blackish. |
| Size | From the growth study it was observed that at the completion of the $4^{th}$ year Koth should attain 95 cm. Koth is growing to a size over 160 cm. |

## Growth of Koth

During the growth study, Kutty (1961) and Jayaprakash (1976) observed that the average length of *O.brunneus* with 1 to 13 rings in the scales to be 41.8, 60.0, 74.1, 83.1, 107.9, 127.0, 134.7, 136.1, 138.5, 142.0, 148.2, 149.0 and 152.0 cms respectively.

'K' value of Koth based on 11 years growth data was 0.14 cm and L value was 206.4 cm.

## Distribution

It is reported that the maximum density of distribution of 'Ghol' in Gujarat and Maharashtra waters lies inside the 20 fathom line.

## Food and Feeding

Prawns and fishes formed major food items. Other crustaceans like crabs, *squilla* spp. alpheids and molluscs like *Sepia* spp., *Loligo* spp., gastropods were observed occasionally in small proportion.

*Acetes indicus, Parapenaeopsis stylifera* were important prawn species that were often seen in the stomach contents.

The number of juvenile "ghols" landed by "dol" nets gradually decrease from January onwards and by March they practically disappeared from the catch. This suggests that the post larvae and juveniles enter the inshore waters in June-July and remains there till January or February during which time they actively feed mostly on prawns. The peak landing of prawns from inshore waters (dol nets) was observed to be high from September to

November and from January onwards the prawns landings gradually decline reaching minimum in March. The juveniles of Ghol, perhaps move to deeper waters due to scarcity of prawns, their major food item, from February-March.

## Sex and Fecundity

### *Fecundity*

The estimation of fecundity becomes a difficult task in fish which spawn in more than one batch in the same season, However, as per the reasonable rough estimate of specimen 86.6 cm in length is low (20,91,220) while the ova count per gram sample showed maximum (12,222). Ova estimation per gram varies from 5,000 to 12,250.

### *Sex Ratio*

The sex composition of the random samples estimated each month for male: female showed average range between 38:62.

## Maturity

### *Maturation (P. indicus)*

Males: Three different stage of testes can be seen.

*Immature*

The testes is long, narrow, thread or ribbon like structure extending up to 2/3 or 3/4 of the body cavity width 0.1 to 0.5 cm. No milt expressed by pressure on the gonad.

*Mature*

Testis white or creamy in appearance, broader and thicker than the immature, extending up to 3/4 body cavity. Milt can be expressed by pressure on the gonad.

*Spent*

Testis broad, flaccid, blood shot extending up to 2/3 body cavity. No milt expressed with pressure.

Females: Development of ova are as under. Table 9.1

**Table 9.1 : Development of Ova**

| | Stage | Detail | Size of ova |
|---|---|---|---|
| 1. | Immature | Polygonal, completely transparent, Nucleus visible | 0.02 to 0.2 mm in diameter |
| 2. | Maturing | The central portion is opaque due to deposition of yolk. Nucleus faintly visible the larger ova fully opaque. | 0.2 to 0.4 mm. |
| 3. | Mature | Completely opaque yolk deposition complete. | 0.4 to 0.8 mm. |
| 4. | Ripe | Completely transparent and fully ripe. A single oil globule 0.2 to 0.3 mm. in diameter could be made out. | 0.8 to 1.0 mm in diameter |

## Minimum size at Maturity

It was observed that only 5% of Ghol were mature in 75-80 cm. group. 36.8% in 80-85 cm group and 91% in 85-90 cm group 100% in 90-95 cm and above group. Hence 85 cm may be taken as the average size of the fish at first maturity.

## Spawning season

It is observed that *P. diacanthus* spawns during the period June to August. October to March is the rest period, April to June is the period of active maturation. Spawning taking place throughout the year.

## Craft and Gear

### *Boat*

The craft used earlier in the Bedi fishery were non-mechanised Gujarat "Vahan". The description of Gujarat "Vahan" is a sturdy plank built boat with overall length of 32 to 45 ft, depth of 6 to 7.50 ft, beam of 9.5 to 12.5 ft and a draft of 3.5 to 6 ft this type boat was used between 1950 to 1980. Now a days mechanised boats are also in operation. Generally 32 ft OAL fiberglass boat with inboard engine are used. (Bhatt *et al,* 1964).

### *Craft*

Usual dimensions of the boats used in Ghol-Dara fishing was as follows:

Over all length–9 m to 14 m. (32′ to 45′)

Beam–2.89 m to 3.81 m. (9.5′ to 12.5′)

Depth–1.83 m to 2.29 m. (6′ to 7.5′)

Draft–0.914 m to 1.83 m (3.5′ to 6′)

The vessels are manned by 6 to 7 people.

The vessel are carvel built and without decks their 'U' bottom and lesser draft enable.

### Net

Saurastra and Bombay waters are the richest, contributing about 80% of the polynemides of the country. Polynemids are caught using different types of gear viz. Seine nets, gill nets, long-lines and hand lines which are operated by the country craft having engine for propulsion or sails. For this fishery, live baits and dead baits are used. In the Northern part of the country where they are found in abundance and fished with stake nets (Dol), bottom set long line (Khanda) and bottom drift gill net (Waghrajal). Among these gears, bottom drift gill net is a specialized gear employed almost exclusively for *Polynemus indicus* (Dara) during the season. These special types of gear are in use only in Gujarat and Maharashtra. This also forms a minor quantity in the trawl landings.

In Jamnagar area bottom set gill nets the *Halar Raach* gill nets of different mesh sizes used for catching Pomfrets. Mullets etc. and *Add,* a barrier a net are used. In Tuna-Bhadreshwar regions of Kutch Dist. *Add* a net made up of Malbar Coir rope is used. At sachhana near Jamnagar, Wada a boulder built tidal fish trap is used for capturing Koth, Ghol and Dara. At Rupen (Dwarka) the sole method employed is the widely practiced and most effective tackle of bottom-set gill nets. The pattern of the net is almost the some as that described by Bhatt *et. al,* for Ghol-Dara fishery of Bedi port. Formerly the material used for making the net was Italian hemp. Now it has been substituted by high density polythene twine of 2 to 2.5 mm diameter. Each beat carries 60 to 80 shots of length of 12 to 17 meters (7 to 9 fathoms) and the mesh size varies between 17 to 20 cm. Sole is slightly thinner. It is generally mounted on mono filament rope size of 0.63 cm to 1.27 cm, the foot being slightly thicker than the head rope one shot has about 15 wooden floats and 8 to10 stone or cement sinkers.

Generally bottom set gill net fixed usually at depths ranging from 15 to 30 meters. Many pieces of nets are tied together, the number of pieces in one net or unit varying from 80 to 90. Each pieces of net has 12 or more floats attached to the upper margin at an interval of 1250 mm.

During 1950's 'Ghol' 'Dara' and 'Koth' were captured using "Wadas". At Sechna about 18 km from Jamnagar this type of most primitive method of capturing fish were used. This type "Wadas" were bounder-walled tidal fish traps. these are circular enclosures with a diameter of 100 to 150 meters, which let in water during high tide. The single opening facing the sea is closed during ebbing tide when water flows out through the stone-walls entrapping the fishes.

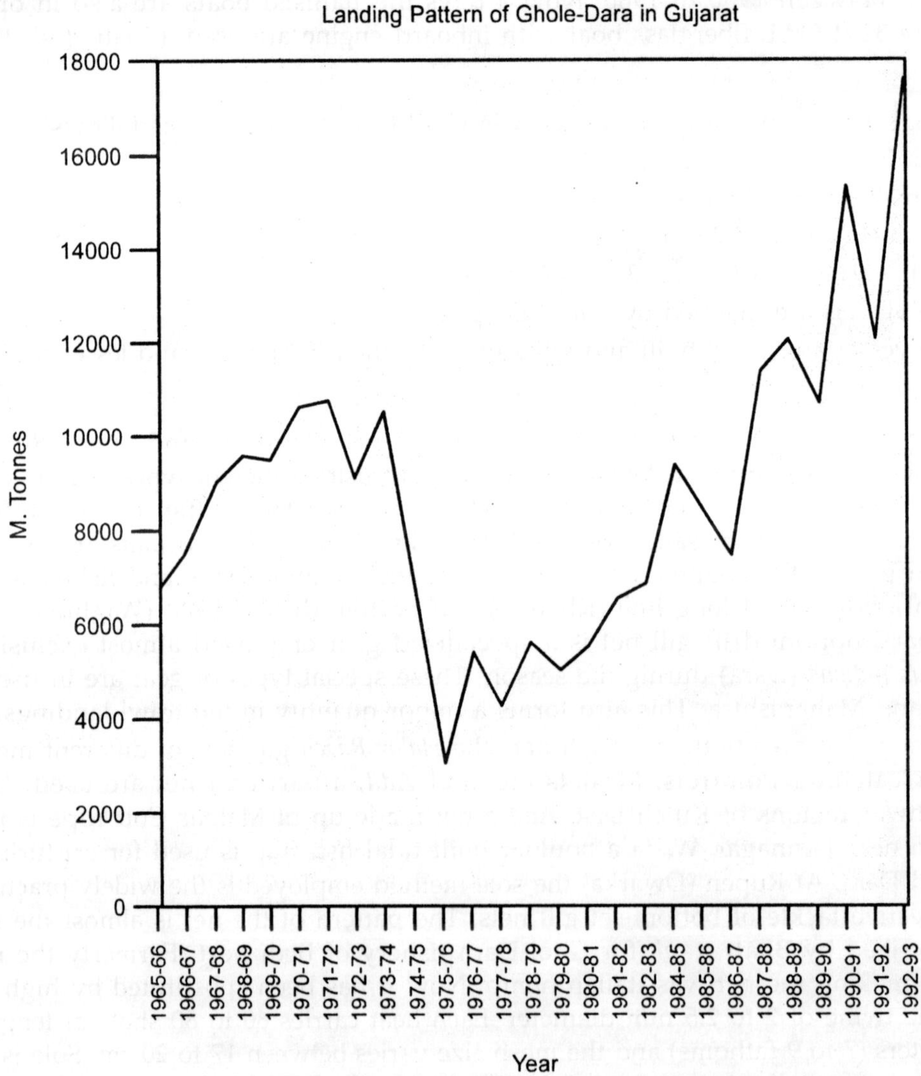


**Fig. 9.4 : Landing Pattern of Ghole-Dara in Gujarat**

## Fishery

Ghol-Dara fishery lasts for about six weeks from March to May and is peculiar in that it is made up exclusively of adut fishes (1000 to 1500 mm) and are being caught by highly selective, large-meshed gill net operated at the bottom.

The main area of activity of this fishery is off Bedi (Jamnagar) port (Long. 70° 2′ E lat. 22° 31′ N) situated inside creek near Jamnagar in the Gulf of Kutch.

The fishery at Bedi port is a spring tide fishery. It was observed that the catches tends to be very good around the new moon and full moon days. (Bhatt. *et. al,* 1964) Similar observation was made by Jayaramen *et. al.,* (1955) that the catch rate for 'Dara' in the trawling grounds off Dwarka (mostly between 31 and 50 meters) was higher in the spring tide period during January and May and in the neap tide period in all other months.

'Koth' are generally caught in trawl fishing. Maximum abundance being reported from off Dwarka in the month of December, January and February (Jayaramen, 1959).

This is significant when considering the fact that the trawler catches of Ghol, Dara and Koth from Dwarka region are mainly represented by Juveniles. While reporting Ghol-Dara fishery of Bedi port (Jamnagar), Bhatt. *et.al,* (1964) observed that 90% of the catches were composed of Dara. But observation at Rupen port shows that the catches are mainly formed by Ghol, followed by Koth and Dara in the order of abundance.

## *References*

Azeez, M.A. and K.G.S. Nair. 1975. Ghol-Dara fishery of Rupen Port (Gujarat) India. Sea Food Export Journal, 6:25-29.

Bhatt, Y.M. Narayanan Kutty, M. Venkata Subha Rao K. and D.M. Punwani 1964. "Ghol-Dara" Fishery off Bedi port in the Gulf of Kutch. *Indian J. Fish.* 11 (i):135-156.

Chhaya, M.M., Abdul Azeez and V.J. Thacker. Exploratory survey of trawl fish resources off Kutch (Gujarat) India.

Dholakia, A.D. 1980. List of fish available in Gujarat with scientific, English and local name in "Fish Procesing Preservation and quality control." Published by D.P. Dholakia, Ahmedabad pp. 134-144.

Dholakia, A.D. 1987. Method of utilizing fish skin from Ghol and Koth. Presented at National Seminar on Diversification of Post Harvest Technology for low coast fish. During 10-11 March 1987 at CIFT, Cochin.

Dholakia, A.D. 1987. New way of utilizing fish skin from Ghol and Koth. Fishing Chimes, 6 (10):14-15.

Fransis Day., 1958. The Fishery fo India Published by William Dawson & Sons Ltd. London. pp.179-189.

Jaya Prakash, A.A., 1976. Age and growth of the juveniles of Koth *Otolithoides brunneus* (Day) in Bombay waters. *Indian J. Fish.* 23 (1&2): 86-96.

Jayaraman, R., Seshappa. G., Mohamad, K.H. and Bapat, S.V. 1959. Observation on the trawl fisheries of the Bombay and Saurashtra waters. 1949-50 to 1954-55. *Indian J. Fish.,* 6 (i): 58-144.

Kagwade P.V. 1970. The Polynemid fishes of India., *Bull. Cent. Mar. Fish. Res. Inst.,* 18: 1-69.

Kagwade P.V. 1970 , Polynemid fishery resources of India. *Proc. Symp. Living Resources of the seas around,* sp. Publication CMFRI, pp. 424-433.

# 10
# Indian Mackeral Fishery

## Introduction

The Indian mackeral, *Rastralliger kanagurta* is a very important pelagic resources along the west coast of India, showing wide fluctuation from year to year. According to Yohannan 1982, the existing quantity of the average standing stock can give an average annual catch of 1,98,870 tones instead of 96,451 tones calculated by other workers. On an average it accounts for about 8% of the country's marine fish landings.

### *Classification*

| | |
|---|---|
| Order | Parsiformis |
| Sub-Order | Acanthuroidei |
| Family | Scombridae |
| Genus | *Rastralliger* |
| Species | *Kanagurta* (Cuvier) |

### *Synonyms*

*Scomber grychysoma (?)* Bleeker, Markr. P. 356.
*Scomber microlepidotus* Rupp. N.W. Fish. P. 38.
*Scomber kanagurta* Russel. Fish Vizag (ii), p. 987.
*Scomber reani*, Day. Proc. Z. S. 1870. P. 690.

### *Names of Other Languages*

| | |
|---|---|
| English | Indian Mackeral |
| Gujarati | Bangda. |
| Oriya | Karan kita. |
| Telugu | Kangurta, Kanangadtha. |
| Tamil | Kumla, Kanangaluthi. |
| Malayalum | Ayila, Ayla. |
| Kanad | Bangda. |
| Marathi | Bangda, Keulagader. |

*Mackeral*

## *Identification*

B. vi, D. 8-10/ $\frac{1}{11}$ + v vi, P. 21, V. 1/5, A.$\frac{1}{11}$ + v vi, C. 24, L.i 130 to 150 L.Tr. 10/28, Vert. 13/16.

| | |
|---|---|
| Head | Length of the head 3.75 to 4.25 of the caudal 4.75 to 5, height of body 4 to 4.66 the total length. |
| Eye | With a broad anterior and posterior adipose lid extending 0.33 across the eye. Diameter of the eye 4 to 4.5 in length of head, 1 to 1.25 diameters from end of snout, and also apart. |
| Jaw | Lower jaw slightly the longer, cleft of mouth deep, the maxilla reaches to below the hind edge of the eye. Length of preorbital 0.5 that of the head, its greatest depth equals half the diameter of the eye. Interobital space flat. |
| Teeth | Minute in both jaws, teeth in the upper jaw more or less disappear with age, none on vomer or palate. |
| Fins | Dorsal spines weak and receivable into a groove, the second to the eighth subequal and above 0.5 the height of the body, posteriorly they rapidly decrease in length of the base of the latter, the upper edge of which last is concave. Finlets commence just behind the fin and are opposite to the similar to those the anal. Pectoral short, nearly equaling half the length of the head. Caudal with deeply pointed lobes. |
| Scales | Smaller above than below the lateral-line, and largest just beneath the pectoral fin the first few rows beneath the lateral - line are parallel with it, those below the pectoral are irregularly concave. Minute scales on second dorsal and anal fins. |
| Lateral-line | Very slightly curved. Air - vessel present. Pseudobranchiae - largely developed. |

Colours Back greenish, side and abdomen descent, a row of sixteen spots along the summit of the back, close to the base of the dorsal fin : summit of head spotted : side shot with bluish -purple. Dorsal fins yellowish, tipped with black : caudal bright yellow, stained with black at its extremity. Pectoral bright yellow, with a dark mark on the body below it which show through the fin : ventral and anal finely dotted with black, but which fades soon after death when all the shot colours also disappear, and it becomes of a dull green with the abdomen of a lighter colour. Posterior edge of caudal some times blackish with a outer margin. In large specimen the colours differ, there are from five to eight dark longitudinal bands along the back and upper half of the body, the superior of which is occasionally broken up into spots. There are usually two golden bands below the lateral - line and one a long its course. In this stage it appears to closely resemble the description of *Steindloo* which is said to grow to a larger size than *R. kanagurta*, and to be destitute of teeth. (Source : F. Day, 1958, 251.)

## Food and Feeding

The Indian mackeral is a surface feeder in the inshore waters. They eat phytoplankton as well as Zooplanktons. The phytoplankton elements comprising the food are the diatoms represented cheifly by *Coscinodiscus, Pleurosigma, Chaetoceros, Nitzschia, Rhizosolenia, Fragillaria, Skeletonema, etc.*

Zooplanktons are from varied groups of crustaceans, mainly with copepods in high percentage. Different genera observed from stomach contents are *Eucalans, Pseudodiaptomus, Acartia, Oithona Centropages.*

However, *mysis, Acetes and Lucifer* are the main larger groups of zooplankton organisms can be visually seen. (Kutty, 1962).

On the West Coast of India two period of feeding intensities have been observed, one during October-December and the other during March-April. Among the adults the feeding intensity was high during maturing stages, low in mature and spawning individuals and again high in the spent and spent-recovering ones.

Different types of feeding are reported. The post-larvae (5-6 mm) are said to be "Herbivorous", feeding on diatoms and algae; the juveniles (25 mm) are "Omnviorous" feeding on all available food organisms in the plankton; and adult (35 to 224 mm) "Carnivorous", feeding on larval and juvenile fish. (Rao, K.V. 1962).

## Sex

R. kangurta is hetro sexual, but the two sexes are not distinguishable from their external characters. In rare cases occurrence of hermaphroditism has been reported (Prabhu and Antony Raja, 1959). From the gonadal examination, the two sexes can be differentiated in fish ranging from 12 cm onwards. (Bal & Rao, 1984).

## Age and Size Composition

The Commerical fishery begins to explorit mackeral from about a size of 18 cm., which they attain within six months. About 80-90% of fish in the commercial catch comes from size below 11 cms., which they attain in the first year and its life.

## Maturity

It has been observed by many workers that the minimum size at first maturity observed vary between 19 and 22 cms. According to Pradhan (1956), the first maturity of the Mackeral is attained when they have completed or are about to complete the second year of their lives in Karwar region. However, near Andaman Island it was reported that R. *Kanagurta* attains first maturity at 25 to 25.0 cms. When the fish is in the third year of its life (Luther, 1973). Pradhan and Palekar (1956) has published maturity Key and is given in Table 10.1.

**Table 10.1 : Key to the stages of sexual maturity (female)**

| Extent of in body cavity and general appearance, nature of Ova | Ova Diamater (mm) | Sate of maturity | Maturity stage |
|---|---|---|---|
| Ovary less than half the length of body and reddish-pink. Ova transparent without yolk and invisible to naked eye | 0.38 – 0.13 | Immature | I |
| Ova semi transparent, yolk in trace, oil globules minute and close to nucleus. | 0.14 – 0.27 | Immature | I-a |
| Ovary slightly more than half the length of body, pinkies and granular, Ova small whitish and opaque. | 0.28 – 0.37 | Maturing | II |
| Ovary extending to about two third of body cavity, pinkies, creamy to yellow and granular. Ova of medium size, opaque and whitish. | 0.38 – 0.46 | Maturing | III |
| Ovary extending to little over two-third length of body cavity, creamy and granular. Ova large white and opaque. | 0.47 – 0.56 | Maturing | IV |
| Ovary extending to entire length of body cavity with network of superficial blood vessels, yellowish and granular. Ova slightly translucent loosely attached to ovarian wall, with 1-3 oil granular of 0.17 to 0.23 mm. | 0.57 – 0.81 | Mature | V |
| Ovary extending in body cavity as in stage V "plum pudding" Appearance and yellowish with translucent all over gonad but not in the lumen, oil globule large. | 0.57 – 0.81 | Mature | VI a |
| Same as in VI a, but lume with transparent ova. | 0.57 – 0.81 | Mature | VI b |
| Ovary shrunken to about half the length of the abdominal cavity, blood-red in appearance, mature ova unhealthy and opaque. | 0.57 – 0.81 | Spent | VII |

Source : Pradhan and Palekar. 1956.

## Spawing Season and Reproduction

The fish spawns in the shelf waters in the areas of its distribution for extended periods. The individual fish releases eggs in batches. majority spawning is poticed around and during the south-west monsoon season in the West Coast. One female lays about 70,000 eggs at a time. It is generally considered that the makeral spawns during night time.

## Migration and Shoaling Behaviour

The mackeral and different fish group move in separate shoals. They move in semicircular or arrowhead formation and their speed is about 8-10 mile per hour. They scatter, when pursued by seer fish. But when sharks or porpoise chases the shoals, the mackeral submerges with the head downward into a compact mass. When the mackeral dive, a pitch & muddy water is seen at the surface, which is due to churnings & water by large mass & fish. The luminescences caused by mackeral shoal passing through a patch & Phosphorescent noctiluca off Ratnagiri coast are not unusual. (Biswas. U.P. 1989).

## Fishing Season

The fishing season on the West Coast begins in July-August and ends by April, peak season being November-December. It starts early in the south and lasts longer in the north. The fishery is sustained by immature fish, mostly 16 to 20 cm in size beieved, sustains the fishery to be mostly of 0-year class fish. As the stock consists mainly of a single year class it is susceptible to large variation in recruitment and mortality. This explains to a great extent the wide annual fluctuation in the stock size.

## Distribution

Mackeral is widely distributed along the Indian coast including Andamans.

## *References*

Anon. C.M.F.R.I. 1979. The Indian Mackeral in 1978. *Mar. Fish Infor. Serv.*, T & E Ser., No. 8 : 1-11.

Anon M.P.E.D.A. 1979. Indian Fisheries 1947-1977. The Fifth Session of the Indian Ocean Fishery Cochin. 19-26 Oct. 77.

Bal, D.V., K.V. Rao. 1984. The Fisheries of India, Published by McGraw Hill Publishing Company, New Delhi. P. 103.

Biswas, K.P. 1989. Text Book of Fish, Fisheries and Technology. pp. 286-290.

Chhaya, N.D. 1990. Fisheries Management and Science. Published by Gujarat Granth Nirman Board, Govt. of Gujarat. p. 260.

Dholakia, A.D. 1986. Scientific, English and local names of fishes available in Gujarat in "Fish processing, Preservation and Quality Control". Published by D.P. Dholakia, Ahmedabad. Pp 133-145.

Fransis Day. 1958. Fishery of India. Published by William Dawson & Sons Ltd., London. pp 251.

Kutty, M.N. 1962. Observation in the Indian Mackeral *Rastrelliger kanagurta* (Cuvier) from the trawl catches along the Bombay coast. *Indian J. Fish.*, 9(2) : 590-603.

Luther, G., 1973. Observation on the Biology and Fishery of the Indian Mackeral *Rastrelliger kanagurta* (Cuvier) from the Andaman Islands. *Indian J. Fish.*, 20(2) : 425-474.

Pradhan, L.B., 1956. Mackeral Fishery of Karwar. *Indian J. Fish.*, 3(1) : 141-185.

Pradhan, L.B. and V.C. Palekar, 1956. Key to the stages of sextual maturity of *Restrelliger kanagurta* (Cuvier). *Indian J. Fish.*, 3(1) : 183-185.

Prabhu, M.S. and B.T. Antony Raja. 1959. A note on hermaphroditic Gonad in the Indian mackeral *Rastrelliger kanagurta* (Cuvier) *J. Mar. biol. Ass., India*, 4(2) : 241-243.

Vijayaraghavan, P., 1962. Some observation on the spawning behaviour of mackeral. *Indian J. Fish.*, 9(2) : 647-652.

Yohannan, T.M. 1982. Population dynamics of Indian Mackeral based on data from Manglore during 1967-75.

11

# Lobster Fishery

Total 18 species of rock lobsters representing to nine genera in family Palinuridae have been reported from Indian ocean.

The lobster fishery of India is supported by 6 species of shallow water spiny rock lobster belonging to the genus *Panulirus* namely *P. homarus, P. polyphagus, P. ornatus, P. penicillatus, P. versicolor* and *P. longipes*, Of these, three species (*P. homarus, P. polyphagus* and *P. ornatus*) are commercially exploited. maximum lobster landing in India was realised mainly from Gujarat; Maharashtra and Tamil Nadu. Sandlobster *Thynus orientalis* is also found in commercial catch of Gujarat.

In Gujarat water following species of lobsters were found.

| Sr. No. | Species | Maximum Size |
|---|---|---|
| 1 | *Panulirus polyphagus* (Herbst) | 45 cm. |
| 2 | *Panulirus penicillatus* (Olivier) | 45 cm. |
| 3 | *Panulirus ornatus* (Fabricius) | 46 cm. |
| 4 | *Thynus orientalis* (Lund) | NA |

## Fishery

Some aspects of the biology and fishery of *P. homarus* have been studied. The species breeds during November-December. It has been estimated that one and two years old lobsters largely support the commercial fishery.

In year 1997-98 exports reached a high Rs 478 million for about 1290 tones of lobsters. The annual catch is in the range of 2500 tones. Out of which lobster landing from Gujarat was recorded as 442 tones.

High prices in the international market, in the range of Rs 370 to Rs 400/kg. particularly in Japan, Singapore and the U.S. have led to the large scale exploitation of India's lobsters.

*Classification*

| | |
|---|---|
| Phylum | Arthropoda |
| Class | Crustacea |
| Order | Decapoda. |
| Family | Palinuridae |
| Genus | Palinuridae |

*Name in Other Languages*

| | |
|---|---|
| English | Rock lobster, Spiny lobster, Sand lobster. |
| Gujarati | Titan, Lobster, Kako. |

*Taxonomy (Identification)*

*Panulirus polyphas* (Herbst)

## Synonyms

*Cancer* (Astacus) *Polyphagus.* (Herbst)

*Polinurus faciatus* (Fabricius)

*Panulirus faciatus* (Milne-Edwards)

*Panulirus Polyphagus* (Nobili).

Greenish gray or muddy brown in colour with numerous indistinct white spots. A row of six large white spots are present on either side of the carapace. These are a large oblique patch on either side of the first abdominal Semite. The hind margin of each abdominal Semite has a brown band with a white line in the middle. Yellowish white spots are present on the walking legs.

The antennuar plate with two principal spines situated far interiorly and the rest of the plate is unarmed. The median spine of the fused coxicerites of the antennae is large and situated in front of the lateral spines. There are no spinules. The exopod of the second maxilliped has a many jointed flagellum, where as the third maxilliped is without and exopod. The abdominal Semites are without transverse groove.

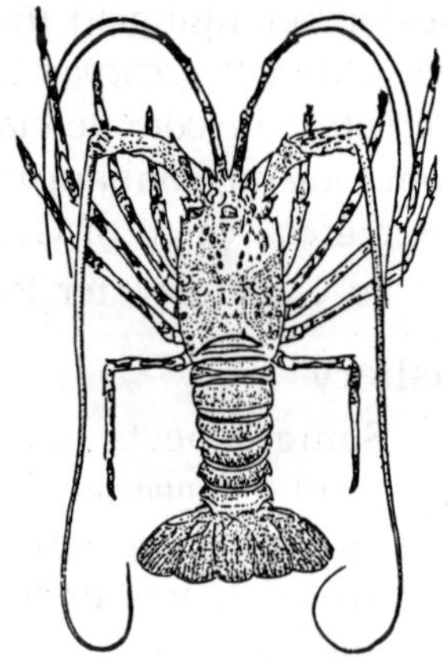

**Panulirus Polyphagus**

*Panulirus penicillatus* (Olivier)

## Synonyms

*Astacus penicillatus* Olivier

*Palinurus penicillatus* Oliver

*Panulirus penicillatus* White

Dark brownish green or brownish pink in colour with numerous orange yellow spots. The entire abdomen is speckled with minute white spots and on top of the plate D a conspicuous white spot is present. There are pale yellow straight or wavy lines along the length of the walking legs.

The antennular plate has four principal spines united at the base, the anterior pair being smaller than the posterior. Rest of the antennular plate is unarmed. The median spine of the based coxicerites of the antennal is large than the lateral spines and there are one or two spinules in between the median and lateral spines. The second maxilliped has an exopod with many jointed flagellum white the exopod of the third maxilliped is without flagellum. The abdominal somites have an uninterrupted transverse groove, the anterior margin of which is non-crenulated and without setae.

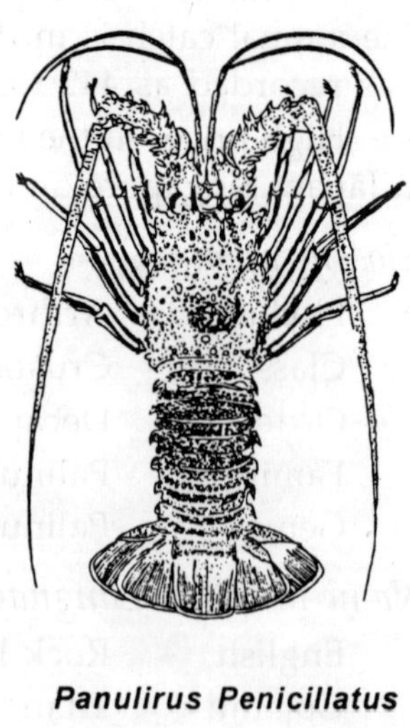

**Panulirus Penicillatus**

*Panulirus ornatus* (Fabricius)

## Synonyms

*Cancer* (Astacus) *homarus* (Herbst)
*Palinurus ornatus* (Fabricius)
*Palinurus sulcatus* (H. Milne Edwards)
*Palinurus sulcatus* (White)
*Palinurus ornatus* (Stimpson)
*Palinurus* (*Senex*) *brevipes* (Pfeffer)
*Palinurus* (*Senex*) *Sulcatus* (Pfeffer)
*Palinurus homarus* (Pfeffer)
*Palinurus polyphagus* (Borradaile)

Bluish green in colour, the carapace has pale cream vermiculite lines and bright orange spots. Each abdominal somite has a dark blue transverse band with one or more two cream coloured oblique spots on the sides. Walking legs alternately banded or mabled with cream and maroon.

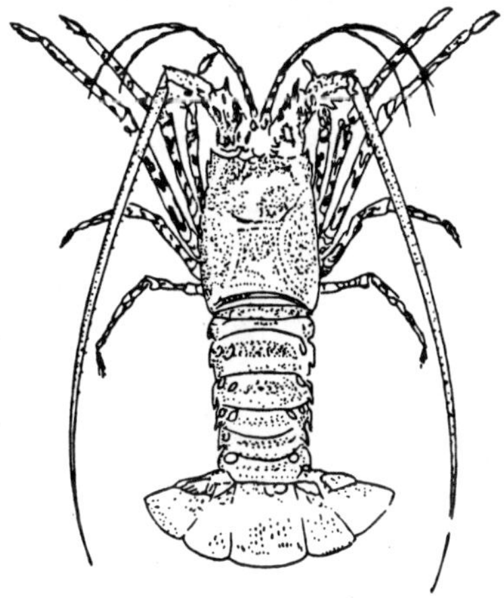

*Panulirus ornatus*

Four principal spines, the anterior pair being larger than the posterior are present on the antennular plate. Small spinules may be present in between the two pairs of spines. The median spines of the fused coxicerites of the antennae is larger than the lateral spines and the three spines are situated almost on the same line. There are no spinules. The second maxillaped has an exopod without a flagellum but with turf of setae, where as the third maxillaped is without an exopod. Abdominal somites without a transverse groove. (R. Raghu Prasad 1986)

*Panulirus homarus* Linnaeus.

## Identification characters

Common name : Scalloped Spiny Lobster, Green Lobster

The carapace of the animal is cylindrical with spines of varying sizes. There is no rostrum but two frontal horns are present. Eyes are not embedded in sockets. Bases of antennae are separated by antennular plate bearing 2 equal well-separated pairs of principal spines and scattered smaller spines in between. Each abdominal segment is with transverse grove, sometimes interruped in the middle. Its anterior margins are formed into shallow scallops. First pair of legs is without pincers. Colour of this animal varies from dark greenish to blackish with numerous small white spots and without transverse bands. This is one of the important Lobster, which form good fishery in India. This is one of the important culturable species, growing to a maximum of 30 cm in body length and its carapace length 12 cm. It is suitable for culture in pen and cages erected in the coastal waters.

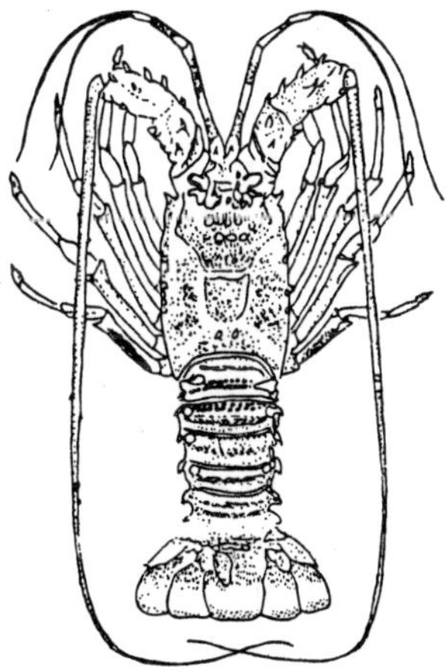

*Panulirus homarus*

This species breeds throughout the year and carry eggs for short period. Larval life is long *P. homarus* is a shelter dweller, lives in den and crawl among corals, lime stones, rocks and other hard objects on the sea bottom. (Sundararaj C. and B Srikrishnadhas, 2000).

### *Thenus orintalis*

| | |
|---|---|
| Phylum | Arthropoda |
| Order | Decapoda |
| Class | Crustacea |
| Family | Scyllaride |
| Common name | Sand Lobster or Slipper lobster |

***Thenus sp.***
**Sand Lobster**

Sand lobster or slipper lobster has extremely flattened body, especially the carapace and antennual plates, and attains about 25 cm in length. Carapace broader than long and strongly narrowing posteriorly. Eyes situated at outer angle of carapace and protected by sockets. Carapace and abdomen covered with dark brownish tubercles. Body colour uniformly muddy gray.

### Food and Feeding

The main food item for lobster is mussel *Perna perma* which directly related with the occurrence of lobsters. They also take fish, warms, crabs, clams, archinodermis, and animals which are habitated in the bottom of the sea.

### Growth Rate

It was observed that growth in ablated females juvenile show an overage increase of 1.02 gm/day while the increase was only 0.35 gm/day on control. Maximum weight gain of 4.6 gm/day ahs obtained in an ablated mature lobster weighing 256 gm (Redhakrishnan 1982). It is also reported that salable lobsters of about 500 gm can be grown in 2 to 2.5 years. From 30 gm to 100 g. (which is minimum marketable size) it takes about 80 to 90 days.

### *Habitate*

Most of the species of lobsters prefer particular regions or hibitat. There is a great variation in the habitat preference of adult lobsters and it may vary from the inter tidal zone to great depths of oceans characterised by mud-ooze substract and perpetual darkness. It is generally believed that the adult lobsters do not undertake long migrations. It is reported that *P. ornatus* undergoes mass migration between habitats which explains the significant contrast between habitats of the juveniles and adult of this species.

The most characteristic feature of the habit of many of the palinurids is the residence area or den. Dens are not usually constructed by the palinuridae but are opportunistically chosen from the surrounding habitat. The burrowing ability of most palinuridae is usually assumed to be minimal (Kancrick 1980).

Adult lobsters prefer from intertidal zone to great depths of the ocean characteritised by mud-ooze substrate and perpetual darkness. They are concentrated in Den. Dens are not constructed but chosen from surroundings. The burrowing ability of most *Palinuridae* is unusually assumed to be minimal. (Reghuprasad, 1986).

### *Habit*

According to Kancrick 1980 "Interactions between environment like temperature, light, photoperiod, salinity etc. and Palnurid behavior are complex, and evidence indicates that such environmental parameters strongly affect activity patterns, migratory movements, reproduction, growth and regional as local distribution. Based on current research, the most important of these factors seems to be light and temperature.

Spiny lobsters are shelter dwellers and include singular or multiple den residency and show definite homing patterns. Most of the species do not undertake extensive migration. They generally begin to feed at dusk, are gregarious and return to their dens before down.

From the difference spectrum of the photopigment one would predict that lobsters maximum visual sensitivity in near 515 m. that its ability to see beyond 600 m. (orange red and red colour) is negligible and that after minimum 420 m. its sensitivity increase slightly into the ultraviolet region. Provided that eye has no cut off filters for this spectral region. (Talbot Waterman, 1988).

## Breeding and Spawning Season

### *Breeding*

Breeding pattern is not understood but maximim numbers of larvae are found at about 50 m. depth (Raghuprasad 1986). No female lobsters observed with eggs were from the landing stock near Bhavnagar Dist. in Gujarat (Thaker & Nimavat 1987). Female of lobster carries about 50,000 eggs. (Edward 1989). More abundance of larvae were found at 100m. depth.

## Craft and Gear

Hand, traps and gill nets catch lobsters. Boys drive in shallow rocky areas where lobsters occur in sufficient numbers. Skilled boys catch some half a dozen lobster within a couple of hours.

## Gill Nets

Gill nets are kept vertically at bottom set. Lobsters get entangled in the nets. Stake net are also used in Bhavnagar District.

## Landing

Table 11.1 gives the detail of lobster landing in Gujarat and its share in Gujarat and in India from 1975-76 to 1998-99.

## Traps

These are box-like, about 0.6 m. long and 0.45 m. broad flat at bottom and round or conical at the top. Some of them are made of bamboo splints, which last for longer time. The trap has a one-way entrance on one side, and another opening with doors on the other side. These traps are kept on the seabed with muscles as bait inside.

One fisherman may keep even 50 traps in one day. They are lifted periodically by diving from a boat, and lobsters are taken through the doorway. 2 to 3 lobster may enter a trap.

## *Chemical Composition of Lobster Meat*

A critical analysis of meat from the tail portion showed following chemical composition

| | | | | | |
|---|---|---|---|---|---|
| (1) | Moisture content | 71.87 to 78.95% | (2) | Proteins | 15.71 to 24% |
| (3) | Fat | 0.24 to 01.35% | (4) | Ash | 0.94 to 02.94% |
| (5) | Carbohydrate | 0.29 to 08.08% | (6) | Calcium | 0.02 to 0.63% |
| (7) | Phosphorous | 0.16 to 0.51% | (8) | Food energy | 91 K.cal/100 gm |

(Source : 'Indian Lobsters' Published by M.P.E.D.A.)

## Lobster Culture

In Gujarat lobster culture has been taken up in Bhavnagar District. Lobster are being landed at Ghogha, Osara, Hathab, Mithi Virdi, Sartanpar, Ralia, Rajpara, and Katpra landing centers. Maximum availability is between August to October.

At present lobster culture has started at Sekhadia near Mahuva light area preparing pit known as "Bandhan" 7 m.(L) x 4.6 m.(w) x, 1 m. (h) in rocky area. The rocks are made of corals. Big coral stone are kept in pit and pit is covered with monofilament net. Small lobsters are kept in pit to grow to marketable size.

**Table 11.1 : Lobster landing in Gujarat and its share in Gujarat and India**

| Year | Total marine Fish landing | Landing of Lobster | Per cent of Total Guj. | Marine fish Landing of India | Per cent Share of Guj. in India |
|---|---|---|---|---|---|
| 1975-76 | 208300 | 424 | 0.20 | 2991 | 14.18 |
| 1979-80 | 206749 | 209 | 0.10 | 1136 | 18.40 |
| 1980-81 | 218872 | 235 | 0.11 | 679 | 34.61 |
| 1981-82 | 220607 | 265 | 0.12 | 1481 | 17.89 |
| 1982-83 | 192669 | 249 | 0.13 | 1764 | 14.12 |
| 1983-84 | 223291 | 320 | 0.14 | 1301 | 24.60 |
| 1984-85 | 290708 | 871 | 0.30 | 3223 | 27.02 |
| 1985-86 | 306577 | 899 | 0.29 | 4082 | 22.02 |
| 1986-87 | 315942 | 783 | 0.25 | 3074 | 25.47 |
| 1987-88 | 327560 | 967 | 0.30 | 2530 | 38.22 |
| 1988-89 | 414075 | 1162 | 0.28 | 1587 | 73.22 |
| 1989-90 | 432384 | 1169 | 0.27 | 1590 | 73.52 |
| 1990-91 | 500462 | 875 | 0.17 | 1834 | 47.71 |
| 1991-92 | 530017 | 1141 | 0.22 | 2254 | 50.62 |
| 1992-93 | 609103 | 939 | 0.15 | 2095 | 44.82 |
| 1993-94 | 619836 | 832 | 0.13 | 1671 | 49.79 |
| 1994-95 | 645261 | 578 | 0.09 | 2692 | 21.47 |
| 1995-96 | 598351 | 592 | 0.10 | 2631 | 22.50 |
| 1996-97 | 660058 | 542 | 0.08 | 2787 | 19.45 |
| 1997-98 | 702355 | 476 | 0.07 | 2620 | 18.17 |
| 1998-99 | 441660 | 442 | 0.10 | 2094 | 21.11 |

Source : (1) Commissionerate of Fisheries, Govt. of Gujarat, Gandhinagar, (with permission)
(2) CMFRI Published Annual Reports.

## Need for Protection

From the analysis it can be seen that quantity of lobster landing is decreasing every year. Sudden increase in 1985 landing may be indication of over exploitation. It requires protection from conservation point. It is necessary to also start lobster culture on a large scale in Gujarat. Dholakia envisaged this in 1985. (Dholakia, 1985). It can be seen that after 1991-92 lobster landing is decreasing, in spite of heavy demand in International market.

### *References*

Anon, 1986. Indian Lobsters Published by the Marine Products Export Development Authority. Extention Section, Cochin - 16.

Dholakia, A.D. 1985. High necessity of protection of lobster, Fishing in Gujarat "Matsya Vikas" 88-89-90 : 29-31.

Dholakia, A.D. 1986. Scientific, English and local names of fishes found in Gujarat in "Fish Processing preservation and quality control" Published by D.P. Dholakia, Ahmedabad.

Eric Edwards (USA) 1989. Sea food International, July : 33-34.

Kanciruk, P. 1980. Ecology of juvenile and adult Palinuridae. In the Biology and Management of Lobster : 59-96. Academic Press. Inc., New York.

Radhakrishnan, E.V. and M. Vijay Kumaran. 1982. *Mar. Fish. Infor. Serv.* T&E Ser., 43: 6-8.

Raghu Prasad, R. 1986. Distribution habits and habitats of Palinurid lobsters and their larvae. *Mar. Fish. Infor. Serv.* T&E Ser., 70: 8-15.

Sarvaiya, R.T. 1987. Possibilities of lobster culture near coast of Gujarat. "Matsya vikas" 105 : 12-16, Published by Dept. of Fisheries, Govt. of Gujarat.

Sundararaj, V. and B. Srikrishnadhas, 2000. Cultivable Aquatic organisms. Published by Narendra Publishing House, Delhi.

Talbot, H. Waterman. The physiology of crustacea Vol II Dept. of Zoology yale Uni : pp 44.

Talbot H. Waterman. 1988. The physiology of crustacea Vol II Dept. of Zoology yale Uni : pp 44.

Thakar, D.B. and D.M. Nimawat, 1987. Initial stap in lobster culture "Matsya vikas" 102 : 18-21 published by Dept. of Fisheries, Govt. of Gujarat.

# 12
# MULLET FISHERY

## Introduction

The gray mullet, locally known as "Mankan", forms a seasonal fishery along the Dwarka coast on the North-western part of the Arabean Sea.

### *Classification*

Class : Teleostel
Order : Acanthopterygil
Family : Mugilidae
Genus : *Mugil*
Species : *cephalus*

*Mugil cephalus* (Cuv. and Val.) Day 355

Synonyms *Mugil ocur : Day*
*M. japonicus :* Tem and Sschlog. Fauna. Japan p.134, PI 72
*M. macrolepidotus* : Richards Ich. China P. 24-PI.9

### *Distinguishing Characters*

B. Vi, D. 4/1, P. 15, V, 1/5, A3, C. 15, L : 1, 422-44, L. tr. 14

Head : Length of head 4.50 to 4.66, of caudal 4.75, height of body 5.33 to 5.66 in the toal length.

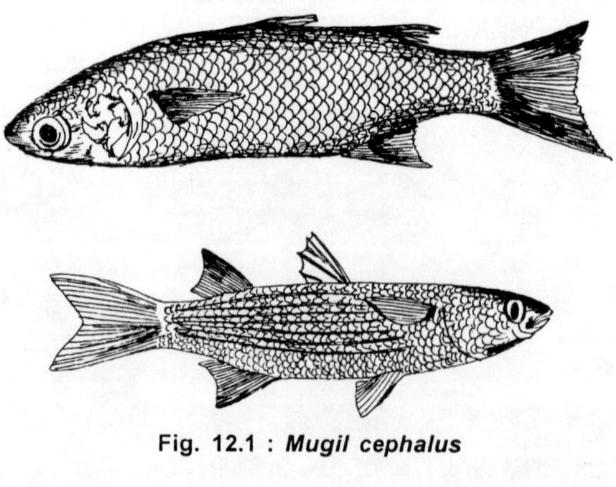

**Fig. 12.1 : *Mugil cephalus***

Eyes : With broad anterior and posterior adipose lips, both of which reach the pupil, diameter 4.66 in length, 1 diameter from end of snout and 2.33 apart. Inter orbital space flat and its width equal to half the length of head; the dorsal profile from the snout to the base of the dorsal fin is nearly horizontal scaled, neither notched nor bent, is lower and posterior edges serrated, the maxilla only visible as a narrow strip or even quite concealed.

Lip : Upper lip thin, and in the nesial line forms the front of the snout. The mandibles join at right angle, are notched at their point of junction and have two small open gland beneath them.

Uncovered space on chin rather broad and rounded anteriorly narrow and calceolate posteriorly. The extent of the cleft of the mouth is little more than half the width of its gape.

Teeth : Distinct in both jaws.

Fins : The first and second dorsal fins commence above the twelfth and twenty-fifth scales of the lateral line. The origin of the first dorsal is midway between the end of the snout and the base of the caudal, its spines are rather strong. The first is as long as the head behind the eye, and slightly exceds the height of the second dorsal, the extent of the base of which last fins equal three-forth of its height. Pectoral situated above the middle of the depth of boy, as long as the head behind the eye and extending to the eleventh scale of lateral line, Anal arises apposite.

## *Mugil Persia* (Hamilton) Day-350

Synonyms *Mugil cantoris,* Bleeker, Beng. En, Hindi; p 100

Distinguishing characters

B.V:, D. $4/\frac{1}{8\text{-}9}$, P.14, V. 1/5 A, $\frac{3}{8\text{-}9}$, C. 14, 14, L1 34-35, L.tr. 11-12

Head : 5 to 5.25 height of body 4.33 to 4.5 in the total length.

Eye : 3.50 in length of head ¾ of a diameter from end of snout. End of maxilla visible. Broad posterior eyelid.

Fins : Anal commences slightly anterior to the second dorsal.

Scale : No long axillay scale, 21 or 22 rows before dorsal fin, eleventh and twenty second scales of lateral line correspond to origins of first and second dorsal fin.

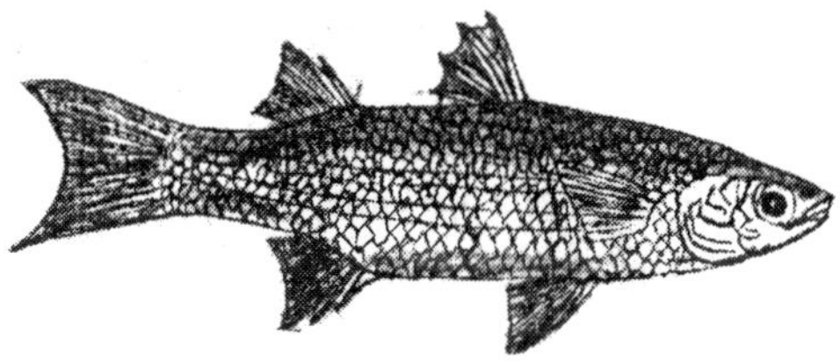

Fig. 12.2 : *Mugil parsia*

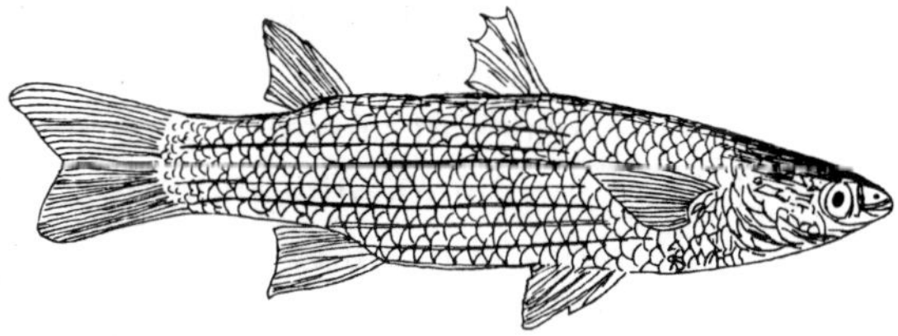

Fig. 12.3 : *Mugil tade*

## *Mugil tade* Forscal Day-346

Synonyms : *M, planicveps* Day

Distinguishing characters

B. vi, D. $4/\frac{1}{8}$, P. 17, V 1/5, A/$\frac{5}{9}$ C. 17, L.1, 33-35, L. tr. 11

Head : 4.50 to 5, height of body 5.50 to 6 in the toal length.

Eye : Broad posterior eyelid. Eyes 6.5 to 7 in length of head, 1.50 diameter from end of snout. End of maxilla visible.

Fins : Anal commences slightly anterior to the second dorsal.

Scale : No long axially scale : 20 rows before dorsal fin : tenth and twenty first scales of lateral line correspond to origins of first and second dorsal fins. A dark line along each rows of scales.

*Mugil seheli* Forscal Day-347

Synonyms : *Mugil axillaries,* Cuv. and Val. Xi, P. 148

Distinguishing characters

B. vi, $D.4/\frac{1}{8}$, P. 15, V. 1/5, $A.\frac{3}{9}$, C-16, L.1, 42-43, L.tr. 13-14

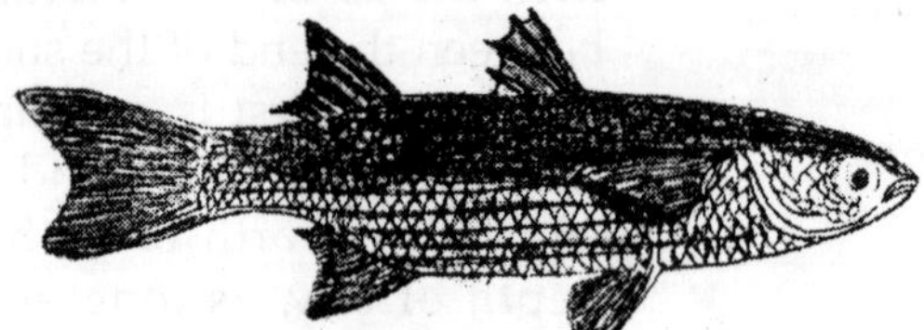

Fig. 12.4 : ***Mugil cunnestus***

Head : 4.66 to 5 height of body 4.66 to 5.50 in the total length.

Eyes : 3.50 to 4 in total length of head ½ of a diameter from end of snout. Extremity of maxilla usually concealed.

Fins : Anal commences slightly before second dorsal.

Scale : A long axially scale 21 or 22 rows before the dorsal in, twelfth and twenty forth scales of lateral line correspond to the origins of the first and second dorsal fins.

*Mugil cunnesius* Val Day-346

*Synonyms : Mugil kunnesae* Russel, Fish. Vizag. li. P. 65. Pl 181

Distinguishing characters

B. vi. D. 4 1, P. 16, V. 1/5, A.3, C.14, L.1, 33-34, L tr. 12.

Head : 5.25, height of body, 4.50 to 4 in length.

Eye : Broad posterior eyelid, 3.50 to 4 in length of head, 2/3 of diameter from end of snout. Extremity of maxilla just visible.

Fins : Anal commences rather anterior to second dorsal.

Scale : A long axially scale. Eighteen rows before dorsal fin; tenth and twentieth scales of lateral line correspond to origins of first and second dorsal fins.

Field Key for the identification of adult Mullet. (Sundaran, 1951)

1. With adipose eye lids ---------- *Mugil*

(A) Eye lids thick and fleshy

(a) Eye large, 4.33 to 5 parts in the length of head; dark blue band across half the base of the pectoral ---------- *Mugil cephalus.*

(b) Eyes lids small, 6 to 6.50 parts in the length of head; long and abnormally flat head ---------- *Mugil tade*

(B) Eye lids thin and very transparent.

(a) Dark spot at the upper edge of the base of the pectoral.

(i) Pectoral equals to or longer than head ---------- *Mugil Speigleri*

(ii) Pectorals markedly shorter than head---------- *Mugil cannesiu*

(b) No dark spot at the upper edge of the pectoral.

(i) Head flat 3 to 6 faint brownish lines along the body ------------------ *Mugil subviridis*

(ii) Head convex : No line along the body ---------------------------------------- *Mugil jerdom*

## Identification of fry of Mullet

### *Mugil cephalus*

*M. cephalus* fry can be distinguished by the characteristic ventral profile and the comparatively thicker head. The dorsal profile from D, to behind the orbit is only vey slightly declivitous, but is steeply so from the orbit to the snout. The prportionate height of the body is more than in any of the above species. The colouration is dark gray dorsally, grayish silvery on the sides and silvery on the ventral aspect. The lines are gray.

### *Mugil tade*

The fry of *M. tade* are smaller, and not quite so round in body as those of *M. corsula*. Dorsal and ventral profiles of the body are similar. The colouration is silvery with a tinge of gray on the back. The fins are uniformly tinged with gray. The highy compressed head and pointed snount of the adult are clearly distinguishable in the fingerling stage.

### *Mugil parsia*

The fry of the species have no specially outstanding characteristics. The body is of medium height and thickness. The ventral profile of the head body is arched. The colouration is light gray dorsally and silvery on the sides and on the sides and on the ventral aspects. The edges of the fins are gray.

### *Mugil cunnesius*

The characteristics features of the fry of this species are the laterally compressed body and the ventral profile of head, which is steep and angular from tip of snout to orbit, and arched thereafter. The height of body is also more than for any of the other species described here. The dorsal profile from D to nape of head is very gradually declivitous, after which it is somewhat more stooping to the tip of the snout. The colour is green olive dorsally extending to about ¼ the sides and silvery below with a tinge of yellow. The fins are light and transparent. Fins pigment spots are arranged in rows along the height on the body, these spots are clearly visible in the living condition, can be discerned on freshly caught specimens and become very prominent after death and remain so when preserved in formalin.

The fingerlings have the characteristic features of the adults such as the pointed axillary scale and the dark spot on the base of the pectoral. The arrangement of pigment spots along white myotomes is clearly evident in the fingerlings and at this stage the unpaired fins assume dark margins as seen in the adult.

Fild Key for Identification of Fry of Mullet (Sarojini)

(1) Ventral profile of head arched ............................................................ 2
Ventral Profile of head not arched ........................................................ 4

(2) Dorsal and ventral profiles of body similar .............................................. 3
Ventral profile of body distinctly more convex than the dorsal profile .................................................................................. *M. corsula*

(3) Dorsal profile from first dorsal fin to tip of snout gradually declivitous ........................................ *M. parsia*
Dorsal profile from first dorsal fin to orbit horizontal and from orbit to tip of snout declivitous slightly convex ........................................ *M. tade.*

(4) Ventral profile of head steeply declivitous from tip of snout to opercle ........................................ *M. cephalus*
Ventral profile of head from snout to orbit steep and angular and arched there after ........................................ *M. cunnesius*

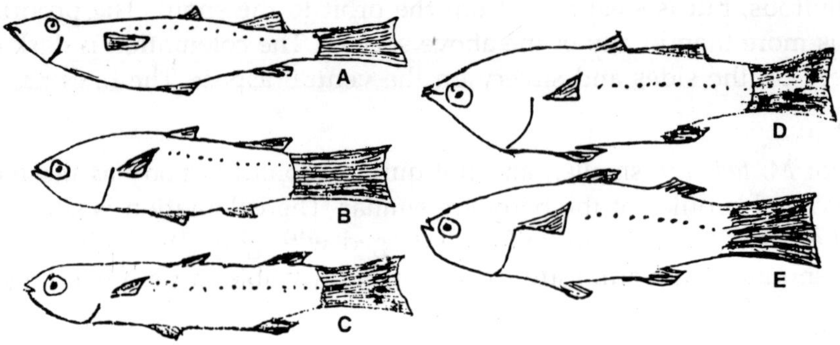


**(a) *Mugil corsula* 27 mm (b) *Mugil parsia* 20 mm (c) *Mugil tade* 20 mm (d) *Mugil cephalus* 26 mm (e) *Mugil cunnesius* 23 mm.**

## Resources

### *Mullet Varieties in Okhamandal (Gujarat)*

Mullet fry resources of Okhamandal was studied by Narayanan *et al.* (1962). According to them following varieties of Mullet fry are available in Okhamandal.

| Sr. No. | Scientific Name | Local name | Per cent availability |
|---|---|---|---|
| 1. | *M. parsia* | Karochi, Cheri, Boyee | 55% |
| 2. | *M. cunnesius* | Pamia, Pamiyo | 15% |
| 3. | *M. cephalus* | Bodki, Bodka | 08% |
| 4. | *M. Macroienpis* | Karochi | 06% |
| 5. | *M. tade* | Thromsi,Thramchi | 05% |
| 6. | *M. sehli* | Mankan | 03% |
| 7. | *M. wieginsis* | Motda | 05% |
| 8. | *M. carnatus* | Cheri | 03% |

Source : Narayanan *et al.*, (1962)

## Maturation

The mature *M. cephalus* has been reported from various parts of the country. Ranged from the length of female fishes was 445 to 460 mm weighing 1000 to 1200 gm. The ovary measured 106 to 165 mm in length and weight 50 to 270 gm. The ova diameter was 0.210 to 0.629 mm and the gonadosomatic index was found to be 4 to 20.

The length of male specimens ranged from 355-380 mm and weighed 450 to 560 gm. The testis measured 40 to 85 mm weighing 6 to 10 gm. Maturation was measured during June-July. (Lal Mohan and Nandu Kumaran, 1987).

## Growth Rate

The growth rate per month ranged between 2.7 mm and 27.8 mm in length and 0.8 g and 25.4 g in weight with a monthly average of 16.9 mm and 10.6 gm in respect of length and weight.

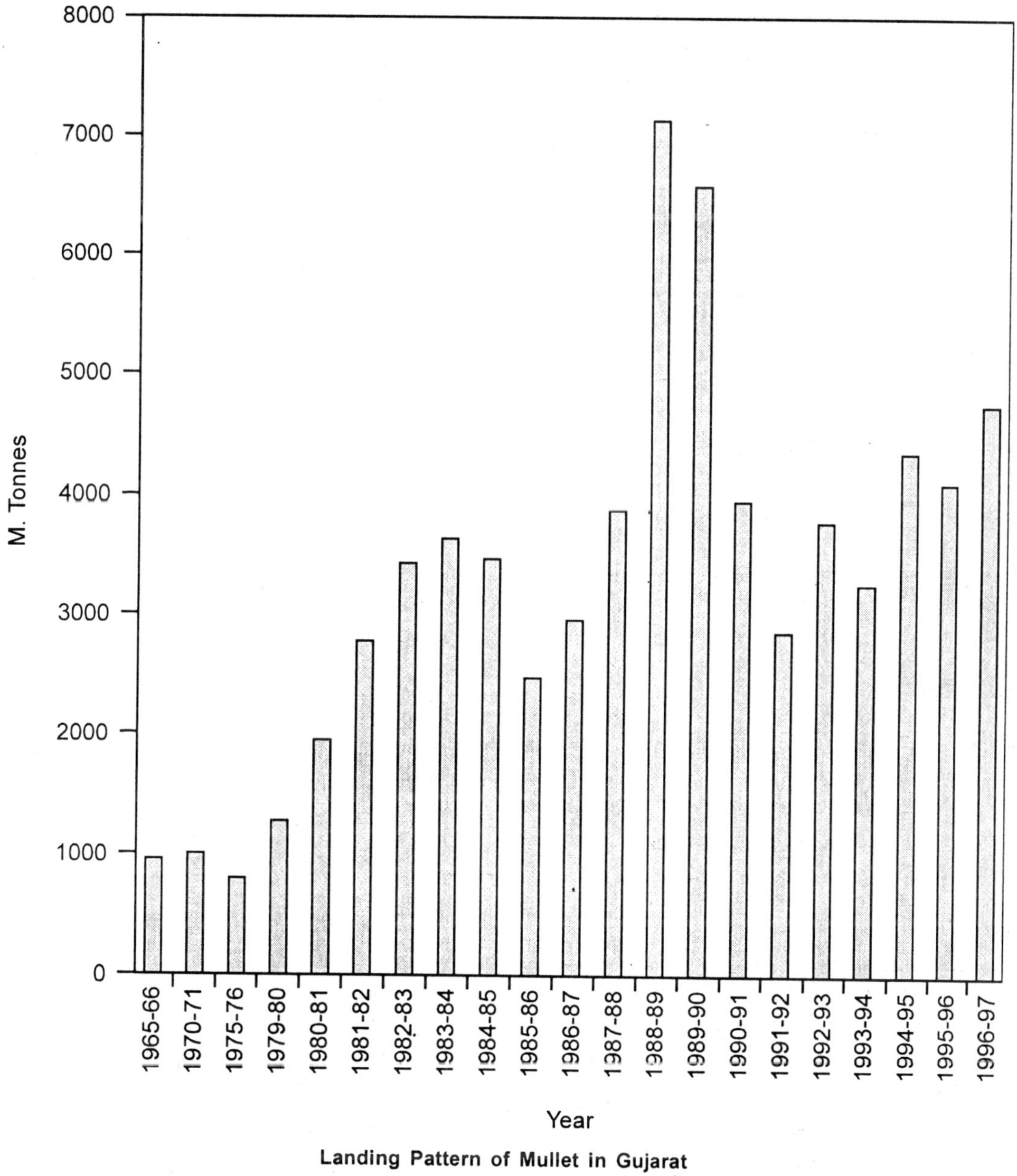


**Landing Pattern of Mullet in Gujarat**

## Spawning Season

The catch of *M. sheli* consist mostly of sexually mature specimens, with ripe gonads which indicates that the shore ward movement of this species during winter could be for spawning.

The species such as *M. parsia* appears to breed within estuarine areas. Most of mullets breed in the open sea, not far from the shore, and there is a tendency for the fries to ascend the comparatively shallow estuarine areas along the high tide.

Mullet fries of one species or other are always available throughout the year in varying quantities, this indicate that they are spawning throughout the year. However, the rate of availability goes on increasing from August onwards will January-February.

## Fishing Season

The fishing season is from November to March. The fishermen participating in this fishery migrate from Beyt Balapur and Warwala, (villages North of Dwarka of Gujarat) with their boats and camp at Rupen Bandar.

## Fishing Operation

Fishing is conducted in the inshore areas from Sivrajpur to Harshad (Gujarat) at an approximate depth of 3 m and less. The number of boats participating in fishing a single shoal varies depending upon size of the shoal and its intensity. Each boat carries 3-member crew.

While fishing two boats approach the shoal on its seaward side, secure the ends of the fleet of net carried on each vessel and move in the opposite direction quickly, paving the nets until the shoal is encircled. The boats then cross each other so that two end of the net overlap. The fishes trapped thus are vigorously disturbed with bamboo poles driving them in to the nets and the nets are then hauled in. The entire operation does not take more than 30 minutes. If the shoal is dense, more vessels participate and put additional rings of nets inside the first. (Bhaskaran *et al.*, 1970).

## Craft and Gear

### *Craft*

The boat used in the Mullet fishery is plank-built, non-mechanized and propulsion is by single sail. Measurements of boat are as under.

OAL; 6.3 m Beam : 1.2 m

Depth : 0.9 m Draft : 0.6 m

### *Gear*

The gear used in Mullet fishery of Gujarat is an encircling net, locally called "Rachha" made of 210/2/3 or 210/3/3 nylon. A unit gear is 35 to 45 m long and has a hung depth of 3 m. The head rope is of 3 mm. Cotton and foot rope 6 mm standard size. Cork or polythene floats are rigged at intervals of 6 meshes. Lead sinkers of 100 gm each are driectly mounted on the foot rope at intervals of 48 cm. Each unit of gear varying 30 sinkers on an average. The selvedges of the webbing are some times made of thicker twine. Each boat carries an average 5 units of gear. Different units have varying mesh size of 7.8, 8.8 and 9.7 cm.

## *References*

Bhaskaran, M., P. Gopalkrishnan and Mohan Chandy, 1970. "Munkan" Fishery of the Dwarka Coast. *Indian. J. Fish.;* vol 17.; 169-172.

Chayya, N.D. Bhaskaran M. and A.U. Bush, 1979. Observation on fisheries of mullet and scope for its culture in gulf of Kutch. *Indian. J. Fish,* 26 (1&2) : 47-57.

Dholakia, A.D., 1980. List of important marine fishes of Gujarat in "Fish processig preservation and quality control" Published by D.P. Dholakia, Ahmedabad, P.P. 134-144.

Dholakia A.D., D.V. Nandasana, P.H. Patel, D.M. Makwana, D.V. Bhola, 1996. Growth study of two species of *Mugil cephalus* and *Mugil tade* from Gulf of Kutch. Annual report, fisheries research station, Gujarat Agriculture University Sikka.

Dholakia A.D., N.G. Akolkar, 1992. A study on identification characters and availability of Mullet seeds for culture. Annual report 1991-1992, Dept. of Fisheries, Govt. of Gujarat. GFASRI, Okha.

Francis Day. 1959. The Fisheries of India. Published by William Dawson & Sons Ltd. London-PP. 346-355

Gopalkrishnan, P., and Surendra, R. Thakar, 1987. Studies on maturation and swpaning of gray mullets of reservoir fishery in Okhamandal (Gulf of Kutch). *Nat. Symp. Res. & Dev.* in *Mar. Fisheries.* Mandapam. CMFRI Bull. 44 : 367-383.

Lal Mohan and Nandakumaran, 1987. Occurrence of mature *Mugil cephalus* along calicut coast. *Mar. Fish. Infor. Serv.* T&E Ser., 48; 20.

Mohanraj, A. Raju, V. Gandhi, 1983. Fish culture in marine fish at mandapan, *Mar. Fish. Inf. Serv.* T&E Ser. 48

Narayan, K.R., A.U. Buch, P.K. Epan. Observation on the Mullet Resources of Okhamandal.

Sarojini, K.K. On the collection acclimatization and transport of Mullet seed in West Bengal (India). *J. Bomb. Nat. Hist. Soc.,* (1) : 42-55

Sundaraman, M. Pater Deva, 1951. Systematic of Chilka Mullet with a key for their identification *J. Zool. Soc.* India; 3 (1) : 19-25.

# 13
# Pomfret Fishery

## Classification

| | | |
|---|---|---|
| Order | : | Perciformes |
| Sub order | : | Stromateoidei |
| Family | : | Stromateidae |
| Genus | : | *Pampus* |
| Species | : | *argennteus* (Euphrasan 1788) |
| Synonym | : | *Stromateus linereus* (Day). |

## Strematеus argenteus

### *Name in Other Languages*

| | | |
|---|---|---|
| English | : | Silver, Pomfret |
| Gujarati | : | Vichuda, Paplet, Chamna (Small size) |
| Marathi | : | Chandwa, Paplet Saranga. |
| Tamil | : | Vella, volvel |
| Telugu | : | Tella, Chanduva |
| Bengali | : | Chandi |
| Orissa | : | Chandi, Gheya |
| Malyalum | : | Aboli |
| Kannada | : | Manji, Thondriti |

Silver Pomfret : *Pampus argenteus* (Euphrasan 1788). (Day I-247)

B. vii, D 5-9, $\frac{1}{38\text{-}43}$, P. 27, A. 5-6 $\frac{1}{32\text{-}41}$, C 0.19.

Head : Length of head 4 to 4.5, of pectoral 2.33 to 2.66 height of body, 1.5 to 1.66 in the total lengh excluding the caudal fin.

Eye : Diameter 0.25 to 0.20 of length of head, 1 diameter from end of snout, and 1.5 to 2 apart. Snout projecting over the mouth; the mexilla reaches to below the first third of the orbit.

Teeth : In a single raw in either jaw.

Fins : The dorsal spines appear in truncated from above the skin, as do also those of the anal : anterior portion of soft dorsal elevated but not to so great an extent as the anterior part of the anal : which in the immature reaches to below the middle of the caudal fin but as age advances it gradually becomes shortened. Pectoral rather pointed. Caudal deeply forked, the lower lobe much the longer in immature some times being twice as long as upper.

Scales : Small and very deciduous

Colour : Upper surface of the head and back as low as the lateral-line of a grayish neutral tint with purplish reflections, sides of head and body silvery-gray, fading to white on the abdomen, and everywhere covered with minute black dots : a dark spot on upper portion of the opercle. Dorsal and anal gray and minutely dotted with black, the outer half being the darker. Caudal and pactoral yellowish-white, also minutely dotted with black, the outer being the darker, Iris silvery. The young are much darker, the vertical fins being nearly black.

(Source : F. Day 1958)

***Stromateus argenteus* Silver Pomfret**

## *Classification*

| | | |
|---|---|---|
| Order | : | Acanthopterygii |
| Sub order | : | Stomateoidaei |
| Family | : | Stromateidae |
| Genus | : | Stromateus |
| Species | : | Niger (Day) |

## *Synonym*

*Apolectus niger* (Block)
*Apolectus stromateus* (cuv. and val).
*Parastromateus niger* (Block)

## *Names in Other Languages*

| | | |
|---|---|---|
| English | : | Black Pomfret/Brown Pomfret |
| Gujarati | : | Halwa, Adadiya |
| Marathi | : | Halwa |
| Telugu | : | Thella, Chanduva, Nalla chandura |
| Orissa | : | Bal |
| Tamil | : | Karwnvaval, Vellai-vaval |
| Malayalum | : | Karu, Abhuli, Aboli, Vallaoli |
| *Stromateus Niger* | : | Black Pomfret Day-I-247 |

B. vii, D. $\frac{5}{42\text{-}44}$, P. 22, A. $\frac{5}{35\text{-}39}$, C. 19.

Head : Length of head 0.22 of caudal 0.25 height of body 0.43 to 0.3 in total length.

Eyes : Diameter 4.5 to 5 in length of head, 1.25 diameter from end of snout, and nearly 2 apart. Dorsal and abdominal profiles equally angular. The lower jaw longer; the maxilla extends to beneath the anterior margin to the orbit.

Teeth : In a fine single series in both jaws.

Fins : The dorsal and anal are much elevated anteriorly, very low posteriorly, with concave external margins. Pectoral 1.5 longer than the head, and falciform : ventral only apparent in the young, in a specimen 3.33 inches in length they are jugular, and equal 1.125 of the total length. The spines, before the dorsal and anal fins, which are concealed in the adult fish, are apparent in the young specimen.

Scales : Small, and extended over the vertical fins.

Lateral-line : Gently curved down words, and in its last fourth passes straight to the centre of the caudal, in the form of araised keeled line with lateral shields as perceived in the genus Caranz.

Colour : Deep brown or grayish with blue reflections : cheeks, opercales, and abdomen pale-neutral or brownish-neutral. Dorsal and anal graying-brown stained black towards their margins : pectoral and caudal brownish, edged with black. Irish brownish-blue : in the young it is gray. The dorsal and anal fins black, and the tail yellow, with three brown cross bands.

## *Classification*

Order : Perciformes
Suborder : Stomateoidei
Family : Stromateidae
Genus : *Pampus*
Species : *Chinensis* (Euphrasen 1788)

***Stromateus niger*** **(Black Pomfret)**

## *Synonym*

*Stromateus sinensis* (Day)

## *Names in Other Languages*

English : White pomfret, Chinese pomfret
Gujarati : Paplet, Pathu, Kafri
Marathi : Saranga, Khalwad, Kapri
Tamil : Mogang, Vovel
Malyalam : Vella, Avuli
Telugu : Chamba

## *White Pomfret Stomateus Sinensis*

*Pampus chinensis* (Euphrasen 1788)

B. vi, D. 43-50, P. 25, A. 39-42, C. 19

Head : Length of head 4.33 to 5, of pectoral 3.5 to 4, of caudal 4.33 to 5, height of body 1.66 to 1.5 in the total length.

Eyes : Diameter 3.5 to 4.5 in length of head, 0.66 to 1 diameter from end of snout, and 2.66 apart. Dorsal and labdominal profiles about equally convex. The lower jaw longer, especially in young specimen; the maxilla reaches to beneath the first third of the eye.

Teeth : In a fine single row in each jaw, becoming lost with age.

Fins : First four or five dorsal rays very short, and not appearing above the skin, its anteiror portion is similar to that of the anal and equals the length of the pectoral fin. First five rays short but not appearing above the skin. Caudal with equal lobes, in the young it is slightly emarginate. Even in the young there is no trace of any ventral fins.

Scales : Small and very deciduous, specially in the immature, in the adult they cover the vertical fins.

Colour : Upper surface of head and body as far as the lateral-line of deep neutral-tint, the rest of the body with a mixture of brownish-grey, having metallic reflections, becoming lighter and silvery towards the abdomen; it is dotted all over with brown, the large spots having a silvery point in their centres. Fins silvery-grey marginal half blackish. Cavity of mouth and tongue pale bluish-grey with brown dots, silvery in the centre. Iris reddish-silvery or copper coloured, minutely dotted with brown. The young are grey, the head covered with irregularly star-shaped spots, and the fins nearly black, especially at their edges.

## *Field Identification and Salient Feature of Pomfret*

(A) Silver Pomfret : Dorsal and anal fins falcate caudal fin strongly forked and lower lobe the longest, 5-10 blade like spines before dorsal and anal fin maximum attainable size 25 cm.

(B) Chinese Pomfret : Body very deep compressed Gill opening restricted to vertical slit on side of the body. Single dorsal and ventral fin. No pelvic fins, Anal fin present and not palate. Caudal fin only slightly forked. Maximum size 35 cm.

***Pampus chinensis*** **Chinese Pomfret**

(C) Black Pomfret : Dorsal and anal fins are much elevated anteriorly, very low posteriorly, scales amall, leteral line gently curves down wards. Maximum size 50 cm.

## Food and Feeding

The stomach contents of the young immature fish often show couarse paste like material representing macerated fish food, while those of adult fish have a gelatinous consistency due to *Thaliaceans* and *Hydromedusae* which form a major portion of the food of the fish *Pompus argenteus.* (Raje and Bal 1963).

It is reported that food content of stomach was found chiefly to be copepods and other smaller crustaceans. However, some seasonal trends related to the fluctuations in the components of diet were noticed. Like the availability of the respective food components in the environment. Feeding intensity was more during mature and maturing stages of the gonad than in fish with ripe gonads (Kuthalingam 1963).

In black pomfret, many species of salps like *Jasis, Zonerea, pagea, salpa,* and *Thalia* spp. formed major food items of fish (42.8 per cent). Next in abundance were crustaceans (8.8 per cent) consisting of prawns, *Acetes, Lucifer, Mysids, Ostacods, Copepods, Zoea* and megalopa etc are about 1.1 per cent. Not much of seasonal variation in the in take of food items was seen (Bal and Rao).

### *Composition of the diet of adults*

It is observed that the percentage composition of the food of adult silver pomfret was as under.

(1) Smaller Crustaceans 38.8%
which includes the Copepods.

(2) Large crustacea 20.8%
includes *Penaeus* spp, *Acetes* spp, *Squilla* spp.

(3) Fish remains 20.2%
Fish scales, ventebra, and *Cynuglossus* spp.

(4) Molluscus 6.6%
Larval decapods, Foraminifera *Sagitta*

(5) Miscllaneous 13.6%

It was observed that maturity has a direct bearing on the feeding intensity of *Pampus argenteus.* Feeding intensity was found to be more during the early stages of maturity and less in fishes with mature gonads. Spent fishes were observed to have a feeding intensity higher than in the mature stages. (Kuthaliangam, 1963). It was also observed that feed percentage varies with the depth at which fish was found. Between 5 & 15 fathoms, the percentage of crustacean items was found to be more where as between 15 and 40 fathoms fish, Polychactes and Forminifera were the dominant items of diet.

## Sex and Fecundity

In Pomfret, sexes are separate. However, they are not distinguishable from external characters. It is found that number of eggs varies with overy size of fish. Observation for *P. argenteus* was as under :

(1) 65,702 eggs were found from 278 mm sized overy.
(2) 1,71,794 eggs were found from 320 mm size overy.

## Maturity

The gonads could be distinguished as ovaries or testes in fish of about 15 cm and more in length. The ovary first becomes apparent as two small, compact lobes : wine coloured and united at anterior end, and the testes as two thin, long strips of tissue, white in colour.

Maturity stages for female and male were defined by Sivaprakasam 1965 for *P. niger* as follows :

**Table 13.1 : Maturity Stages for Female**

| Stage | Description | Largest size of ova |
|---|---|---|
| Immature I | Ovaries thin, small piece of tissue, wine-coloured, microscopic tranparent ova, not visible to necked eye. | 0.21 mm in diameter |
| Immature II | Ovaries with compact lobes, right lobe often shorter than the left, pale yellow in colour. Ova with trace of yolk, visible to necked eye. | 0.32 mm in diameter |
| Maturing III | Ovaries large, yellow in colour. Ova large, Semi opaque | 0.42 mm in diameter |
| Maturing IV | Ovaries large, bright yellow in colour. Large ova completely opaque, fully laden with yolk. | 0.63 mm in diameter |
| Mature V | Ovaries large and jelly like, "Speckied" appearance due to large transparent ova. Ova largest, still retained inside the follicles. | 0.95 mm in diameter |
| Ripe VI | Ovaries very much distended ripe ova shed into the cavity of the ovary and oozing out through the oviduct. | |
| Spent VII | Ovaries shrunken, bag like, and blood shot. A few large ova may be present. | |

**Table 13.2 : Maturity Stages for Male**

| Stage | Description |
|---|---|
| Immature I | Testes thin, long strip of tissue, white in colour. |
| Immature II | Testes slightly large in size, compact, right lobe shorter than the left. White in colour. Traces of milt in the central core. |
| Maturing III | Testes longer. Transverse grooves and wave margins appear. Milt is being formed. |
| Maturing IV | Testes much longer, white in colour, ovary margins and transverse grooves. Milt oozes out when pressed hard. |
| Mature V | Testes very long and broad, milky white in colour, turgid due to milt, but not oozing. |
| Ripe VI | Testes as above, but oozing. |
| Spent VII | Testes hard contracted and dull white in colour. Without any milt. |

## Sex Ratio

Out of a total of 600 fish taken for observation 285 were males and 315 were females, showing a more or less equal distribution of sexes.

In 15-25 cm group males and females were more or less equal. Males were predominant, about twice as many as the females in the 25-35 cm group. Females were remarkably more, about four times, the males, in the size group 35-45 cm. While in size group 45-55 cm. There were no males at all.

As regards the distribution of the sexes from month to month, females predominated during August to October. The sexes were more or less equal in November to March. Males were predominant in April-May.

**Table 13.3 : Ovary Condition and Size of Ova in Different Stages of Maturity in Pomfret**

| Stage | Size of Ova | |
|---|---|---|
| | *P. argentius* | *S. niger* |
| Immature (microscopic) | 0.24 mm | 0.22 mm |
| Maturing-I Ovary swallen yellowish | 0.5 to 0.56 mm | 0.32 mm |
| Maturity-II Ovary crift yellow | 0.7 to 0.75 mm | 0.63 mm |
| Mature | 0.86 to 0.94mm | 0.95 mm |
| Ripe and running condition | 1.2 to 1.3 mm | 1.2 mm |
| Maturity size length | | |
| (1) Immature | Less than 110 mm | less than 250 mm |
| (2) Maturity (ripe) | 190-210 mm | 320 mm |
| Depth at which they lay eggs | 90 to 120m. | 90 to 120 m. |

## Spawning Season

Experiments have shown that in case of *P. niger* females, on the Saurashtra coast early stage I and II occur, throughout the year (August to May). Mature fish in stage III and IV start appearing in April and continue till November. Ripe fish (Stage V) were recorded in September. Actual spawners stage VI were found in September and October.

Spawning starts from July and continued till October, with the peak of activity in August-September.

In males similar picture was observed. The early stages occurred throughout the period. Stage III and IV occurred during April to November stages V was recorded in September. Oozing male stage VI were not recorded in October. This also suggests that spawning period extends from July to October.

Mature fish in stage IV and juvenile fish, in size range from 3 to 15 mm were recorded on Saurashtra coast from (Div-Head) Madhwad to Porbandar. This indicates that the fish spawn throughout Saurashtra coast.

Different spawning season could be observed for east and west coast of India. For silver pomfret it was reported that spawning season on Maharashtra coast is between October and December (Raje 1958), on Gujarat coast it is between February and August (Gopalan 1958), and on east coast it is between January and February (Kathalingum 1963).

For black pomfret it is reported that spawning takes place between September to December. (Shiva Prakashan 1963).

## Fishing Season

The fishing season of Pomfret is generally from September to May.

## Fishery Protection

During 1962-63, when fishing with trawlers were not in operation fish below 15 cm in length, i.e. indeterminate, were landed in very small quantities. However, now a days fisherman are reducing mesh size of gill net and trawling operation have reduced fish size. Due to over fishing depletion in catch ws also recorded. Regulation in mesh size for Pomfret is necessary to preserve stock.

## Resources

As estimated by Marine Product Export Development Authority Govt. of India, resources of Pomfrets are as under

(1) Silver Pomfret 60,000 T/annum
(2) Chinese Pomfret 74,000 T/annum
(3) Black Pomfret 62,000 T/annum

It is observed that for getting mroe quantity fishermen are reducing mesh size of Gill net.

It is observed from the landing of catch that day-by-day smaller size pomfrets are caught. Fishermen are not allowing pomfret to mature and stock is reducing. This gives a red warning signal for future fishery.

## Vertical Distribution

Highest contribution of Silver pomfret *P. argenteus* was found between 380 cm and 424 cm. in the bottom set gill net. (Sulochan and Rao, 1967)

## Craft and Gear

Gear : The two main gears currently used for pomfrets are surface drift gill net (Tarti) and bottom drift gill net (Tagli) in a Depth 30-50 M. They are also caught in Trawl net.

Craft : Generally pomfret fishery is carried out in gill netter boat with OBM, IBM and Trawlers of standard size.

**Table 13.4 : Council of Scientific Industrial Research, New Delhi published chemical composition of Pomfret, is as under.**

| Chemical component | White Pomfret (*Pampus argenteous*) | Black Pomfret (*Pampus niger*) |
|---|---|---|
| Moisture% | 71.00 | 70.60 |
| Protein% | 16.69 | 18.40 |
| Ash% | 1.48 | 2.19 |
| Calcium mg/100g | 250.60 | 158.00 |
| Phosphorus mg/100g | 21.24 | 16.93 |
| Iron mg/100g | 3.76 | 4.48 |

Source : The Wealth of India, 1962 pp99

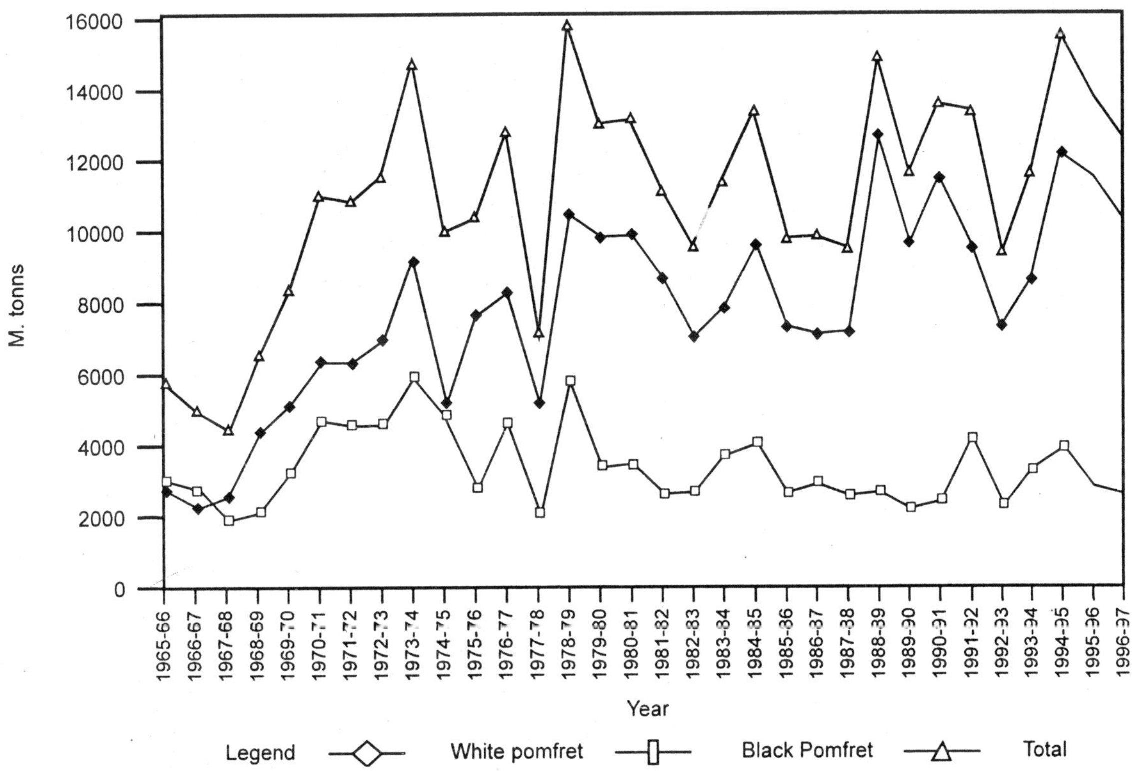


**Graph 13.1 : Landing Pattern of Pomfret fish in Gujarat**

*Source* : Department of Fisheries, Govt. of Gujarat (with permition)

### *Landing*

Graph 13.1 gives the details of landing pattern of all three variety of Pomfret in Gujarat from 1964-65. (Dept of Fisheries, Govt of Gujarat).

## References

Anon, 1962. The Wealth of India. p.p. 99. Council of Scientific & Industrial Research, Govt. of India. New Delhi.

Anon, 1980. Indian Fishery Atlas, Marine Products Export Development Authority, Government of India, Cochin.

Anon, 1987. Gujarat Fisheries Statistics, Commissionerate of Fisheries, Govt. of Gujarat, Gandhinagar.

Bal, D.V. and K.V. Rao. 1984. Marien Fisheries, book publishing by Tata Mac Graw Hill Publishing Company Ltd, New Delhi.

Dholakia, A.D. 1986. List of important fishes available in Gujarat, in Fish Processing Preservation and Quantity Control Publishing by D.P. Dholakia, Ahmedabad. P.P. 134-144.

Fransis, Day. 1958. The fishery of India. Published by William Dawson and Sons Ltd., London.

Gopalan, U.K., 1967. Pomfret fishery resources. *Proc. Symp* on *Ocean. Bul. Nat, Inst, Sci.* 38 : 785-96.

Gopal Krishna Pillai, N., and R. Sathiadhas 1982. Pair trawling strikes good ground for white pomfret in palk Bay. *Mar. Fish. Infor.* Serv, T&E Ser., 39; 1-6.

Kgwade, P.V. 1989. Pomfret resources along North-West coast of India National Symposium on the utilization of living resources of the Indians Seas, Abst : 25

Kuthalingam, M.D.K., 1963. Observation on the fishery and biology for the silver Pomfret *Pampus argenteus* (Euphrasen) from the bay of Bengal. *Indian J. Fish,* 10(1); 59-74.

Raje, M.S., 1958. A study of stromatied fisehes of Bombay Ph.D. thesis, University of Bombay.

Sivaprakasam, T.E. 1965. Observations on the maturation and spawning of Brown Pompfret *Parastromateus niger* (Bloch) in Saurashtra waters. *Jar. Bomb. Nat. hist. Soc.,* 62(2) : 245-253.

Sulochan, P. and K. Krishna Rao, 1967. On the Vertical distribution of silver pomfret (*Pompus argenteus* Euphr) in the bottom drift gill nets off Veraval and Porbander. *India J. Fish,* 11(1); 9-14.

Talwar, P.K., and R.K. Kacker, 1984. "Commercial Sea fishes of India" Publishing by Zoological Survey of India Calcutta.

# 14
# Prawn Fishery

## Introduction

The prawns constitute a large group of Crustaceans varying in size from microscopic to 35 cm long. Although nearly 2500 species are known, less than 300 are economically harvestable; and of these about 100 comprise much of the annual world prawn and shrimp catches.

Prawns are widely distributed, occurring in marine, brackish and fresh waters from the equator to the polar regions. They are found from nearly shallow waters to depth of nearly 5700 m. However, most of the commercial prawns are taken on the continental shelf at depth of less than 100 m.

The commercial prawns of India can be grouped in to penaeid and non-penaeid. Penaeid form a little over 50% of the total marine prawn catch. The rest being non-penaeid. (Dholakia 93).

There was a dispute about the use of word Prawn & Shrimp. The term "Prawn" used in India is identical to term "Shrimp" used in Western Countries.

As decided in Indo-Pacific Fisheries Council meeting held at Tokyo in 1955, only Penaeid, Pandalids and Palemonids varieties will be property known as "Prawn" while other small varieties will be identified as "Shrimp". Penaeid and Non-penaeid varieties can easily be identified in field as under.

Penaeid (Prawn) Exoskeleton of second segment of the body does not over lap on first segment but over laps only on third segment.

Non-penaeid (Shrimp) Exoskeleton of second segment of the body over laps on both first and third segment.

## Classification of Penaeid Prawns is as under:

| | |
|---|---|
| Class | Crustacea |
| Phylum | Arthropoada |
| Order | Decapoda |
| Sub order | Mecrura |
| Family | Penaeidae |

Main Genus and species found in Important Commercial varieties are as under.

| | Genus | Species | |
|---|---|---|---|
| (1) | Penaeus | indicus | H.M.Edward |
| (2) | Penaeus | monodon | Fabrisious |

| | | |
|---|---|---|
| (3) Penaeus | merguiensis | de Man |
| (4) Penaeus | semisulcatus | de Haan |
| (5) Metapenaeus | monoceros | Fabrisious |
| (6) Metapenaeus | kutchensis | Jeorge and Rao |
| (7) Metapenaeus | afinis | H.M. Edward |
| (8) Parapeneopsis | Stylefera | Haler |
| (9) Parapeneopsis | hardwiki | Morse |

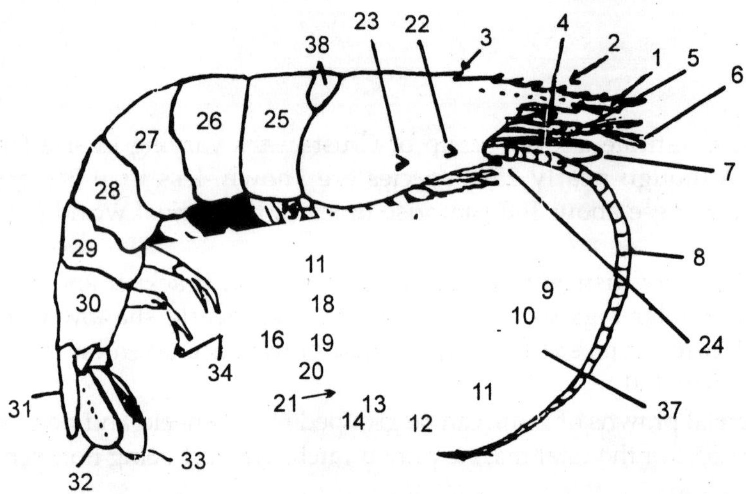


**Fig. 14.1 Diagrammatic view of Penaeid prawn shown taxonomic characters.**

**Diagramatic view of Penaeid prawn showing taxonomic characters.**

| | |
|---|---|
| 1. Rostrum | 2. Rostral spine |
| 3. Epigastric spine | 4. Eye |
| 5. Antennule | 6. Antennular flagellum |
| 7. Antennal scale | 8. Antennal flagellum |
| 9. Third maxilliped | 10. First pereopods |
| 11. Second pereopods | 12. Third pereopods |
| 13. Fourth pereopods | 14. Fifth pereopods |
| 15. Coxa | 16. Basis |
| 17. Ischium | 18. Merus |
| 19. Carpus | 20. Propodus |
| 21. Dectylus | 22. Post orbital spine |
| 23. Hepatic spine | 24. Pterygostomian spine |
| 25. First Abdominal segment | 26. Second Abdominal segment |
| 27. Third Abdominal segment | 28. Fourth Abdominal segment |
| 29. Fifth Abdominal segment | 30. Sixth Abdominal segment |
| 31. Telson | 32. Endopodite |

33. Exopodite
34. Plepods
35. Cephalothorax
36. Abdomen
37. Petasma
38. Arthroidal membrane

Source : A.D. Dholakia, "Fish processing preservation and quality control":1986: 3-4.

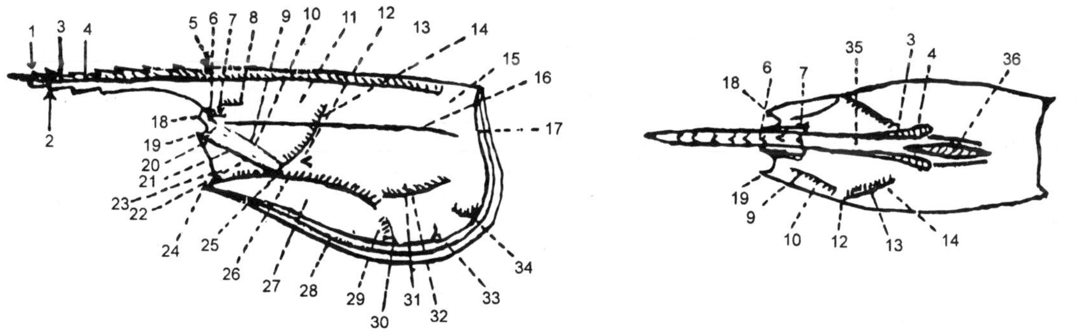


*(A) Side view of the carapace* *(B) View from the above of the carapace*

**Fig. 14.2 : Identification characters using carapace of penaeid prawn.**

## Identification Characters using carapace of a penaeid prawn.

1. Rostral tooth
2. Ventral rostral tooth
3. Adrostral carina
4. Adrostral sulcus
5. Epigastric spine
6. Gastrofrontal carina
7. Gastrofrontal sulcus
8. Postocular sulcus
9. Orbito-antennal sulcus
10. Gastroorbital carina
11. Gastric region
12. Hepatic spine
13. Cervical sulcus
14. Cervical carina
15. Cardiac region
16. Longitudinal suture
17. Pterygastomian sulcus
18. Orbital or supra orbital spine
19. Antennal
20. Postorbital spine
21. Antennal carina
22. Postantennal spine
23. Brachiostegal spine
24. Ptery-gastamian spine
25. Hepatic carina
26. Hepatic sulcus
27. Ptery-gastomian region
28. Marginar region
29. Inferior carina
30. Inferior sulcus
31. Branchiocardinal carina
32. Branchiocardinal sulcus
33. Transverse suture
34. Stridulatin organ
35. Postrostral carina
36. Postrostral or median sulcus.

*Source* : Kurian and Sebastian. Prawns and Prawn Fisheries of India, pp. 77.

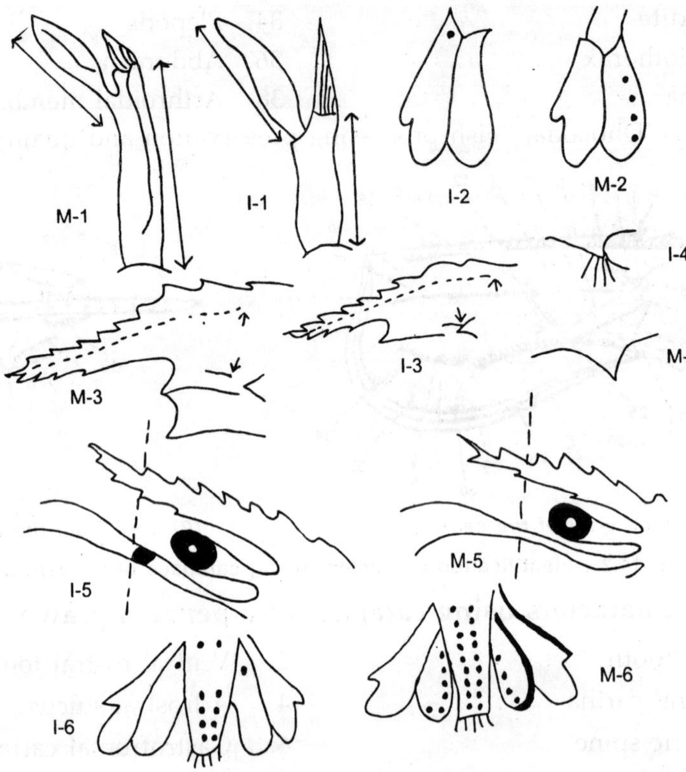


**Fig. 14.3 : Identification difference between *Penaeus merguiensis* and *P. indicus*.**

## Identification difference between *P. merguiensis* and *P. indicus*

M=*Penaeus merguinsis*; I=*Penaeus indicus*

(1) Third maxilliped
(2) Chromatophores on inner uropod
(3) Position of Gastro orbital crest
(4) Mid-Ventral prominence
(5) Position of anterior most dorsal tooth.
(6) Chromatophores on telson

*Source* : Dholakia, A.D. 1994. Ph.D. Thesis. Saurashtra University.

## Identification of juveniles of different Prawn species.

PMo-1 = Rostrum of *Penaeus monodon*
PMo-2 = Inner and outer uropod of *P. monodon*
PMo-3 = 6th abdominal segment of *P. monodon*

PS-1 = Rostrum of *P. semisulcatus*
PS-2 = Inner and outer uropods of *P.semisulcatus*
PS-3 = 6th abdominal segment of *P.semisulcatus*
PJ-1 = Rostrum of *P. Japonicus*

PJ-2 = Inner and outer uropods of *P.Japonicus*
PJ-3 = $6^{th}$ abdominal segment of *P. Japonicus*

MA-1 = Rostrum of *Metapenalus affinis*
MA-2 = Inner and outer uropods of *Metapenalus affinis*

MD-1 = Rostrum of *M.dobson*
MD-2 = Inner and outer uropodds of *M.dobson*

MM-1 = Rostrum of *M. monoceros*
MM-2 = Inner and outer uropods of *M.monoceros*

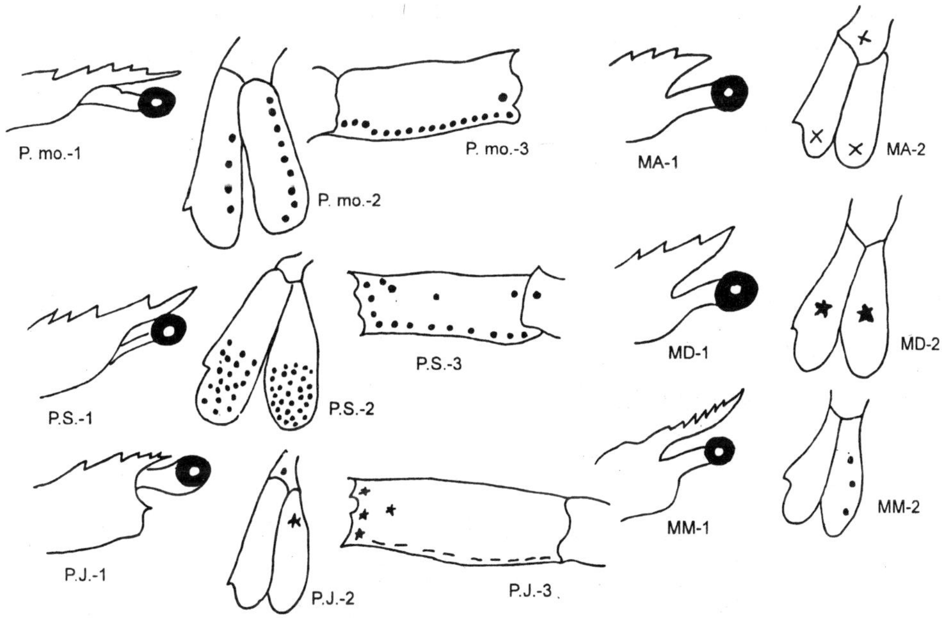


**Fig. 14.4 : Identification of juveniles of different prawn species.**

While important non-penaeid varieties coming under four different families are as under.

| | | | |
|---|---|---|---|
| (1) | Family | *Hippolitidae* | |
| | Genus | *Hippolysmata* | |
| | Species | *acirostos* Camp. | |
| (2) | Family | *Palamonidae* | |
| | **Genus** | **Species** | |
| (i) | *Palamon* | *tenucpis* | Handerson |
| (ii) | *Palamon* | *styleferus* | H.m. Edward |
| (iii) | *Macrobranchium* | *rosenbergii* | de-Man |
| (iv) | *Macrobranchium* | *malcomsi* | H.M. Edward |

| | | | |
|---|---|---|---|
| (3) | Family | *Sargastidae* | |
| | **Genus** | **Species** | |
| (i) | *Acetus* | *indicus* | H.M. Edward |
| (ii) | *Acetus* | *arethaeus* | Nobily |
| (4) | Family | *Pandalidae* | |
| | **Genus** | **Species** | |
| (i) | *Parapandapus* | *spineus* | Bet |
| (ii) | *Hitorocarpus* | *gobosus* | Bet |
| (iii) | *Plosonica* | *martia* | H.M. Edward |

Gujarat state has longest coast line and widest continental shelf area in India. It also ranks second in having potential backwater area next only to West Bengal.

74 species of penaeid prawns are known to exist in India, but only 21 of them form significant portion of the commercially expoiled resources. Following are few penaeid prawn species which grow faster and can be cultured in India.

| | *Prawn species* | Maximum size |
|---|---|---|
| (1) | *Penaeus monodon* | 320 mm |
| (2) | *Penaeus semisulcatus* | 250 mm |
| (3) | *Penaeus merguiensis* | 240 mm |
| (4) | *Penaeus indicus* | 230 mm |

Details of only important prawn varieties are given below.

## (A) *Peaeus indicus Milne Edw.*

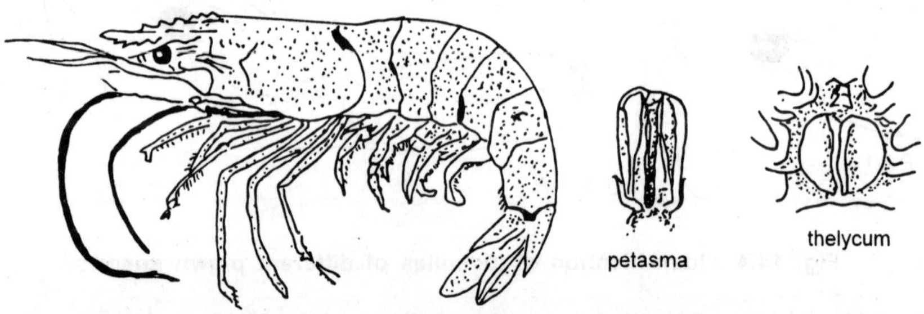


**Fig. 14.5 : *Penaeus indicus* with petasma and thelycum**

Adrostral carina not reaching behind middle of carapace; gastro-frontal carina absent. Hepatic carina is also absent. Gastro-orbital carina occupying the posterior 2/3 distance between hepatic spine and orbital angle; rostral crest must be elevated but not triangular in profile.

Rostrum with ventral teeth. On dorsal side of rostrum 7 to 9 spine are present while on ventral side they are 4 to 6; rostrum crest feeble. Post orbital spine absent; carapace without a median dentate crest except occasionally over the eyes; No distinct median tubercle on ocular peduncle; upper antennular flagellum sub-equal to the lower one attached to apex of third antennular segment.

## (B) *Penaeus monodon* (Fabricius.)

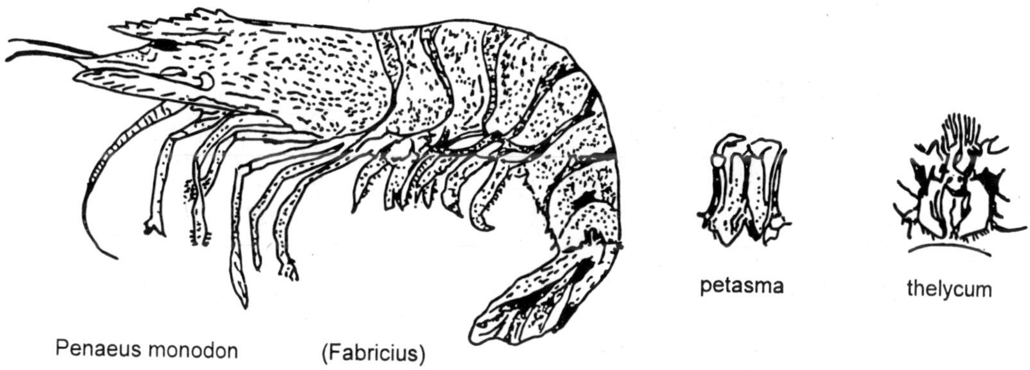


**Fig. 14.6 : *P. monodon* with petasma and thelycum**

Post orbital spine absent; carapace without a median dentate crest except occasionally over the eyes. No distinct median tubercle on ocular peduncle; upper antennular flagellum sub-equal to the lower one attached to apex of third antennular segment. Rostrum with ventral teeth. Adrostral carina not reaching behind middle of carapace: gastro-frontal carina absent. Hepatic carina horizontally straight; $5^{th}$ pereopod without exopodite.

## (C) *Penaeus semisulcatus* (de-Haan.)

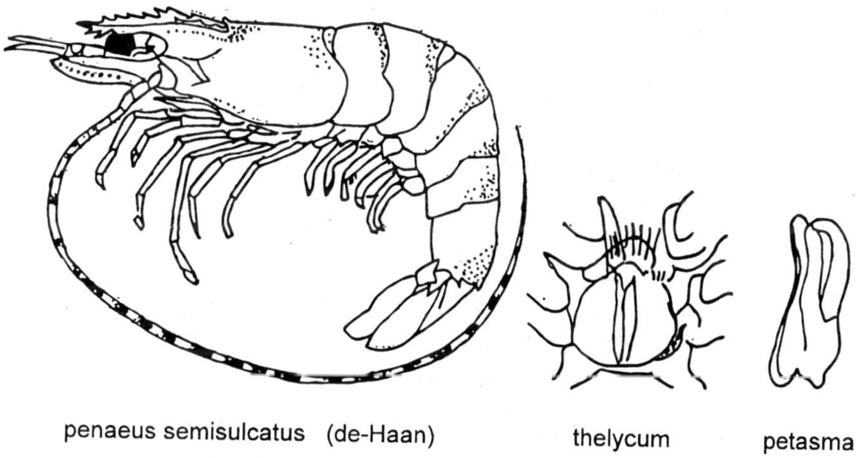


**Fig. 14.7 : *P. semisulcatus* with thelycum and petasma**

Carapace without median dentate crest except occasionally over the eyes. Post-orbital spine absent. No distinct median tubercle on occular peduncle; upper antennular flagellum sub-equal to the lower one attached to apex of third antennular segment.

Adrostral carina not reaching behind middle of carapace; gastro-frontal carina absent: Hepatic carina inclined at an angle of 20° anteroventrally; $5^{th}$ pereopod with small exppodiete.

## *(D) Penaeus merguiensis* (de-Man)

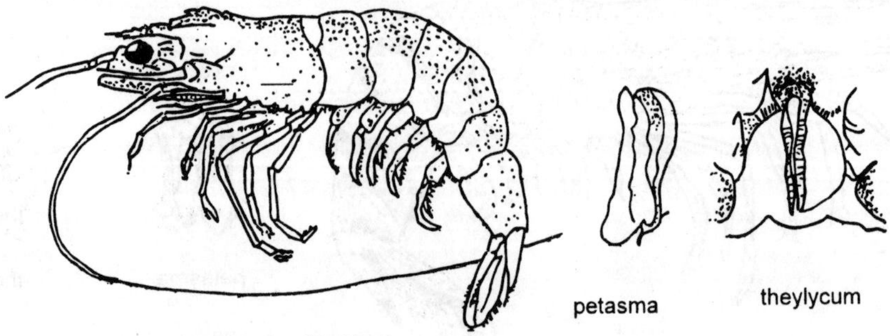


**Fig. 14.8 : *P. merguiensis* with petasma and thelycum**

Adrostral carina not reaching as far as epigastric tooth; gastro-frontal carina absent; Hepatic carina absent; Gastro-orbital carina absent or not reaching hepatic spine and occupying the middle 1/3 distance between hepatic spine and orbital angle. Dectyl of 3rd maxilliped of adult male 1/2 size of propodus; Rostral crest triangular in profile post orbital spine absent; carapace without a median dentte crest except occasionally over the eyes. No distinct median tubercle on occular peduncles upper antennular flagellum sub-equal to the lower one attached to apex of third antennular segment.

## (E) *Pendeus penicillatus* (Alcock)

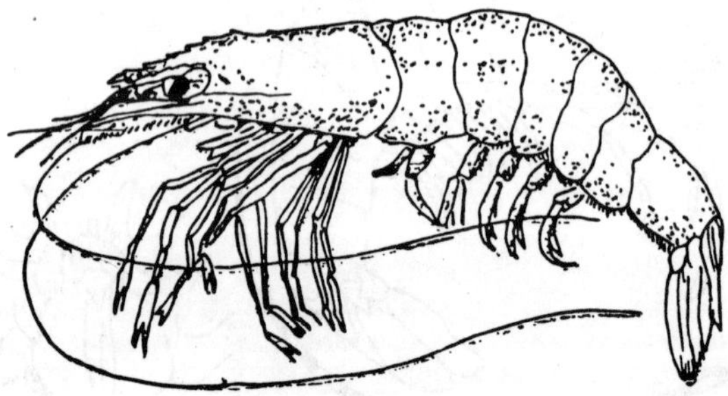


**Fig. 14.9 : *P. penicilatus* with petasma and thelycum**

Adrostral carina reaching just beyond epigastric tooth; Gastrol-frontal carina as well as Hepatic carina bsent. Gastro-orbital carina not reaching hepatic spine and occupying the middle 1/3 distance between hepatic spine and orbital angle. Detyl of 3rd maxilliped of adult male much longer than propodus. Rostral crest markedly elevated.

## *Metapenaeus monoceros* (Fabricius)

Rostrum without ventral teeth. Post-orbital spine absent. No distinct median tubercle on occular peduncle; upper antennular flagellum sub-equal to the lower one attached to apex of third antennular segment. No distal fixed pair of spine on the telson; lateral mobile spines present. No exopod on 5th pereopod; pleurobranch on 7th thoracic somite present.

Distomedian petasmal projection without apical filament looks as hood-like and directed anterolaterally. Thelycum of impregnated females without white conjoined pads. Brachiocardiac sulcus distinct in at least posterior thelycal plate tong like lateral thelycal paltes with salient and paralleled ear shaped lateral ridges.

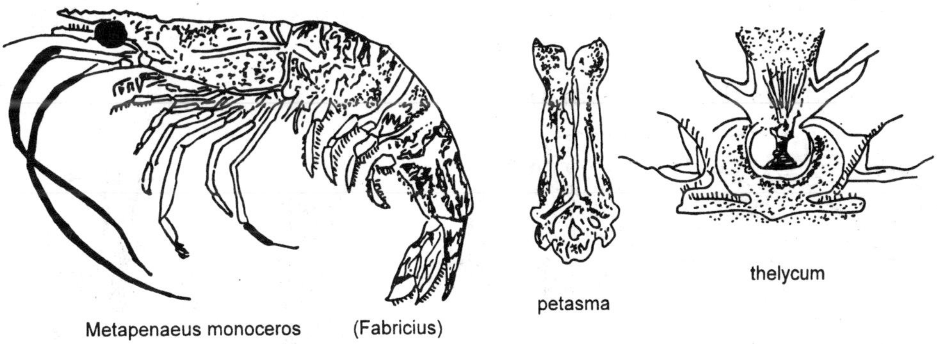


**Fig. 14.10:** ***Metapenaeus monoceros***

## *Metapenaeus dobsoni (Mires)*

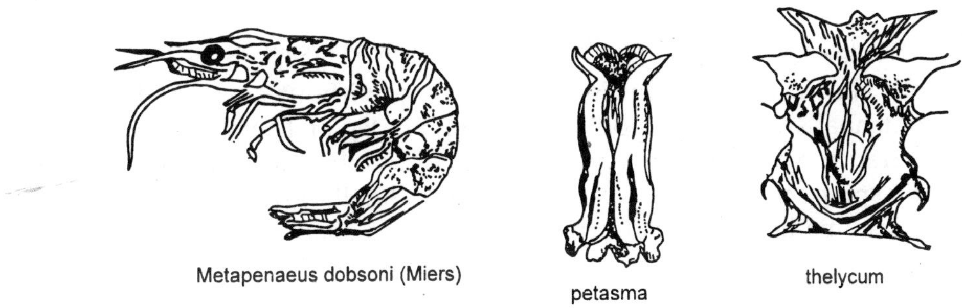


**Fig. 14.11 :** ***M. dobsoni***

Ventral teeth absent on rostrum. Rostrum more curved, with basal crest; last pair of pereopods fall short of middle of antennal scale; Exopodite absent on $5^{th}$ leg of Thorax. Basal spine on male $3^{rd}$ preopod long and barbed; apical petasmal filaments not really visible; anterior thelycal plate tong-like. Pleurobranch on $7^{th}$ thoracic somite present.

## *Metapenaeus Kutchensis* (George and Rao)

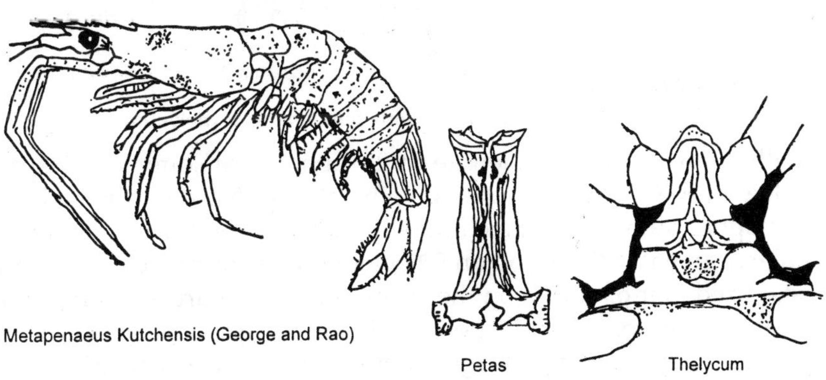


**Fig. 14.12 :** ***M. kutchensis*** **with petasma and thelycum**

Rostral without ventral teeth. Exopodite absent on 5th leg of thorax. Rostrum straight and extends the 2nd segment of antennular peduncle.

Dorsomedian abdominal carina from 4th segment. Rostal tip more straight than in *M. affinis*: posterior extension of anteromedian thelycal plate not bond by oval plate on either side. Epigastric teeth 7-8. Adrostral carina ending near epigastric: sulcus extending to middle of carapace.

## *Metapenaeus affinis* (H.M. Edward)

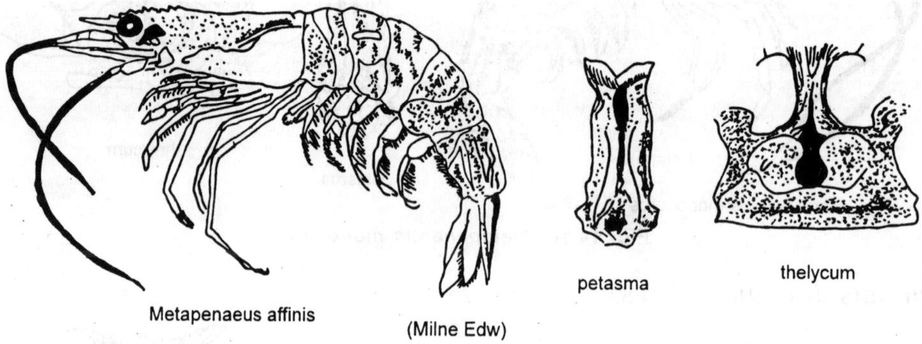


**Fig. 14.13 : *M. affinis* with petasma and thelycum**

Ventral teeth absent on rostrum. Rostrum curved without crest. 5th leg of Thorax surpassing the antennal swcale. Expopodite absent on 5th leg of Thorax. Ischial spine on 1st pereopod small. Branchio cardiac carina distinct extending from posterior margin of carapace almost to hepatic spine; anterior Thelycal plate longitudinally grooved, wider posteriorly than anteriorly; distomedian petasmal projections crescent-shaped.

## *Metapenaeus brevicormis* (Milne Edward)

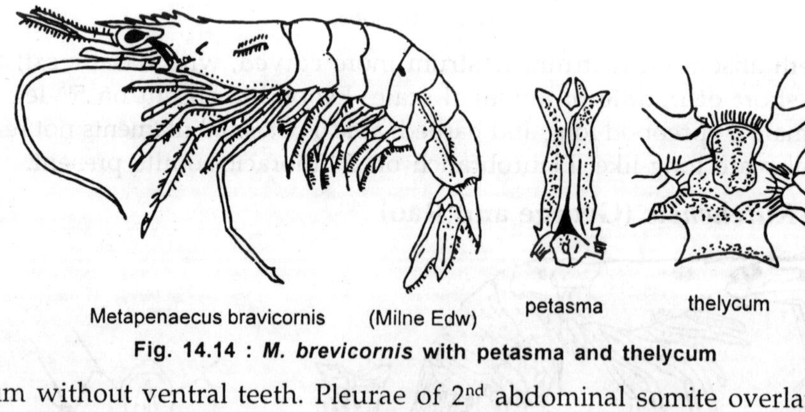


**Fig. 14.14 : *M. brevicornis* with petasma and thelycum**

Rostrum without ventral teeth. Pleurae of 2nd abdominal somite overlapping those of 1st segment, 3rd leg with a chela. Post-orbital spine absent. No distinct median tubercle on ocular peduncle; upper antennular flagellum sub-equal to the lower one attached to apex of third antennular segment. No exopod on 5th pereopod; Pleurobranch on 7th thoracic somite present. No distal fixed pair of spines on the telson. Posterior part of rostrum with distinctly elevated crest; basial spine on male 3rd pereopod simple; apical petasmal filament slender slightly converging; thelycum with a large anterior and small lateral plates.

## *Parapenaeopsis stylifera* (Milne. Edward)

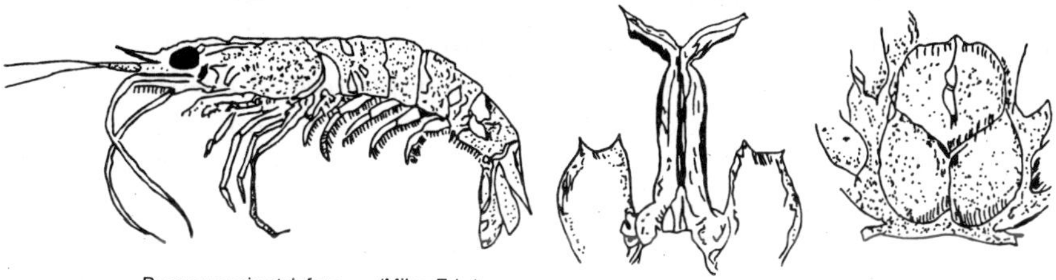


**Fig. 14.15 : *Parapenaeopsis stylefera* with petasma and thelycum**

Pleurae of 2nd abdominal somite overlapping those of 1st segment, 3rd leg with a chela. Last two pairs of walking legs will developed; gills many. Rostrum without Ventral teeth. No distal fixed pair of spines on the telson; lateral mobile spine may be present. Exopode on 5th pereopod present; Pleurobranch on 7th thoracic somite absent. Carapace with longitudinal sutures; ischial spine on 2nd pereopod absent. 3rd pereopod without epipodite. Epipodite present on 1st and 2nd pereopods. 2nd percopods with basial spines. Telson with pair of fixed sub apical spine; at least distal 1/2 free portion of rostrum unarmed.

## *Acetus indicas* (H. Milne)

Pleurae of 2nd abdominal somite overlapping those of 1st segment, 3rd leg with a chela. Last one or two pairs of walking legs reduced or absent; gills few. Procurved spine present between 1st pair of pleopods. Basis of the third pereopod with tooth on inner free margin; petasma without membraneous coupling folds.

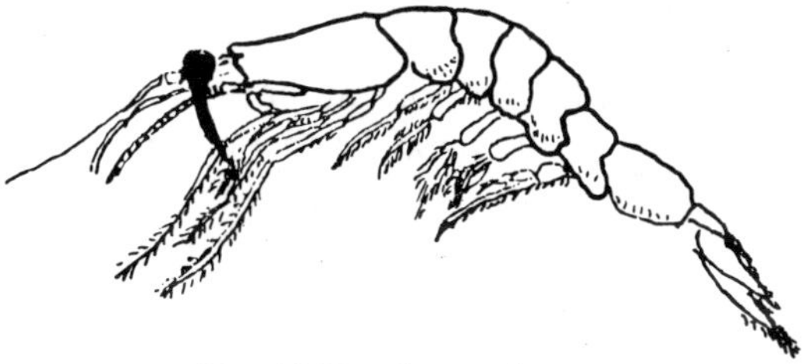

**Fig. 14.16 : *Acetus indicus***

## *Hippolysmata ensirostris* (Kemp.)

Chelae of 1st pair of pereopods distinct on both sides; ends of fingers of this chela usually dark coloured; eyes free, never extremely elongate. Pleurae of second abdominal somite overlapping those of first and third segments; no chela on third pereopods. Rostrum longer than carapace, with an elevated dentate basal crest; lateral margins of telson concav, apex acute and unarmed.

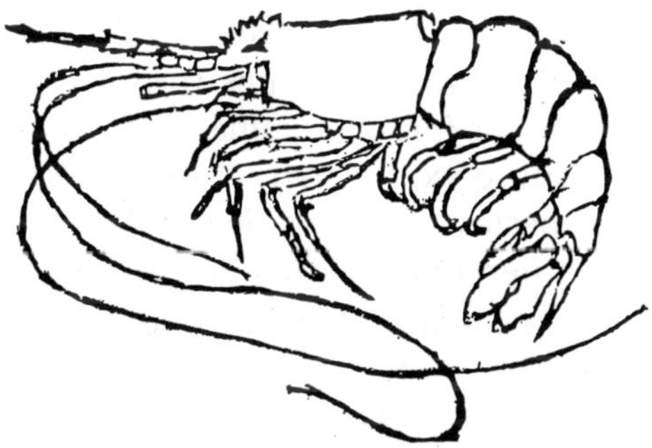

**Fig. 14.17 : *Hippolysmata ensirostris***

## *Parapenaeopsis sculptilis* (Hellar)

Pleurae of second abdominal somite overlapping those of first and third segment; no chela on third pereopods. Carpus of 2nd pair of pereopods divided into two or more articles; if not, 1st pair of pereopods not chelate. Chelae of 2nd pair of pereopods small and slender. Rostrum without ventral teeth. No distal fixed pair of spines on the telson; No exopod on

5[th] pereopod; Pleurobranch on 7[th] thoacic somite absent. Carapace with longitudinal surtures; ischial spine on 2[nd] pereopod absent, but basal spine is present. 3[rd] pereopod without epipodite. Epipodites present on 1[st] and 2[nd] pereopods. Telson without fixed subapical spine, with or without lateral movable spine; 1/3 or less free portion of rostrum unarmed. Petasma with a pair of disto-lateral projections usually short and spout like and pair of cap like distal projections; Post-rosteral carina reaching almost to posterior border of carapace; Antennular flegellum 0.5-0.6 length of carapace; thelycum with median tuft of setae on posterior plate.

### *Parapenopsis hardwickii* (Morse)

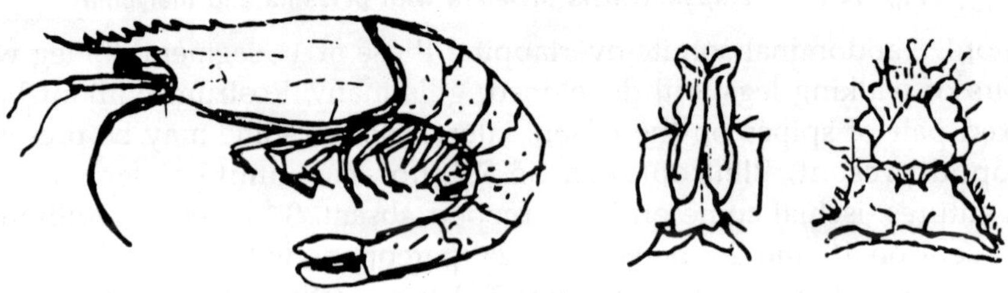

Fig. 14.18 : *P. sculptilis*

Pleurae of second abdominal somite overlapping those of first and third segments; bo-chela on third peropods. Gill phyllobranchiate. Corpus of 2[nd] pair of pereopods entire; no epipods on legs; upper antennular flagellum bifid; 3[rd] maxilliped normal. Rostrum without ventral teeth. A distal fixed pair of spines onm the telson and 1-3 pairs of mobile spines. Petasma symmetrical; 3[rd] maxilliped without basial spine. Carapace with longitudinal sutures extending from orbital margin to almost posterior margin. Epipodites present on 1[st] and 2[nd] parepods. Telson without fixed subapical spines, with or without lateral movable spines; 2[nd] pareopods with basial spine; 1/3 or less free portion of rostrum unarmed. Petasma with a pair of distro-lateral projections directed laterally or distro-laterally, usually short and spout-like and pair of cup-like distal projections. Antennular flagella 0.7 length of carapace or longer; thelycum without median tuft of setae on posterior plate.

Of the total earnings from export of marine products, earnings through prawns and crustaceans form a dominant percentage. It is above 70% of total earning in foreign exchange. To capture export market and good price per unit, it is necessary to export the qualities and species of prawn demanded by buyers abroad. It is, therefore, necessary to identify and differentiate different species of prawns at different stages.

The identification work of such different species of prawns from nauplii stages to adult stage were done by many workers like George and Rao (1963), Raje and Ranade (1072) (I) & (II), Kurian and Sebastian (1976) Muthu, Pillai, and George (1978), Dholakia (1993) etc. Identification keys are given below.

A genetic key for the various larval stages are given below:

### Nauplius

1. (a) The proximal inner lateral seta of antennule longer than 2 anterior ones and filamentous in N IV to VI .................................................................. *Parapenaeopsis.*
1. (b) The proximal inner lateral seta of antennule shorter than & anterior ones in N IV to VI .................................................................................................... (2).

2. (a)The anterior most inner lateral seta in antennule very long: the posterior most inner lateral seta absent or rudimentary in N-I to N-III. In the N VI the 3 aesthaetes on the outer distal margin of antennule equally spaced i.e. the middle one is equidistant from the distal and proximal ones ................................................ *Penaeus.*
2. (b) The size of the 3 inner lateral seta on antennule decreases gradually from the anterior to the posterior one. In N VI the 3 aesthaetes on the outer distal margin of antennule not equally spaced, the 2 anterior ones closer to each other than to the posterior one .................................................................................................. *Metapenaeus.*

## Protozoea-I

1. (a) Antenna exopod with 11 setae along inner and distal margin, endopod with 1+1+2 lateral setae; outer most furcal setae dorsally disposed ................................ *Penaeus.*
1. (b)Antenna exopod with 10 setae along inner and distal margin, endopod not with 1+1+2 lateral setae. Outer most furcal setae venrally or laterally disposed...... (2).
2. (a) Antenna endopod with 1+2+3 lateral setae. Frontal horns usually present above frontal organs. Caudal furcae separated by shallow semicircular space, outermost furcal setae slightly ventrally disposed ..................................................... *Metapenaeus.*
2. (b) . Antenna endopod with 2+2 lateral setae. No frontal horns above the frontal organs, caudal furcal separated by deep A shaped space; outermost furcal setao laterally disposed and widely separated from the penultimate outer seta ............ .................................................................................................................. *Parapenaeopsis.*

## Protozoea-II

1. a. Supraorbital spine absent ............................................................... *Parapenaeopsis.*
1. b. Supraorbital spines present ............................................................................... 2.
2. a. Supraorbital spines bifid ............................................................................ *Penaeus.*
2. b. Supraorbital spines simple ................................................................... *Metapenaeus.*

## Protozoea-III

1. a. Telson with 7+7 furcal setae ............................................................ *Metapenaeus.*
1. b. Telson with 8+8 furcal setae.................................................................................. 2.
2. a. Antennule 4 segmented, supraorbital spine absent ......................... *Parapenaeopsis.*
2. b. Antennule 3 segmented, supraorbital spine present .................................. *Penaeus.*

## Mysis stage

1. a. $5^{th}$ abdominal segment with a pair of posterolateral spines. Dorsal spines present on $4^{th}$, $5^{th}$ and $6^{th}$ abdominal segments; sometimes on $3^{rd}$ segment also..... *Penaeus.*
1. b. $5^{th}$ abdominal segment devoid of posterolateral spines. Dorsal spines present only on $5^{th}$ and $6^{th}$ abdominal segment ............................................................................ 2.
2. a. Telson with 7+7 spines. Hepatic spine preent in later mysissubstage ................. *Metapanaeus.*
2. b. Telson with 8+8 spines, Hepatic spine absent in all mysis substages .................. *Parapenaeopsis.*

## Early Postlarvae

### *(Up to Pl-2)*

1. a. 5th abdominal segment with posterodorsal spine .......................................... *Penaeus.*
1. b. 5th abdominal segment without posterodorsal spine .......................................... 2.
2. a. Telson with 7 pairs of spines .......................................... *Metapenaeopsis.*
2. b. Telson with 8 pairs of spines on either side of a median spine ... *Parapenaeopsis.*

Source : Muthu, Pillai and George (1978).

## A key for the Identification of mysis stage of five commercial penaeid prawns of India

1. a. Telson with 8+8 spines .......................................... 2.
1. b. Telson with 7+7 spines .......................................... 3.
2. a. Supraorbital spines present .......................................... *Penaeus indicus.*
2. b. Supraorbital spine absent .......................................... *Parapenaeopsis stylifera.*
3. a. Median dorsal spine on 5th abdominal segment conspicuous .......................... 4.
3. b. Median dorsal spine on 5th abdominal segment absent ........ *Metapenaeus dobsoni.*
4. a. Rostrum extending to 3/4 eye; larvae conspicuously brownish ....... *Metapenaeus monoceros.*
4. b. Rostrum reaching tip of eye or slightly beyond it; larvae not brownish .......................................... *Metapenaeus affinis.*

## A key for the Identification of first postlarve of fine Commercial penaeid prawns of India

1. a. Telson with 7+7 spines .......................................... 1.
1. b. Telson with more than 7+7 spines .......................................... 2.
2. a. Rostrum with 4 spines (2 large and 2 small); long setae on distal lateral aspect of 6th abdominal segment .......................................... 3.
2. b. Rostrum with 2 spines; no setae on distal lateral aspect of 6th abdominal segment .......................................... *Metapenaeus dobsoni.*
3. a. Epigastric spine on carapace absent; larvae brownish ..... *Metapenaeus monoceros.*
3. b. Epigastric spine on carapace present; larvae not brownish ... *Metapenaeus affinis.*
4. a. Telson truncate, with 8+8 spines, median posterior spines equal; median dorsal spine present on 5th and 6th abdominal segment .......................... *Penaeus indicus.*
4. b. Telson tapering between posterolateral spines, with 8+1+8 spines, median posterior spine very much larger than adjacent spines; median dorsal spine on abdominal segment absent .......................................... *Parapenaeopsis stylifera*

Source : Vedvyas Rao (1973).

## Key for identification of juveniles of four species of *Metapenaeus*

1. a. Rostrum long, without basal crest, extending to middle of eye in younger individuals and beyond eye in older forms; abdominal carination commences from 3rd segment; ground colour of body buish brown to brownish; chromatophores closely spaced, forming short band on the lateral region of carapace .................. 2.

1. b. Rostrum short, with basal crest, extending to base of eye in younger individuals and tip of eye in older forms; abdominal carination commences from 4th segment; ground colour of body cream white; chromatophares widely spaced, not forming bands.............................................................................................. *Metapenaeus dobsoni.*
2. a. Rostrum reaching middle of eye or little beyond it; the short band on carapace less conspicuous; 'M'-like patch on mid-dorsal region of 4th abdominal tergum absent.............................................................................................. *Metapenaeus affinis.*
2. b. Rostrum reaching tip of eye or beyond it; the short band on carapace conspicuous; a distinct 'M'-like patch on mid-dorsal region of 4th abdominal tergum present .............................................................................................. *Metapenaeus monoceros.*
2. c. Rostrum straight with small crest; Mid-dorsal carination of the abdominal segment commences from 4th segment; ischial spine present on 1st pereiopod. Body partly covered with harsh and very short tomentum ...................... *Metapenaeus kutchensis.*

Source : George and Rao.

It is difficult to differentiate prawn species in early stages specially when species are very nearer to each other. In Gujarat as it is reported, *Penaeus merguiensis, Penaeus indicus* and *Penaeus penicillatus* are present. They are very near to each other. Dholakia (1993) has given the identification difference between these species in his Ph.D. Thesis (pp 66-69) (Fig. 14.3). (Fig. 14.4). Similarly field identification of major commercially important species available in Gujarat are as under:

## Identification characters of juveniles of *P.monodon*

Details of characters

1. Rostrum reaches just beyond the eye stalk. (Fig.__PMo-1).
2. Chromatophores on inner uropod ramus in raw along median aspect & on outer uropod ramus 3 to 5 in the middle on median aspect. (Fig___PMo-2).
3. Anterolateral chromatophores absent on 6th abdominal segment and on Ventral side 13 to 18 chromatophores present. (Fig___PMo-3).
4. Chromatophores on telson present from base to distal end.
5. Colour of chromatophores in live specimen is maroon.
6. Body very slender and elongated appears like brownish stick and a red steak is present along the entire ventral side of the body of post larvae (5 to 6 mm in size).
7. In the specimen over 15 mm the reddish steak turns pink and then green.
8. It swims rapidly with straight movement along the edge to cling over fleating twinges, leaves & algal matters.
9. Juveniles are pale green with dark transverse band on first, third and last abdominal segment.
10. Posteromedian dorsal spine on 5th abdominal segment absent after two rostral spine stage.

## Identification characters of juveniles of *P.semisulcatus.*

Details of characters

1. Rostrum long, surpasses end of eye. (Fig___P.S.I.).

2. Chromatophores on telson present from base to distal end. Chromatophores are numerous on inner & outer uropod ramus, on distal half only. (Fig___P.S.2)
3. On 6th abdominal segment anterolateral chromatophore present & on ventral side 9 to 10 chromatophore can be seen. (Fig___).
4. Abrostral creast and grove extending beyond epigastric tooth.
5. Juveniles are creamise with irregular mettled brown bands on all abdominal segment.
6. Abdomen has brown-gray and pale yellow dorsal transverse bands.
7. Antenuals are banded white and brown.
8. Posteromedian dorsal spines on the 5th abdominal segment present up to 4 rostral spine stage.

## Identification characters of juveniles of *P.japonicus.*

Details of characters

1. Rostrum short, reaching only half length of eye. (Fig___PJ-1).
2. One chromatophores present near base on median aspect of inner uropod ramus and absent on outer uropod ramus CL 7 to 13 mm. (Fig___PJ-2).
3. On Anterostral side on 6th abdominal segment chromatophores absent, and on ventral side 10 to 11 chromatophores present, CL 7 to 13 mm. (Fig___PJ-3).
4. Chromatophores present on telson from base to distal end.
5. Colour of chromatophores in live specimen is crimson.
6. Spinules on dorsal carina of 6th abdominal segment present on all along length of carina.
7. Prominent antennal spine on carapace is found.
8. Telson with 3 pairs of lateral spinules.

(CL = Carapace length).

## Identification characers of juveniles up to 5 mm of carapace length of *Metapenaeus affinis*

Details of characters

1. Rostrum very short and pointed. (Fig MA-1).
2. One-chromatophores present on each inner and outer uropod ramus at distal end. (Fig MA-2).
3. On 6th abdominal segment one anterolateral chromatophore is present and on ventral side two chromatophores are present.
4. Chromatophores present on telson from base to distal end.
5. Telson bears 7 pairs of spines on side and distal margin.
6. Colour of chromatophores in live specimen is brown.
7. 1 to 2 spinules on dorsal carina of 6th abdominal segment may be present near posertior end.
8. Antennal spine on carapace present.
9. 'M' like patch on mid-dorsal region of 4th abdominal tergum absent.

## Identification characters of juveniles up to 5 mm of carapace length of *Metapenaeus dobsoni.*

Details of characters.

1. Rostrum long with adrostral crest reaching as far as epigastric tooth. (Fig.__MD-1).
2. Two prominent chromatophores present in the middle on telson (Fig___MD-2).
3. One prominent chromatophore present in the middle of each inner and outer uropod ramus.
4. Tip of the rostrum is actually blunt.
5. Body colour is pale yellow with red brownish or greenish specks.
6. Swims with jerky movement.
7. No setae on distal lateral aspect of 6$^{th}$ abdominal segment.
8. Median dorsal spine on 5$^{th}$ abdominal segment absent.
9. 2 to 3 spinules may be present near posterior end on dorsal carina of 6$^{th}$ abdominal segment.
10. Telson bears 7 pairs of spine on side and distal margin.
11. Abdominal carination commences from 4$^{th}$ segment.
12. Ground colour of body cream white.
13. Chromatophores on body widely spread, not forming bands.

## Identification characters of juveniles up to 5 mm of carapace length *M. monoceros*

Details of characters.

1. Rostrum extending to 3/4 eye. (Fig.___MM-1).
2. 2-3 chromatophores on inner uropod ramus present and absent on outer uropod ramus. (Fig___MM-2).
3. Chromatophores on 6$^{th}$ abdominal segment present and 4$^{th}$ & 5$^{th}$ on ventral side only chromatophores are absent.
4. Tip of the rostrum is pointed and serrated on its upper margin.
5. Rostrum with 4 spines (2 large & 2 small).
6. Epigastric spine on carapace absent in post larvae.
7. Long setae on distal lateral aspect of 6$^{th}$ abdominal segment.
8. Swims in a straight line direction without any jerky movements.
9. 'M'-like patch on mid-dorsal region of 4$^{th}$ abdominal tergum present.
10. The short band on carapace conspicuous.

### Maturation

In mature, maturation time and size are different for different prawn species. Different environmental conditions also affects maturation. Maturation in prawn can be obtained using unilateral eyestalk ablation technique under controlled condition in *P. indicus, P. monodon* and *P. dobsoni.* (Muthu and Laxminarayan 1977). Artificial maturation through eyestalk ablation is possible. Eyestalk ablation can be done by placing a cut (with a sharp sterilized blade) at the base of one eye stalked. The fluid of the eye is then squeezed out. Eyestalk ablation shortens maturation period. (Browdy and Samocha 1985.). After ablation, the minimum time for a female with fully developed ovaries ready to release eggs is three days, and the maximum time is 3 to 4 weeks.

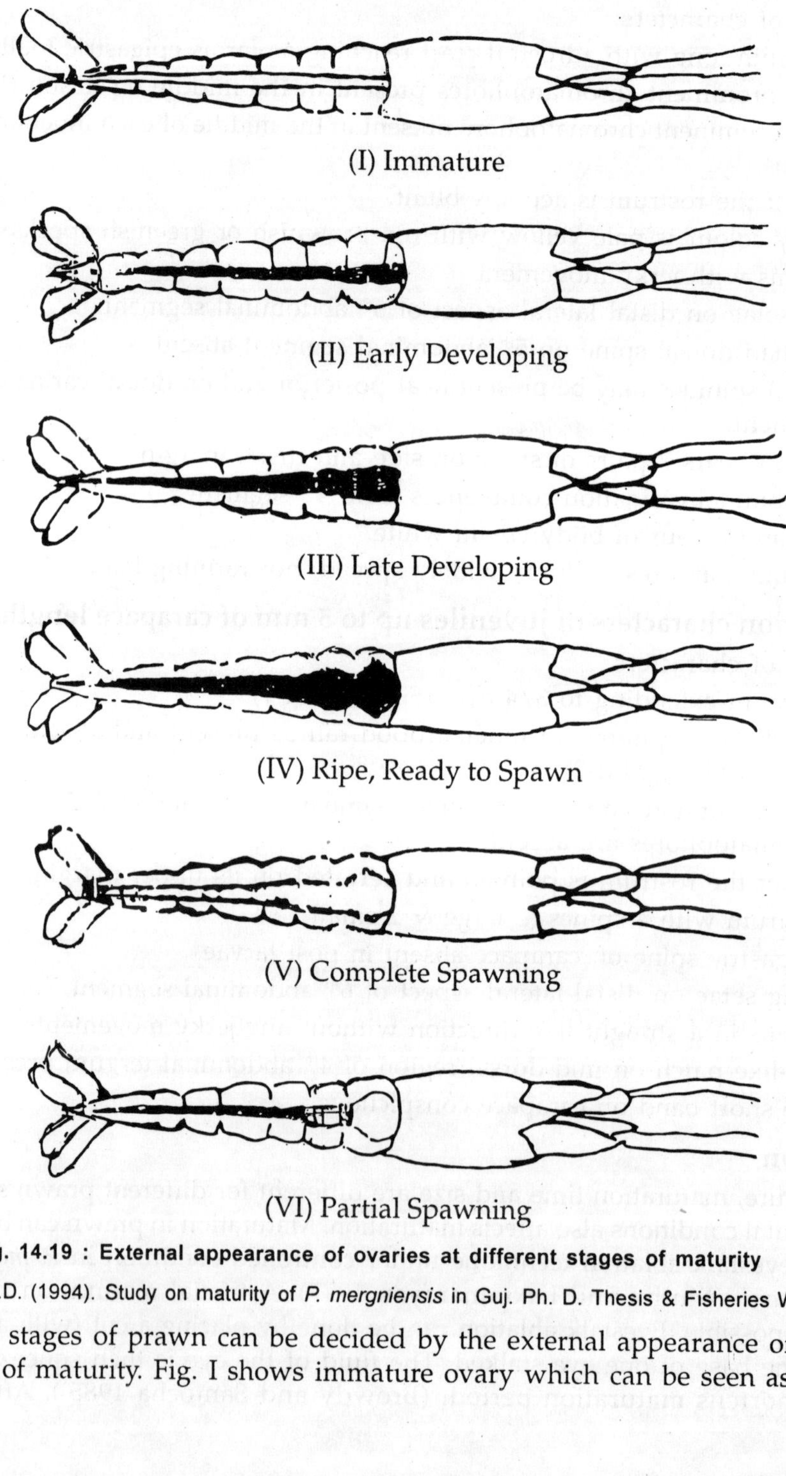


**Fig. 14.19 : External appearance of ovaries at different stages of maturity**

*Source* : Dhokkia, A.D. (1994). Study on maturity of *P. mergniensis* in Guj. Ph. D. Thesis & Fisheries World (1996).

Maturation stages of prawn can be decided by the external appearance of ovaries at different stages of maturity. Fig. I shows immature ovary which can be seen as white line on dorsal abdominal part. Fig.–II shows early developing appearance of ovary. White line becomes dark and little thick. In Fig.–III thickness of ovary is increasing which shows late development.

When ovary is ripe and ready to release eggs, its appearance is like Fig.–IV dark olive green and first abdominal segment diamond shape can be seen. Fig.–V shows same appearance like ripe ovary, but colour is white. This shows that prawn has released eggs recently and full spawning has occurred. Some times full spawning does not take place. Fig. VI represents this stage-half spawning condition, part of ovary will be dark green and part will be white. Outer appearance thus is an indication of ovary will be dark green and part will be white. Outer appearance thus is an indication of ovary condition. In case of *P. merguiensis* Maximum number of fertilised eggs (2.30 lakh) were observed from a female weighing 42 gm at Okha (Dholakia, 1993, 1996.)

## Spawning

It was reported that spawners of *P. merguiensis* were available in great quantity from October to December as well as March to April. In February 50% prawns were on way to maturation. In April, it was seen that about 30% were mature and remaining were spent or immature. It can thus be inferred that spawning occurs during March-April phase.

The ablated brood stock spawned more and produced more than non-ablated controls. Shortening in maturation time led to an increase in the number of spawners per moult cycle. A maximum of four spawners per moult cycle was observed for ablated female Vs maximum of two from non-ablated female. (Browdy and Samocha, 1985).

From the Kn value of *M. kutchensis* it was found that an individual prawn may spawn about six times during its life. Similarly the Kn value of *P. merguiensis* shows higher than one, indicating the high relative robstness and suitability of prawn in Saurashtra environment.

**Table. 14.1 : Monthwise Maximum and Minimum size of prawn found on Saurashtra coast**

| Prawn sp. | Male | | Female | |
|---|---|---|---|---|
| | Max. | Min. | Max. | Min. |
| *P. penicillatus* | 169 mm March. | 110 mm Dec. | 180 mm April. | 117mm April. |
| *P. semisulcatus* | 167 mm Feb. | 128 mm March. | 205 mm Feb. | 120 mm March. |
| *P. merguiensis.* | 164 mm March. | 123 mm Dec. | 204 mm Nov. | 116 mm March. |
| *P. japonicus* | 142 mm March. | 107 mm Dec. | 145 mm Dec. | 115 mm Feb. |
| *P. monodon* | 218 mm Oct. | 216 mm Oct. | 298 mm Oct. | 296 mm Oct. |
| *M. monoceros* | 158 mm Feb. | 111 mm March. | 206 mm Feb. | 106 mm March. |
| *M. affinis* | 138 mm April. | 137 mm April. | 170 mm April. | 132 mm April. |
| *M. brevicornis* | 98 mm March. | 84 mm Dec. | 145 mm March. | 101 mm Dec. |
| *P. hardwickii* | 65 mm Oct. | 57 mm Oct. | 123 mm Feb. | 70 mm Jan. |
| *P. stylifera* | 88 mm Nov. | 62 mm Oct. | 110 mm Nov. | 67 mm Nov. |

## Growth

It is well known fact that the growth of prawn is a complex process which is controlled by many factors like salinity, temperature, water depth, tropic condition of the ecosystem, age of prawns. density of prawns, food conservation efficiency of prawns etc. Taissersion (1960) has stated that "the growth in size in crustaceans is essentially discontinuous since increase in linear dimension takes place abruptly at echdysis and length within the rigid exoskeleton in the interval between moults remain contant."

According to Rajyalakshmi (1961), there is direct correlation between seasonal fluctuation in Kn values and rate of growth in prawn.

Different workers have tried to study the effect if salinity on growth and survival in different prawn species which are as under:

| | Salinity | Growth | Survival | Reference. |
|---|---|---|---|---|
| *Peneus monodon* | 7 to 10 ppt. | 0.6 mm/day | 70.9% | Rajyalakshmi, 1982. |
| | 11.6 to 19.48 pt. | 1.43 mm/day | 86% | Venketesan & Bose, 1982 |
| *Penaeus indicus* | 45 ppt | 4 mg/day | 68% | Paul Raj & P.J.S. Raj |
| | 35 ppt | 10 mg/day | 70% | 1982. |
| | 25 ppt. | 4 mg/day | 95% | |
| *P. merguiensis* | 39 to 42 ppt | 32 mg/day | | Dholakia, *et. al.*, 1991. |

Different workers have studied prawn species available on Gujarat-Saurashtra coast. George and Suseedan (1982) have found distribution in Narmada and Tapti estuary as well as Rann of Kutch. Dholakia (1993) has observed near Okha, Dwarka, Harshad and Saurashtra coast, Joseph (1991) has found monthwise male and female near Okha and Dwarka. Mansuri *et. al.*, 1986 near Navibunder, Gosabara, Chhaya and Nandasana (1990) have observed juveniles in Navibandar creek. Dholakia and Nandasana (1996) have noted availability in Kutch. Consolidated picture of availability of prawn species are given in Table-14.1.

## Prawn Fishery of the Little Rann of Kutch

The Little Rann of Kutch has an area of approximately 3000 sq. km. bordering Rajkot and Surendranagar districts on the Southern side and Kutch district on the northern side. The fishery for prawns exists only in an area of 1200 sq. km.on the southern border of the Little Rann, as the depth in muddy, making it favourable for operating drag nets.

The Little Rann is dry from October to June except for Adesar area (300 sq. km.) With the onset of monsoon in the third week of June, the rivers like Banas, Sarswati, Rupan, Umai, Chandrabhaga, Phulka, Kankavati, Bambhan. Godadhroi, and Machhu, discharge in the Little Rann. By the middle of July, generally, the monsoon picks up momentum and the Rann gets flooded. During this period the Little Rann with its fresh water is connected to the Gulf of Kutch, facilitating entry for fish and prawns into the Rann. However, during onset of monsoon (July) the salinity is very low and hence migration of prawns is at a minimum. But at this time there is a good fishery for Hilsa Spp. in the Rann for about 10 days. With the cessation of heavy rains, salinity increase in the Rann and an estuarine environment is established. Juvenile prawns migrate into this water in large quantities and are caught. Generally the prawn fishery in this area starts by the first week of July and ends by the middle of October, (Sudhakar Rao, 1983).

## Craft and Gear

### *Craft*

The craft employed is a Malia type of Plank-built ribbed boat locally known as "odie". The dimensions vary from 5.4*0.9*0.5 m. to 6.6*1.33*0.6 m. The bottom of the boat is flat to facilitate easy beaching on the estuarine mud.

### Gear

In this area "Gunja" net (a small bag net) is used for prawn fishing. At Surbari and Lakhiaser the net is used as stationary net.

### Fishery

It is reported that *Metapenaeus kutchensis* formed about 99% of the Prawn catches of the area. *Macrobranchium malcomsonii, Palaemon stylefera* and *Parapenaeopsis sculptilis* contributed to the rest of catch. Data have shown that in August maximum catch was reported compare to July and September.

### Size Composition

*M. kutchensis* of 41-115 mm. in total length were represented in the catch.

### Sex Ratio

At Surbai, in the beginning of the season (July). Males and females were observed in more or less equal proportion; later the proportion of male increased gradually.

### Food and Feeding

The study of stomach contents of *M. kutchensis* collected from Little Rann of Kutch have shown prawns (7.61%), copepods (7.11%), other crustaceans (10.28%), fish (5.01%), algae (31.07%), Anagiosperm matter (5.83%), molluscan shells (1.95%), sand (6.32%) and detritus (24.81%). This indicates that *M.kuchensis* is an omnivores with possible preference to crustaceans and algae. (Sudhakar Rao, 1983).

### Prawn Fishery of Gulf of Khambhat.

Prawn fishery of Gulf of Khambhat, consists principally of *Parapenaeopsis sculptilis* (Heller) and *Metapenaeous dobsoni* (Miers) throughout the year. Generally they are landed in the Bhavnagar-Ghogha coast of Bhavnagar District, and Dehej-Kavi coast of Bharuch District. (Patel and Balapatel 1982.)

It was reported that *P. sculptilis* were relatively in abundance in November and December 78, January June-September 79 and May-July in 1980. The largest size observed was 147 mm. and juveniles of the species was 20 mm.

### Maturity stage

| | *Stage* | *Observation* |
|---|---|---|
| 1. | I | Opaque and Slender |
| 2. | II | Light green and slightly elongated |
| 3. | III | Dark green and distended |
| 4. | IV | Dark green, granular and fully distended |
| 5. | V | Orange yellow and flaccid |

Observation of large number of live female of *P. sculptilis* revealed that some female had a noticeable expansion of the posterior lobes of ovary in region of the first & third abdominal segments. The females belonging to maturity stage III and IV were considered for determination of spawning season. The majority of the female attains the ovarian development at the size of 70 mm. The majority of the females of *P. scultilis,* measuring 70 mm to 140 mm. The spawning is associated with high temperature seasons. Salinity and pH showed no influence on spawning.

## Sex Ratio

Month wise sex ratio (male/female) was calculated and it was found to be in the range of 0.16 to 0.72 for *P. sculptilis.*

## *M. dobsoni*

In prawn fishery of Gulf of Khambhat, one of the largest quantity of prawn species is *Metapenaeus dobsoni.* On analysis of the prawn catch from the area revealed that *M. dobsoni* contributes 51.6% on an average and occurs throughout the year. Abundance was found in August 78 February to May 79 and October to February 86. The largest size was 124 mm. juveniles was measuring 20 mm to 40 mm in most of the months. This is suggestive of prolonged spawning of *M. dobsoni.* The individuals of *M. dobsoni* measuring more than 110 mm. were less in number. (Sudhakar Rao, 1983).

## Maturity

In *M. dobsoni* four maturity stage were found.

1. Opaque and slender
2. Dirty and slightly distended
3. Pigmented and fully distended
4. Whitish yellow with less pigments.

Observation of a large number of five female of *M.dobsoni* revealed that some developing females had a noticeable abdominal segments. The females belonging to maturity stage III were considered for detemination of the spawning season.

## Spawning season

The females of *M. dobsoni* attains the ovarian development at the size of 55 mm. Females measuring 55 mm to 120 mm collected during April, May and June had the developed ovary indicating their spawning season during these months.

## Sex Ratio

Month wise sex ratio (male/female) was calculated and it was found to be in the range of 0.14 to 0.77 for *M. dobsoni.*

**Table 14.2 : Maximum Attainable Size of Prawns found in Gujarat.**

| Sr. No. | Genus and Species | Maximum size |
|---|---|---|
| 1 | *Penaeus indicus* | 230 mm |
| 2 | *Penaeus monodon* | 320 mm |
| 3 | *Penaeus merguiensis* | 240 mm |
| 4 | *Penaeus semisulcatus* | 250 mm |
| 5 | *Penaeus japonicus* | 270 mm |
| 6 | *Penaeus penicillatus* | 210 mm |
| 7 | *Metapenaeus affinis* | 180 mm |
| 8 | *Metapenaeus dobsoni* | 125 mm |
| 9 | *Metapenaeus kutchensis* | 140 mm |
| 10 | *Metapenaeus brericornis* | 125 mm |
| 11 | *Parapenaeopsis stylifera* | 140 mm |
| 12 | *Parapenaeopsis sculptilis* | 165 mm |
| 13 | *Parapenaeopsis hardwickii* | 120 mm |
| 14 | *Macrobranchium rosenbergii* | 320 mm |
| 15 | *Macrobranchium malcomsonii* | (male) 230 mm<br>(female) 200 mm |
| 16 | *Palaemon styliferus* | 90 mm |
| 17 | *Hippolysmata ensirostris* | 80 mm |
| 18 | *Acetus indicus* | 40 mm |

Source : Kurian and Sebastian, 1976: 92-102.

**Table. 14.3 : Estimated Resiource potential of Prawns in tonnes**

| Sr. No. | Species | Estimated Resources in Gujarat | Estimated Resources in India |
|---|---|---|---|
| 1 | *Penaeus indicus* | 8,500 | 32,000 |
| 2 | *Penaeus monodon* | 1,000 | 8,000 |
| 3 | *Penaeus semisulcutus* | 500 | 5,600 |
| 4 | *Penaeus merguiensis* | 300 | 4,200 |
| 5 | *Metapenaeus effinis* | 2,000 | 16,000 |
| 6 | *Metapenaeus dobsoni* | 2,000 | 56,000 |
| 7 | *Metapenaeus monoceros* | 2,000 | 14,000 |
| 8 | *Macrobranchium rosenbergii* | 100 | 4,600 |
| 9 | *Parapenaeopsis stylifera* | 3,000 | 31,000 |
| 10 | *Parapenaeopsis sculptilis* | 2,000 | 3,500 |
| 11. | *Parapenaeopsis hardwickii* | 1,200 | 4,200 |
| 12 | *Panadalus sp.* | 3,500 | 27,000 |

Source : M.P.E.D.A. Altas, 1982.

## Prawn Fishery

Table 14.2 gives the maximum attainable size of prawns found in Gujarat. This will be helpful in selection of species for prawn culture. Table 14.3 gives the species-wise estimated potential resources of prawns (in tones) in Gujarat and India. Table 14.4 gives th details of quantity wise and location-wise availability of various cultivable prawn species on the Gujarat coast.

**Table 14.4 : Table showing location of quantitative availability of various species of cultivable prawn pecies on the Gujarat Coast**

| Name of location | Cultivable prawn Species | Availability of prawn seeds |
|---|---|---|
| 1) Mundra | *Penaeus merguiensis*<br>*Metapenaeus dobsoni* | More than 450 No/haul<br>Between 250-450 No/haul |
| 2) Jodia | *P. merguiensis* | Less than 75 No/haul |
| 3) Port Okha | *P. meguiensis*<br>*P. semisulcatus* | Between 75-100 No/haul<br>Loss than 75 No/haul |
| 4) Harshad | *P. merguiensis*<br>*P. monodon.*<br>*P. affinis* | More than 450 No/haul<br>Less than 75 No/haul<br>Less than 75 No/haul |
| 5) Navibandar | *P. merguiensis* | More than 450 No/haul |
| 6) Mangrol Bara | *M. monoceros*<br>*P. japonicus* | Beween 100-250 No/haul<br>Less than 75 No/haul |
| 7) Sartanpur | *P, indicus*<br>*M. affinis* | Between 250-450 No/haul<br>More than 450 No/haul |
| 8) Dandi and Kapasi | *M. dobsoni* | More than 450 No/haul |
| 9) Vansi Borsi | *M. monoceros* | Between 100-250 No/haul. |

Source : Guj. Agri. Uni. Okha.

In recent years capture prawn landing is not sufficient to meet increasing international demand. Hence it is necessary to culture prawn. All counries of the world are undertaking prawn culture activity to get maximum production. Table 14.5 give district wise details of brackish water area, cultivable land, allotment and shrimp culture farms in Gujarat. Prawn landing pattern of small, medium and Jumbo size and its percentage of total prawn landing of Gujarat since 1978-79 is calculated and shown in Table 14.6. Shrimp farming in Gujarat is compared with India and the same is given in Table 14.7.

Different workers have calculated relationship of total length and weight and carapace length and total length of some of the important cultivable prawn species. The same is consolidated and shown in Table 14.8.

In India almost all-maritime state have started shrimp farming. Table 14.9 give the details of area under culture, estimated production and per hector production in Kg for each state since 1990-91 to 1998-99. From this table we can see that per hectare average shrimp produuction is 582 kg, which is very less compared to Thailand (14 tons/ha). In Gujarat only 256 ha land is utilized for shrimp farming up to 1998-99 from total 3,76,000 ha suitable land for culture. This shows that there is good scope of shrimp culture in Gujarat. Table 14.10 gives the penaeid and non-penaeid composition of different states of India.

**Table. 14.5 : District wise details of Brackish water area, Cultivable land, allotment and Prawn culture farms in Gujarat (Up to 1994-95)**

| Name of District | Potential cultivable land (Ha) | Land allotted for prawn farming | No. of Farms. |
|---|---|---|---|
| Valsad | 20,235 | 1,428 | 129 |
| Surat | 24,300 | 30 | 6 |
| Bharuch | 47,500 | 505 | 63 |
| Amroli | 80,095 | 25 | 5 |
| Junagadh | 19,020 | 62 | 2 |
| Jamnagar | 22,300 | 50 | 10 |
| Kutch | 68,400 | 425 | 26 |
| Rajkot | 2,000 | 20 | 4 |
| Bhavnagar | 5,200 | 46 | 4 |
| Kheda | 20,735 | – | – |
| Ahmedabad | 49,780 | – | – |
| Total | 3,76,000 | | |

Source : M.P.E.D.A.

**Table 14.6 : Prawn landing pattern of Gujarat with its percentage share from 1978-79 to 2000-01**

| Year | Total marine landing | Shrimp Landing | Its % share | Medium Landing | Its % share | Jumbo landing | Its % share |
|---|---|---|---|---|---|---|---|
| 1978-79 | 229971 | 5322 | 2.32 | 5700 | 2.47 | 945 | 0.41 |
| 1979-80 | 206749 | 4482 | 2.15 | 4997 | 2.42 | 1373 | 0.66 |
| 1980-81 | 218872 | 8192 | 3.74 | 6745 | 3.08 | 1805 | 0.82 |
| 1981-82 | 220607 | 8708 | 3.94 | 5436 | 2.46 | 1800 | 0.82 |
| 1982-83 | 192669 | 9764 | 5.07 | 4505 | 2.34 | 1002 | 0.52 |
| 1983-84 | 223291 | 10557 | 4.73 | 5918 | 2.65 | 1091 | 0.49 |
| 1984-85 | 290708 | 12028 | 4.14 | 7109 | 2.45 | 1452 | 0.52 |
| 1985-86 | 306577 | 10546 | 3.44 | 5564 | 1.82 | 1946 | 0.63 |
| 1986-87 | 315942 | 14575 | 4.61 | 4870 | 1.53 | 982 | 0.31 |
| 1987-88 | 327560 | 15968 | 4.87 | 5991 | 1.83 | 1367 | 0.42 |
| 1988-89 | 414075 | 20146 | 4.87 | 9892 | 2.39 | 2004 | 0.48 |
| 1989-90 | 432384 | 18347 | 4.34 | 6851 | 1.61 | 1722 | 0.41 |
| 1990-91 | 500462 | 18463 | 3.69 | 4360 | 0.87 | 1364 | 0.27 |
| 1991-92 | 530017 | 27560 | 5.20 | 8759 | 1.65 | 1648 | 0.31 |
| 1992-93 | 609103 | 31815 | 5.22 | 10117 | 1.66 | 2056 | 0.34 |
| 1993-94 | 619836 | 29483 | 4.76 | 9576 | 1.54 | 2246 | 0.36 |
| 1994-95 | 645261 | 38063 | 5.90 | 15794 | 2.45 | 1988 | 0.31 |
| 1995-96 | 598351 | 35096 | 5.86 | 9711 | 1.62 | 2333 | 0.39 |
| 1996-97 | 660068 | 39096 | 5.91 | 10349 | 1.57 | 1225 | 0.19 |
| 1997-98 | 702355 | 36676 | 5.22 | 11147 | 1.58 | 1000 | 0.14 |
| 1998-99 | 551660 | 39038 | 7.08 | 7284 | 1.94 | 878 | 0.16 |
| 1999-00 | 670951 | 48032 | 7.16 | 7284 | 1.09 | 1126 | 0.18 |
| 2000-01 | 620474 | 44970 | 7.25 | 7905 | 1.27 | 1108 | 0.18 |

Source : Commissioner of Fisheries, Govt. of Gujarat, Gandhinagar. (with permission).

**Table. 14.7 : Shrimp Farming in Gujarat compared to India. Since 1990-91 to 1995-96**

| | Gujarat | | India | |
|---|---|---|---|---|
| Estimated potential area in ha. | 3,76,000 | | 11,90,900 | |
| Year | Area under culture in Ha. | Estimated production in M.T. | Area under culture in Ha. | Estimated production in M.T. |
| 1990-91 | 125 | 125 | 65,100 | 35,500 |
| 1991-92 | 231 | 170 | 68,227 | 40,000 |
| 1992-93 | 360 | 200 | 70,700 | 47,000 |
| 1993-94 | 475 | 500 | 82,540 | 62,000 |
| 1994-95 | 700 | 700 | 1,00,700 | 82,850 |
| 1995-96 | 884 | 546 | 1,18,983 | 70,573 |

Source : MPEDA, Kochi.

From the above it can be seen that prawn culture activities in Gujarat is very limited. It can be increased. Secondly though total area under prawn cultured in all India base is increased, production per ha is reduced. This is due to disease attack at many farms.

**Table. 14.8. : Total Length-Weight and Carapace length-Total length relationship of some important prawn species**

| Sr. No. | Name of Prawn species | Total length – Weight | Carapace length –Total length | Reference |
|---|---|---|---|---|
| 1 | *Penaeus merguiensis* | Male: Log w = -4.8579 + 2.9062 log TL<br>Female: Log W = -3.5055 + 2.33097 log TL | Male: Log CL = 1.1824 + 0.6549 log TL | Dholakia A.D., 1993 (Ph.D. Thesis) Saurashtra Uni. |
| 2 | *Penaeus monodon* | Male: Log W = -6.1269 + 3.4554 Log TL<br>Female: Log W = -5.3480 + 3.0776 Log TL | CL = -2.5339 + 0.2493 TL | Rao P.A.V. 1967. Indian J. Fish. 14 (2): 251 – 270 |
| 3 | *Penaeus indicus* | Male: Log W = -6.1262 + 3.4554 Log TL<br>Female: Log W = -5.3480 + 3.077 Log TL | CL = 0.9408 + 0.2073 TL | Rao P.A.V. 1967. Indian J. Fish. 14 (2): 251 – 270 |
| 4 | *Penaeus semisulcatus* | Male: Log W = 0.9093 x L 2.9540<br>Female: Log W = 111.5492 x L 2.4242 | CL = 3.2231 + 3,4802 TL | M.M. Thomas, 1975, Indian J. Fish No.1 (22) 1 & 2 : 133 – 142. |
| 5 | *Penaeus ponisiloatus* | Male: Log W = -4.50 + 2.67 log TL<br>Female: Log W = -4.87 + 2.85 Log TL | Male. Log CL = 0.81 + 1.20 Log TL<br>Female: Log CL = -0.95 + 1.28 Log TL | Mansuri, A.P., M.N. Prasad, R.I. Patel, ICAR Project; 4(4)/82-AS(F) |
| 6 | *Metapenaeus monoceros* | Male: Log W = -4.88 + 2.88 Log TL<br>Female: Log W = -4.87 + 2.88 Log TL | Male: Log CL = -0.11 + 0.76 Log TL<br>Female: Log CL = 0.13 + 0.75 Log TL | Mansuri, A.P., M.N. Prasad, R.I. Patel, ICAR Project; 4(4)/82-AS(F) |
| 7 | *Metapenaeus dobsoni* | W =0.0037936 L 2.9078891 | N.A. | Sreekumaran Nair S.R., H Krishna Iyer, C.B. Lalithambika Devi and M. Krishman Kutty. 1982, Mahasagar No 10, 15(2) : 95 – 104. |
| 8 | *Palaemon malcolmsonii* | Male: Log W = -5.91 + 3.45 Log TL<br>Female: Log W = -5.32 + 3.16 Log TL | Male: Log CL = -0.81 + 1.21 Log TL<br>Female: Log CL = -0.63 + 1.13 Log TL | Mansuri, A.P., M.N. Prasad, R.I. Patel, ICAR Project; 4(4)/82-AS(F) |

**Table 14.9 : State-wise details of Shrimp Farming in India**
**(A-Area under culture in ha; P - Estimated Production in MT; H - Per ha production in Kg)**

| State | Estimated Potential Area (ha) | | Annual levels of Area brought under culture (ha) 1990-91 | 1991-92 | 1992-93 | 1993-94 | 1994-95 | 1995-96 | 1996-97 | 1997-98 | 1998-99 |
|---|---|---|---|---|---|---|---|---|---|---|---|
| West Bengal | 1,05,000 | A | 33,815 | 33,918 | 34,050 | 34,150 | 34,400 | 34,660 | 42,605 | 42,525 | 42;067 |
| | | P | 12,500 | 13,800 | 16,300 | 16,500 | 25,000 | 23,445 | 19,949 | 15,121 | 18,326 |
| | | H | 369 | 407 | 472 | 483 | 726 | 676 | 457 | 356 | 435 |
| Orissa | 31,600 | A | 7,075 | 7,417 | 7,760 | 8,150 | 8,500 | 11,000 | 11,332 | 11,332 | 8,000 |
| | | P | 4,100 | 3,800 | 4,300 | 3,300 | 4,800 | 6,000 | 6,805 | 5,000 | 6,000 |
| | | H | 579 | 512 | 554 | 505 | 565 | 545 | 600 | 441 | 750 |
| Andhra Pradesh | 1,50,000 | A | 6,000 | 8,100 | 9,500 | 19,500 | 34,500 | 50,000 | 60,249 | 66,290 | 71,000 |
| | | P | 7,350 | 9,700 | 12,800 | 26,000 | 34,000 | 27,140 | 30,577 | 34,075 | 44,856 |
| | | H | 1,225 | 1,197 | 1,347 | 1,333 | 985 | 543 | 507 | 574 | 631 |
| Tamil Nadu | 56,000 | A | 250 | 480 | 530 | 1,050 | 2,000 | 2,879 | 640 | 670 | 1,087 |
| | | P | 450 | 700 | 1,100 | 2,000 | 3,000 | 1,092 | 1,129 | 1,197 | 1,820 |
| | | H | 1,800 | 1,458 | 2,075 | 1,904 | 1,500 | 379 | 1,764 | 1,780 | 1,674 |
| Pondicherry | 800 | A | – | – | – | – | – | 37 | 22 | 22 | 22 |
| | | P | – | – | – | – | – | 10 | 27 | 20 | 27 |
| | | H | – | – | – | – | – | 270 | 1,227 | 909 | 1,227 |
| Kerala | 65,000 | A | 13,000 | 13,145 | 13,400 | 13,860 | 14,100 | 14,657 | 14,658 | 14,595 | 14,705 |
| | | P | 8,925 | 9,500 | 9,750 | 11,500 | 12,000 | 9,000 | 8,225 | 7,290 | 7,660 |
| | | H | 686 | 723 | 727 | 830 | 851 | 614 | 561 | 499 | 520 |
| Karnataka | 8,000 | A | 2,500 | 2,542 | 2,570 | 2,600 | 3,500 | 3,500 | 3,500 | 3,540 | 3,564 |
| | | P | 1,000 | 1,100 | 1,150 | 1,500 | 2,500 | 2,050 | 2,300 | 2,640 | 2,690 |
| | | H | 400 | 433 | 447 | 577 | 714 | 585 | 657 | 746 | 755 |
| Goa | 18,500 | A | 525 | 525 | 550 | 575 | 600 | 650 | 680 | 650 | 650 |
| | | P | 245 | 300 | 350 | 400 | 450 | 550 | 580 | 590 | 590 |
| | | H | 466 | 571 | 636 | 696 | 750 | 846 | 892 | 907 | 907 |
| Maharashtra | 80,000 | A | 1,800 | 1,869 | 1,980 | 2,180 | 2,400 | 716 | 929 | 970 | 426 |
| | | P | 800 | 930 | 1,050 | 300 | 400 | 740 | 523 | 700 | 409 |
| | | H | 444 | 498 | 530 | 137 | 166 | 1,033 | 563 | 722 | 960 |
| Gujarat | 3,76,000 | A | 125 | 231 | 360 | 475 | 700 | 884 | 997 | 997 | 316 |
| | | P | 125 | 170 | 200 | 500 | 700 | 546 | 572 | 235 | 256 |
| | | H | 100 | 736 | 555 | 1,052 | 1,000 | 618 | 574 | 236 | 810 |
| **Total** | **11,90,900** | **A** | **65,100** | **68,227** | **70,700** | **82,540** | **1,00,700** | **1,18,983** | **1,35,582** | **1,41,591,** | **1,41,837** |
| | | **P** | **35,500** | **40,000** | **47,000** | **62,000** | **82,850** | **70,573** | **70686** | **66,868** | **82,634** |
| | | **H** | **545** | **586** | **664** | **751** | **819** | **593** | **521** | **472** | **582** |

Source : Fishing Chimes, VOl 20, (10 & 11) : 15

From the table 14.6 it can be seen that landing percenmtage of medium and jumbo variety of shrimps are decreased since 1985-86 and small size shrimp landing is increased, which proves that juveniles are killed before they are matured. This condition is very alarming condition. Our fishermen are destroying potential stock. Government should take proper action for over fishing of small size shrimps. So as to once again after maturing and breeding sufficient natural stock can be obtained.

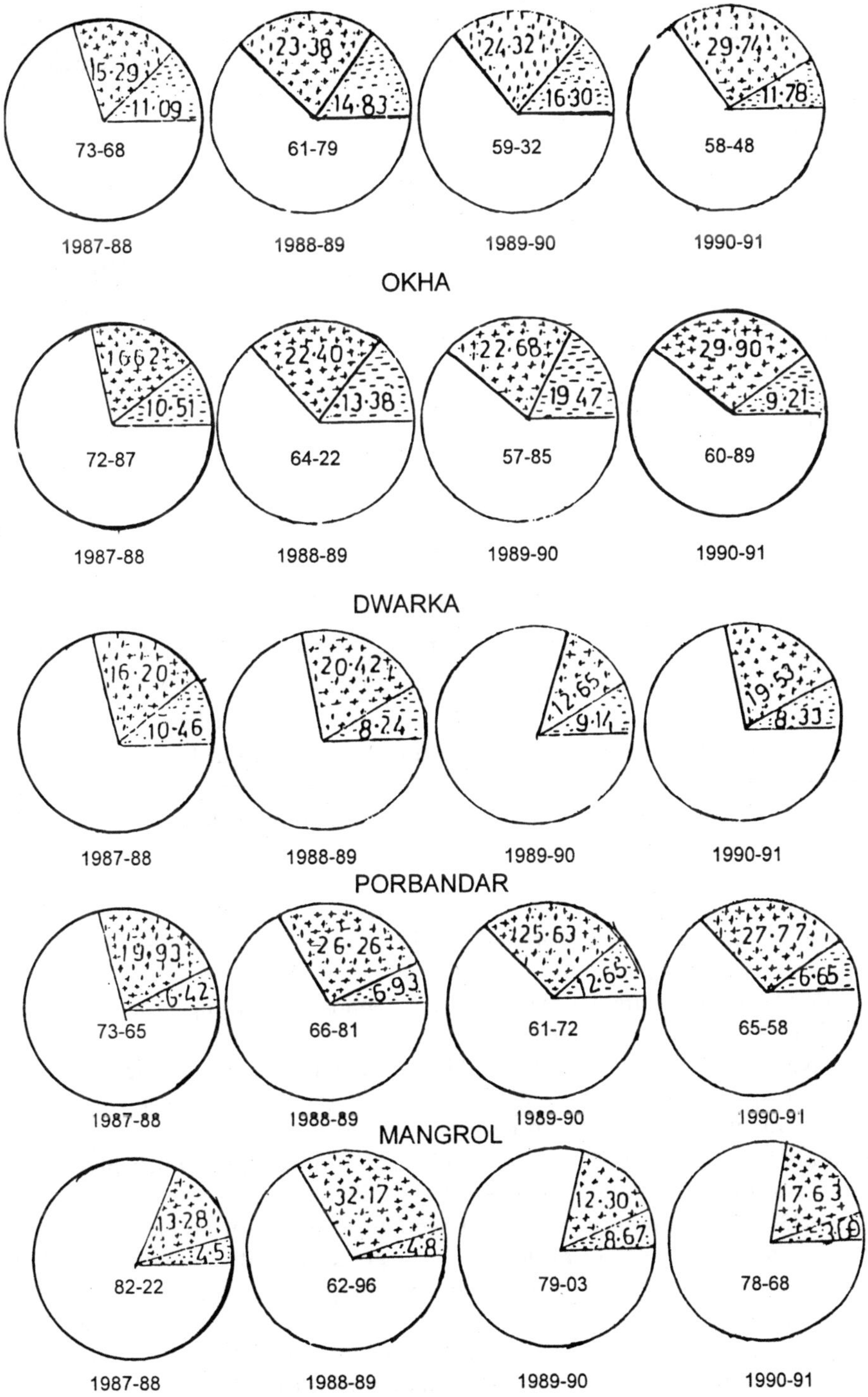


Source : Ph.D. Thesis of Dr. A.D. Dholakia

**Fig. 14.20 : Prawn landing group-wise at major centers in Saurashtra during 1987-88 to 1990-91**

Table. 14.10 : Penaeid and Non-penaeid composition of different State of India

| State | Penaeid | Non-Penaeid |
|---|---|---|
| Gujarat | 71.9% | 29.1% |
| Maharashtra | 48.1% | 51.9% |
| Goa | 98.4% | 1.6% |
| Karnataka | 99.9% | 0.2% |
| Kerala | 99.1% | 0.9% |
| Tamil Nadu | 95.8% | 4.2% |
| Pondichary | 77.5% | 22.5% |
| Andhra Pradesh | 84% | 16% |
| Orissa | 99.5% | 0.5% |
| West Bengal | 47.5% | 28.0% |
| All India | 72.00% | 28.00% |

Source : *Mar. Fish. Infor. Serv.* T & E Ser. No 10 1979, P. 15

## *References*

Anon. 1978, Symposis of Marine Prawn Fishery of India-1978. *Mar. fish. Infor. Serv.* T. & E. Ser., 10

Anon. 1980, Symosis of Marine Prawn Fishery of India-1979. *Mar. Fish Infor. Serv.*, T & E. Ser.,: 25.

Anon. 1982, Symposis of Marine Prawn Fishery of India-1980. *Mar. Fish Infor. Serv.* T & E. Ser., : 35.

Anon. 2001. An over view of the present state of shrimp aquaculture in India. Fishing Chimes, Vol 20m (10 & 11) : 11 - 16

Chhaya, N.D. and D.V. Nandasana 1990. Some observation on availability of culturable penaeid prawn juveniles in Navibandar creek in Junagadh District - Gujarat. *Fishing Chimes.* 10 (2) : 45 - 46.

Dholakia, A.D. 1986. List of important Fishes available in Gujarat with scientific, English and local names. in Fish Processing Preservation and Quality Control. pp 133 - 148.

Dholakia, A.D., M.I. Patel and N.D. Chhaya 1991. Observation on growth of banana prawn *Penaeus merguiensis* (de-Man) using different feed. National Seminar on Shrimp production and farming, 12 - 13 Feb, held at Berhampur, Orissa.

Dholakia, A.D., N.G. Akolkar, M.I. Patel and N.D. Chhaya 1991. Constraints and solution of transportation of prawn seeds of *Penaeus merguiensis* (De-Man) in tropical areas. National Seminar on Productivity, constraints in coastal areas, held at Calcutta.

Dholakia, A.D. and N.D. Chhaya 1993. (i) Some information on artemia which plays an important role in prawn culture. "Krushi go Vidya" 46 (4) : 26-28.

Dholakia, A.D. 1993. (ii). Availability of prawn seed from different creeks of Saurashtra area. Workshop organised by College of Fisheries, Guj. Agri. University, Veraval on 20 June.

Dholakia, A.D. 1993. (iii). Details of feeds to be given during different stages of growth in prawn culture. Workshop organised by College of Fisheries, Guj. Agri. University, Veraval on 20 June.

Dholakia, A.D. 1994. Marine prawn fishery and culture in Saurashtra with special reference to *Penaeus merguiensis* (de-Man)-Ph. D. Thesis, with Saurashtra University.

Dholakia, A.D. and N.D. Chhaya 1975. Model size grioup of banana prawn *Penaeus merguiensis* (de-Man) *GAU, RES, J.*, 20 (2) : 94-97.

Dholakia, A.D. and D.V. Nandasana 1996. Information on commercial production of *Metapenaeus kutchensis* available in Kutch and Gujarat Marine Water. Sovenior at Guj. Agri. University Junagadh on 16th January, pp 146.

Dholakia, A.D. 1996. Study on maturity of *Penaeus merguiensis* (de-Man) banana prawn in Gujarat-Saurashtra coast. *Fisheries World,* 4 (4) : 13-15.

George, M.J. 1962. On the breeding of Penaeids and recruitment of their Post larvae in to the back waters of Cochin. *Ind. J.Fish.,* 9 A (1) : 110-116.

Kagwada, P.V. 1967. Prawn catches by mechanised vessel in the trawling groups of Bombay and Saurashtra. *Proc. Symp, Crust. Mar, Biol. Ass. of India.,* : 4 : 1348-1381.

Kurian, C.V. and V. Sebastian, 1976. Prawns and Prawn Fisheries of India, Published by Hindustan Publishing Corporation Delhi,

Mansuri, A.P., M.N. Prasad, R.I. Patel, A.Y. Desai, V.V. Prasad 1986. Assessment of Prawn Fishery resources of River Ojat, Bhadar, Mindsar and their esturies. Their impact on Socio economic condition of the fishermen village (Saurashtra). A case study-Final Report ICAR Report 82-AS (F) : 68, Saurashtra University.

Marichamy, R. and John Motha. 1987. Prospects of prawn culture in salt pan areas. *Mar. Fish. Infor. Serv.,* T & E Serv., No 70 : 1-7.

Muthu, M.S. and G. Sudhakar Rao. 1973. On the distinction between *P. indicus & P. merguiensis* with special reference to juveniles. *Indian J. Fish.,* 20 (1) : 61-69.

Muthu, M.S. 1978. Larval development-specific identity of Penaeid post larvae found in brackish water areas. C.M.F.R.I. Bulletin No 28 : 86-90.

Ramamurthy, S. 1963. A note on prawn fishery of Kutch. *J. Mar. Biol. Ass. India. 5 (1)* : 146-148.

Ramamurthy, S. 1967. Studies on the prawn fishery of Kutch. *Proc. Symp. Crustacea* ; 4 : 1424-1436.

Sarvaiya, R.T. 1978. Prawn Fishery of Kutch with special reference to Sukhper and Lakhpat. *Indian J. Fish.,* 25 (1 & 2) : 35-40.

Sudhakar Rao, G. 1983. Observation on the seasonal prawn Fishery of the little Runn of Kutch during 1980. *Indian J. Fish.* : 30 (1) 124-134.

Vedavyasa Rao, P., 1973. Identification of juveniles of *Metapnaeus. J. Mar. Biol Asso. India.* 15 (1) : 45-55.

Vedavyasa Rao, P., 1973. Larval development of Penaeid prawns. *J. Mar. Biol/ass. India.,* 15 (1) : 95-124.

Vyas, A.A., A.Y. Desai, A.D. Dholakia, and N.D. Chhaya 1993. Unavoidable live feed *Chaetoceros* and *artemia* for seed culture in hatchery. Workshop on Marine Prawn Culture, 20-6-93. Organised by Guj. Agri. Uni, Veraval : 22-25.

# 15

# PEARL OYSTER FISHERY

The Indian pearl fisheries have been famous since ancient times for the most beautiful pearls they yield. Following are the main species of pearl oysters occurring in the Indian waters.

| *Species* | *Main area of occurrence* |
|---|---|
| 1) *Pinctada fucata* (Gould) | Main area of occurrence. |
| 2) *P. Vulgaris* | Gulf of Kutch. |
| 3) *P. maargaritifera* (Linnaeus) | Andaman and Nicobar. |
| 4) *P. chemnitazii* | Palk Bay and small in Gulf of Kutch. |
| 5) *P. sugilla* (Reeve) | Rarely available |
| 6) *P. arropurpurae* | Rarely available |
| 7) *P. aanomioides* (Reeve) | Rarely available |

Outer appearance of some of above species are shown in Fig. 15.3. Of those. *P. fucata and P. vulgaris* occurs in extensive beds in Gulf of Kutch and Gulf of Mannar. It is commercially very important being the source of the "Oriental pearls". The other five species of pearl oysters on the Indian coasts are not of any importance as their pearl yield is low or the pearls produced by them are not of high quality.

World wide also main variety of oyster cultivated for pearl is *Pinctada fucata.* The scientist Kokichi Mikimoto who, in July 1893, is believed to have been the first person to succeed in artificial culturing pearls from this species only.

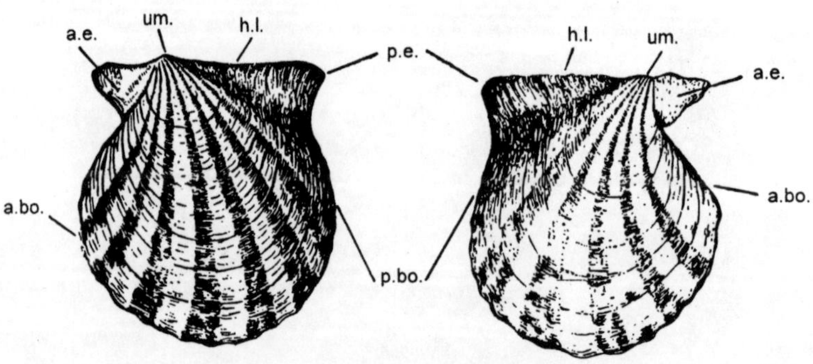


**Fig. 15.1 : Taxonomy of Pearl Oyster**

## Synonyms

*Pteriaa vulgaris* Jaameson, 1901.
*Margaritiferaa vulgaris.* Hornel 1922.
*Perlamater vulgaris.* Schumacher, 1817.

## Taxonomy

The following features characterize members of Pinctada. The hinge is long and straight; the king axis of the shell is at right angle to thee hinge; thee left valve is a little deeper than right; there is a byssal notch on each valve at the base of the anterior lobe. (Rao, 1970).

## Description of *P. fucata*

### *Shell*

The hinge is fairly long, its ratio to the broadest region of the body of the shell is about 0.85 and its ratio to the longest dorso-ventral measurement is about 0.76. In both the vlaves there are hinge teeth, one at the anterior and posterior ends of the ligament. The anterior ear is larger than in other species and the byssal notch at the junction of the body of the shell and the ear is slit like. The posterior ear is fairly well developed. The posterior border of the shell shows a small or moderately large sinus. The adductor impression is large and subcentral. There are 12 to 15 small scars caused by the insertion of pallial muscle between the umbo and the anteroventral border. The convexity off the valves is greater than in other species of the genus. The shell of the pearl oyster of Tuticorin pearl banks are reddishbrown or yellowish brown in colour with non-nacrcous broder on the inner surface of the valves possesses brownish or reddish patches coinciding with the external rays. The younger shell have thin, fat, marginal radiating blunt projections. (Rao, 1970).

**Table. 15.1 : Analysis of Oyster gm/100 gm**

| Name of chemical | Window-pane oyster | Pearl oyster | Edible oyster |
|---|---|---|---|
| Calcium Ca. | 40.08 | 41.05 | 36.68 |
| Calcium carbonate $CaCo_3$ | 90.95 | 91.51 | 86.85 |
| Calcium Oxide Cao | 50.046 | 51.25 | 46.046 |

### *Body*

The soft body of the pearl oyster consists of a viscero-pedal mass covered by the right and left mantle lobes which are free anteriorly, ventrally and posteriorly but fused dorsally. The mantle edge has two thin, folds with pigmented, papillate edges. The outer fold is parallel to the inner surface of the shell and the inner fold called the pallial veil or velum projects at right angle from the mantle edge. In life the pallial is of the opposite sides are in contact with each other except in the regions of the inhalant aperture about min-ventrally and the exhalent aperture at the posterior end. The foot is elongated and muscular arising about mid-way between the mouth and the intestinal lobe and has a grove ventrally. The byssus is at the proximal end of the foot and bears byssus fibers with which the pearl oyster attaches itself to the substratum.

The adductor muscle consists of a narrow region formed of white, glistening muscle fibers and a broad region of colourless semi-translucent fibers. The alimentary canal consists of the mouth provided with two pairs of labial palps, the oesophagus, the stomatch and in healthy oyster a crystalline style is present in the intestine. The rectum passes through the pericardium, curves ventrally and lies around the posterior aspect of the adductor muscle terminating in the anus.

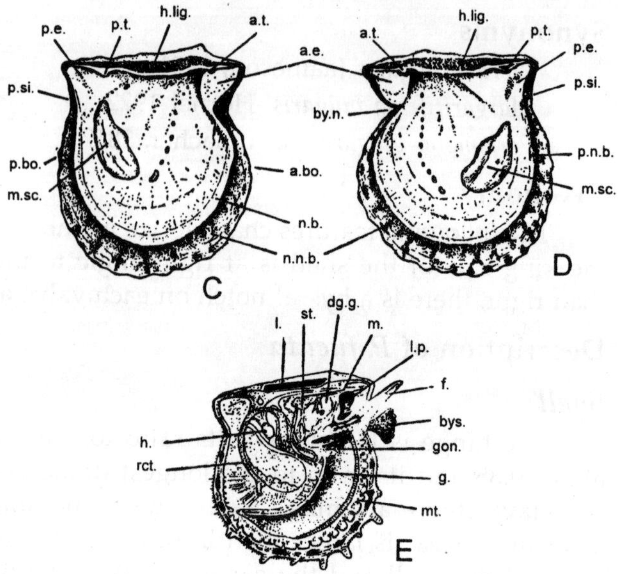

Fig. 15.2 : Anatomy of Pearl Oyster

The gills are paired, stickle-shaped structures lying on either side of the viscero-pedal mass. The vascular system consists of a heart with paired auricles and a ventricle enclosed in a pericardium, and a series of arteries and venous sinuses in the visceral organs. The nervous system has the paired cerebal, pedal and visceral or paricto-splanchnic ganglia from which nerves run to different parts of the body. There are two nephridia one on either side of the posterior half of the viscero-pedal mass. The gonads are paired, asymmetrical and creamy yellow in colour. The gonadal follicles cover the stomach and part of the intestine. (Fig. 15.2) (CMFRI Bulletin No 25, 1974).

## Food and Feeding

Pearl oysters are filter feeders, feed mainly on phytoplankton. Minute food organisms present in the water enter inside the pearl oyster along with water current passing through the narrow slit formed by the inwardly directed edges of the pallial lobes and they are carried towards the branchieae which act as finestrainers arresting every particle in the water current. The food particles collected thus are carried by the cilia to the crest of the branchial lamellae and from there they are directed by the labial palps into the mouth. The labial palps have the ability to reject unwanted materials like mud particles etc.

## Sex

The pearl oysters are hermaphrodite animals. Hence it is not possible to distinguish sex from outer appearance. The gamotogonic activity commences when they are 78 months old and about 17-25 mm in size Pandya (1972) found that during the period from December and January and from June to July about 63% of the oyster (above three years of age) were with ripe gonads in Gulf of Kutch. In the rest of the months majority of the oysters were inactive gondial condition.

## Early Development

Fertilization is external. The eggs when discharged is pyriform or flaskshaped but becomes spherical on fertilization. Segmentation is complete but unequal. The trochophore larvae is formed about 20 hours after fertilization, having a pre-equatorial tuft and a patch of circlet of cilia at the opposite pole. The body becomes elongate and the equatorial band moves forwards and becomes a premolar circlet. By the end of the second day the velgir larvae is formed with a shell and velum. A little later the shell grows larger, the umbones are visible, rudiments of brachial filaments and otocysts appear and the anterior and the posterior adductor muscles are prominent. In older larvae the umbones of the shell are larger, the digestive gland is bilobed and there is an increase in the number of brachial filaments. Later the velum is reduced, the foot is prominent and is actively protruded and moved. The byssal gland is formed about the middle of the posterior edge and octocysts lie near the base of the foot. The larvae then settle down as spat. In the earlier stage of spat that measures 0.1 mm in diameter the shell shows growth at the margin. At a size of 0.175 mm prismatic shell substance is deposited marginally. There is asymmetrical growth and the umbones are more anterior in position. A byssal sinus and delicate byssus are developed for attachment. Further growth depends on the availability of suitable substratum and other favorable environmental conditions.

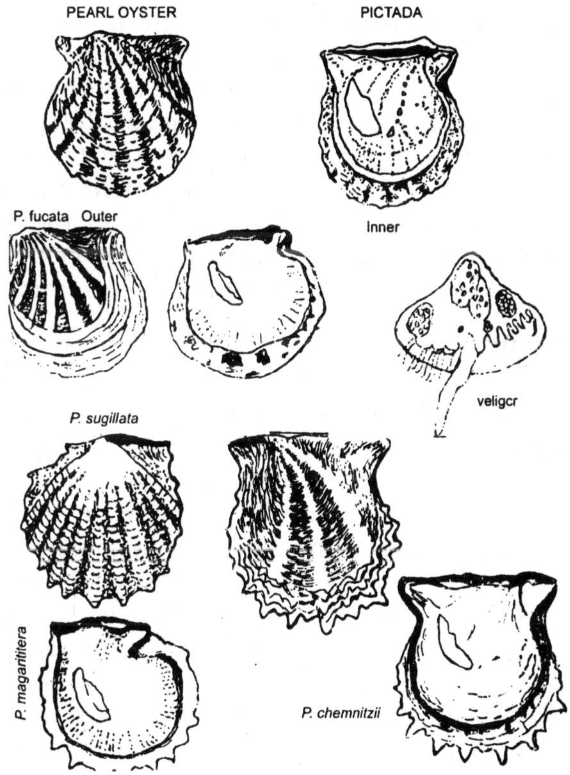

Fig. 15.3 : Different species of Pearl Oyster

**Table. 15.2 : Protein Percentage of Body of some Molluscans**

| Name of molluscan | Protein % of body weight (On dry weight base) |
|---|---|
| 1) Clams | 10-15 |
| 2) Edible Oyster | 11-12 |
| 3) Pearl Oyster | 19-20 |
| 4) Scallops | 14-15 |
| 5) Blood clams | 12-13 |
| 6) Squid meal | 81-82 |
| 7) Mussels | 08-09 |

Table No 15.1 and 15.2 gives chemical analysis of some nutrients of oysters and some moluscans of Gulf of Kutch. (Sarvaiya and Chhaya, 1993).

## Spawning season

In Indian waters, two seasons of spawning, one in June-September and other in November-December were observed in an year.

Fig. 15.4 shows life cycle and duration of each stages of pearl oyster *P. fucada* (Gould). Favorable temperature and salinity for breeding and culture is 20-25° C. and 28-35 ppt respectively. (Dholakia, et al., 1997)

## Growth

The growth rate of the pearl oyster (*P. fucada*), species occurring in the Gulf of Kutch, as denoted by increase in length (height), breadth (width) and hinge length is not proportionate to age as they get retarded after sixth year. The increase in thickness and hinge width is more or less uniform during growth. (Narayanan and Michael, 1968, Pandya 1976) Faster growth was observedin witner months October to February than February to August. To find out age group and growth of the pearl oyster, *P. vulgaris* (Schumachar) of the Gulf of Kutch. Narayan and Michel (1968) has mentioned about the oyster, *P. vulgaris* growing faster from November to February and sowing down of growth during summer.

## Fishery

In the Gulf of Kutch, pearl fishery used to be conducted on a relatively moderate scale. The season generally commenced just after the south-west monsoon. The Department of Fisheries were organizing the fishery by engaging fishermen who were paid according to the number of oyster they collected. The pearls in the oysters became the property of the Government. The fishery used to be held almost every year or alternate years from 1913 to 1939. Subsequently, it was held after every 3-4 years; but after 1996-97 there has been no fishery at all; as the oyster population became depleted. (Alagarswami and Quasim, 1973).

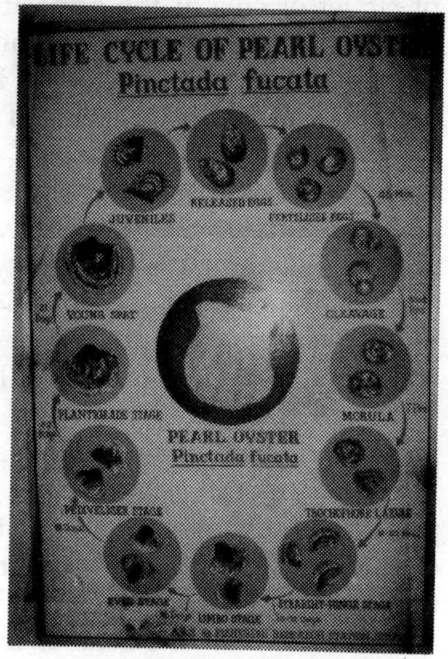


Fig. 15.4 : Life cycle of Pearl Oyster *Pinctada frucata.*

Table 15.3 : Exploitation of pearl Oyster from Gulf of Kutch

| Year | No of Pearl Oyster Harvested |
|---|---|
| 1916-17 | 76,658 |
| 1920-21 | 52,306 |
| 1926-27 | 33,816 |
| 1931-32 | 20,829 |
| 1937-38 | 44,655 |
| 1946-47 | 31,059 |
| 1950-51 | 32,441 |
| 1956-57 | 35,389 |
| 1961-62 | 17,208 |
| 1966-67 | 30,000 |

Source : Wonderful world of Gulf of Kutch. Reprinted with the permission of Commissionerate of Fisheries, Govt. of Gujarat.

## Habitat

The pearl oyster reefs of the Gulf of Kutch-locally known as "khaddas" are scattered all along the southern coast, bordering the coast line of the Jamnagar District. There are about 42 important reefs covering an area of 60,000 acres from Jodia (22° 42′ N., 70°16′ E) in the east and Ajad (22°23′ N., 69°20′ E) in the west. These reefs are not continuous but separated by sand patches, mud flats and mangrove forests, some of them exposed during the ebb tide. (Eswaran *et. al.*, 1969).

A typical bed consists of a hard bottom of coral and rocky frame work, both an admixture of mud and sand, Pearl oysters of Gulf of Kutch inhabit at depth from 1 to 20 m, while the strain of Gulf of Mannar inhabit in deep waters.

In 1969 Eswaran *et. al.*, found that the pearl oyster beds at Pirotan, Movada. Chhad, Chicheya, Deeda, and Sirwal appeared to be comparatively more productive than Betwara, Jindhada, Kalubhar, Narara. Deposition of sand reduces population. Table 15.3 shows the area of *different* locations and number of pearl oysters found per sq. m. during 1993 to 1996 (Raval *et, al.*, 1998, Dholakia, 1996)

Pearson (1927) have reported that large bed of oyster in Ceylon was lost in 1925, due to action of sand. Many island and reefs in Gulf of Kutch is not affected with heavy sedimentation resulted in depletion of population.

It was reported in survey report (Raval *et al.*, 1998) that the population was reduced to 6 to 7 No. per sq. km. Hence Indian Council of Agricultural Research (ICAR), Govt. of India have approved a project in 1998 on establishment of hatchery at Sikka, under Dr. A.D. Dholakia as Principal Investigator to increase the population of pears and edible oysters in Gujarat.

## Resources

The Pearl oyster resources show wide fluctuations which indicate needs for the development of pearl culture industry in the country. Kutch region is very rich in the resources of clam, cockle oyster, chank, seaweed coral etc. (CMFRI sp. Pub. no 38, Narsimhan, 1973).

## Marketing of Pearls

In Bombay markets four varictics of Indian pearls are recognized viz.

1) Jeevan : In which to the pearl is perfectly spherical in shape, has bright luster and the colour is rosy, pink and the pinkish-white.
2) Gholwa : The pearl is nearly spherical lustrous with colouration of varying shades.
3) Ghat : Small, irregular pearls.
4) Mosi : Very small sized one.

It is reported (Eswaran *et. al.*, 1969) that, the Gulf of Kutch pearls are costlier and their pearl yielding capacity is more than those of Gulf of Mannar.

## Effect of Different Parameters on Survival of Oyster

For commercial culture of oyster, it is necessary to know optimum limits of different water parameters like temperature, Oxygen, slainity, feeding preference etc.

Effects of temperature, oxygen, feeding preference were studied for oyster of out side Gujarat. While salinity tolerance limit for survival of pearl oyster *Pinctada fucata* and edible oyster *Crassostrea gryphoides* were studied by Dholakia *et al.*, in 1997. They found that edible oyster can tolerate wide range of salinity (0 to 70 ppt). While salinity tolerance for pearl oyster is between 12 to 76 ppt if they are properly acclimatized. (Dholakia, *et.al.*, 1997)

Stemps to be followed in undertaking a Pearl Culture Venture.

### *A) Selection of suitable sites*

Area with

- Salinity above 30 ppt
- Good phytoplankton production
- Mild moderate current
- Low siltation
- Depth 2 to 3 m and above

### *B) Selection of Oyster for operation*

- Age of Oyster - Above 1.5 to 2 years
- Weight - more than 25 gm (40 mm)

- Stage of maturity - Spent resting stage
- Overall health - Good, free from polycheates/sponges trematods infections.

### C) Conditions

- Arranging oyster in container with hinge pointing downwards.
- Narcotization of selected oyster by sprinkling Menthol in the sea water.
- Insertion of a small wooden leg between the two valves to facilitate nucleus implantation.

### D) Preparation of Graft tissue

- Selected healthy non narcotized oyster.
- Cut mental in to thin strips of 5 cm length and 0.5 cm width.
- Remove mucous and muscle from the mantle.
- Cut the mantle strip in to 20-25 pieces of 2-3 mm square.
- Keep cells live by adding Azumin/Eosin solution in sterilized sea water and use within 15 min.

### E) Implantation

Mounting oysters in the stand with the valves facing upwards.

Making incision at the right Place and placing the graft metal piece

Single implantation in the gonad near the intestinal loop.

Double implantation additionally close to hepato-pancreas.

Multiple implantation more than double implantation.

Placing the sterilized nucleus on the graft metal piece

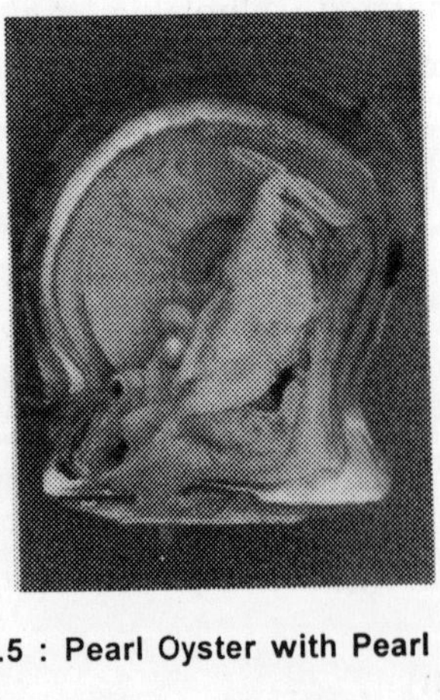

**Fig. 15.5 : Pearl Oyster with Pearl inside**

**Fig. 15.6 : Pearl Oyster *Pinctada fucata***

## F) Concalescence

- Placing the implanted oyster in fresh sea water with mild circulation for two or three days.
- maintaining water quality and water exchange.
- Removing dead oyster and shifting the healthy implanted oyster in to the natural environment.

According to Department of Fisheries, Govt. of Gujarat Pearl Oyster production from Gulf of Kutch was maximum in 1916-17 (76658 Nos.) Table 15.3 with an average of about 30,000 nos. in last 50 years. Up to independence pearl oyster harvesting was conducted by King of Jamnagar "Jam saheb", only once in a year. And he was giving labor charge to fishermen. Same practice was continued up to 1966-67. After 1966-67 no special efforts were carried out for such harvest. Hence uncontrolled collection of pearl oyster were started. Secondly affluent from established big factories on the coast line have changed ecosystem and thus population was reduced to 6-7 no/K.M. to 17 no/km Table 15.4. It was considered that pearl oysters are available in 130 KM area of Gulf of Kuth. i.e. total stock can be calculated as 2230 nos in 1992-93. This reduction may be due to pollution by factories and over exploitation of pearl oyster by local fishermen.

Because of the efforts of the scientists of Fisheries Research Station, GAU, Sikka, it is reported that total numbers are increasing to 29 no/km i.e. total stock can be calculated as 3770 no in 1996-97.

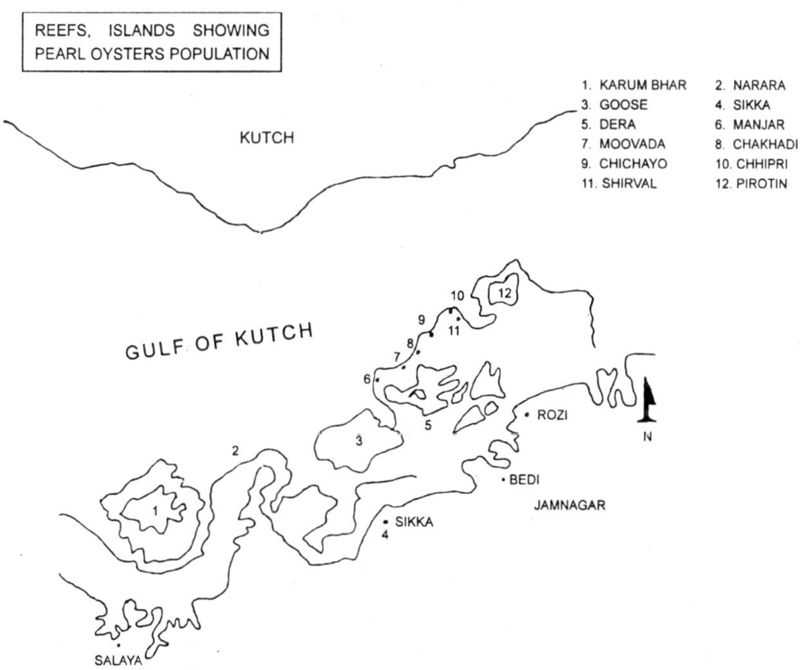


**Fig. 15.7 : Reefs, islands showing Pearl Oyster population in Gulf of Kutch**

Table.15.4 gives the number of pearl oysters found on different island from 1984-85 to 1997-98. From the table it can be seen that Pataria island which is near to Sikka was having popultion 88/km in 1990-91 was reduced to 5 No/km in 1997-98.

**Table 15.4 : Details of survey of pears oyster grounds during 1984-85 to 1997-98 number of oyster/sq. km.**

| Sr. No | Name of ground | 84-85 | 88-89 | 89-90 | 90-91 | 92-93 | 94-95 | 97-98 |
|---|---|---|---|---|---|---|---|---|
| 1. | Narara | 20 | 15.5 | 16 | 12 | 17 | 37 | 28.30 |
| 2. | Sikka | 03 | 12 | – | – | – | 23 | 19.57 |
| 3. | Goose | 10 | 04 | – | – | 48 | 16 | 23.75 |
| 4. | Movada | 14 | 22.4 | 03 | 05 | – | 6 | 21.90 |
| 5. | Chichaya | 12 | 08 | 01 | – | – | – | 10.00 |
| 6. | Pirotin | 01 | 6.6 | 18 | 15 | 15 | 12 | 11.60 |
| 7. | Dantala | 06 | 01 | 01 | – | – | – | 14.20 |
| 8. | Manzer | – | 07 | 04 | 1.5 | 6.3 | 16 | 35.00 |
| 9. | Othad | – | – | 40 | 12 | 32 | 10 | 16.26 |
| 10. | Bhansli | – | – | 56 | 60 | – | 20 | 18.48 |
| 11. | Pataria | – | – | 56 | 84 | 40 | 4 | 5 |
| 12. | Aghor | – | – | 14.4 | 12.8 | – | 12 | – |
| | Average of 12 island | 5.5 | 6.37 | 17.45 | 16.86 | 16.52 | 13.00 | 16.98 |

Source : Annual Report and GAU Sikka.

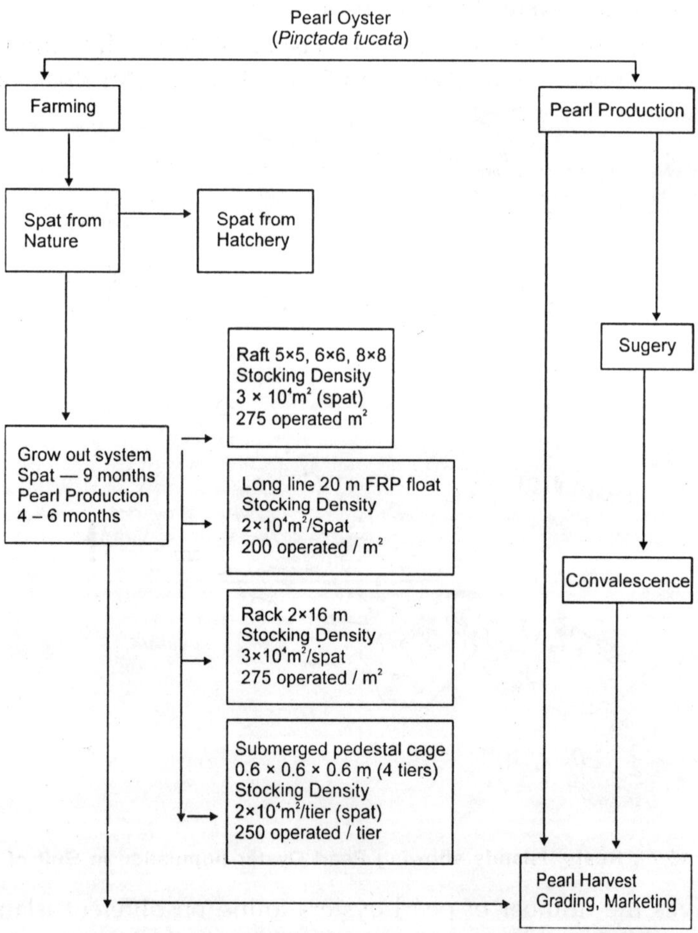


**Fig. 15.8 : Protocol for Pearl Oyster farming and Pearl Production in India.**

For increasing population of pearl oyster, Govt. of India, I.C.A.R., New Delhi have approved a project on "Establishment of pearl oyster hatchery" at Sikka worth Rs. 9.14 lakh. The project was submitted by Dr. A.D. Dholakia, Research Officer, GAU, Sikka.

## *References*


Algarswami, K. and S.Z. Quasim 1973. Pearl culture-its potential and implications in India. *Indian J. Fish.* 20 (2): 533-550.

Anon. 1978. Mariculture Research and developmental activities, CMFRI Sp pub. No 2.

Anon. 1992. The Indian Edible Oyster Technology of seed production and Farming Booklet published by CMFRI Cochin.

Dholakia, A.D. 1996. Pearl Oyster culture in Gujarat. Guj. Agri. Uni. S.K. Nagar.

Dholakia, A.D. and Y.B. Raval 1996. Edible Oysters in Gujarat Guj. Agri. Uni., Fisheries Research Station, Sikka.

Dholakia, A.D., D.V. Bhola, P.H. Patel, D.M. Makwana. 1997. Salinity tolerance limit for survival of pearl oyster, *Pinctada fucata* (Gould) and Edible Oyster *Crassostrea gryphoides* (Schlotheim) Fishing chimes, 17 (5) : 17-18.

Durve, V.S. 1973. On breeding of mussels.

Eswarn, C.R., K.R. Narayan and M.S. Michael. 1969. Pearl fisheries of Gulf of Kutch. J. *Bombay. nat. Hist. Soc.,* 66(2) : 338-344.

Hornel, 1909. The Marine Zoology of Okhamandal in Kathiawar-Part I Published by Williams and Norgate London-pp 148.

Jones, S. 1968. The Molluscan Fishery Resources of India. *Proc. Symp. Mollusca Mar. Biol., Asso of India. Pt-III* : 906-918.

Narayan, K. and Michael, M.S. 1968. On relation between age and linear measurements of Pearl Oyster, *Pinctada vulgaris* (Schumacher) of the Gulf of Kutch. J. *Bombay. Nat. Hist. Soc.* 65 (2) : 442-252.

Narsimhan, K.A. 1973. Clam, Cockle and oyster Resources of the Indian coasts. *Symp. Living resources of the seas around India,* CMFRI Cochin ; 648-658.

Pandya, J.A. 1972. Age and growth of the pearl oyster, *Pinctada volgaris* (Schumacher), of the Gulf of Kutch. *J. Indian Fish, Ass.* 3 (1 & 2) : 47-54.

Pandya, J.A., R.T. Sarvaiya, G.L. Gajera and K.R.Chaudhary. 1976. A quick, economic and hygienic method of extracting pearls from the window pane oyster. *C.I.F.E. Souvenir,* (6) : 22-26.

Pandya, J.A. 1996. Influence on temperature on growth ring formulation in the pearl oyster *Pinctada fucata (Gould)* of the Gulf of Kutch. *Indian J. Mar. Sci.,* 5 (2) : 249-251.

Raval, Y.B., A.D.Dholakia, D.V. Nandasana, 1998. Stock assessment of *Pinctada fucata* in the Gulf of Kutch, causes of Depletion and Measures to protest their population. "Current and Emerging Trends in Aquaculture" Published by Daya Publishing House, New Delhi. 234-237.

Sarvaiya, R.T. and N.D. Chhaya, 1993. "Saurashtra Molluscans" Book published by Smt. P.R.Sarvaiya, Ahmedabad.

# 16
# Ribbon Fish Fishery

Ribbon fish contributes about 7 to 8% of total marine fish landing of Gujarat. It can be seen that during 1975-76 Ribbon fish landing was 0.11%, fish landing was increased to 1.8 to 2.4%. After 1986-87 Ribbon fish landing increased from 6.6% to 13.71% in 1996-97.

## Classification

| | |
|---|---|
| Class | Pises |
| Sub Class | Teleostomi |
| Order | Acanthopterinii |
| Family | Trichiuridae |

Following species are available in Gujarat.

## Species available in Gujarat

In Indian waters following species of ribbon-fish are found.

(a) *Trichiurus lepturus* (linn)
(b) *T. Naumela*
(c) *Lepthracanthus savala* (cuv)
(d) *Eupleurogrammus intermedius* (gray)
(e) *E. muticus* (gray)

Out of these (a) (b) and (c) are commonly found in Gujarat.

## Local names in other Language

| | |
|---|---|
| English | Ribbon fish, cutlass, |
| Gujarati | Patty, Lapdi |
| Marathi | Wagti, balu |
| Kannada | Baale, Pambaale |
| Malayalam | Thylayana, Pambabu |
| Tamil | Savalani, sona-ka-whala |
| Telugu | Savalla, Sawala |
| Oriya | Puttiah |
| Andaman | Pape. |

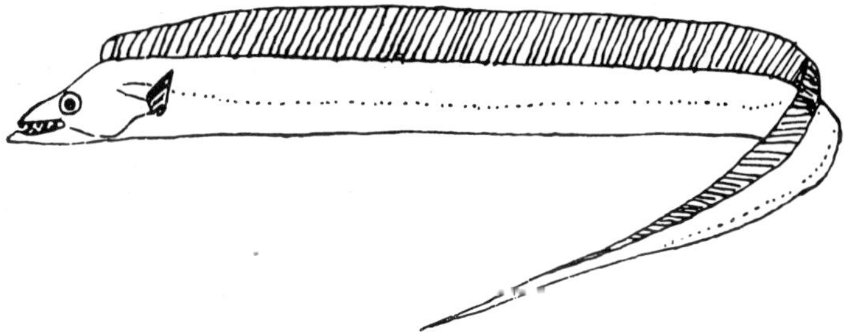

Fig. 16.1 : *Eupleurogrammus intermedius*

## Taxonomy (Identification) Day 1958:201

### *Trichiurus haumela*

### *Synonyms*

*Trichiurus Malabaricus*, Day, Fish, *Malbar* P. 65 P1 V

*Enchelyopus haumela*, Bleeker; Bintang 1868 P.4.

B. Vii, D. 127-133, P-11.

| | |
|---|---|
| Head | Length of head from 0.13 to 0.12 part of the total length. |
| Height | Height of body from 0.08 to 0.06 part of the total length. |
| Eye | Diameter 4.75 to 6.5 part in length of head, 1.66 to 2.33 diameter from the end of snout, and 0.75 part of a diameter apart: it's height is at least 0.50 part of that of the head where it is situated. |
| Jaw | Lower jaw considerably the longer, the posterior extremity of the maxilla reaches to below the middle of the orbit. Dorsal profile, between the end of the snout and the eye, rather concave. |
| Teeth | Upwards of 10 to 12 sharp compressed ones laterally in either side of upper jaw, usually increasing in size posteriorly whilst anteriorly in the premaxillaries are two pairs of large curved and barbed canines, there are two similar but smaller teeth above the symphysis of the lower jaw (and in advance of the end of the snout when the mouth is closed), most distinct in the adult, and 8 lateral ones also of less size than those in the upper jaw, and some of which are occasionally barbed, teeth also in the palatiness. |
| Fins | The first few dorsal rays are of less height than those near the middle of the fin which about equal that of the body. Pectoral nearly or quite as long as the height of the head. Behind the anus the anal fin is seen in the form of short spines, often entirely concealed or else blunted at their external extremities: in one specimen it was 74, while in another 80. |
| Lateral-line | Gradually descends until above the commencement of the anal fin where it is in the lower third of the body. |
| Colour | Greyish along the back, becoming silvery on the side and beneath: a dark mark along the edge of the pre-orbital. Fines of a pale yellow, the upper half of the dorsal dark, due to numerous fine black dots. |

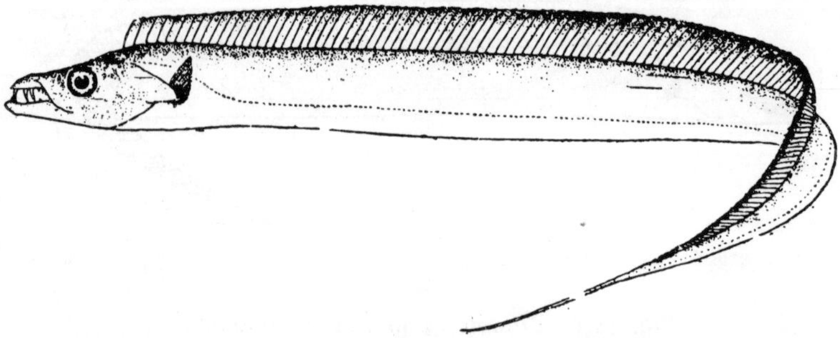

**Fig. 16.2 : *Trichiurus haumela***

## *Trichiurus savala* Day 1958 : 201

### *Synonyms*

*Trichiurus savala,* Cuv. and val. Viii, p. 251. P1 244
*Trichiurus armatus,* gray, zool misc., p. 9; Gunther, catal, ii, p. 347;
*Enhelyopus savala,* bleeker, Bitang, 1868, p. 14
B, vii, D, 122-120, p,11

| | |
|---|---|
| Head | Length of head 7.5 part of the total length, Weight equals 2.5 to 2.66 part in it's length. |
| Height | Height of body 0.06 part of the total length. |
| Eye | Diameter 0.15 to 0.14 part of length of head, 2.75 to 2.5 part in the length of snout, and 1 apart: it's height is above 0.33 of that of the head where it is situated. |
| Jaw | The lower jaw considerably prolonged beyond the upper; the maxilla reaches to below the middle of the eye. Inter-orbital space nearly flat; dorsal profile between uper surface of eye and of snout slightly concave. The distance from the eye to the upper jaw nearly equals to the diameter of the orbit. |
| Teeth | About 8 shapes and compressed ones laterally in the upper jaw,whilst anteriorly are two three pairs of large: there are two similar and rather smaller ones above the symphysis of the lower jaw and which are anterior to the snout when the mouth is closed: laterally there are about 9 similar to those in the upper jaw but smaller. |
| Fins | The dorsal fin commences over the hind edge of teh pre-opercle, it's first rays are short, being about equal to 1 diameter of the orbit, the longest rays are about equal to the height of the body. |
| | Pectoral 0.36 of the length of the head, no rudiment of ventral fins. Anal in the form of about 76 to 82 free spines which may be concealed in the skin but are generally distinct, especially the first, which is twice as long as seen in *T. houmela.* |
| Lateral-line | It passes downwards to the lower third of the side |
| Colour | Silvery, fins yellowish white. |

Fig. 16.3 : *Trichurus savala.*

## Food and Feeding Habit

The food of *E. intermoedius* consists of fish followed by crustaceans. The diet of adult *E. intermedius* consists chiefly of *thrissocles, stolephorus, sphyraena hemirhamphus, sardinella, tetrodon reiognathus, dussummieria, atherina* and in crustaceans. They eat *Acetus*, prawns *Squilla*, Octopus, *Mysids, Isopods Lucifer*, crab larvae, copepods, sepia etc. However it is observed that *lucifer*, is most preferred diet.

Unlike in other fishes, the intensity of feeding does not seem to be related to spawning activity. From the fish items of the diet it can be seen that Ribbon fish is surface feeder. On stomach analysis of mature fish shows that the fish is also cannibalistic. It is observed that all the species are pisivorous through out their life, and their diet is more or less common. There may be seasonal and regional variation in composition of the food (Jams 1967 and Pillai 1974).

## Sex/Fecundity

## Fecundity

Many workers have noted different fecundities. The potential stock of eggs in *E. intermedius* varied from 2249 in fish of 40.9 cm. to 9950 in a fish of 45 cm length. In *E.muticus*, it varied from 1327 (49.5 cm) to 2087 (55.1 cm) in L. *salva* from 9178 (37.0 cm) to 17347 (54.0 cm) and in *T. lepturus* from 4000 (42 cm) to 16,000 (60.0 cm) (Bal and Rao 1984).

## Sex Ratio

In *E. Intermedius* the percentage of male was low throughout the year (28.94) but was considerably lower during April, May, August and November, varying between 21.7 and 23.1 in *L. salva* the male: female ratio is 1:1.4 practically throughout the year. The lower ratio of the male during the peak spawning in April, May and November. Observation shows that Ribbon-fish breed in deeper water beyond the 35 m line (Bal and Rao 1984)

## Length Weight Relationship

The length-weight relationship ribbon fish is as under.

*E.muticus*

Male $\log W = -4.2705 + 3.5560 \log L$

Female $\log W = -4.1491 + 3.4848 \log L$

*Lepturus*

Male $\log W = -3.9468 + 3.4169 \log L$

Female $\log W = -3.9552 + 3.4367 \log L$

(Narsimhan, 1970 and 1974)

Narsimhan (1976) have studied relation between age and growth of ribbon fish *T. lepturus* and found that the fish attains an average length of 41.6, 69.0 and 88.5 cms in $1^{st}$, $2^{nd}$ and $3^{rd}$ year respectively.

**Table 16.1. Distinctive Characteristics of Four Common Species of Ribbon-Fishes**

| | Character | E.*intermedius* | E.*muticus* | T.*lepturus* (*T. haumela*) | L.*savala* |
|---|---|---|---|---|---|
| 1. | Dorsal fin | III, 123-129 | III, 139-147 | III, 131-136 | IV, 108-123 |
| 2. | Anal fin | i + I + 113-121 aggregate 115-124 | i + I + 113-122 | i + I + 103-109,I | i + I-77-93,I |
| 3. | Head | - | - | 6.18-8.04 in length | 7.1-10.0 in length |
| | Depth | - | - | 12.89-18.60 in length | 14.8-22.1 in length |
| | Eye | - | - | 5.0-7.0 in head | 5.6-9.7 in head |
| 4. | Vertebrae | 31.32+126-131=158-162. | 39-42+150-159=189-201. | 38-40+127-137=167-175. | 33-37+131-152=164-189. |
| 5. | Teeth (main series) | 7-12 in both jaws | 12-25 in upper; 16-23 in lower. | 9-14 in both jaws. | 7-15 in both jaws. |
| 6. | Gill rakers (main series) | 10-23 | 13-22 | 10-12 | 3-13 |
| 7. | Post anal portion of body | Ends abruptly | Ends gradually in whip-like Filament. | - | - |
| 8. | Snout | Elongated | Short | - | - |
| 9. | Dorsal profile of head | Slopes gradually | Slopes abruptly | - | - |
| 10. | Parallel line to long axis of body from snout | Passes below margin of eye of eye | Passes through eye | - | - |
| 11. | Ventral fins | Scale-like, relatively more conspicuous | Scale-like, less conspicuous | Absent | Absent |
| 12. | Ventral profile | Behind vent rough indicating vestiges of anal fin | Behind vent smooth indicating suppression of anal fin is extreme | Spines rudimentary, fin suppressed | Spinules behind vent clearly visible |
| 13. | Caudal fin | Absent | Absent | Absent | Absent |
| 14. | Teeth | Canine-like teeth distinctly barbed; other smooth, widely spaced | Canine-like teeth faintly barbed, other smooth close set | Canine-like teeth barbed, 9-14 teeth in both jaws | Canine-like teeth barbed, 7-15 teeth in both jaws |
| 15. | Operculum | Lower hind margin of operculum convex | Lower hind margin of operculum convex | Lower hind margin concave, partly overlaps pectoral base and fin | Lower hind margin concave, partly overlaps pectoral base and fin |

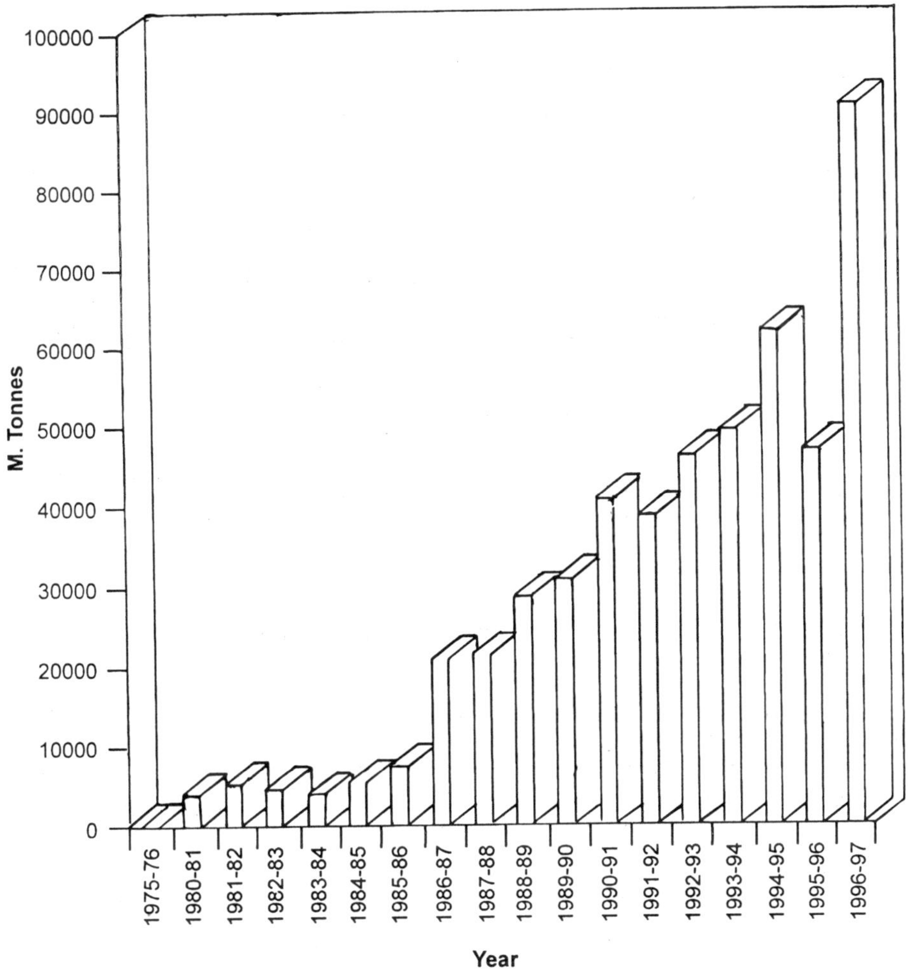


**Fig. 16.4 : Landing pattern of Ribbon fish in Gujarat since 1975-76**

## Reproduction

The gonads lie just above the alimentary canal. The testes are thin and ribbon like. They become thick and milky white when mature. The ovaries are asymmetrical in length and broader. The lobes are fused along the dorsal edge.(James 1967).

## Spawning periodicity

It is reported that *E. intermedius* spawns twice a year however, it is difficult to determine the interval between the two spawning. Observation have shown two distinct modes of maturing and ripe ova clearly separated from each other and immature stock of ova supports two spawning a year. Data of mature fish shows two peaks in March-April and August-November.

The minimum size at maturity is 30 cm in *E. intermedius* 47-48 cm in *T.lepturus* and about 43 cm in *E.maticus* (Bal and Rao 1984).

## Craft and Gear

### *Craft*

Trawlers, mechanised vessels are being used in Gujarat for Ribbon fish fishing.

### *Gear*

Ribbon-fish generally caught in trawl Net and in 'Dol' Net in Gujarat and Maharashtra, shore seines in Karnataka, Andhra Pradesh, Gill Net in Tamil Nadu, beg Nets in Kerala and in West Bengal.

**Table 16.2 : Key to the Stage of Sexual Maturity of Ribbon fish E.*intermedius***

| Stage of maturity | Extent of ovary in body cavity and general appearance. Nature of ova | Ova Dia. |
|---|---|---|
| Immature | The ovaries are small, transparent, occupying about one third of the body cavity ova are not visible to the naked eye. Immature ova are transparent, irregular in shape. | Max. 0.19 mm. Dominant range 0.01 to 0.09 mm. |
| Maturing | Ovaries occupy slightly more than one third of body cavity, with one batch of eggs, separated from the immature stock. Maturing ova is opaque rounded. | 0.34 to 0.47 mm. |
| Mature | Ovaries occupy half to three forth, of body cavity yellow colour, ovarian wall thick, ova spherical opaque with yolk. | Max. 1.14 mm Range 0.6 to 0.81 |
| Late ripe | Ovaries turgid, transparent, white in colour, occupying entire body cavity by displacing viscera to one side; ovarian wall thin easily ruptured; eggs semi-transparent; fully formed oil globule. | Max. 1.71 mm. average 1.57 mm. |
| Spent | Ovary shrunken, flaccid, not fully packed with ova most of which are small few whitish granular scattered about. | |

**Table. 16.3 : Ribbon fish landing in Gujarat and its share in Gujarat and India**

| Year | Total marine Fish landing | Landing of Ribbon fish | Percent of Total Guj. | Marine fish Landing of India | Percent share of Guj. in India |
|---|---|---|---|---|---|
| 1975-76 | 208300 | 225 | 0.11 | 75330 | 0.39 |
| 1980-81 | 206749 | 4101 | 1.87 | 62690 | 6.55 |
| 1981-82 | 218872 | 5367 | 2.43 | 41566 | 12.91 |
| 1982-83 | 192669 | 4771 | 2.47 | 48875 | 9.76 |
| 1983-84 | 223291 | 4152 | 1.86 | 39071 | 10.63 |
| 1984-85 | 290708 | 5576 | 1.92 | 5308 | 10.5 |
| 1985-86 | 306577 | 7483 | 2.44 | 84403 | 8.86 |
| 1986-87 | 315942 | 21030 | 6.66 | 95872 | 21.93 |
| 1987-88 | 327560 | 21564 | 6.58 | 81729 | 26.0 |
| 1988-89 | 414075 | 28941 | 6.99 | 68496 | 38.0 |
| 1989-90 | 432384 | 30947 | 7.16 | 65150 | 42.25 |
| 1990-91 | 500462 | 40907 | 8.17 | 74362 | 47.50 |
| 1991-92 | 530017 | 38684 | 7.30 | 95321 | 55.01 |
| 1992-93 | 609103 | 46167 | 7.60 | 114729 | 40.58 |
| 1993-94 | 619836 | 49362 | 7.96 | 91314 | |
| 1994-95 | 645261 | 61818 | 9.58 | — | — |
| 1995-96 | 598351 | 46925 | 7.84 | — | — |
| 1996-97 | 660058 | 90518 | 13.71 | — | — |

It can be observed from the above data that up to 1985-86 share of Ribbon fish in total fish catch of Gujarat was less than 2.5% while after increasing trawling activity in the State, Ribbon fish landing is increased and in 1996-97 it was 13.71%. Secondly after 89-90 share of ribbon fish in landing of Gujarat fish production was increased and reached to about 40 to 50% in total ribbon fish landing of India.

## *References*

Bal, F.V. and K.V. Rao. 1984. Marine Fisheries. McGrow Hill Publishing Co. New Delhi, pp. 251-52.

Chhaya, Niranjan D., 1990. Fisheries Management and Science. Gujarat Granth Nirman Board Ahmedabad, pp. 100-101.

Dholakia, A.D. 1986. List of fishes available in Gujarat. In Fish processing, reservation and Quality control. pp. 139.

Francis Day, 1958. The Fishes of India. By William Dawson & Sons Ltd. pp. 200-201.

James, P.S.B.R. 1967. The ribbon fishes of the family *Trichiuridae* of India. *Mem, I. Mar. Biol. Asso. of India.*

Narsimhan, K.A. 1970. Ribbon Fishes of India, *Indian, J, Fish.,* 90-96.

Narsimhan, K.A.1974. *Indian, J, Fish.,* 21(2): 479-94.

Narsimhan, K.A., 1976. Age and growth of Ribbon Fish *T. kepturus. Indian J, Fish,* 23 (1&2) : 174-182.

Pillar Mahadevan, P.K., 1974. *Indian J, Fish.,* 21(2): 597-600.

Rao, K.V. 1973. *Symp. Living Resources of Sea around India, CMFRI.* pp. 18-101.

17

# Scianeid Fishery

## Introduction

All scianeides are known as "Jew fish" and crockers. Total 30 species are recorded in Indian waters. Out of which 12-14 species are found in bulk landing.

Following main species are found from Indian Seas.

(1) *Otolithus rubber*

(2) *O. argenteus*

(3) *Johnius dussumieri*

(4) *Pseudosciaena sina*

(5) *P. axillaries.*

*Otolithus rubber* (Schneider) contribute to an important demersal fishery along the Indian coast. Though caught by most of the indigenous units all along coast, their landings are mainly by the trawlers. Prior to 1980 *Otolithus rubber,* in Gujarat, were considered as trash fish. However, thereafter export to Gulf countries were started. (Dholakia)

## *Classification (As per Day)*

| | |
|---|---|
| Class | Pisces |
| Sub Class | Teleostei |
| Order | Acanthoterygill |
| Family | Sciaenidae |
| Sub Family | Otolithinae |
| Genus | *Otolithus* |
| *Species* | *Rubber.* |

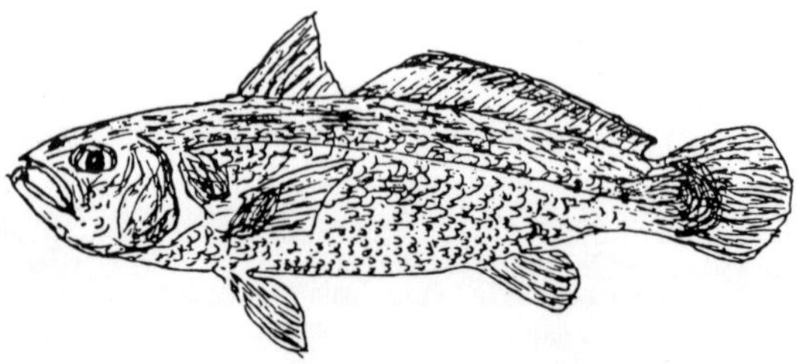

**Fig. 17.1 : *Otolithus rubber***

## *Synonyms*

*Johnius rubber* Bl. Schn. P. 75 t Xvii.

*Otolithus rubber,* Cuv and Val V, p.60, PL 102;

*Otolithus submaculatus,* Blyth., J.A.S. of Beng. 1960, p. 141.

## Distinctive characters

B.Vii, D.10/$\frac{1}{29\text{-}31}$, P. 17, V 1/5, A.$\frac{1\text{-}2}{7}$, C. 17, L.1,$\frac{60\text{-}65}{52\text{-}54}$, Ltr 7/17

Caec. Phl. 4-5

Head : Length of head 0.25 to 0.28 of caudal 0.17 to 0.66, height of body 0.25 to 5.50 in the total length.

Eyes : Diameter 0.25 to 0.20 or even 0.17 of length of head, nearly 1 diameter from end of snout, and 1.25 apart. Greatest width of head equals 2.25 in its length of head, and its length excluding the snout.

Mouth : Cleft of mouth rather oblique, the maxilla reaches to below the last third of the eye : nostrils opposite the upper third of the orbit. Preopercle scarcely denticulate : two opercular spine. Shoulder-flap entire. The distance between the eye and the upper jaw equals two-third of a diameter of the eye in extent.

Teeth : On either side of the symphysis of the upper jaw a pair of large canines, and inner villiform band, also an outer row of conically curved ones, occasionally even between the canines : in the lower jaw a central (some times a second) curved canine having a few villiform teeth behind it, and a lateral row of distantly placed conical teeth.

Fins : Dorsal spines slender the second and third equaling from 2 to 2.25 times in the height of the body and from 0.50 to 0.66 higher than the rays. Pectoral as long as the head excluding the snout. The ventral does not reach quite half way to the vent. First anal spine minute or wanting, the length of the second equals about 0.4 of that of the first ray, the length of the base of the anal fin about 0.17 in that of the soft dorsal. Caudal wedge shaped.

Scales : Cycloid

Lateral-line : Gradually curves to above the middle or end of the anal : the tubes arborescent posteriorly.

Air Vessel : Somewhat contracted at its first fourth and having about 34 branching processes on either side.

Colour : Brownish - red, shot with silvery and white, sometimes glossed with gold in the lower third of the body. First dorsal stained black at its edge, soft dorsal and anal with grayish outer margins : pectoral ventral and anal yellow.

## *Otolithus argenteus* (Cuvier)

D.x 27-31 : A 11/7; L about 50; L tr. $\frac{7\text{-}8}{10\text{-}12}$

Fig. 17.2 : *Otolithus argenteus*

Head about 3.5, height 4-4.5 in length, eye 4-5 in head, 1.4 in snout, without pores, about 3.5 in head. Anterior nostril rounded, posterior oval, mouth oblique, lower jaw operculum with two soft spines, one or two strong, curved long canines on each side of mandibular symphysis, scale cycloid in adult, ciliate in young. Even in adult scales on head before ventral and on lower half of tail more or less ciliate. Origin of anal opposite the 12th dorsal ray, its base nearly one fifth of that of soft dorsal, its spines slender and weak. Caudal wedge - shaped air bladder with 25 lateral processes on either side.

Colour : Colour silvery, dark on back, dark spot on opercle, fins yellowish, border of dorsal grayish. (Bal & Rao, 1990)

### *Johnius dussumieri (Cuv & Val)*

It is classified in sub family Scianinae. The geneus *Sciaena* is having barbal at mandibular symphysis and a wide pore as its base.

### *Synonyms*

| | |
|---|---|
| *Corvina dussumier* | Bleeker, 1874 |
| *Sciaena dussumieri* | Cuv & Val |
| *Johnius dussumier* | Flower, 1933 |
| *Sciaena glaucus* | Day, 1889 |

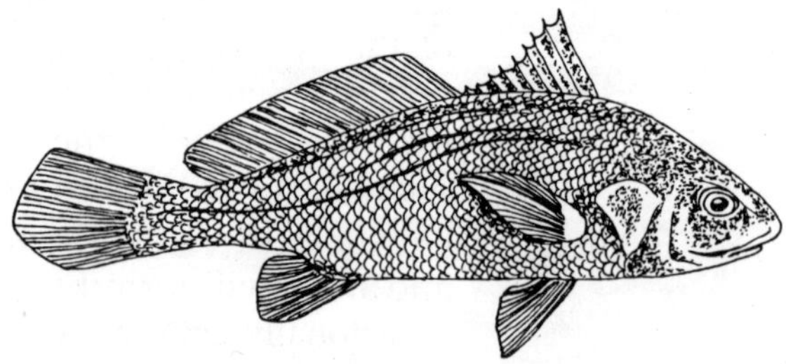

Fig. 17.3 : *Johnius dussumieri*

### Distinctive Characters

D.x1/28-30; AII/8; L.1.50; L. tr. $\frac{5}{10\text{-}12}$

Dorsal profile slightly convex from snout peduncle, ventral profile more conves, height 3.25, head 3.3, Eye 4.2 in head three fourth of diameter from upper jaw, snout bluntly rounded and swollen, three pores across its base and five along free border, a median pore below mandibular symphysis and two others behind them, usually all fissure-like. Anterior nostril rounded, valvate, the posterior nostril much larger, close to middle border of eye, interiorly with a flap.

Opercle with two flat weak spines, scales cycloid on cheek, preparcle and opercle, otherwise ciliate. Lateral line slightly arched to middle of soft dorsal tubes distally bifid. Dorsal fin deeply incised, anal below middle of soft dorsal, dorsal and anal rounded, pectorals, pointed, caudal rounded or wedge-shaped.

Colour : Gray along back, silvery below, shot with golden-orange on belly, diffused blotch on opercles and a dark spot at base of pectorals; first dorsal nearly black, fine dark spots on upper half of second dorsal, caudal and half of pectoral grayish. (Bal and Rao, 1990).

Maturity : The bigger sciaenids attain maurity late in their life. There are no external secondary sexual characters which differentiate the sexes. In *P. sina* and *P. axillaris* females attain maturity when they reach 11.5 - 12.5 cm size. In *Johneous dussumeireis* maturity reaches at 14 cm length and breed more than once. *O. argentius* attains maturity at 15 cms in length, and breeds only once in a year. Details of maturity, fecundity and spawning season of some of sciaenids are given in Table 17.1

### *Pseudosciaenasina (Cuv. & Val.)*

### Distinctive characters

D.X 1/27-29; A II/7; L.I. 48-50; L. tr. $\frac{6\text{-}7}{12\text{-}13}$

Dorsal profiles from snout to dorsal strong convex, ventral profile straight, height 3.3-3.41, head 3.15-3.3 in length, eye 4-4.6, equal to or somewhat shorter than convex inter-orbital space. Five pores below mandibular symphysis. Posterior nostril ova, anteiror smaller and rounded. Opercles with flat spines. Outer series of teeth in upper jaw enlarged and somewhat caninoid. Scales on opercles, cheeks, such orbitals and snout cycloid, rest ciliate, lateral line arched up to middle of soft dorsal, sensory scales arborescent; spinous dorsal deeply incised, its origin opposite that pectoral, origin of anal before middle of soft dorsal; soft dorsal and anal rounded, pectoral falciform; caudal rhomboidal.

**Table 17.1 : Table showing Maturity, Fecundity and Spawning of some Sciaenides**

| Name of Sciaenid | Size of 1st maturity (mm) | Fecundity | Spawning Season | Spawning periodicity | Reference |
|---|---|---|---|---|---|
| *Otolithus ruber* | 220 | 49,216 to 1,19,932 | July to Sept. | Once | Vaidya (1960) |
| *O. argeneus* | 151 | 28,159 to 1,19,625 | Peak May-July | Only once | Basu (1975) |
| *Johnius dussumeri* | 130-140 | 39,630 to 83,100 | Oct-July Peak June-July | More than once | Bhusari (1971) |
| *Pseudosciaena sina* | 115-125 | 32,174 to 60,840 | Oct-July Peak June-July | Intermittent | Bhusari (1971) |
| *P. axillaris* | 115-125 | 30,885 to 62,414 | Peak June-July | More than once | Bhusari (1971) |

## *Colour*

Silvery, grayish or brownish above, opercle with dark bluish patches, dorsal dark, other fins hyaline to yellowish.

**Fig. 17.4 : *Psedosicaenasina***

## *Food and Feeding*

They are carnivorous and active predator feeding on crustaceans, fish, molluscan, achnoderm and annelides. *Otolithus ruber* prefers crustaceans. It is an active surface and mid water feeder. *O. argentius* prefers acetes, squilla along with *leander spp. Hippelysmata penaeus, meta penaeus, paarapeopsis,* etc. varies from 32 to 83% in stomach analysis. The crustacean diet varies from 3.7 to 90.5% in *O. ruber.* The presence of polychetes, crabs, bivalves and gastropods in food, through in small quantities, indicates that they are bottom feeder also. *O. ruber* seems to restrict itself to surface and mid water feeding. *O. argenteus* is a surface feeder, occasionally exhibiting cannibalism. Juveniles of *Pseudo sciaena sina, J. dussumieri* are mainly plankton feeder. Their foods are *Acetes,* copepods, Lucifer, small prawns etc (Bhusari, 1971).

The juveniles as they grow in size gradually change their diet from crustaceans to teleosts. Favourite food of *J. dussumieri* are prawns being consumed maximum of 78 per cent (January). Fish form its next preferential item food during most of the months. *O. argenteus* feeds occasionally on echuroids, polychaetes and mysids in addition to its main items of food viz. crustaceans and fish. Among crustaceans *Acetes* and *squolla* are the main species favoured by the fish. (Basu, 1975)

## *Age and Growth*

Age and growth has been studied in very few sciaenids. *J. dussumeri* grows at the rate of 9.6 mm per month in the first year and 6.7 mm per month in the second year reaching a length of 115 mm by the end of the second year. The maximum size recorded has been 250 mm. *O. argenteus* grows 9 mm during the first two years of this life. (Bal and Rao, 1990).

**Talle 17.2 : Craft and Gear used in different parts of India**

| Sr. No. | State/area | Craft | Gear |
|---|---|---|---|
| 1. | North Gujarat | Vahan | Bottom set Gill net. |
| 2. | Saurashtra | Vahan and Trawler | Bottom set Gill net - Trawl net |
| 3. | Kutch | – | Gill net "Malar Rachhi" |
| 4. | Jamnagar Dist. of Gujarat | – | Trape type net called "Wadas" |
| 5. | Tamil Nadu | Catamaran | Gill net "kalluvela" |
| 6. | Andhra Pradesh | Masula Catamaran | Gill net "Koduvavala" and "Valavani" |
| 7. | Orissa | Masula Catamaran | Gill net "Koduvavala" and "Valavani" |
| 8. | Karnataka | – | Drift net, Cast net. |
| 9. | Kerala | Odam or dugout canoe | Gill net |
| 10. | Maharashtra | Ratnagiri boat | "Dol" net and "Jot" net |

Source : Bal and Rao, 1990 : 141-145

## Fishery

The sciaenids constitute about 5% of the total all India marine fish landings for the 22 years. The smaller sized species like *O. ruber; O. argenteus* etc which are collectively known as "Dhoma" are caught all along both coasts throughout the year. It has been observed that, mostly juveniles of this fish are landed on the western and southern coasts of Saurashtra while the larger size adults are landed along the northern coast of Saurashtra along the Gulf of Kutch. (Kutty, 1967) In the Gulf of Kutch around Bedi big meshed drift nets are used while small meshed bag net (dol net) is used in the Bombay region. *O. argenteus* is most widely distributed in the estuaries of Bengal and Orissa also. The sciaenid catch along the Gujarat and Maharashtra coast is pretty high, each contributing about 25% of the total sciaenid catch of India. Landing pattern of sciaenid in Gujarat from 1974-75 to 1998-99 (25 years), and its share in total marine fish landing is given in Table 17.2.

**Table 17.3 : Landing Pattern of Sciaenid Fish in Gujarat**

| Year | Total Marine Fish landing in MT | Landing in Gujarat in M.T. | % share to Total sciaenid landing |
|---|---|---|---|
| 74-75 | 157400 | 20793 | 13.21 |
| 75-76 | 208300 | 38665 | 18.65 |
| 76-77 | 126879 | 59130 | 46.60 |
| 77-78 | 176895 | 44446 | 25.13 |
| 78-79 | 229901 | 61959 | 26.95 |
| 79-80 | 296749 | 64974 | 31.43 |
| 80-81 | 218872 | 59999 | 27.41 |
| 81-82 | 220607 | 67431 | 30.57 |
| 82-83 | 192669 | 51807 | 26.89 |
| 83-84 | 223291 | 60101 | 26.92 |
| 84-85 | 290708 | 81396 | 28.00 |
| 85-86 | 306577 | 107818 | 35.17 |
| 86-87 | 315942 | 108248 | 34.26 |
| 87-88 | 327560 | 105832 | 32.32 |
| 88-89 | 414075 | 134513 | 32.48 |
| 89-90 | 432364 | 155112 | 35.88 |
| 90-91 | 500462 | 184851 | 36.94 |
| 91-92 | 530017 | 196500 | 37.07 |
| 92-93 | 609103 | 210176 | 34.51 |
| 93-94 | 619836 | 221728 | 35.77 |
| 94-95 | 645261 | 235934 | 36.56 |
| 95-96 | 598551 | 203807 | 34.06 |
| 96-97 | 660088 | 192775 | 29.21 |
| 97-98 | 702355 | 216890 | 30.88 |
| 98-99 | 551660 | 178442 | 32.35 |

Source : Commissioner of Fisheries, Govt. of Gujarat, Gandhi Nagar with permission.

## Fishing Season

These species are caught through out the year on the both the coasts of India. However, the landing is less in monsoon season, which is closing seasondue to rough sea.

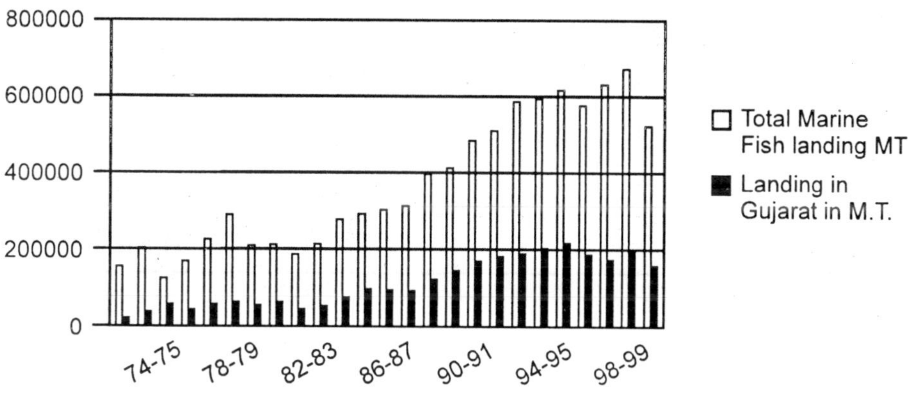


**Fig. 17.5 : Landing pattern of Sciaenid in Gujarat**

## *Craft and Gear*

Traditional Craft and gear used for Sciaenids in different parts of India is summarized in Table 17.2. However, in recent age of motorization more fishing is done by Trawl net, Pair Trawling, Boat Seine, or by using echosounder etc.

## *References*

Anon, 1998. Statistical branch of Commissioner of Fisheries, Government of Gujarat, (With permission)

Bal, D.V. and K.V. Rao. 1990. Marine Fisheries of India. Tata McGrow Hill Publishing Company Ltd, New Delhi : 126-145.

Butt, Y.M. Narayanan Kutty, K.V.S. Rao, and D.M. Punwani. 1964. *Indian J. Fish.* 11(1) : 135-56.

Dholakia A.D. 1986. List of Fishes available in Gujarat with English and Gujarati name. "Fish Processing Preservation and quality control", D.P. Dholakia, Ahmedabad. 134-144.

M.P.E.D.A. 1986. Statistics of Marine Products Exports 1986. Marine Products Exports Development Authority, Cochin.

# 18
# SEER FISH FISHERY

## Introduction

Seer fish contribute 1.47% of total marine landing. This is prime table fish, very much in demand in fresh and cured forms. White flash has high fat content. There is good demand in U.K. and other foreign market. They are available on both the coasts of India. They are ideal game fish.

### *Classification*

Family : Scombridae
Sub-family : Scomberomorinae

### *Specie available*

(1) *Scomberomorus coinnerson* (Lacepede)
(2) *S. guttatus* Bloch & Schenider
(3) *S. lineolatus* (Cuvier)

## Identification

(1) *Scomberomorus commerson* (Lecepede)

First dorsal, with XII-XXII spines. Second dorsal and anal with 15-21 rays. Gill rackers 3-4, Lateral line with deep inflection below posterior end of D2 to second of third dorsal fin let behind it, distinct dark vertical irregular stripe in adults. Caudal deeply forked. Caudal peduncle with three lateral keels, five to ten dorsal and four to ten anal fin lets.

**Fig. 18.1 : *Scombermorus commerson***

### *Synonyms*

*Scomberomorous konun* Bleekar 1851

### *Local names*

English Seer fish, Barred Spanish Makeral.
Gujarati Surmai, Chapri
Marathi Sormai
Tamil Konam

Malayalam Chum-bum
Bengali Champa.

(2) *Scomberomorus konun guttatus* Bloch & Schneider

Five to ten dorsal and four to ten anal finlets are observed. Distance from anterior insertion of second dorsal to caudal fork distinctly longer than second pre-dorsal length. Gill rackers 7-10 on lower limb of outer arch. More than three rows of black spots (round or oval) on side of the body. Caudal-keel not conspicuously broad.

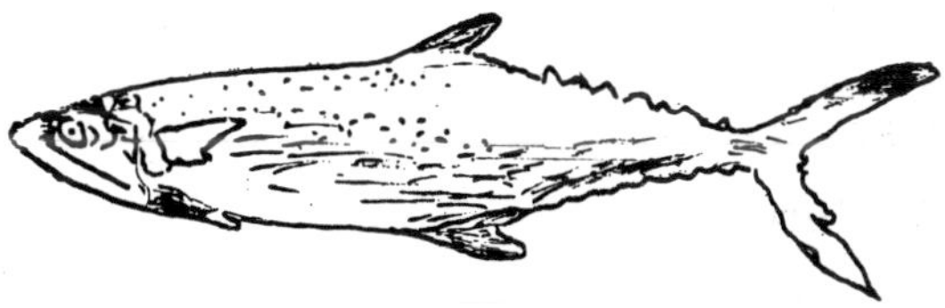

**Fig. 18.2 : *Scombermorus guttatus* (Bloch & Scheider)**

### *Synonyms*

*Scomber guttatus* Bloch and Schneider 1801
*Cybium guttatus* Bleeker 1845.

### *Local names*

English Seer Fish, spotted Spanish Mackeral
Gujarati Surmai, Chapri
Marathi Toower
Kanada Anajal
Telugu Vanjaramu
Bengali Birjram.

(3) *Scomberomorus konum linrolstus* (Cuvier)

Distance from anterior insertion of second dorsal to caudal fork shorter than or rarely equaling second pre-dorsal length; 9 gill rackers on lower limb of outer arch; side of body with irregular rows of horizontal black streaks or broken bars; Caudal-keel conspicuously broad.

**Fig. 18.3 : *S. linolatus***

### *Synonyms*

*Cybrium lineolatus* Cuvier 1831
Local names
English : Streaked spanish mackeral
Gujarati : Surmai
Tamil : Pulli arakula

## Food and Feeding

All the three species are carnivorous. They can chase and capture the moving prey by sight. They are predominantly piscivorous, feeding occasionally, on prawns and squids. *Saurida* are eaten *S. commerson* and *S. lineolatus*. They also consume *percoids, decapods* floating feeds near the surface.

During gut inspection *Anchoviella* sp is recorded in almost all the months; where as *Trichiurus sp.* is present only during June to October. *Alma* larvae are reported to be significantly high from August to February and are found in 41 per cent of the total food content. (Bal and Rao, 1990) Seer fish below 50 mm length are observed feeding more actively than the bigger individuals. The perior of intensive feeding in *S. guttatus* is from June to December. *S. Commerson* takes in a large number of fish where as *S. guttatus and S. lienolatus* takes only a few of them. *S. guttatus* eat teleosts (71%) and crustaceans (24%).

### *Maturity*

Minimum size at which first maturity obtained is as under :

(1) *S. guttatus* 480-520 mm
(2) *S. commerson* 750 mm
(3) *S. lineolatus* 700 mm

(Devraj, 1977)

## Spawning

Females are larger than male in all species. *S. commerson* span in coastal waters. There are two major spawning acts in all the species. Duration from the onset of maturity to the first spawning act is 112 days in *S. guttatus,* 73 days in *S. lineolatus.* The period between first and second spawning act is about 30 days in *S. Commerson,* 92 days in *S. guttatus* and 21 days in *S. lineolatus* (Devraj, 1977)

### *Spawning Season*

Spawning season for different species is different in seer fish. It is as under:

(1) *S. guttatus* May to August with a peak in July
(2) *S. commerson* May to July
(3) *S. lineolatus* January to August with a peak in April-May in Mandapam area.

## Life History

The eggs and larvae of *S. guttatus* have been recorded in the inshore water off Madras. The Coromandal cost and Gulf of Mannar. Eggs of *S. commerson,* measure 1.05-1.12 mm in diameter with an oil globule of 0.31-0.39 mm (Vijayaraghavan, 1955). The four-celled stage is reached nearly an hour after fertilization and subsequent cell divisions take place at intervals of about 10 minutes each. After 13.5 hours of fertilization, the embryo has 8-9 somites and pigments begin to appear about 1.5 hour later. Complete investiture of the yolk by the blastoderm takes place after about 16.5 hours, when 17 somites are discernible. Hatching takes place 3-4 hours later. The newly hatched larvae of *S. guttatus* is 2.5 mm in total length with an oil globule of 0.26 mm in diameter. (Bal & Rao 1990).

*Distinguishing characters of early stages of seer fish are as under.*

1 (a) Gill rackers 7-11 on lower limb of first arch ............................................................ 2

(b) Gill rackers 3-6 on lower limb of the first arch longest opercular spine projects beyond the operculam in the individuals up to 25 mm length. Pigmentation scarce up to 40 mm after which vertical bands appear. Origin of anal fin recedes behind that of second dorsal fin from 25 mm onwards .................................... *S. commerson*

2 (a) Preopercular spines decrease in size from above; upper two sub-equal; non-projects beyond operculum. Pigmentation dense on snout, over hind brain and above meddiaam longitudinal axis of the body. At 40 mm size, a dark wide lateral bend near the dorsal edge and fainter one along the mid-lateral line have already formed. Origin of anal fin recedess behind the origin of second dorsal fin from 25 mm onwards ........................................................................................................ *S. guttatus*

(b) The first and third preopercular spines from the upper end subequal; second comparatively longer never project beyond the operculum. Pigmentation on the body very scanty, without any longitudinal streaks on bands even at 50 mm. A faint dark band appears along the dorsal edge at about 100 mm. Origin off the anal fin below the origin of second dorsal fin throughout the development stages ..... *S. lineolatus*

Source : Bal & Rao, 1990

**Table 18.1 Landing pattern of seer fish in Gujarat and India and it share**

| Year | Total Marine Fish landing in MT | Landing in Gujarat in MT | Its share in Gujarat | Landing in India in MT | Its share in India base |
|---|---|---|---|---|---|
| 75-76 | 208300 | 7101 | 3.41 | 20159 | 35.22 |
| 76-77 | 126879 | 2127 | 0.94 | 21119 | 10.07 |
| 77-78 | 176895 | 2073 | 1.17 | 20779 | 9.98 |
| 78-79 | 229901 | 2698 | 1.17 | 29547 | 9.13 |
| 79-80 | 296749 | 1991 | 0.96 | 25986 | 7.66 |
| 80-81 | 218872 | 4169 | 1.9 | 27553 | 15.13 |
| 81-82 | 220607 | 3108 | 1.4 | 33197 | 9.36 |
| 82-83 | 192669 | 4305 | 1.76 | 35615 | 9.56 |
| 83-84 | 223291 | 5153 | 2.31 | 36651 | 14.06 |
| 84-85 | 290708 | 5720 | 1.97 | 34078 | 16.78 |
| 85-86 | 306577 | 5089 | 1.66 | 33157 | 15.35 |
| 86-87 | 315942 | 7269 | 2.3 | 30672 | 23.7 |
| 87-88 | 327560 | 5469 | 1.67 | 36380 | 15.03 |
| 88-89 | 414075 | 6888 | 1.66 | 40062 | 17.19 |
| 89-90 | 432364 | 6880 | 1.59 | 29101 | 23.64 |
| 90-91 | 500462 | 6331 | 1.26 | 37037 | 17.09 |
| 91-92 | 530017 | 7225 | 1.36 | 42140 | 17.14 |
| 92-93 | 609103 | 1703 | 0.28 | 41718 | 4.08 |
| 93-94 | 619836 | 1561 | 0.25 | 40794 | 3.83 |
| 94-95 | 645261 | 1093 | 0.17 | 45912 | 2.38 |
| 95-96 | 598551 | 1034 | 0.17 | 37394 | 2.76 |
| 96-97 | 660088 | 8705 | 0.89 | 41889 | 2.37 |
| 97-98 | 702355 | 7269 | 1.04 | 54876 | 1.91 |
| 98-99 | 551660 | 6278 | 1.14 | 45040 | 1.63 |
| 99-00 | 670951 | 7133 | 1.06 | 50375 | 1.42 |

Source : 1. Commissionerate of Fisheries, Govt. of Gujarat (With permission)

2. C.M.F.R.I. Year Book Published data

## *Shoaling Behaviour*

Very large shoal of *S. lineolatus* occurs near esturies, *S. commerson* makes small shoals.

## *Fishery*

Fishing grounds were found in varying degree along the Indian Coasts. Fishing activities are confined to 25 km from the coast and 20 to 50 meters depth.

Percentage-wise landing is diferent in different maritime States. They are as under:

(1) Andhra Pradesh 20.4%

(2) Gujarat 11.4%

(3) Maharashtra 11.0%

(4) Karnataka 08.1%

(5) Kerala 17.4%

## *Landing*

Table 18.1 shows the detail of landing of seer fish in Gujarat from 175-76 to 1990-00 and its share in Gujarat and India. (Dept of Fisheries).

From the seer fish landing of last 26 years, in Gujarat it was observed that it is about 1 to 2% of total marine fish landing upto 1991-92. Since then declining trend has been observed. In 1998-99 it was only 0.13%. Though seer fish landing of Indian is almost steady with little increasing trend. It requires finding out reasons for such fall in Gujarat.

## *References*

Anon, Annual Reports. Department of Fisheries, Govt. of Gujarat, Gandhi Nagar.

Anon, CMFRI Annual Reports.

Bal, D.V. and K.V. Rao, 1990. Marine Fisheries of India, Tata MacGraw Hill Publishing Company Ltd., New Delhi. 185-199

Devraj, M. 1977. The Biology of and Fishery for the Seer fishes in India. Ph. D. Thesis Madurai Kamaraj University, Madras.

Dholakia, A.D. 1986. List of Marine Fishes of Gujarat with local names. D.P. Dholakia, Ahmedabad.

Rao, K.S. 1962. *Proc. Symp on Scombroid Fishes. Mar. biol. Asso. India.* 2 : 591-98.

Statistical Branch Commissionerate of Fisheries, Govt. of Gujarat.

Vijayaraghavan, P. 1955. Sear fish Fisheries of India. *Indian J. Fish* 2 : 360-72.

# 19
# TUNA FISH FISHERY

## Introduction

Tuna is one of the least exploited resources of the Indian seas. It is high sea fish occurring in shoals in tropical and temperate seas. On the other hand countries such as Japan, Korea and Taiwan from the Indain Ocean have exploited tuna resources. It is recommended in utilization and conservation of Tuna resources held at Mandapam (1963) that "Scientist in the countries of the would primarily concerned with the high seas fishery for Tunas and bill fishes devise and publish a suitable system for gathering biological useful statistics of yield and efforts in these fisheries that will be capable of being employed on world wide bases."

The coastal tuna fishery of Kerala constitute 50% of the total tuna landings of the country, Korea and China (Taiwan) have emerged as important producers.

## Classification

| | |
|---|---|
| Order | Perciformes |
| Sub-order | Scombridae |
| Family | Scombroidae |
| Sub-family | Thunninae |

## Species occurring in the Indian Sea

| | Scientific Name | Common Name |
|---|---|---|
| 1. | *Euthynmus affinis* | Little Tunny |
| 2. | *Auxis thazard* | Frigate Tuna |
| 3. | *A. rochei* | Frigate Tuna |
| 4. | *Sarda orientalis* | Oriental bonito |
| 5. | *Thunnus tonggol* | Northern blue fin tuna |
| 6. | *T. albacares* | Yellowfin Tuna |
| 7. | *T. obesus* | Big eye Tuna |
| 8. | *Katsuwonus pelamis* | Skipjack Tuna |

### *Common Names*

| | |
|---|---|
| English | Short corseletted Frigate |
| Gujarat | Mackeral, Bangda. |
| Tamil | Kutteli, Churai |

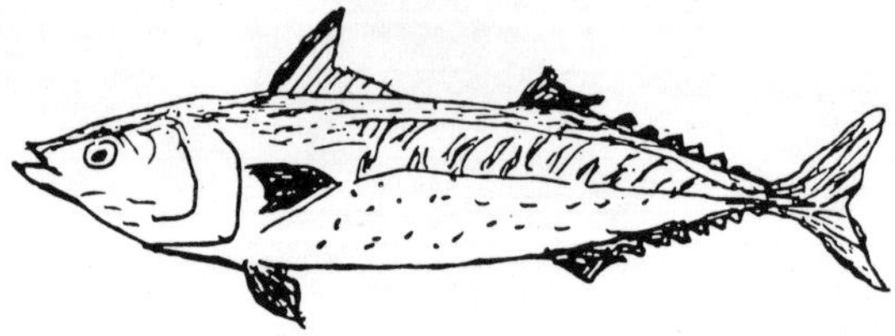

**Fig. 19.1 *Auxius thazard***

### *Synonyms*

*Auxis vulgaris* Cuvier, 1831

*Scomber thazard* Lacepeda 1802

### *Identification*

D1; X – XI D – 2, 11-13 + 8 P1 24, A – 13 + 7, G.R. 9 – 10 – 29 – 32 = 38 – 42.

Head 3.2 to 3.8; depth 3 – 9 to 4.5 is standard length; snout 3,62 to 4.0 and eye 5.0 to 5.8 in head. Body robust and little laterally compressed; mouth oblique; teeth in both jaws minute and in single series; palatines edentulous and vomers sometimes with scattered teeth; scales only on corselet; corselet short, abruptly tapering as narrow band on lateral line; first dorsal larger than the second, the two being widely separated by distance shorter than length of head; caudal widely forked; peduncle with poorly developed lateral keel; air bladder absent.

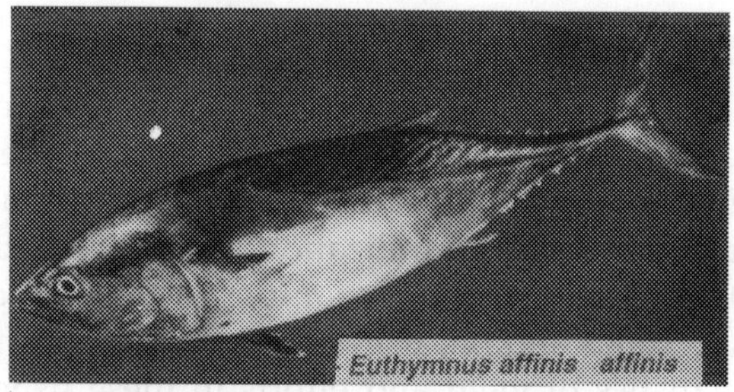


**Fig. 19.2 *Euthymnus affinish* (Cantor 1860)**

### *Colour*

Head dark, black bluish, oblique black bars behind corselet dorsal to lateral line; pale or light yellowish – brown below.

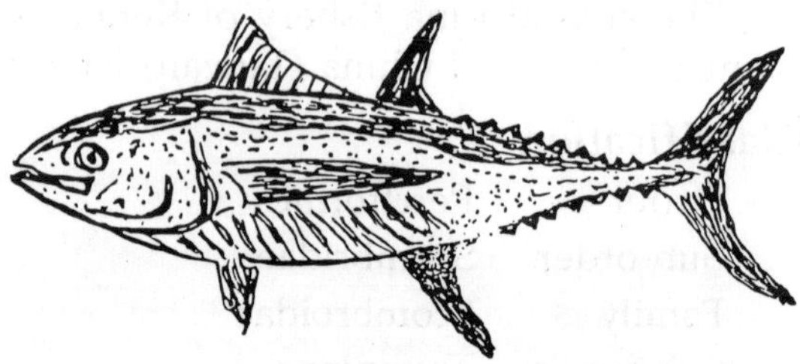

**Fig. 19.3 *Thunnus* (Neothunnus) *albacares* Bonnaterre, 1788.**

### *Common Names*

| | |
|---|---|
| English | Little Tunny or Mackeral Tuna |
| Tamil | Suraly |
| Malayly | Kembel-mas |

### *Synonyms*

*Thynnus affinis 1850,*

*Euthynnus alleteratus affinis* Munro 1956,

*Euthynnus affinis affinis* Fraser – Brunner 1949; Nair 1970.

### *Identification*

D-1'XIV; D-2 – 11-14 + 8 – 9; A – 12 – 15 + 6 – 8; P-1 24 – 28; G,R, 7 – 10 + 22 – 25

Mouth oblique, kaws with minute conical teeth, palatines and vormers also provided with teeth; scales on corselet better developed on dorsal hind position; first dorsal is separated from second dorsal by narrow space; caudal peduncle with median keel and two small keels on either side.

### *Common Names*

| | |
|---|---|
| English | Yellow fin Tuna. |
| Gujarati | Tuna |
| Tamil | Soccer and Kalavai |

### *Synonyms*

*Thynnus albacore* Lowe 1839
*Neothunnus itosibi* Jardan and Evermann 1926
*neothunnus macropterus* Kishinouye 1923.

### *Identification*

D-1 XIII; D-2 – 14 + 9 : A – 14 – 15 + 8-9 : G.R. 9 + 21.

Head length about equal to body depth; body scaled with larger scales on corselet; tail elongated; pectoral reaching origin of second dorsal; in larger specimens second dorsal and anal much elongated.

### *Colour*

Dark bluish above, faint gray oblique lines alternating with rows of silvery spots on abdomen; Pectoral greenish – yellow; other fins and finlets bright yellow with narrow dark border.

### *Common Names*

| | |
|---|---|
| English | Oceanic skipjack stripped tuna |
| Gujarati | Gedra |
| Lakshdweep | Kali Philama |
| Sri Lanka | Alaguduva |

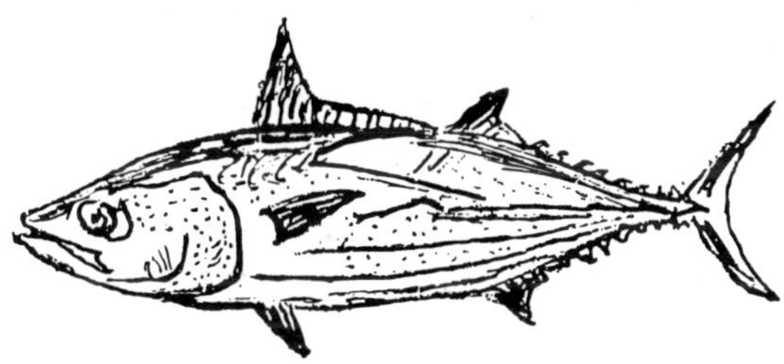

Fig. 19.4 *Katsuwonus pelamis*

### *Synonyms*

*Scomber pleamis* Linnaeus 1758
*Thynnus pelamis* Temminck and Schegal 1844
*Euthynnus (Katsuwonus) pelamis* Brunner 1950

### *Identification*

D-1 XV – XVI; D-2 – 13 – 14 + 7 – 8; P-26 – 28, G.R. 14–18 + 34 – 39 + 48 – 57

Body rubst and round tapering posteriorly well developed corselet, teeth single series only on both jaws, pectoral on a level with eye, gill rackers 50 and over, vertebrae 41 (20+21) and longitudinal bands on lower half of body. Head laterally compressed, snout long and pointed, mouth oblique, anterior spine of first dorsal long and posteriorly concave, lateral line arched below second dorsal and then passes straight to caudal base, caudal with strong lateral keels.

## *Colour*

Steel blue on back, slightly tinghtly violet, abdomen yellowish - white with 4-6 longitudinal blackish strips below lateral line.

## *Common Name*

English Oriental Nonito

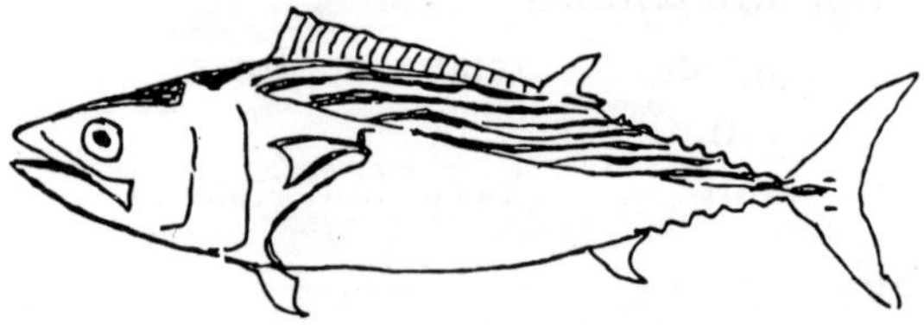

Fig. 19.5 *Sarda orientalis*

## *Synonyms*

*Pelamys orientalis* Tamminck and Schlegel 1842,

*Pelamys chilensis* Day 1889

## *Identification*

D-1; XVII – XIX; D-2. 14 – 16 + 7 – 9; P – 123 – 25; A-13 – 16 + 5 – 7; G.R. 1 – 4 + 8 – 10 = 9 – 14.

Head large, maxilla over half the length of head; corselet not extending beyond tip of pectoral; lateral line arched over pectoral and slightly undulated over body; teeth in upper jaw more than in lower one; finlets 7 – 8 / 5–7; gill rackers small in number usually around 3 + 9 in fish from Indian Coast.

## *Colour*

Adults have eight longitudinal blue lines on the lateral line and silver white below. The juveniles have a varying number of transverse dark bands, which, as growth advances get divided into longitudinal strips which ultimately join together to give the characteristics longitudinal bands of the adults.

## *Size Range*

**Table 19.1 : The size range and weights of different species as per available record is as under**

| Name of species | Fork length (Cm.) | | Weight (Kgs.) | |
|---|---|---|---|---|
| | Min. | Max | Min. | Max |
| *E. affiniis* | 21 | 71 | 0.9 | 5.2 |
| *A. thazard* | 33 | 48 | 0.6 | 2.0 |
| *T. tonggol* | 38 | 64 | 09 | 3.5 |
| *T. albacare* | 63 | 78 | 4.8 | 7.1 |
| *S. orientalis* | 11 | 52 | 0.1 | 1.8 |

Source : Bal and Rao 1990

## *Species wise abundance*

| | |
|---|---|
| *E. affinis* | April-June and September-November with peak during May and October |
| *A. thazard* | February-April with peak in May. |
| *A. rochei* | May and June to December with peak in October. |
| *S. orientalis* | May to October with peak in September. |
| *T. albacares* | January and October. |
| *T. tonggol* | June, October and November |
| *K. oelamis* | January and October. |

## Craft and Gear

The chief gear employed for tuna are drift net and hooks and lines. Which are operated either by catamaran or dugout canoe. Now a days fiberglass boats are being used. Exploratory fishing surveys conducted along Indian coast have drawn attention to the effectiveness of using gears such a Purse seines for Tuna fishing in the potential fishing grounds in the shelf wates. (Silas E.G. 1969) in south India they are caught near the surface and gear used is the pole and line with lead coated iron hooks of simple type without any darb. The pole used is of light but strong bamboo of 3-4 meters in length. The length of line is equal to that of the pole and the hook is kept fixed to the base of the later, when not in use. The firs 2/3 of the line is of cotton and the rest portion closed to the hook is of steel wire formerly, but now of nylon. (Biswas, K.P. 1989). In Gujarat, Maharastra and Karnataka they are caught in drift gillnets operated by the mechanised boats known as the "Machwahs" during September to December with peak in October (Bal and Rao, 1990).

**Table 19.2 : Showing Maturity, Fecundity Spawning Period and Food Feeding of some of Tunas**

| | *A. thazard* | *S. orientalis* | *E. affinis* | *T. albacares* | *K. pelamis* |
|---|---|---|---|---|---|
| Size at first maturity | Between 416-442 mm | 386 mm | 550 mm | 700 mm in some while 1200 mm in all | 400-500 mm |
| Fecundity | 197223 to 1056468 eggs | 24000 to 115000 eggs | 21000 to 68000 in single spawing and 79000 to 250000 in entire season | 23700 to 859000 eggs | 150000 to 197000 eggs |
| Size of eggs | 0/97 mm with an oil globule of 0.22 mm | N.A. | 0.99 mm with oil globule of 0.25 mm | 0.76 to 1.22 mm with oil globule of 0.26 mm | 0.85 to 1.25 mm India. With single oil globule of 0.22 to 0.455 mm |
| Spawning period | N.A. | April to September with peak period in July-August | April & September with peak in May-August | October-April | Probably round the year with a peak in Septermber to April. |
| Food and Feeding | Feed mainly fishes, cephalopods, crustaceans etc. | Fishes, juveniles of *S. orientalis.* | Fish, crustaceans cephalopods, Tunicated etc. diet changing with available pray. | Fish crustaceans Cephalopods, Tunicated etc. diet changing with available pray. | Crustaceans dominated by stomatopods, cephalopods, larval & juveniles of fish like bate fish. |

### *Species wise distribution*

*Ausix Thazord* — The West coast of India, Lakshadweep, Sri Lanka, Indian and Pacific Ocean

*S. orientalis* — West coast of India, in the Indian Ocean, China, Japan, Ghana.

*E. affinis* — West and east coast of India, Lakshadweep and Maldive Islands, Pakistan, East coast of Africa.

| | |
|---|---|
| *T. Tonggol* | West and east coast of India, Lakshadweep and Maldiv Island, Sri Lanka, Andaman and Nicobar Island, throughout Indo-Pacific Ocean. |
| *T. albacare* | Lakshadweep, Andaman Islands, Gulf of Mannar, and Arabian Sea of the Maharashtra and Gujarat coast. |

## Migration

They are denizens of the high seas, but most species frequent the coastal waters. Their migration into the coastal waters are quite regular. There seems to be an optimal temperature range of occurrence for each species as has been observed for yellow fin Tune 21 – 24°C and *T. albacares* is 17 – 19°C. The thermo cline ridges and places the convergence and divergence of water masses are preferred for tuna migration.

### *Seasonal Pattern of Tuna Fishery*

It was observed that peak fishing season were May and September-November period. However, it varies from place to place.

| *Place* | *Peak Period* |
|---|---|
| Cakicut | January |
| Vizhinjam | May |
| Neondalcara | May-June |
| Goa | September - October |
| Maharashtra | October |
| Cochin | June - July. |

### *References*

Bal, D.V. and K.V. Rao, 1990. Marine Fisheries of India Tata MacGraw - Hill Publishing Company Ltd., New Delhi. 158-184.

Biswas, K.P. 1989. A Text book of Fish, Fisheries and Technology.

Dholakia, A.D. 1986. Availability of Marine Fishes in Gujarat with Scientific, English and local names, D.P. Dholakia, Ahmedabad, 133 - 148.

Gopakumar, G. and P.S. Sadasiva Sharma. 1989. The present status of coastal Tuna Fishery at Vizhinkam, Trivendram Coast. *Mar. Fish. Infor. Serv. T & E Ser.* 97 : 1-7.

Rao, K.V. N. 1962. *Proc. Symp. Scimbroid Fishes,* Marine Biological Asso. of India, Pt-II : 733 - 743.

Silas, E.G. 1969. *Bul. Cent. Mar. Fresh. Res. Inst.* 12 : 1 - 86.

Silas, E.G., M.S. Rajagopalan and P. Parmeswaran 1979. Tuna Fisheries in India : Recent trends. *Mar. Fish. Infor. Serv. T & E Ser.* 13 : 1-6.

# 20

# Window Pane Oyster Fishery

## Introduction

Large number of marine and few fresh water molluscs are known to produce pearls. Pear produced from oyster is considered as best quality and size. The window-pane oyster *Placenta placenta* is one of the important molluscs found in Gujarat (Anon. 1962). Window-pane oyster contributes one of the biggest fishery along Gujarat coast, particularly near Poshitra, Pindara and other villages in the Gulf of Kutch. This is perhaps the second biggest window-pane oyster fishery on the Indian coast. In Kakinada Bay, where the window-pane oyster fishery is the biggest recorded one, it is exploited only for purpose of preparing lime. (Narshmhan, 1973). In the Gulf of Kutch it is exploited almost exclusively for mother of pearls. (Jones, 1968). This is traditional fishery. Since this oyster is not edible is comparatively a stranger to layman. This less known oyster is popularly called as window-pane oyster because the shell of the animal were made use of in window panes by people in the erstwhile Portuguese Colonies in India.

### *Classification*

| | |
|---|---|
| Phylum | Mollusca |
| Class | Pelecypoda |
| Order | Eulamellibranchina |
| Sub-order | Ostracea |
| Family | Ostreidae |
| Geneus | *Placenta* |
| Species | *placenta* |

Following two species of window-pane oysters are represented on Indian coasts

(1) *Anomia achaeus* (Gray)

(2) *Placenta placenta* (Linnaeus)

### *Synonyms*

*Anomia placenta* Linnaeus 1758, *Placuna placenta* Hornell 1909, *Placenta placenta* Rao 1969.

## Taxonomy

### *Shell*

The shell of adult oyster is free, very much compressed, slightly unequal, sub-orbicular in shape, the height and length approximately equal. The shell valves are slightly unequal, very flat, rounded and translucent. The inner surface of the valves is pearly. The adductor impression is at about center. The umbo is small. Two thin ridge like teeth diverge from the umbo making a characteristic inverted V shape angle. The shell is fairly large with a diameter of 14 cm or even more. It bears numerous concentric line of growth on the exterior consisting of slightly projecting lamellae the margins of which are minutely uneven with finger-like or spatulate processes. The shell of the adult window-pane oyster is white in colour (Fig 20.1A) Fig. 20.1B shows anatomy with the left mantle and distal half of left gill having been removed.

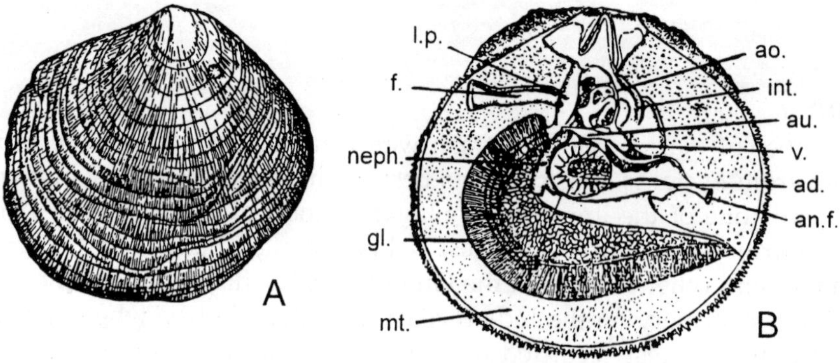


**Fig. 20.1 : Window pane oyster**

(A) *Placenta placenta*

(B) Anatomy with the left mantle and distal half of left gill having been removed.

| | |
|---|---|
| l.p. = Labiai palps | ad. = Adductor muscle |
| An.f. = anal fu | a.o. = aorta |
| f. = foot | gl. = gills |
| mt. = mantle | neph = nephridia |
| int. = intestine | v. = ventricle |

Source : The Commercial Molluses of India CMFRI Bulletin No. 25. pp. 142-143.

## *Body*

The body is very much latterly compressed and is covered by two folds of mantle. Anteriorly on either side of the visceral mass is a pair of elongated, narrow labial pulps between which is the mouth. There is a pair of gills (gl) on either side of the visceral mass. The foot (f) is cylindrical, flattened laterally and attached to the anterior surface of the visceral mass and its tip has a deep cuplike sucker, which facilitates dispersal of foreign matter settling on the body. The alimentary canal consists of a slit-like mouth, esophagus, a large stomach surrounded by the digestive gland, well-developed pyloric caecum with crystalline style, a short intestine a rectum which ends by anus in the center of the broad ventricle (v) membranous collar. The vascular system consists of the heart with a ventricle (v) and two auricle (au), being bigger than the left, a signle aorta, the anterior aorta, arteries and venous sinuses. The nervous system consists of a pair of widely separated and asymmetric cerebral ganglia, a pedal ganglion formed by the fusion of paired, pedal ganglia and single parieto-splanchnic ganglion, from which nerves lead to different parts of the body. The excretory system is composed of a pair of asymmetric nephridia (neph) which are connected dorsally by a short, transverse channel; the right nephridium has a long caecum posteriorly while the left one has only a short caecum. The rephiridia open separately to the exterior at the ventral extremity of each renal tube near side is a highly irregular mass composed of a large lobe covering the stomach, digestive gland and coil of the intestine and number of other lobes spread out in various directions in the mantle. The gonad opens in to the inter nephridial passage by a small aperture. Sexes are separate (K. Satyanarayana Rao, 1974).

## Distribution

Windowpane fishery in Saurashtra is located near Pindara Bay located between Poshitra and Ajad Island is encircled on the three sides by land, some of its coastal area has a characteristic association with mangrove swamps. Its northern side opens in to the Gulf of Kutch by a wide mouth. The Bay mouth is protected by various islands like Paga, Mangunda, Boria, Dona, Dabdaba, Pashu, Bural chank, Vudda-kudda, Ajad, Gunja-kudda etc. which are at some distance from the coast line. The bottom of the Bay is mostly muddy. Thus Bay is a natural habitat for windowpane oyster. The Bay is shallow in the coastal sections with up to 1 m depth, but the depth is about 3 m in some areas in the northern section. During low tide vast stretches of coast line are exposed. Windowpane oyster also occurs in lesser quantity in Balapur Bay. (Sarvaiya and Chhaya, 1983) The right valve is flat.

## Habitat

The windowpane oysters inhabit muddy bottom of bay and creeks, which are more or less land, locked. Usually the habitat is subjected to wide changes in salinity due to the opening of streams or rivers in to them. The species can tolerate large variations insalinity. The internal organs confine themselves to the superior shell i.e. towards the right valve. The oyster lying with convex surface down on the seabed. This characteristic of the animal is attributed to the muddy bed of its habitat which to animal chooses and at the same time would want to keep away the seedments of mud, sand and detritus away from its internal organs. Absence of a byssi's or siphon to keep the body in a vertical position or attached is of significance, with reference to its habitat.

## Sex

Windowpanes are dioecious i.e. the sexes are separate. While attributing these characteristics to this animal Hornel compares this is similar dioecious nature of the true pearl oyster. But the latter however, he recorded in subsequent studies to be prone to sex-reversal. The sex of the oyster is determined by the age and sea condition. Such phenomenon of sex reversal is known to exist in some marien fauna.

## Fecundity

It is cited by American Hard-clam in Symposium series 3 of Marine Biological Association of India, which was held at Cochin that windowpane releases 24 million eggs at a single spawning and 250 million eggs during a spawning season. The eggs diameter is given as 70 micron.

## Spawning Season

The species appears to spawn at the onset of the northeast monsoon in October. It is believed that the fall in the specific gravity of the witner in October following rains acts as a stimulant for spawning to take place (Hornel, 1909). Mosses (1947) has observed that windowpane oyster has two spawning seasons. Viz. October-December and April May. The seasons for spawning therefore appears to be identical to that of the edible oyster whose spat fall occurs at the same time as recorded by Commissioner of Fisheries, Govt. of Gujarat.

## Growth

The size of about 3 months old oyster observed were 15 mm. The shells of young oysters upto about one year age are almost transparent with the soft parts seen through the shells. When oyster grow further the shells are thicker and usually become white and translucent.

## Migration

A significant characteristic noted is that though windowpane oyster is fairly mobile, tagging experiments indicate that population do not migrate and although normal exigencies of larval and juvenile life take their toll, predation parasitism and diseases do not seem to impair the recruitment of the species.

## Economic Importance

The windowpane oyster presently is valued for its seed pearls. Since "Ayurveda" the traditional Indian Medicinal Science is being revived the traditional usage of seed pearls for pearls medicine finds an increased use. Usage of seed pearls in preparation of medicines is not confined only to India. Seed pearls are valued for similar usage in Middle-East countries, and also in scientifically advanced countries like Japan and China, Germany even today. Pearl medicines are believed to be effective against a number of human ailments. Hence the importance of the windowpane oyster fishery of the Gulf of Kutch. Possibilities of putting this mollusc for edible use, as the sea-scallop in America where the animal is relished for its abductor muscle opens up another dimension is repsect of the economic use of the windowpane oyster. The value realization of windowpane oyster from the Gulf of Kutch (since 1947-48) is given in Table 20.1. We can observe from the table that value of seed pearl is increasing every year. Previously Government was giving lease every year for lifting windowpane oyster from the Gulf of Kutch. However, because of over exploitation of this species government has stopped giving lease for lifting windowpane oyster since last 5 years.

## Fishery

Windowpane oysters are collected manually. Collection rights are leased out for a fixed period by the Department of Fisheries, Government of Gujarat. Fishing is started from the end of March and continues till the end of October. Oysters are collected from various centers like Asaba, Sidibava, Nagsura, Juda, Divdi, Kurchali, Mongra, Jagania, Dhoribhut, Sarhad, Pindharabeds, Ghungra and Gadu, situated in the Pindhara Bay.

Normally fishermen collect the windowpane oyster during daytime but a few also catch them during night. One man can collect 50 to 200 kg of oysters during the collection of two tides. Fishing is carried out from the 12th phase of moon to 5th phase of moon every fortnight. Oyster in deeper water upto 3.3 m are hand picked by divers without any diving aid. However, major part of collection is made at the depth of 1.5 to 2 m. The duration of diving is up to 30 sec. and depends up on the endurance capacity of fishermen. Oysters are lept in a bag (approximately 2′ to 2.5′ in length and 2.5 to 3′ width mesh size 3″ to 4″) locally called "Gumbha".

## Parasites and Pests

Larval cestodes and trematodes parasites the window-pane oyster and lie encysted in the mantle edge. Fishes are infected when they attack the oysters and ingest fragments of the mantle with the parasites. The parasites became sextually maturate in the vertebrate host. The polycahets *Pinnotheres placunae* with a body highly compressed dorsoventrally is common in the mantle cavity of window-pane oyster in Okha. (Hornell and Southwell, 1909).

### *Conservation*

Hornell (1909) has made some suggestion for the conservation of the window-pane oyster of the Gulf of Kutch as under :

(1) Keeping watch for large beds of mature oysters and then attempt to lease them to contractor.

(2) Stipulating that oysters below 11.4 cm (4.5 inches) in length should not be fished.

(3) Oyster below this size should be relaid with the convex side downwards in the fishing ground if captured.

(4) The lease of the beds should be given for a term of three or five year rather than for one year since in the former case the lessee will have more interest in the prosperity of the fishery and overcome loss incurred in a bed season. Sarvaiya and Chhaya (1983), have suggested that Pindhara Bay window-pane oyster fishery needs a closer and serious review. More and more small size window-pane oysters are noted in the catches.

**Table 20.1 Statement showing exploitation of Windowpane oysters in the Gulf of Kutch (Gujarat) since 1947-48**

| Sr. No. | Year | Value realized from lease |
|---|---|---|
| 1. | 1947-48 | 2500.00 |
| 2. | 1948-49 | 2816.00 |
| 3. | 1949-50 | 2201.00 |
| 4. | 1950-51 | 625.00 |
| 5. | 1951-52 | 2500.00 |
| 6. | 1952-53 | 4001.00 |
| 7. | 1953-54 | N.A. |
| 8. | 1954-55 | 5001.00 |
| 9. | 1955-56 | 7001.00 |
| 10. | 1956-57 | 7001.00 |
| 11. | 1957-58 | 7001.00 |
| 12. | 1958-59 | 11002.00 |
| 13. | 1959-60 | 4001.00 |
| 14. | 1960-61 | 8501.00 |
| 15. | 1961-62 | 21001.00 |
| 16. | 1962-63 | 11001.00 |
| 17. | 1963-64 | 14001.00 |
| 18. | 1964-65 | 12001.00 |
| 19. | 1966-69 | 60501.00 |
| 20. | 1969-70 | 45000.00 |
| 21. | 1972-75 | 60000.00 |
| 22. | 1975-78 | 75561.00 |
| 23. | 1978-79 | 85000.00 |
| 24. | 1980-81 | 115252.00 |
| 25. | 1981-82 | 291000.00 |
| 26. | 1982-83 | 1131111.00 |

Source : Commissionerate of Fisheries, Govt. of Gujarat. (With permission)

### *Resources*

According to Murthy, *et al.*, (1979), the total biomass of the window-pane oyster is estimated to be 8,945 tonne.

### *Utilization*

Window-pane oyster can be utilized as under :

(1) The thin, flat, indescent shells of the window-pane oysters are used for glazing windows and doors.

(2) The shells are also used as decorative pieces of gardens.
(3) Pearls can be obtained, though of poor quality. They lack hardness and have poor luster.
(4) In China, they were used for making medicines.
(5) After proper cutting window-pane shells can be used as glass cover dish.
(6) In Philippines, extremely fine lamp shades and lamp stands are made out of this shell.

## *References*

Anon., 1962. The Wealth of India, Council of Scientific and Industrial Research, New Delhi. pp 126.

Anon., 1983. Wonderful world of Gulf of Kutch. Commissionerate of Fisheries, Government of Gujarat.

Gokhale, S.V. 1963. Shell fisheries of Saurashtra region, Gujarat State. Department of Fisheries, Government of Gujarat.

Hornel, J. 1909. Report on the Marine Zoology of Okhamandal in Kathiawar. Government publication of State of Saurashtra.

Mosses, S.T. 1946. Pearl and Pearl Fisheries of Baroda State. *J. of Natural Hist. Soc. Vol XII.*

Nayan Nagapan, K. and K. Satyanarayan Rao 1985. Molluscan Fisheries of India. *Mar. Fish. Infor. Serv.* T & E Serv. No. 61, pp. 4-5.

Sarvaiya, R.T., and N.D. Chhaya 1993. "Saurashtra Molluscans" Smt. P.R. Sarvaiya, Ahmedabad.

Satyanarayan, K. Rao 1974. The Commercial Molluscs of India, CMFRI Bulletin, No. 25, pp. 142-143.

# 21
# CATLA FISHERY

## Introduction

This is the fastest growing major carp fish in India. It is considered excellent eating, especially when the fishes are of moderate size. The head is very fatty and those habituated to fish head preparation will appreciate its flavor. Hence people prefer this fish for culturing in fresh water pond. It is on of the most productive food fishes in India.

## Classification

| | |
|---|---|
| Phylum | Chordata |
| Sub-phylum | Vertebrata |
| Class | Pisces |
| Sub class | Teleostomi |
| Super Order | Ostariophysi |
| Order | Cypriniformes |
| Family | Cyprinidae |
| Sub family | Cyprinae |
| Genus | *Catla* |
| Species | *catla* |

## Names in other languages

| | |
|---|---|
| English | Catla |
| Gujarati | Catla |
| Kutch | Tambra |
| Bengali | Katla |
| Telugu | Botchee, Krishna-botcha |
| Tamil | Theppn-menu, Thoppa, Kooro-kandi, Kanavi. |
| Malayali | Karakatla |
| Oriya | Barkur, Karkura |
| Bombay | Tambra |
| Punjabi | Theila |
| Uttar Pradesh | Chepti, Catla |

**Fig. 21.1 : *Catla catla***

## Identification

D iii-iv 14-16; A iii 5; P1 20;VI (8).

B III C 19; L.1 40-30, Ltr. 7.5/9

Body deep, its depth 2.5 to 3 times in standard length. Head enormously large. Mouth wide and upturned with a prominent protruding lower jaw. Pectoral fins long, extend to pelvic fins. Scales conspicuously large lateral with 40 to 43 scales. No barbell; eyes with 3 orbital margins, Dorsal fin rather long without osseous ray. Some pores on the snout are visible 'Dorsal fin is inserted above tip of pectoral fin and has 1-19 rays.

Colour in life, grayish on back and flanks, silvery white below. Fin dusky. Generally those inhabiting weedy or turbid ponds have a darker colour. (Talwar and Jhingran, 1991).

## Reproduction

Heterosexual sexes are difficult to distinguish externally except breeding season when females develop a buldging abdomen and a swollen reddish vent. Pectoral fin in mature male is slightly stouter and longer and having rough dorsal surface, while female, it is very smooth. The rough surface of pectoral fin helps in gripping the female during courtship. (Chakraborti, N.M. 1998) Generally three or more males may chase female catla, but the reverse may also happens occasionally. A single couple may be seen very rarely. Chakraborti (1998) has noticed an interesting phenomenon that guttural sounds are produced by the breeders for calling the other partners especially in a case where inmates are separated from the congregations.

## Mating

The males chase the female and the former are found to held the latter. A male bends his body round the female and starts rubbing, knocking and nudging the female. The pair seen to be locked in embrace, their body twisted round each other with fins erect and caudal fin quivering. Mating occurs with vigorous splashing of water. Many fishes loose their scales and sustain minor injury. The duration of sexual play is short, the coiling and interwisting of male and female exerts pressure on the abdomen resulting in the extrusion of ova and milt.

## Fecundity

The fecundity of catla vary from 230,831 to 4,202,250 eggs depending upon the length and weight of the fish and weight of ovary.

## Eggs and Larval developments

Fertilized eggs of Catla are mostly spherical some are oval in shape, transparent and light red in colour. They are non-floating and non-adhesive. The shape of yolk is spherical without any oilglobule and pre-vitalline space in 3 mm in width. Fully ripe ovarian eggs may measure up to 2.2 mm (Mukharjee 1945). After spawning they swell to 5,3 to 6.3 mm.

## Diagnostic Characters of spawn and fry of *Catla catla*

| Time | Characters |
|---|---|
| Hatching | - Average size is 4.68 mm<br>- Bulbus part of the yolk and narrow part are equal in length.<br>- 26 pre-anal and 14 post anal myotomes present |
| 24 hrs after hatching | - Average size 5.80 mm<br>- Narrow end and yolk does not end in a point sharply. |
| 36 hrs after hatching | - Average size 6.4 mm<br>- Anterior profile of yolk sac convex. |
| 72 hrs after hatching | - Average size 7.3 mm<br>- Dark somewhat triangular spot in caudal peduncle |
| 96 hrs after hatching | - Average size 7.56 mm<br>- Semicircular mark formed of black chromatophores, seed, on ventral side of the caudal towards end of the notochord.<br>- Lip margins are very thick<br>- Dorsal fin separating from the embryonic fin fold. |
| 5th day after hatching | - Average size 9.0 mm.<br>- Caudal rays about 18, branching distally. |
| 7th day after hatching | - Average size 12 mm.<br>- Dorsal fin separated from embryonic fold. |
| 10th day after hatching | - Average size 16.5 mm<br>- Lips are thick.<br>- Mouth somewhat upturned. |
| 15th day after hatching | - Average size 23 mm.<br>- Margin of the dorsal fin dark. |

Source : Reddy and Rao, 1999.

## Food and Feeding

*Catla catla* is a surface feeder and plankton feeder. It is non-predatory. Fry of C.*catla* feeds water fleas and animalclues. Fingerling feeds on water fleas, few planktonic algae and some vegetable debris. Adult fish feed on water fleas, vegetable debris, and some algae.

In analysis of stomach content it was found that zooplankton forms about 65 to 72% of the total content. However, it depends up on their size. They are:

| Size | Gut length/fish length ratio | Feeding habit % by volume estimates. |
|---|---|---|
| Short | 1:7.4 | Phytoplankton 49.1%<br>Zooplankton 94.1% |
| Medium | 1:7.6 | Phytoplankton 79.4% |
| Long | 1:6.7 | Zooplankton 72.6% |

Source : Biswas, 1996.

## Growth

*Catla catla* attains a length 31 cm in 6 months, 38-46 cm in a year growth of 65 to 70 cm in the first has been recorded (Talwar and Jhingran, 1991). It is reported that it grows about 7.5 to 10 cm per month. Maximum growth reported is 1.8 meter in length. Best size for consumption is when not more than 61 cm. (Biswas, 1996).

## Maturity and Spawning

*Catla catla* becomes mature when 56 cm long and the age of two years.

It migrates to river in the plains for spawning but in the higher reaches, migration is not so marked. Uterine eggs are 2-3.2 mm in diameter, but after fertilization, they swell up to 4.1 mm. They are spherical, transparent and demersal with no oil globule. Hatching takes place in 16-18 hours and newly hatched larvae are 4.4-5.3 mm in length.

## Spawning season

The spawning season of catla in Gujarat is between June to September. While in Bangladesh it is between May to August. In the rivers of south India the spawning season appears to be somewhat variable. (Talwar and Jhingran, 1991)

## Stocking density and stocking ratio

The stocking density primarily depends on the natural fertility of pond. Stocking combination of several comptible species and increasing the carrying capacity of the pond by proper fertilization and supplementary feed can considerably increase this.

**Table. 21.1 : Landing pattern of *Catla catla* in Gujarat and is share in Gujarat (M. tones)**

| Year | Total Inland fish landing | Landing of Catla | Percentage of total inland catch of Gujarat |
|---|---|---|---|
| 1984-85 | 23814 | 1543 | 6.48 |
| 1985-86 | 25172 | 1715 | 6.81 |
| 1986-87 | 24451 | 1703 | 6.96 |
| 1987-88 | 22551 | 1421 | 6.30 |
| 1988-89 | 22315 | 1397 | 6.26 |
| 1989-90 | 27146 | 1333 | 4.91 |
| 1990-91 | 45687 | 1461 | 3.20 |
| 1991-92 | 39870 | 1716 | 4.3 |
| 1992-93 | 51154 | 3686 | 7.21 |
| 1993-94 | 65019 | 4888 | 7.52 |
| 1994-95 | 70100 | 8132 | 11.60 |
| 1995-96 | 60158 | 6753 | 11.23 |
| 1996-97 | 65278 | 7653 | 11.72 |
| 1997-98 | 70450 | 8329 | 11.82 |
| 1998-99 | 80068 | 13114 | 16.38 |
| 1999-00 | 70328 | 11091 | 15.77 |

Source : Commisioner of Fisheries, Govt of Gujarat Gandhinagar. (with permission)

Generally stocking density for Calta, Rohu and Mrigal at the ratio of 4:3:3 and stocking rate is maintained at 3750/ha. If more varieties of species are stocked, the ratio for Catla, Rohu, Mrigal, Silver Carp, Grass Carp Common Carp, Calbasu may be kept as 2.5:5:2.5:2:2.5:0.5 at stocking rate 5000/ha. In pond, recording highest production of 4210 kg per hectare per year.(Biswas, 1996).

Catla landing in Gujarat is between 1333 to 11716 mt upto 1991-92. Thereafter fresh water culture activity was increased which directly reflects on total landing. Landing from 1716 mt in 1991-92 was increased upto 7653 mt in 1996-97.

Percentage wise catla was contributing about 5 to 6% of total inland catch prior to 1991-92. While it was increased to 11.72% in 1996-97.

## *References*

Anon, 1999. Statistical information on Gujarat Fisheries, Commissionerate of Fisheries, Govt. of Gujarat. Gandhinagar. (With the Permission)

Biswas, K.P. 1996. A Text Book on Fish and Fisheries and Technologies.

Chakraborti, N.M. 1998. Biology, culture and production of Indian Major Carps. Narendra publishing House, Delhi.

Mukharjee, H.K. 1945. Lifehistory of some major carps of Bengal *Sci. Cult.* 10(9): 400-2.

Pandey, A.K. and G.S. Sandhu. 1992. Encyclopaedia of Fishes and Fisheries of India. Anmol publications, New Delhi.

Reddy, M.Srinivasulu and K.R.S. Sambasiva Rao. 1999. A Text Book of Aquaculture, Discovery Publishing House, New Delhi.

Talwar, P.K. and Arun G. Jhingran. 1991. Inland Fishes of India and Adjacent countries, Vol. 1:163. Oxford & IBH Publishing Co. Pvt. Ltd.

# 22

# ROHU FISHERY

Rohu is considered testiest and most common of the Indian carps and is suitable for pond culture. Rohu is regarded as excellent game fish and seems to put up a better growth in a river than in a tank.

## Classification

| | |
|---|---|
| Phylu | Chordata |
| Class | Pisces |
| Sub Class | Teleostomi |
| Super Order | Ostariostomi |
| Order | Cypriniformes |
| Sub Order | Cyprinoidel |
| Family | Cyprinidae |
| Sub Family | Cyprininae |
| Geneus | *Labio* |
| *Species* | *rohita* |

## Names in other languages

| | |
|---|---|
| Gujarati | Rohu |
| Bengali | Ruee |
| Marathi | Tambada-massa |
| Oriya | Rohu |
| Punjabi | Rohu, Tapra, Dhambra |
| Assam | **Rohiti, rai.** |

**Fig. 22.1 : *Labio rohita***

## Distinguishing Characters

D iii-iv 12-14, A ii-iii 5 ; PI 16-18; VI 8.

Body moderately elongate, its dorsal profiles more arched than the ventral profile. Snout fairly depressed, Projects beyond mouth, devoid of lateral lobe. Eyes large, not visible from underside of head, the diameter 4 to 6 times in head. Mouth small and inferior; lips thick and fringed, with a distinct inner fold to each lip. Dorsal fin inserted midway between snout-tip and base to caudal fin. Pectoral fins shorter than head. Caudal fin deeply forked scales moderate; lateral line with 40 to 44 scales; lateral transverse scales - rows 6 to 6.5 between lateral line and pelvic fin base; Pre-dorsal scale 12 to 16.

Colour in life, bluish along back, becoming of silvery on the flanks and beneath, with a reddish. mark on each scale during breeding season; eyes reddish. Fins grayish or dark; Pectoral fins dusky. The colour tends to very in fishes living among weeds exhibiting greenish black on back. (Talwar and Jhingran, 1991).

## Distribution

This fish is found in North and Central India, Pakistan, Bangladesh, Tarai region of Nepal, Burma. This graceful riverine fish is one of the mjor carps in India. It is reported in Narmada and Tapi in Gujarat.

*Labio rohita* is found in following rivers of India.

| | |
|---|---|
| Ganga River System | Ganga, Yamuna, Ghaghra, Gomti, Tapi, Koshi, Son, Damodar, Chambal, Betwa. |
| East Coast | Mahanandi, Godavari, Krishna, Cauvery. |
| West Coast | Narmada, Tapti, Mahi, Sabarmati. |

## Reproduction

Rohu is heterosexual and sex is recognized when ripe females show significantly buldging and soft abdomen with a reddish vent. In mature female a gentle pressure on the abdomen of fish release eggs and males exude milt. In males during breeding season the pectoral fin is found to be better developed and greater than anal fin.

In male the pectoral fin reached $8^{th}$ or $9^{th}$ lateral line scale, while in female it touches $6^{th}$ or $7^{th}$ lateral line scale. In breeding seaon the dorsal surface of pectoral fin shows roughness, while in female it is smooth to touch. (Chakraborti, 1998)

## Maturity

Rohu usually attains maturity towards the end of second year of life in riverine condition and after one year in pond condition, (Alikhni ; 1957)

In case of male the first maturity is observed at size of 462 mm and in length group of 650 mm all were matured. In females the smaller mature fish was 515 mm and all attains maturity in 700 mm length group.

Fecundity increases with age group up to $6^{th}$ year and later decreased in subsequent age group. (Chakraborti, 1998).

## Mating

Rohu is polygamous and promiscuous fish. Spawning activities followed by spawning and frequent jumping. Males first chase the female and the female is hold by a male and bends its body round the female and starts rubbing knocking and nudging the female. During this period pairs are seen to be locked in embrace, body twisted round each other. The fins are erect and quivering of the caudal fin. Mating occurs vigorous splashing of water, during that time many fishes loose number of scales and sustain minor injuries and the sexual play lasts for 5 to 10 seconds. Several such mating takes place before it totally spent indicating that all eggs are not laid at one time. The bending activities exert pressure on the abdomen of mating pairs as a result of which extrusion of ova and ejection of milt takes place.

## Spawning

Rohu spawns once a year in nature. It was noticed that a single batch of ova started developing at late winter months and they increase gradually in size to maturity at post summer months and were finally discharged in monsoon. Spawning takes place in flooded rives. Spawnig occurred during different times of the day. Flooding, drop of temperature, stimultae spawning, sufficient depth of water for to and fro movement of fishes, moderate to fast current at breeding site, temperature range 22-30° C occasional rain with thunder storm are conductive factors influencing the breeding. Rohu does not breed in confined water.

## Spawning season

Spawning season of Rohu generally coincides with the South west monsoon but it varies in different parts of India as under.

| | |
|---|---|
| Western U.P., Punjab and Gujarat | July-August. |
| Orissa, West Bengal | June-August. |
| Ganga river in Bihar and U.P. | April-August. |
| Assam | April-May |

## Fecundity

The fecundity of Rohu varies from 2,26,000 to 2,794,000 depending upon the length and weight of the fish and weight of the ovary. (Talwar and Jhingran, 1991).

Fecundity is reported to be greater than that of *Catla*.

## Maturity and spawning

Sexual maturity is attained toward the end of the second year. Spawning takes place in flooded rivers. The female lays two million eggs at a time. Fertilized eggs are reddish. The fry has three small spots on the caudal region, which later merge and disappear in the adult.

## Diagnostic characters of larval stages of *Labio rohita*

| Time after Hatching | Characters |
|---|---|
| Hatching | Newly hatched larvae measures 3.5-4.5 mm. It has transparent laterally compressed body and is characterized by the presence of gill slits, Pectoral fin and median fin fold. |
| 24 hours | Average size 5.5 mm. A few chromatophores are seen on the head above the eyes. Embryo pale yellow in colours. Yolk sac sharply ending distally to a point. Striations seen in the caudal. Notochord turned upward only at the very end. |

Contd.........

| Time after Hatching | Characters |
|---|---|
| 36 hours | Average size about 5.9 mm.<br>Lower lip clear. Anterior profile of yolk sac more or less straight. A few black chromatophores on dorsal margin of the yolk sac through out its length, a few on the dorsal fin. No chromatophores on the caudal region.<br>Notochord slightly upturned at the tip. Mouth appears as a slit |
| 48 hours | Average size 6.2 mm.<br>Yolk sac. Convex anteriorly<br>Embryo yellow in colour<br>Air bladder distinct<br>Pectoral fin prominent<br>Head dark and body is faint yellow. |
| 4th day | Average size 7.6 mm<br>Distinct black chromatophores seen behind the eyes on the head region.<br>Notochord bent at the tip. |
| 7th day | Average size 11 mm<br>Lips are very thick.<br>Ventral embryonic fin fold stretching from abdominal region up to anus<br>Caudal fin with 22 rays<br>Pelvic fin with 2-3 rays, Dorsl fin with 2-11 rays.<br>Two black chromatophore clusters at the peduncle and two crescent shaped patches on the caudal fin. |
| 12th day | Average size 19 mm<br>Maxillary barbell present<br>Body golden coloured on the dorsal side and dirty yellow ventrally<br>Dorsal fin with 3-13 rays and 1 fin with 7 rays, except for anterior one. The rest are branched.<br>Ventral fin with 7 rays, except two anterior ones, the rest are branched.<br>Caudal fin rays 34, a broad triangular block band on the caudal peduncle across its entire with apex facing the head. |

Fry can be identified as spots on caudal peduncle not enclosing a whitish area, its lips fringed, body without longitudinal bands, fins marked with vermillion red.

## Survival, Growth of Spawn, fry, Fingerlings of Rohu.

### *Spawn*

Stocked in nursery pond at length 20-25 mm in 15 days with supplementary feeding resulting survival of 50% or more.

### *Fry/Fingerlings*

In Fry/Fingerlings (23-314 mm) and average weight 0.15-0.3 g and stocking density 62,000-93,000 survival of 72-80% was observed. Fingerlings attains 150 mm with an average weight 41g and survival of 95%.

Rohu grows quickly. A growth of 35 to 45 cm can normally be expected in the first year. It grows to over 1 meter in length.

## Food and Feeding

This fish is herbivorous fish. Rohu is a bottom and column feeder fish. Very young fry feed on zooplanktons. Fingerlings feeds vegetable debris, minute plants, Adults feeds vegetable debris small plants, detritus and mud. The plant material in their food is more than 75% of the total gut content.

Spawn of *L.rohita* start feeding on plankton after 5th day of hatching. The initial food items have been observed to be mainly zooplankton (90%) and remaining phytoplankton (10%). Cladocerans supplemented with copepods and insect larvae are the main content of zooplanktonic food. (Kamal, 1967) From the larval stages onwards in the rohu fry and fingerlings, the phytoplanktonic feeds are mainly constitutes of Microcystis. As the fry grows preference for zooplanktonic food items is noticed cladocerans forms the chief items of food initially on 5th day while copepods dominated the gut contents from 17 to 19th day old fry. Insect larvae, rotifers and nauplii were noticed considerably in small quantities.

Initially the length of the alimentary canal is less than 50% of the body length indicating carnivorous nature and the gut length becomes more than 100% on seventh day indicating herbivorous character.

## Fishing Season

Fishing in tanks, ponds is done almost throughout the year. In river fishing period is generally accentuated during winter months and summer months (November-July), but suspended during onset of monsoon (July-October) because of strong currents.

**Table 22.1 : Landing pattern of Rohu in Gujarat and its share**

| Year | Total Inland Fish landing | Landing of Rohu | Percentage of total Inland catch of Gujarat |
|---|---|---|---|
| 1984-85 | 23814 | 1621 | 6.81 |
| 1985-86 | 25172 | 1519 | 6.04 |
| 1986-87 | 24451 | 1788 | 7.31 |
| 1987-88 | 22551 | 1841 | 8.16 |
| 1988-89 | 22315 | 1302 | 5.83 |
| 1989-90 | 27146 | 1704 | 6.28 |
| 1990-91 | 45687 | 2082 | 4.56 |
| 1991-92 | 39870 | 3091 | 7.75 |
| 1992-93 | 51154 | 4304 | 8.41 |
| 1993-94 | 65019 | 6580 | 10.12 |
| 1994-95 | 70100 | 8462 | 12.07 |
| 1995-96 | 60158 | 7870 | 13.08 |
| 1996-97 | 65278 | 9351 | 14.33 |
| 1997-98 | 70450 | 10048 | 14.26 |
| 1998-99 | 80068 | 12707 | 15.87 |
| 1999-00 | 70328 | 9021 | 12.83 |

Source : Commissioner & Fisheries Govt & Gujarat. Gandhinagar. (with permission)

## *References*

Anon, 1998. Statistical information of Gujarat Fisheries. Published by Commissioner of Fisheries, Govt. of Gujarat. Gandhinagar.

Biswas, K.P. 1996. A text book on Fish, Fisheries and Technologies, Pp 68.

Chakraboarti, N.M. 1998. Biology, Culture and Production of Indian major carps. Narendra Publishing House, New Delhi.

Reddy, M. Srinivasulu and K.R.S. Sambasiva Rao 1999. A Text book of Aquaculture, Discovery Publishing House, New Delhi.

Talwar, P.K. and A.G. Jhingran. 1991. Inland fishes of India and Adjacent Countries Vol I. 219-220.

# 23
# MRIGAL FISHERY

Next in importance to Catla and Rohu for cultural purpose is the Mrigal. It is an important major carp of India and quite a game fish.

## Classification

| | |
|---|---|
| Class | Pisces |
| Sub class | Teleostomi |
| Super Order | Ostariophysi |
| Order | Cypriniformes |
| Family | Cyprinidae |
| Genus | *Cirrhinus* |
| *Species* | *mrigala* |

## Name in other languages

| | |
|---|---|
| Gujarati | Mrugal, Nagri, Morahkee |
| Bengali | Mrigal |
| Telugu | Yerra-masu, Bellala mosu. |
| Oriya | Mirrga |
| Kutch | Morahkee |
| Punjabi | Mori |
| Maharashtra | Mrigal, Mirya |
| Kerala | Mrigalam |
| Andhra Pradesh | Ballalamosa |
| Bihar | Nain, Nainee |
| Uttar Pradesh | Mrigal, Mirki. |

## Identification

B.III; D-10 to 11; (2/8 to 9); P-16; V-9; A-8 (3/8); C-19;L1. 35 to 38; Ltr. 7/9

This carp very much resembles to *Labio rohita* except that it has a wider mouth and thinner lips. Body streamlined its depth about equal to length of head. Snout blunts, often with pores. Mouth broad; Upper lip entire, lowe lip most indistint. Barbells a single short pair of rosterals only. Pharyngeal teeth in three rows, 5.4.2/2.4.5 pattern. Gillrackers 40 to 49 on first arch. Dorsal fin as high as body. Pectoral fins shorter than head. Caudal fin deeply forked. Lateral transverse scale-raws 6-7/5.5 to 6.

Colour : In life, dark gray along back often with coppery tinge, flanks silvery, a yellowish tinge and belly silvery white; eyes golden. Pectoral, Pelvic and Anal fins orange-tipped. Dorsal and Caudal fins dusky. The fry has a black tail spot, which later grows in to a tail 10 cm. long; fins are tinged yellow and orange.

## Diagnostic characters of spawn and fry of *Cirrhinus mrigala* are as under.

| Time after hatching | Characters |
|---|---|
| Immediately | Average size 4.2 mm, yolk more or less club shaped. |
| 24 hrs | Average size 6.21 mm, narrow end yolk does not end in a point sharply. |
| 48 hrs | Average size 6.84 mm, Anterior profile of yolk sac more or less straight. |
| 72 hrs | Average size 7.2 mm, few black chromatophores in the caudal region. |
| 96 hrs | Average size 7.38 mm, Black chromatophores on the dorsal fin rudiment, towards anterior region.<br>Dorsal fin separating from the embryonic fin fold. |
| 5th day | Average size 9.0 mm, Caudal rays about, not branching. |
| 7th day | Average size 13.2 mm, Black chromatophores from a triangle anterior to the origin of the caudal peduncle. Dorsal fin separated from embryonic fold |
| 15th day | Average size is 27 mm, Body is more pigmented. |

## Diagnostic characters of advanced fry and early fingerlings

The spot on the caudal peduncle becomes diamond shaped. Barbells are faintly visible. Lips are not continuous. A few longitudinal lines are visible on the body due to the pigments on the scales.

**Fig. 23.1 : *Cirrhina mrigala***

## Food and Feeding

This is an omnivorous type fish. Adult fish feeds mainly on algae and vegetable detritus and debris. Fingerling feeds on vegetable debris, Unicellular algae, detritus and mud. They also feed rotifers, insects and their larvae, crustaceans, Bryozoans etc. *C.mrigal* is a bottom feeding fish.

Most active feeding period was observed during January. Semidecayed organic matter consisting of leaves twings aquatic vegetation, seeds, spicules etc. was fairly high during July (9.3%) and September (8.5%). Planktonic content in total gut content was only 1.9% and was fairly higher from December to May. Diatomes were dominated and formed 46% of total plankton.

Feeding intensity in general was poor excepting the period from November to January and it improved especially after breeding season. Active feeding was not observed during breeding season and fishes were in starved condition. No significant difference on composition of diet of males and females were observed. It was noticed that mucus secretion in the gut was in abundance during the low intensity of feeding.

## Sex

Sex ratio in population is important in Fishery Management. Sex ratio of Mrigal from rivers indicated that the two sexes appeared to be more or less equal. However, in younger age group more males were found that females. But in case of older fishes dominance of females were observed due to the fact that the males dies earlier than females (Bannet 1962).

## Fecundity

Fecundity varies from 1-2 lakh depending upon the size of the fish.

## Maturity and Spawning

The male matures at the end of the first year; the female takes little longer. Breeding takes place in flooded river during July-September. It becomes sexually matured in the second year.

## Breeding

It breeds during South-West monsoon in shallow pockets in marginal areas and in bundhs. The depth of water at breeding ground is 0.5 to 1 m. (Talwar and Jingran, 1991) Their eggs are non-floating, non-adhesive type round 5.5 mm in diameter, brownish in colour. (Reddy and Rao, 1999) It does not breed in confined waters and is induced to breed by hypophysation technique.

**Table No. 23.1 : Mrigal landing pattern in Gujarat and its share**

| Year | Total Inland Fish landing | Landing of Rohu | Percentage of total Inland catch of Gujarat |
|---|---|---|---|
| 1984-85 | 23814 | 1059 | 4.45 |
| 1985-86 | 25172 | 1033 | 4.10 |
| 1986-87 | 24451 | 901 | 3.68 |
| 1987-88 | 22551 | 791 | 3.51 |
| 1988-89 | 22315 | 948 | 4.25 |
| 1989-90 | 27146 | 888 | 3.27 |
| 1990-91 | 45687 | 1345 | 2.27 |
| 1991-92 | 39870 | 2577 | 6.46 |
| 1992-93 | 51140 | 2868 | 5.61 |
| 1993-94 | 65019 | 3910 | 6.01 |
| 1994-95 | 70100 | 5102 | 7.28 |
| 1995-96 | 60158 | 5094 | 8.47 |
| 1996-97 | 65278 | 5883 | 8.62 |
| 1997-98 | 70450 | 6322 | 8.97 |
| 1998-99 | 80068 | 8236 | 10.29 |
| 1999-00 | 70328 | 7396 | 10.52 |

Source :Commissioner of Fisheries, Govt. of Gujarat.

## Growth

Mrigal grows slower than Catla and Rohu. In fish farms it attains a length of 55-66 cm and weight of 650 to 800 gm after one year. While in tanks and contaminated water it grows, during the same period, to 38-41 cm. They attains maximum length of over 0.75 meter.

## Spawning season

The spawning season depends on the onset of South-West monsoon (June to August).

It can be observed from Table 23.1 that before 1990-91 landing of Mrigal was less than 100 MT. While after that it was increased up to 5883 MT. in 1996-97. This is clear effect of increase in Fresh water culture activities in Gujarat after 1990-91. Percentage wise also increase was recorded from 2 to 4% to 8.62% during above period.

## *References*

Biswas, K.P. 1996. A Text book on Fish, Fisheries and Technology.

Pandey, A.K. and G.S. Sandhu, 1992. Encyclopedia of Fishes and Fisheries of India. Published by Anmol publication.

Rao, M Srinivasulu and K.R.S. Samusiva Rao, 1999. A Text Book of Aquaculture. Published by Discovery Publishing House, New Delhi.

Talwar, P.K. and A.G. Jhingran, 1991. Indian Fishes of India and Adjacent Countries Vol I Published by Oxford & IBH publishing Co. pvt. Ltd.

# 24

# Hilsa Fishery

Hilsa is highly esteemed fish for its culinary excellence. Its popularity as table fish, in spite of numerous fin bones, is probably due to the high fat content (20%) during certain season. It is a pelagic fish, found at depth up to 36 m in fore shore and off shore waters. Essentially marine, it is aeuryhaline type of clupeid which tolerate considerable salinity variations. This capacity of fish enables it to readily adapt to conditions in estuarine and back waters and to undertake long sojourns in fresh water during migration.

## Classification

| | |
|---|---|
| Class | Osteichthyes |
| Sub Class | Actinopterygii |
| Order | Cluperiformes |
| Sub-Order | Clupeoidei |
| Family | Clupeidae |
| Sub Family | Alosinae |
| Genus | *Hilsa (Regan)* |
| Species | *ilisha* |

## Synonyms

*Hilsa ilisha* (Hamilton Buchanan)

*Clupea ilisha* (Day)

Clupanodon ilisha (Ham. Buch.) Fish Ganges, pp 243.

*Alausa plasah,* Blecker, Beng p.74.

## Names in other languages

| | |
|---|---|
| English | Hilsa shed. |
| Gujarati | Palla, chaksi. |
| Hindi | Hilsa |
| Bengali | Ilisha |
| Marathi | Pala, Chakai, Plava (Male). |
| Telugu | Palasah |
| Tamil | Oolum, Ullam. |

| | |
|---|---|
| Kanada | Paliya |
| Malayalam | Paluva |
| Assam | Ilihi |
| West Bengal | Ilishmach. |

## Taxonomy

B. vi, D.18-19 $\frac{3}{15\text{-}16}$, P.15, V.0, A.19-22 $\frac{2}{17\text{-}20}$, C-19, L.1, 46-49, L.tr. 7-19, Vert. 12/32 34.

| | |
|---|---|
| Head | Length of head 4.25 to 4.5, of caudal 5, height of body 3.5 to 3.75 in total length. |
| Eye | With broad adipose lids, situated some distance before the middle of the head and from 1 to 1.5 diameters from the end of snout, and 1.5 apart. |
| Jaw | The posterior extremity of the maxiila reaches to below the middle or even hind edge of the orbit: lower jaw not projecting beyond the upper. The greatest width of the opercle equals about 2/3 of its height sub-opercle rounded posteriorly. |
| Teeth | Absent. |
| Fins | Dorsal commences a little nearer the snout than the base of the caudal fin, its upper edge concave. Pectoral reaches to above the origin of the ventral, which latter fin is inserted beneath the anterior half of the dorsal. Caudal deeply forked. Caudal peduncle as deep as long. |
| Scales | In regular raw: many over the caudal fin. Gill-raker numerous and as long as the eyes: 16 to 17 scutes before and 14 to 15 behind the insertion of the ventral fin. Pseudobranchiae well developed. |
| Colours | Silvery, shot with gold and purple: no spots in the adult, but a raw of them along the upper third of the body in the immature, the most distinct of which is behind the upper third of the opercile. The young are usually of a bronze colour along the back, with silvery sides and a burnished silvery band going from above the eye to the upper half of the caudal, whilst the caudal fin is often deeply edged with black in its entire circumference. |

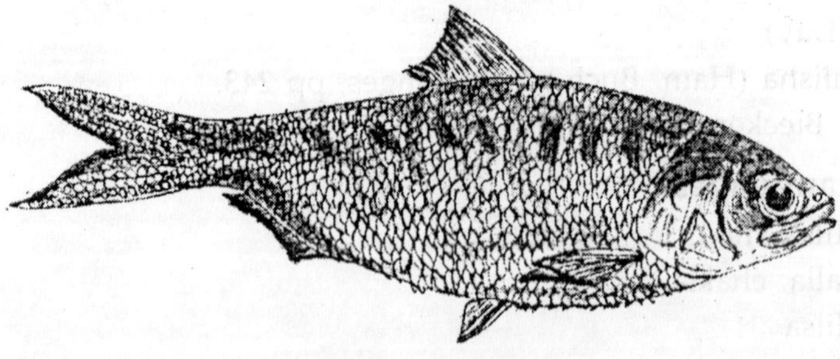

Fig. 24.1 : *Hilsa ilisha*

## Distribution

Along the west coast of India it is distributed on the Saurashtra Coast and the fore shore areas at the mouth of rivers Narmada and Ulhas. In the off shore waters its distribution ranges between 13 km to 16 km away from the coast and at a depth of 36 m. All these areas, therefore, support hilsa fishery to a lesser or greater extent. (Pandey and Sandhu, 1992). It is also available in the inner most part of Gulf of Kutch-around Machhu, Banas, Sarasvati rivers entering the little rann. Good catches during March-June period of good size Hilse.

## Average Size

The average size of an anadromous mature female is 38-46 cm; male generally smaller. The maximum recorded size for female is 60 cm with a weight of 2.49 kgs; for male 43 cm and 0.68 kgs (Talwar and Jhingran, 1991.) They attain maturity size at the age of one or one and half years, (Pandey and Sandhu, 1992).

**Table 24.1 : Landing pattern of Hilsa in Gujarat and in India and its share (in '000 kg)**

| Year | Hilsa landing in Gujarat | Hilsa landing in India | Percent share of Gujarat |
|---|---|---|---|
| 1975-76 | 1608 | 8897 | 18.07 |
| 1979-80 | 3489 | 12068 | 28.91 |
| 1980-81 | 4740 | 69960 | 7.77 |
| 1981-82 | 4155 | 54070 | 7.68 |
| 1982-83 | 3813 | 31680 | 12.04 |
| 1983-84 | 5862 | 38760 | 15.12 |
| 1984-85 | 7097 | 58970 | 12.03 |
| 1985-86 | 4897 | 90951 | 5.41 |
| 1986-87 | 3093 | 40910 | 7.56 |
| 1987-88 | 2737 | 44180 | 6.20 |
| 1988-89 | 2077 | 20060 | 10.35 |
| 1989-90 | 2459 | 14556 | 16.89 |
| 1990-91 | 2705 | 14243 | 18.99 |
| 1991-92 | 2082 | 27728 | 7.51 |
| 1992-93 | 1613 | 22378 | 7.21 |
| 1993-94 | 1899 | 30215 | 6.28 |
| 1994-95 | 2138 | 19518 | 10.95 |
| 1995-96 | 2176 | 25646 | 8.48 |
| 1996-97 | 2598 | 28889 | 8.99 |
| 1997-98 | 2556 | 17206 | 14.85 |
| 1998-99 | 1515 | 10018 | 15.12 |
| 1999-00 | 1710 | 10748 | 15.91 |

Source (1) Commissioner of Fisheries, Govt. of Gujarat, Gandhinagar (With permission)
(2) CMFRI Annual Reports

## Growth rate

Growth rate is estimated at 20 mm per month for one year old hilsa. Hilsa is known to live for 5 to 7 years. It has been found that hilsa from Mahanadi river have the highest fat content (20%) and those from Naramda the lowest (6%). (Pandey and Sandhu, 1992).

## Food and feeding habits

Hilsa is a strict plankton feeder. Its gill and gillrackers are modified in to a very sieve that must filter all microscopic food organisms entering the buccal cavity with respiratory water current. Both phytoplankton and zooplankton make up the food.

## Sex and Breeding

The species is polygamous, the sex ratio in the population is normally 1:1. Breeds mainly in rivers, in some cases far up (400 km) but elsewhere only to about 50 km or less. Younger fishes may breed in the tidal zones of rivers. Fertilization is external. (Talwar and Jhingran, 1991). Percentage fertilization at Bharbhoot on Narmada river varied between 60 to 90. (Malhotra, *et. al.* 1973).

## Spawning Season

Spawning generally occurs during the south west monsoon period but may extend from June to February. (Pandey and Sandhu, 1992).

## Migration

Generally Hilsa fish migrates from sea to river during the winter for spawning. The spent adults of the monsoon migration return to the sea from October to December and those of the post-monsoon migration from April to early June. In addition to this maturing, mature and spent adults, were also found near the mouth of river. (M. Babu Rao, 1969).

## *References*

Chakraborti, N.M. 1998. Biology, culture and production of Indian Major carps. Narendra Publishing House, New Delhi.

F. Day. 1958. Fishes of India.

Malhotra, J.C., P.K. Mathu, S.N. Melhotra, and M.Y. Kamal, 1973. Successful rearing of *Hilsa ilisha* (Hamilton) in fresh water for years. *Proc. Nat. Acad. Sci. of India,* (B), 43 (3) : 207-10.

M. Babu P. Rao. 1969. Some observation of juveniles of *Hilsa ilisha* (Hamilton) pieces from (*Clupeidae)* from Godavari Estuary. *J. Bom. Nat. Hist. Soc.* 66(1) : 116-131.

Pandey, A.K. and G.G.S. Sandhu. 1992. Encyclopaedia of Fishes and Fisheries of India. Vol III. Published by Anmol Publication.

Talwar, P.K. and Arun G. Jhingran. 1991. Inland Fishes of India and Adjacent countries. Vol. I. Published by Oxford and IBH Publishing Co. Pvt. Ltd., New Delhi. Pp 100-101.

# FRESH WATER PRAWN FISHERY

## Introduction

There are about 100 species of Fresh water prawn *Macrobranchium*, identified from various parts of the world of which about 25 species are found in India. Among the Indian prawns, *Microbranchium rosenbergii, M.malcolmsonii, M.brimanicum, M.idella, M.rude, M.equidens, M.seacriculum and M.lamarrei* are suitable for farming in India.

*Microbranchium rosenbergii*, commonly known as "Scampi" is immensely popular from the point of view of commercial aquaculture and because of its ability to grow both in fresh water and low saline waters, compatibility for polyculture with carps, omnivorous feeding habits, hardihood and potential to grow fast. It does not have much of problems of different disease attacks and has good consumer preference. It has ever increasing demand both in the domestic as well as export market.

## Classification

| | |
|---|---|
| Phylum | Arthropoda |
| Class | Crustacea |
| Order | Malcostraca |
| Family | Palaemonidae |

## Identification *Microbranchium rosenbergii (DE-Man)*

Rostrum long with distinct elevated crest, 1-3 teeth on carapace 12-13 teeth on dorsal and 11-13 teeth on the ventral side of the rostrum. Chelipeds equal, spines strong short, widely separated not pigmented.

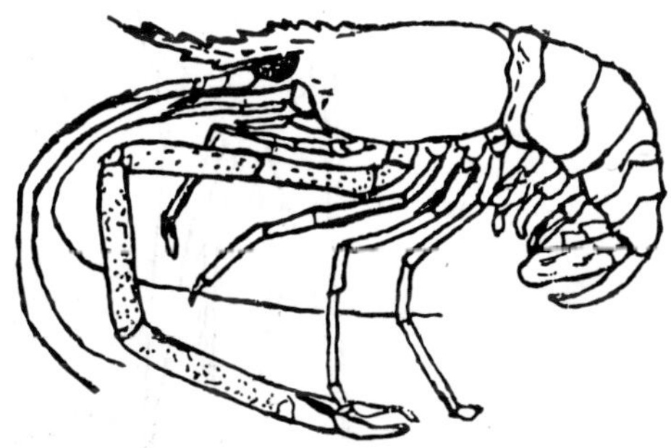

**Fig. 25.1 : *Macrobranchium rosenbergii***

*Macrobranchium malcolmsonii*

## Classification

| | |
|---|---|
| Phylum | Arthrooda |
| Class | Crustacea |
| Order | *Malcostraca* |
| Family | *Palaemonidae.* |

## Identification characters

The rostrum of this species reaches about the end of the antennal scale. The basal part of dorsal margin is high and crust like and armed with 7-11 teeth. The ventral pair of pereopods is enlarged in males. The body and the anterior two pairs of pereopods is are dark grey in colour. The antennae and inner flagella of antennules are grey and the outer flagella are brownish. The maximum size recorded is 24.3 cm in length.

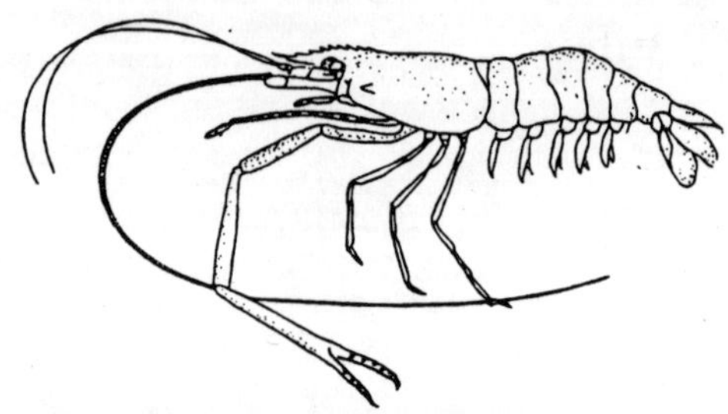

**Fig. 25.2 :** ***Macrobranchium malcolmsonii***

## Fig. 25.3 : Seed identification of *M. molcolmsonii and M. resenbergii*

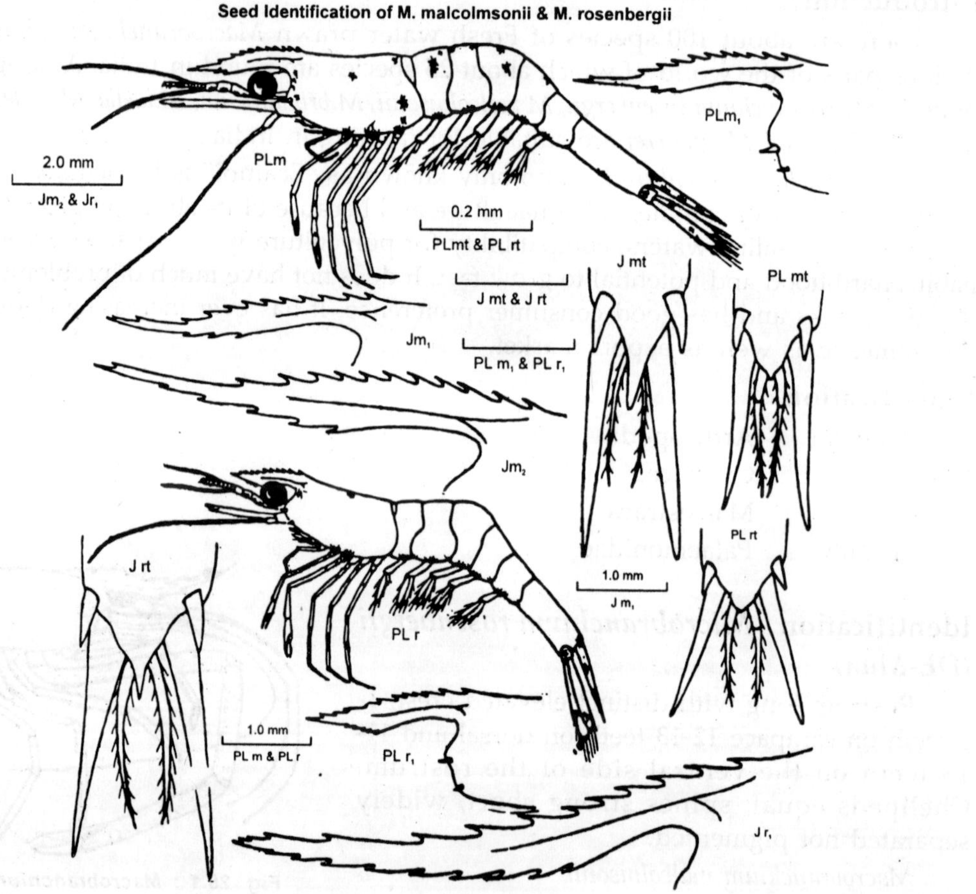


**Explanation of Captions**

PLm— postlarva of *malcolmsonii;* PLmT— anterior part of carapace of PLm; PLmt— telson tip of PLm; $Jm_1$— rostrum of 1st juvenile of *malcolmsonii;* $Jm_2$— rostrum of 45 days old juvenile of *malcolmsonii;* Jmt— telson tip of 30 days old juvenile of *malcolmsonii;* PLi— postlarva of *rosenbergii;* $PLr_1$— anterior part of carapace of PLr; PLrt— telson tip of PLr; $Jr_1$— rostrum of 45 days old juvenile of *rosenbergii;* Jrt— telson tip of 30 days old juvenile of *rosenbergii.*

Source : Fishing chimes Feb. 92, p. 106.

## Distribution

It is distributed in the Indo-west Pacific countries, Pakistan, India and Bangladesh. Its habitats are the freshwaters and brackish waters. In India it is the most common in the river Godavari, Cauvery and also in Chilka Lake. This is the only prawn of commercial fishery in the river Godavari, producing an annual yield of 80 tones. In the Chilka Lake and other places; its fishery is during the monsoon season.

## Biology

This is an omnivore, preferring animal matters and also copraphagous in feeding habit. Spawning takes place throughout the year with a peak during monsoon season.

This is one of the prospective species for culture in freshwater. Natural seeds can be collected from rivers from where it is available. Hatchery seed production is also possible for mass culture of post larvae to take up commercial culture. Juvenile cannibalism and differential growth pattern are observed in culture systems.

## Growth

During the process of culture of *M. rosenbergii*, a major difference in growth sequence of the males and of the females became very apparent (Brody *et. al.*, 1980). In single aged population, the size distribution of female population is rather homogenous as there exists no morphotypic differentiation. On the contrary, males exhibit differential growth rate as manifested by the positive slowness in the size distribution among the male population. Males of *M. rosenbergii* are known to exhibit a complex social and organizational hierarchy comprising dominant, sub-dominant, and subordinate groups of animals which have distinct morphologically distinguishable characteristics. Three such morphotypes and two tansitional stages have now been established on the basis of relative body size, spination on chela, claw characteristics, hierarchical dominance, differential growth patterns and alternative mating strategies. (Kuris *et. al.*, 1987). The most salient distinguishing features and their relative advantage in the growth stages of the three male morphotypes such as Blue Clawed Males (BC), Orange Clawed (OC) and small males (SM). Among the three morphotypes, in the BC males which seem to constitute the final morphotype in the male development process the growth rate is greatly reduced and the moulting is also reported to be infrequent. In OC males, the somatic growth is very fast as they spend very little energy for reproduction and courtship, while SM also represents the slowly growing phase. (Kurup, Madhusudhan B. 1997).

## Feed and Feeding

*M.rosenbergii* is omnivorous in its feeding habit. It feeds on aquatic insects and larvae, worms, small molluscs, crustaceans, grains, nuts, flesh and offal of fish and other animals. Oil cakes, slaughter house waste tapioca root etc. Natural feeds produced through biological cycle in culture ponds serve as a good source of food for the stocked prawns. However, in commercial operations, the natural productivity is not sufficient where the prawns are stocked at higher densities. The feeding rate starts with 10% of the prawn body weight initially and gradually reduces to 2 to 3% towards harvest. It is essential to provide supplementary feed for achieving higher growth. If the prawns are fed with compounded pelleted feed having 25 to 30% protein, higher production could be achieved. Feed may be given twice in a day, in the early morning and in the late evening hours. Food is usually spread around the marginal shallow areas of the pond. Generally the daily ration of the feed is calculated on the basis of estimated total weight of the prawn.

## Mating

Two alternative mating strategies have also been described in males of *M.rosenbergii.* BC males actively court and protect the female prior to mating while OC males demonstrate only a mild reproductive activity especially in the presence of BC males whereas the SM males practice a form of sneak mating consistent with their smaller size and high mobility. (Kurupup M.B. 1997).

## Spawning

Mating takes place immediately after premating moult in matured female and spawning occurs few hours after mating. The release of eggs from ovary to brood chamber is in a tubular form and it becomes rounded while leaving the genital pore. Since the sperms are already present in brood chamber, fertilization takes place immediately after the eggs are free in brood pouch, which comes in contact with the sperm stored within thoracic legs of the female.

Fertilized eggs are slightly oval in shape with 0.4 to 1.0 mm in diameter and deep yellow in colour. The center of egg is darker than periphery and has a thin transparent membrane covering. First division of nucleus occurs at about 4 hours intervals. Embryo is fully developed in 10-15 days after fertilization. During incubation, the female provides aeration to the eggs till hatching. The deep yellow colour of the barried eggs gradually turn to lighter gray with the progress of embryonic development and finally become light slate gray before hatching. Unfertilized and dead eggs are removed by the female through its first chelate legs. During the incubation period the gonadail maturation takes place further and spawning occurs just after hatching of first zoea larvae. This way, incubation period of eggs lasts between 10-15 days depending upon the water temperature. 28-30° C has been found to be optimal. However, at lower temperature, the incubation period is prolonged to more than 21 days (Kanaujia *et. al.*, 1997).

## Seed collection from natural source

the observations made during the last two to three years show declining trend in the nature seed availability, because of the increased esturine pollution, erratic monsoon pattern and indiscriminate fishing of barried females during breeding season. Though natural seeds are available in maritime states like Gujarat, Maharashtra, W.Bengal and A.P., the technique of seed collection are very primitive.

In Gujarat seeds from natural source can be obtained near Bharuch, from 15th September to 15th March. However, maximum availability of *M.rosenbergii* is from 15th September to 1st December. Thereafter, mixed variety of other fresh water prawns are available which affect the final growth and marketable size group. At present there is no fresh water prawn hatchery in Gujarat. However, Department of Fisheries Govt. of Gujarat is planning to establish fresh water prawn hatchery with French Colaboration.

## Prospect of culture and stocking rate

The ever increasing demand for seed is not possible to be met from existing natural resources. Even for 25% of the available ponds and tanks to be brought under freshwater prawn culture, the seed requirement will be 28115 million, when stocked at 50,000 seeds per hectare per crop in monoculture system, at stocking rate of 15,000 seeds per hectare per crop. (Reddy, 1997). Murthy (1996) have suggested to stock 40,000 to 50,000 juveniles/ ha in semi intensive method. In growing period of 5-6 months majority of prawns may attain more than 5g. However, uneven growth within the same batch is the genetic character of the species which could be minimized by providing additional substrates, sufficient food and with low stocking density. On an average, production varying from one to two tonnes/ ha could be obtained.

## Necessity of hatchery establishment

The scientific farming of fresh water prawns was neglected until now but it has been getting tremendous popularity for last few years. The major constraint in the scientific development of fresh water prawn farming on commercial scale is the non-availability of desired quantity of quality seed. At present, if all existing hatcheries work at their optimum capacity, they may be able to produce about 180 million seeds per annum. There is thus a wide gap between demand and supply. Therefore, it is very essential to establish commercial and backyard hatcheries both in the coastal and inland areas.

## Hatchery management technique

The hatchery management techniques for *M.rosenbergii* are discussed here under:

### *Site selection*

Selection of suitable site is one of the most important factors for the establishment of a prawn hatchery. Commercial prawn hatchery requires coastal sites having freshwater free from pollutants. Infrastructural facilities like availability of electricity, accessibility to road, water supply system, aeration system, powerhouse, pump house etc. should be provided.

### *Brood Stock*

Continuous supply of good quality brooders is one of the prerequisites for the successful operation of a commercial hatchery. Therefore, it is eseential to have brood stock ponds located near by the hatchery. Site which is suitable for carp culture is ideal for construction of brood stock ponds.

To obtain good quality, large size and healthy brooders, prawns are to be fed with variety of feeds such as broken rice, oil cakes, trash fish, bivalve meat, offal of fish and other animals and formulated pelleted feeds twice a day.

### *Water quality*

The optimum desired salinity for better growth and survival for *M.rosenbergii* is 14± 2 ppt. this could be maintained by mixing supernatant water from the fresh water and sea water storage tanks in to a mixing tank. The mixed water is treated with 100 ppm chlorine to kill micro flora and fauna. After 10-12 hours of chlorination. the water should be dechlorinated with sodium thiosulphate. The dechlorinated water is treated with 10 ppm of EDTA to eliminate the dissolved heavy metals, if any. Now this water free from any pollutants should be exchanged at regular intervals so as to provide healthy environment to the growing prawns.

### *Selection of Brooders*

Depending on the colour of eggs barried females are collected from brood stock ponds should be separated in three different groups i.e.

1) The females with orange coloured eggs.
2) The females with brown coloured eggs.
3) The females with gray coloured eggs.

And kept in holding tanks.

### Hatching

The size of hatching tanks of FRP Tanks/plastic pools with 0.5 to 1.0 ton capacity should be thoroughly disinfected and filled with fresh water or brackishwater of 5-6 ppt. The barried females with gray colour eggs are carefully removed from holding tanks and released into the hatching tanks, after having treated with 25 to 30 ppm formaline for about half an hour to avoid transfer of infection along with egg mass (Reddy, 1997).

After hatching, the stage-1 larvae are siphoned into harvesting bucket having 250 micron mesh and washed thoroughly with 0.05 ppm Trefian for about 15-20 minutes to avoid any fungal infection.

### Different parameters for culture

For successful rearing of *M.rosenbergii* larvae, (1) temperature (2) suitable feed and (3) maintenance of water quality in rearing tanks are required. 28°C is considered optimum temperature for larval development. However, the larvae can grow in temperature ranging from 26.5 o 31.5°C.

**Table. 25.1 : Average Farm production (Gujarat)**

| Name of Tank | Area | Total nos Juveniles stocked | Prawn production (Kg.) |
|---|---|---|---|
| Umarwada | 4 ha. | 23,000 | 86.750 |
| Muller | 4 ha. | 35,000 | 299.500 |
| Ambhal | 2 ha. | 20,000 | 71.200 |

Source : Biswas, K.P., 1966.

The results of prawn culture in Gujarat State Fisheries Department tanks gavee an average production of 45-75 Kgs per hectare as detailed below:

**Table. 25.2 : Yield in Monoculture and polyculture with Fresh Water Prawn**

| Sr. No. | Type of culture | Yield | Worker |
|---|---|---|---|
| 1. | Polyculture<br>Prawn (M.r.)+Chinese carp | 322.3 kg Prawns/ha/yr<br>2516 kg fish/ha/yr | Malecha et.al. 1981 |
| 2. | (a) Poly culture Prawn + Common Carp and Silver carp<br>(b) mono culture (M.r.) | 580 kg prawns/ha/yr<br>2225.40 kg fish/ha/yr<br>650 kg/ha/yr | Cohen et.al. (1983) |
| 3. | Monoculture Stocking 30,000/ha (Mr.) | (i) 280-700 kgs/ha in six months | Subamanyam (1984) |
| 4. | Monocultue (M.r.) | (ii) 350-725 kg/ha in 6-7 months | Dept. of Tamil Nadu (1990) |
| 5. | Monoculture (M.r.) | 535 kg/ha to 1125 kg/ha in 9-10 months | Raje and Joshi (1991) |
| 6. | (A) Monoculture (M.r.)<br>(B) Polyculture | 105.8 to 254 kg/ha in 160 days<br>297 to 408 kg/ha in 160 days | Susheel et.al. (1992) |
| 7. | (A) Monoculture (m.r.)<br>(b) Polyculture | 630 kg/ha in 6 months<br>411 kg Prawn/ha<br>840 kg Fish/ha (Rohu, Catla and Silver Carp) in 6 months. | Durairaj et. al.(1992) |
| 8. | (A) Monoculture<br>(B) Polyculture | 2.0 to 2.5 tonnes prawns/ha<br>500 to 600 kg/ha | A.K. Reddy (1997) |

M.r. = Macrobranchium rosenbergii.

**Table. 25.3 : Fresh water prawn landing pattern in Gujarat (grade wise and percentage wise) landing in M.T.**

| Year | Total Inland fish landing | Jambo and medium F. W. prawn | Percentage of total inland landing. | Small prawn | Percentage of total inland fish landing. |
|---|---|---|---|---|---|
| 1984-85 | 23814 | 1583 | 6.65 | 2626 | 9.75 |
| 1985-86 | 25172 | 1756 | 6.98 | 3118 | 12.39 |
| 1986-87 | 24451 | 1264 | 5.17 | 2823 | 11.55 |
| 1987-88 | 22551 | 1202 | 5.33 | 2027 | 8.99 |
| 1988-89 | 22315 | 1723 | 7.72 | 1202 | 5.39 |
| 1989-90 | 27146 | 1545 | 5.69 | 1662 | 6.12 |
| 1990-91 | 45687 | 1249 | 2.74 | 1858 | 4.07 |
| 1991-92 | 39870 | 1873 | 4.70 | 1907 | 4.78 |
| 1992-93 | 51154 | 4971 | 9.72 | 6102 | 11.93 |
| 1993-94 | 65019 | 6196 | 9.53 | 8435 | 12.97 |
| 1994-95 | 70199 | 12152 | 17.33 | 11960 | 17.06 |
| 1995-96 | 60158 | 7584 | 12.61 | 1143 | 01.90 |
| 1996-97 | 65278 | 6458 | 9.89 | 2391 | 03.66 |

Source : Fisheries Statistical informtion. Reprinted with the permission from Commissionerte of Fisheries, Govt. of Gujarat.

It can be observed from Table 25.3 that Fresh water prawn jambo and medium variety compared to total inland landing is between 5.17 to 9.89%. while small size prawn landing is between 4.78 to 12.97%. However, sudden increase in landing could be seen in both size during 1994-95. This my be because virus disease have killed many adult as well as small size prawns. Resulting in very poor 1.9% in small size could be seen which confirms the less breeding.

## *References*


Anon. 1997. Fisheries statistical information. Published by Commissionerate of Fisheries (with permission).

Biswas, K.P. 1996. Text Book of Fish, Fisheries and Technology, Published by Narendra Publishing house Delhi, pp. 150-151.

Brody, T., Cohen, D.A. Barnes and A.Sector (1980). Yield characteristics of the prawn *Macrobranchium rosenbegii* in temperate zone acquaculture. *Aquaculture,* 12:375-385.

Dalal, V.M. and G.T. Pandya. (1973). Some land and freshwater molluscans of Ahmedabad. *Biological* VI : 29-41.

Kuris, A.m., Ralanan, Z., Sagi, A. and Cohen, D. (1987). Morphometric differentiation of male Malaysian Giant Prawn, *M.rosebergii. J.Crustacean. Biol.* 7(1):219-237.

Kurup, Madhusoodana. B. 1997. Dynamics of interaction and transformation of male morphotypes of *Macherobranchium rosenbgii* (De Man) as a decisive factor governing economic yield. Fishing chimes 17 (2) : 34-38.

Murthy, H. Sivanand 1996. Farming of the Giant Fresh Water Prawn *Macrobranchium rosenbergii* in India. Fishing chimes, 16(2):43-47.

Reddy, A.K. 1997. Management of fresh-water prawn hatcheries and culture possibilities. Fishing chimes, 16 (10): 32-38.

Roy, Psupendra Prasad 1996. Present status of Fresh water Prawn *M.rosenbergii* culture. Fishing chimes, 15 (10): 17-22.

Sing, C.S. and V.S. Chandrasekharan. 1991. Prospects of Freshwater Prawn seed production and farming in Tari Region of Kumaun Himalaya. National Seminar on Shrimp Seed Production and Farming, 12-13 Feb.

Sree Prakash. 1989. Status of Fresh water Farming in India. Seafood Export Journal, xxi (7) : 5-14.

# 26

# Aquatic Weed (Marine and Fresh Water)

## Introduction

Seaweed or marine algae are a diverse attractive, interesting and useful group of aquatic plants, which help the ecosystem to thrive. The size of different algae varieties from those that are microscopic to those are 80 to 100 ft like 'Kelp'.

In India, the largest amount of marine algae is observed along the state of Tamil Nadu; Gujarat is second to Tamil Nadu. It is believed that in Gujarat the concentration of marine algae is around the Saurashtra coastline, particularly so in the Gulf of Kutch. Kerala is also one of the important maritime States.

Seaweeds are classfied based on their colour in to three divisions viz. Chlorophyta, Phaeophyta and Rhodophyta. The division Chlorophyta includes 550 species of green algae of which ony ten per cent is found in the sea. This green seaweeds contain pigments like chlorophyll 'a' 'b' 'c' and B-Carotenens, Lutein, Zeaxanthin and siphonoxanthin. Second group is brown seaweeds belonging to Phaeophyta division. The largest seaweeds come under this group. Of the 1500 species of brown algae, almost all varieties (99%) are marine. They contain chlorophyll. 'a' 'c' and B-Carotein, Fucoxanthin, Flavoxanthin, and Violaxanthin. The third division Rhodophyta has a total of about 4000 species of red algae. Of these, about 98% is found in the marine environment. They also contain pigments such as Chlorophyll 'a' and 'd' and B-Carotein, Lutein, Zeaxanthin, r-Phycocyanin and r-Phycoerythin. (Sundararaj and Srikrishnadhas, 2000).

The seaweeds form one of the important marine living resources. They are primitive type of plants growing in the intertidal and subtidal regions of the sea. They flourish wherever rocky or coral substratum is available for their attachment with the help of the rhizoids or holdfast. Depending upon the type of pigment present and the other morphological and anatomical structures, the seaweeds are broadly grouped in to green, brown, red and blue green algae. Seaweeds contain portein, iodine, bromine, Vitamin and certain antibiotic substances.

Seaweeds are used as human food, livestock feed and fertilizer for land crops in many countries. Phycocolloids such as agar-agar and corrangeenan are obtained from red sea weeds. Algin, Mannitol and Iodine are extracted form brown seaweeds. In Japan, Malaysia and Phillippines, seaweed such as *Ulva, Entromorpha; Caulerapa, codium and Monostroma* (Green algae); *Sargassum, hydroclathrus, Laminaria, Undaria, Macrocystis* (brown algae); *Porpyra,*

*Gracilaria, Eucheuma, Laurencia and Acanthophora* (red algae) are consumed as vegetables in salad, soup porridge and pickles. (Chennubhotla *et. al.*, 1991).

About 700 species of marine algae have been reported along the east and west coast of India. From the seaweed resources survey carried out by the CMFRI and other research organizations in different parts of India coast, it is estimated that the total standing crop of all seaweeds in Indian waters is more than one lakh tonnes (wet weight) consisting of 6000 tonnes (wet weight) of agar yielding seaweeds and 16000 tonnes (wet weight) of algine yielding seaweeds. They also contain more than 60 trace elements. (Kaliaperumal, N. and S. Kalimuthu, 1993).

In Gujarat chief varieties of alginophytes are *Sargassum tenorrium, S. cinerious* and *S. swartii.* Annually about 4000 tonnes of fresh Sargassum can be harvested from the Gulf of Kutch. According to Fishermen Co-operative Society, Sikka, in the season about 400 persons are earning livelihood out of seaweed collection and marketing (Dholakia, *et, al.*, 1998).

There are about 60 commercially important species of seaweeds in India. A wide variety of Phycocolloid poly saccharides including agar-agar and algin are extracted from seaweed. Considering the standing crop (of about 1,00,000 tons, the present landing of 17,000 tons forming 11% of the standing stock is very minimal. The edible seaweeds form 70% of the standing stock followed by algin 16% Carrageenan 8%, and agar 6% yielding seaweeds. There are about 60, small scale industries using seaweed as raw material in Gujarat, Tamil Nadu and Kerala, but their capacity utilization round the year is very poor primary due to the non-availability of raw material, lack of adequate demand for the products in the domestic markets and products not meeting standards for export. The edible seaweeds are known to be rich in Protein (20 to 25%), Carbohydrates (16 to 24%), lipids (6 to 11%), Vitamins and aminoacids. The agar yielding red seaweeds are known as agarophytes and brown seaweeds, which contain algin, are known as algiophytes. Seaweed culture has immense potential and scope in Inida. The utility of seaweeds in the purification of "aquaculture effluents" in coastal aquaculture is generally associated. In recent days when much more emphases is given to "aqua-friendly ecosystem" the culture of certain species of seaweed may prove to be the best.

## Importance of Aquatic weeds

Aquatic plants reduce the level of carbon dioxide in water and increase oxygen and utilize metabolities of other aquatic organisms present in water for their nourishments. Thus help to keep a balanced ecosystem. Without aquatic ornamental plants aquarium will not give good and attractive appearance. Aquatic plants can offer an ideal place for fishes to refuge and many fishes prefer aquatic plants and parts suitable for egg laying and building nests. Soft aqautic plants are useful as direct food for herbivorous fishes.

They can be divided in two part (1) Freshwater and (2) Marine water plants.

## Fresh water Plants

Fresh water plants, which are ornamental and submerged, are widely used in aquarium tank. They take up the nitrogenous material from water and increase oxygen level of the water by photosynthesis, thus improve water quality. Details of some of the plants are as under.

## *Vallisneria spiralis* Michex

Common name : Tape plant.

*Vallisneria* is dence plant with number of long, ribbon shaped leaves, steam is runner, found underground bearing a rosette of leaves on the nodes at regular intervals. Male flowers are produced near roots and female flowers are seen with very long and coiled stalk. This is found in tropical and subtropical waters. Because of the structure of the leaf, they are commonly known as tape plant. The colour and shape of the leaf are very attractive and flowers add more to the beauty.

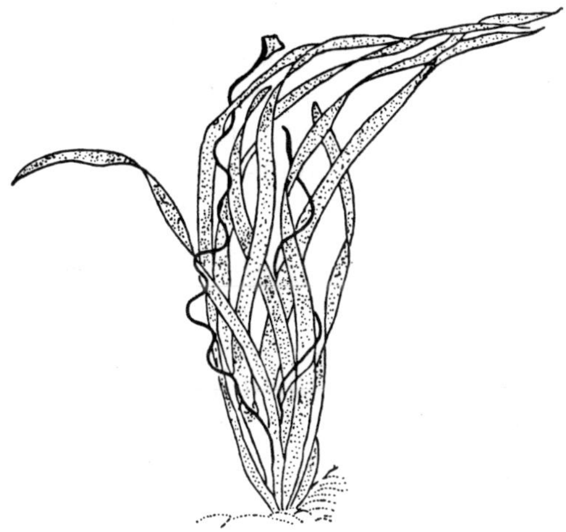

**Fig. 26.1 : *Vallisneria spiralis***

### Biology

Reproduction is mainly vegetative by offshoots. *Vallisneria* requires strong light, neutral pH and soil with mixture of sand and clay for successful propagation. Growth of *Vallisneria* is fast and it can grow up to one meter.

## *Hydrill verticillata*

Common name : Chain weed

**Fig. 26.2 : *Hydrilla verticillata***

### Identification

*Hydrilla* is rooted plant, most suited for aquarium. The leaves grow to a size of about 2 cm and are found in whorls to 3-6 cm. The stem is round, branched and elongated. The roots are fibrous. This plant is found distributed in the tropical and subtropical waters. It grows submerged and mostly found with floating branches. In nature, it is also useful for the fishes to hide. It is an excellent feed for some herbivorous fishes.

### Biology

*Hydrilla* can tolerate wide fluctuations in the growing medium. Reproduction is mainly through vegetative propagation. It is very easy to culture as it is fast growing species.

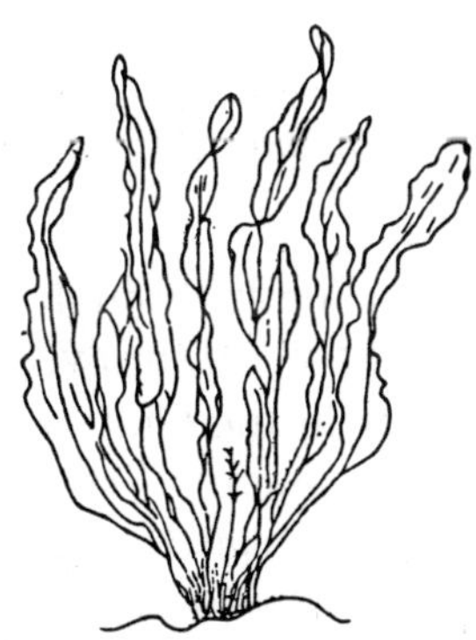

**Fig. 26.3 : *Echinoddorus paniculatus***

## *Echinodorus paniculatus*

Common name : Amazon sward plant.

This is submerged aquatic plant found in

tropical fresh water. Aquarium lovers know this species as Amazon sward plant. It is very beautiful aquarium plant with bright green colour. The leaves are tufted and lanceolate with veins running the whole length. Stem is a runner and the roots are very small.

## Biology

Reproduction is by vegetative propagation through the off shoot which produces plantlets that grow independently after some time. It requires clear water and good light. Growth rate is moderate compared to other aquatic plant. Substratum must be of coarse sand with rich nutrients.

Important seaweeds are as under.

Fig. 26.4 : *Sargassum weightii*

## *Sargassum weightii* Greville

## Identification

Large brown seaweed with higher plants like organization. The branches are flattened and somewhat smooth. Vesicles are separated and are oval shape. Air bladders are single and large. Receptacles are branched and found in clusters. Its habitat is eulittoral and sublittoral. It is most important in tropical reef beds.

## Biology

The reproduction is with oospores. Surgassum grows to a height of 50 cms in 40 days and becomes ready for harvest. Table 26.8 gives the details of algin obtain and viscosity from different species of *Sargassum* with maximum availability in months (Bhandari 81-82).

## Economical Importance

Fig. 26.5 : *Ulva lectuca*

## *Ulva lactuca* Linnaeas

This is a green seaweed with two cell thick thallus. Fround are delicate, flat, foliaceous and grows as large sheets.

## Biology

Reproduction is through vegetative propagation. If culture is done, the propagation is about 2 Kgs (dry wt)/m$^2$/year.

Seaweeds in the Gulf of Kutch was listed by Chhaya (1997), which is given in Table 26.1. Mineral contents like Magnesium, Potassium, Calcium, Zinc, Copper, Nitrogen of seaweed *Sargasum spp.* Using Atomic Absorption spectrophotometer was done by Bio-Science Deparment, Saurashtra University, Rajkot, which is shown in Table 26.2. Mineral constituents of about 23 seaweed was analyzed by Chenaubhatla, et. al., (1991) of CMFRI in mg/100 mg of dry sea weed is reprinted in the Table 26.3. They have given chemical composition like Protein, Carbohydrate, Lipid, lodine, Bromine, Ascorbic Acid, of about 94 seaweeds of India, which is reprinted as Table 26.4. They have also analyzed 12 Agar and 12 Agaroid yielding seaweeds of India. They also informed about availability of such seaweeds in India. Consolidate information is reprinted as Table 26.5. They have also analyzed 24 Algin yielding seaweeds of India, which is reprinted as Table 26.6. Bhandari (1981-82) have found out algin percentage and viscosity of different *Sargasum Spp.* During the month of November and January, which is given in Table 26.7. Kalipaperumal and Kalimuthu (1993) have prepared consolidate information of occurrence of Agar, Carrageen and algin yielding seaweeds of India, which is reprinted as Table 26.8.

Seaweeds have many industrial uses like Food and bakery, Cosmetic, Textile, Rubber Pharmaceuticals, Medical, Dental, Paper Industry. Utilization of seaweed in such industry is consolidated and given in Table 26.9.

**Table 26.1 : List of 209 seaweeds in the Gulf of Kutch of Gujarat**

| | | | |
|---|---|---|---|
| ***Chlorophycea*** | | | |
| 1. | *Enteromorpha clathrata* | 2. | *Enteromorpha flexuosa* |
| 3. | *Enteromorpha gujaratensis* | 4. | *Enteromorpha intestinalis* |
| 5. | *Enteromorpha linza (bicomuta)* | 6. | *Enteromorpha plumosa* |
| 7. | *Enteromorpha prolifera* | 8. | *Enteromorpha Ovata* |
| 9. | *Enteromorpha tubulosa* | 10. | *Ulva beytensis* |
| 11. | *Ulva lactusa* | 12. | *Ulva fasciata* |
| 13. | *Ulva renticulata* | 14. | *Ulva rigida* |
| 15. | *Chaetomorpha antennina* | 16. | *Cladophora facicularis* |
| 17. | *Cladophora crystallina* | 18. | *Cladophora expansa* |
| 19. | *Spongomorpha indica* | 20. | *Willella ordinata* |
| 21. | *Ectohaete leptochaete* (endoderma) | 22. | *Bryopsos plumosa* |
| 23. | *Bryopsis hypnoides* | 24. | *Halicystic borergesenii* |
| 25. | *Caulerpa fastigiata* | 26. | *Caulerpa freycinetii* |
| 27. | *Caulerpa peltata* | 28. | *Caulerpa racemosa* |
| 29. | *Caulerpa scalpelliformis* | 30. | *Caulerpa sertularioides* |
| 31. | *Caulerpa taxofolia* | 32. | *Caulerpa verticillata* |
| 33. | *Caulerpa veravalensis* | 34. | *Acetabularia calyculus* |
| 35. | *Neomeris annulata* | 36. | *Avrainvillia nigricus* |
| 37. | *Codium dwarkense* | 38. | *Codium iyengari* |
| 39. | *Halimeda tuna* | 40. | *Halimeda macroloba* |
| 41. | *Udotea indica* | 42. | *Boodlea composita* |
| 43. | *Chamaedoris auriculata* (auriculata) | 44 | *cladophoropsis sundanesis* (roundansos) |
| 45. | *Cladophoropsis zollingeri* | 46. | *Dictyophaeria covernosa* |
| 47. | *Boergensenia forbessi* | 48. | *Valoniopsis pachynema* |
| 49. | *Valonia aegagrophila* | 50. | *Valonia utricularis* |
| 51. | *Ostroebium remenkei* | 52. | *Vaucheria piloboloides* |
| ***Phaeophycea*** | | | |
| 53. | *Ectocarpus arabicus.* | 54. | *Ectocarpus battersii* |
| 55. | *Ectocarpus columellaris* | 56. | *Ectocarpus elachistae formis* |
| 57. | *Ectocarpus oralis* | 58. | *Ectrocarpus rhpdpchortonoides* |
| 59. | *Giffordia mitchellae* | 60. | *Giffordia sandrina* |
| 61. | *Streblonema turmala* | 62. | *Spacelaria furcigera* |
| 63. | *Spacelaria tribuloides* | 64. | *Dictyopteris acrostichoides* |
| 65. | *Dictyoperis australis* | 66. | *Dictyopteris woodwardii* |

**contd...**

**Table 26.1 *Contd...***

| | | | |
|---|---|---|---|
| 67. | *Dectyota atomaria* | 68. | *Dectyota dichotoma* |
| 69. | *Dectyota rivaricata* | 70. | *Dectyota bartayresiana* |
| 71. | *Dectyota carvicornis* | 72. | *Dectyota ceylonica* |
| 73. | *Dectyota ciliata* | 74. | *Myriactula arabica* |
| 75. | *Dilophus fasciola* | 76. | *Padina gymnospora* |
| 77. | *Padina tetrastomatica* | 78. | *Padina crassia* |
| 79. | *Pocockiella veriegata* | 80. | *Spatoglossium asperum* |
| 81. | *Spatoglossium variabile* | 82. | *Stoechospermum marginatum* |
| 83. | *Zoonaria variegata* | 84. | *Namacystus deepiens* |
| 85. | *Colpomenia sinuosa* | 86. | *Hydroclathrus clathratus* |
| 87. | *Iyengaria stellata* | 88. | *Cystoseria indica* |
| 89. | *Levringia borgensenii* | 90. | *Rosenvingea indricata* |
| 91. | *Cystoceria trinodis* | 92. | *Hermophysa triquentra* |
| 93. | *Sargassum carpophyllum* | 94. | *Sargassum cinetum* |
| 95. | *Sargassum sinereum* | 96. | *Sargassum licifolium* |
| 97. | *Sargassum jhonstonii* | 98. | *Sargassum merifieldii* |
| 99. | *Sargassum plagiophyllum* | 100. | *Sargassum prismaticum* |
| 101. | *Sargassum swartzii* | 102. | *Sargassum tenerrium* |
| 103. | *Sargassum vulgre* | 104. | *Turbinaria omata* |
| 105. | *Turbinaria indica* | | |
| ***Rhodophcea*** | | | |
| 106. | *Goniotrichum flegans* | 107. | *Erythrotrychin carnea* |
| 108. | *Acrocheatium carnea* | 109. | *Acrocheatium crassipes* |
| 110. | *Acrocheatium dwarkense* | 111. | *Acrocheatium erectum* |
| 112. | *Helminthocladia australis* | 113. | *Helminthocladia calvadosii* |
| 114. | *Liagara ceranoides* | 115. | *Actinotarichia fragilis* |
| 116. | *Galaxaura oblongata* | 117. | *Psendoglophloea fasclucularis* |
| 118. | *Scinaia furcellata* | 119. | *Scinaia leotei* |
| 120. | *Scinaia indica* | 121. | *Asparagopsis taxiformis* |
| 122. | *Gelidium puillum* | 123. | *Gelidiella acerosa* |
| 124. | *Gelidiella myrioclada* | 125. | *Gelidiella tenuissima* |
| 126. | *Peyssonella obscura* | 127. | *Amphiroa anceps* |
| 128. | *Amphiroa fragilissima* | 129. | *Athricardia rubens.* |
| 130. | *Athrocardia capensis* | 131. | *Jania rubens* |
| 132. | *Gyptonemia undulata* | 133. | *Gratelopia indiaca* |
| 134. | *Holymenia porphyroides* | 135. | *Holymenia venusta* |
| 136. | *Corynomorpha prismatica* | 137. | *Sebdenia pilydactyla* |
| 138. | *Gelidiopsis variabilis* | 139. | *Gracilnoa cuculia* |
| 140. | Gracilona corticata | 141. | *Gracilona fatiifera* |
| 142. | *Gracilona verrucosa* | 143. | *Gracilona debilis* |
| 144. | *Caulacanthas ustulatus* | 145. | *Agardhiela tenera* |
| 146. | *Agardhiella tenera* | 147. | *Eucheuma spinosum* |
| 148. | *Meristotheca papulosa* | 149. | *Sarconema filiforme* |
| 150. | *Sarconema furcetum* | 151. | *Sarconema surcellatum* |
| 152. | *Solieria robusta* | 153. | *Catenella impudica* |
| 154. | *Hypnea cervicornis* | 155. | *Hypnea musciformis* |
| 156. | *Botryocladia leptopoda* | 157. | *Colearthrum muelleri* |
| 158. | *Colearthrum opuntia* | 159. | *Rhodymennnia australis* |
| 160. | *Rhodymenia dissecta* | 161. | *Rhodymenia palmala* |

***contd...***

**Table 26.11 *Contd...***

| | | | |
|---|---|---|---|
| 162. | *Champia indica* | 163. | *Champia somalensis* |
| 164. | *Champia globulifera* | 165. | *Gastrodonium iyengarii* |
| 166. | Alwothamnion byssoides | 167. | *Centroceros clavulatum* |
| 168. | *Ceramium truncatum* | 169. | *Griffithsia rhizophora* |
| 170. | *Griffithsia tenuis* | 171. | *Holoplegma duperreyi* |
| 172. | *Spyridia aculeata* | 173. | *Spyridia fliamentosa* |
| 174. | *Caloglossa lepieurii* | 175. | *Hypolossum spathulatum* |
| 176. | *Menbraboptera murrayi* | 177. | *Myriogramme bombayensis* |
| 178. | *Mariogramme okhensis* | 179. | *Plastisiphonia miniata* |
| 180. | *Teanioma perpusillum* | 181. | *Acanthophora delilei* |
| 182. | *Acanthophora dendrodes* | 183. | *Acanthophora muscoides* |
| 184. | *Bostrychia tenella* | 185. | *Chondria armata* |
| 186. | *Chondria cornuta* | 187. | *Chindria dasyphylla* |
| 188. | *Layrencia cruciata* | 189. | *Laurencia obtusa* |
| 190. | *Laurencia pamosa* | 191. | *Laurencia pedicularioides* |
| 192. | *Loveillea lungcrmannoides* | 193. | *Lophocladia lallemandi* |
| 194. | *Lynkiella kuruvalensis* | 195. | *Polysiphonia bharadwaje* |
| 196. | *Polysiphonia coacta* | 197. | *Polysiphonia ferulacea* |
| 198. | *Polysiphonia kappannai* | 199. | *Polysiphonia parthasarathyal* |
| 200. | *Polysiphonia platycarpa* | 201. | *Polysiphonia tuticorinensis* |
| 202. | *Polysiphonia unguiformis* | 203. | *Polysiphonia variegata* |
| 204. | *Rochera condensis* | 205. | *Roschere glomerulata* |
| 206. | *Spirocladia baroddensis* | 207. | *Dasya flageliformis* |
| 208. | *Heterosiphonia muelleri* | 209. | *Heterosiphonia wurdemanni* |

*Source* : Chhaya, N.D. 1997, Minding our Marine Wealth. An Apprisal of Gujarat's Coastal Resources.

**Table : 26.2 : Mineral contents of sea weed *Sargasum Sp.* Gujarat.**

| Mineral | Mean of three readings | |
|---|---|---|
| | mg | (Dry weight base) |
| Magnesium | (Mg) | 0.92 |
| Potassium | (K) | 4.02 |
| | Calcium | (Ca) 13.84 |
| Zinc | (Zn) | 0.08 |
| Iron | (Fe) | 0.295 |
| Copper | (Cu) | 0.0168 |
| Total Nitrogen | (N) | 21.00 |

This analysis was done by Atomic Absorption Spectrophotometer.

Source : Department of Bio-Sciences Saurashtra University, Rajkot. (Gujarat). 1996.

## Estimated Harvestable Standing Crop of seaweed in India

Though no overall picture of the estimates of standing crop of marine seaweed of India or each maritime state is available, numerous survey of selected areas of some states was carried out by different workers and estimated stock. This may not give true and total picture. However, it helps to indicate about the general picture. The estimates may vary but are good indicators. Desai (1965) and Kaliaperumal et al., (1987) have estimated harvestable standing corp of seaweed in India's mentioned hereunder.

**Table No. 26.3 : Mineral constituents in seaweed (mg/100 g of dry weed)**

| Sl. No. | Seaweed | Sodium | Potassium | Calcium | Magnesium | Chloride | Nitrogen | Sulphate | Iron | Copper | Manganese | Boron | Zince | Phosphorus |
|---|---|---|---|---|---|---|---|---|---|---|---|---|---|---|
| | ***Green Algae*** | | | | | | | | | | | | | |
| 1. | *Enteromorpha intestinalis* | 1160.0 | 710.0 | 510.0 | 410.0 | 2400.0 | 380.0 | 4000.0 | 14.0 | 0.3 | 13.0 | 0.6 | 4.4 | – |
| 2. | *Ulva lactuca* | 1710.0 | 1580.0 | 630.0 | 1640.0 | 790.0 | – | 12100.0 | 0.4 | 0.9 | 8.2 | 15.6 | 0.7 | 277.6 |
| 3. | *U. rigida* | 1110.0 | 680.0 | 340.0 | 980.0 | 270.0 | – | 7740.0 | 257.2 | 4.7 | 38.4 | 10.0 | 1.6 | 286.3 |
| 4. | *Chaetomorpha linoides* | – | – | – | – | – | – | – | 21.7 | 0.5 | 38.5 | 0.4 | 3.0 | – |
| 5. | *Cladophora fascicularis* | 570.0 | 3590.0 | 520.0 | 70.0 | 2900.0 | – | 2410.0 | 144.5 | 0.5 | 6.2 | 23.5 | 2.3 | 116.2 |
| 6. | *Codium dwarkense* | 10740.0 | 2350.0 | 1190.0 | 180.0 | 15630.0 | – | 5990.0 | 60.6 | 0.7 | 2.3 | 1.1 | 2.0 | 205.7 |
| 7. | *Boodlea composita* | 4820.0 | 4090.0 | 410.0 | 120.0 | 5190.0 | – | 4430.0 | 468.6 | 1.1 | 17.6 | 4.5 | 1.9 | 258.4 |
| | ***BROWN ALGAE*** | | | | | | | | | | | | | |
| 8. | *Padina australis* | 1280.0 | 930.0 | 500.0 | 500.0 | 2400.0 | 600.0 | 1800.0 | 50.4 | 1.1 | 45.0 | 1.1 | 4.4 | – |
| 9. | *P. gymnospora* | | | | | | | | | | | | | |
| 10. | *Colpomenia sinuosa* | 1400.0 | 1060.0 | 160.0 | 20.0 | 870.0 | – | 1390.0 | 456.1 | 2.0 | 24.8 | 3.2 | 3.5 | 28.6 |
| 11. | *Rosenvingea intricata* | 560.0 | 8850.0 | 120.0 | 40.0 | 530.0 | – | 1330.0 | 249.7 | 1.5 | 0.1 | 4.0 | 0.1 | 98.4 |
| 12. | *Cystoseira* sp. | – | – | – | – | – | – | – | 22.4 | 0.5 | 57.5 | 0.7 | 3.2 | – |
| 13. | *Sargassum* | 1200.0 | 1250.0 | 20.0 | 20.0 | 840.0 | – | 2540.0 | 30.1 | 0.1 | 13.8 | 2.6 | 0.7 | 198.0 |
| | *cinereum* var. *berberifolia* | 1670.0 | 7350.0 | 20.0 | 80.0 | 7200.0 | – | 1500.0 | 224.1 | 1.5 | 4.2 | 0.2 | 1.1 | 3.0 |
| 14. | *S. jonhstonii* | 1470.0 | 1670.0 | 20.0 | 10.0 | 1390.0 | – | 1820.0 | 107.4 | 0.6 | 9.1 | 1.6 | 2.1 | 203.6 |
| | ***RED ALGAE*** | | | | | | | | | | | | | |
| 15. | *Porphyra vietnamensis* | 5660.0 | 1110.0 | 300.0 | 450.0 | 3580.0 | – | 110.0 | – | – | – | – | – | – |
| 16. | *Gelidiella acerosa* | 80.0 | 20.0 | 280.0 | 70.0 | 90.0 | 1340.0 | 730.0 | – | – | – | – | – | – |
| 17. | *Gracilaria edulis* | 1230..0 | 2010.0 | 570.0 | 160.0 | 3840.0 | 2140.0 | 4500.0 | 28.0 | 1.0 | 55.0 | 1.4 | 8.3 | – |
| 18. | *Sarconema filiforme* | – | – | – | – | – | – | – | 19.6 | 0.7 | 18.7 | 0.7 | 6.4 | – |
| 19. | *S. furcellatum* | 560.0 | 400.0 | 510.0 | 410.0 | 2400.0 | 930.0 | 2900.0 | 14.0 | 3.0 | 39.0 | 0.9 | 5.8 | – |
| 20. | *Hypnea musciformis* | – | – | – | – | – | – | – | 28.0 | 0.9 | 19.5 | 0.8 | 8.0 | – |
| 21. | *Acanthophora spicifera* | 320.0 | 320.0 | 420.0 | 380.0 | 3060.0 | 740.0 | 2000.0 | 28.0 | 1.2 | 8.5 | 0.4 | 7.0 | – |
| 22. | *Chondria dasyphylla* | – | – | – | – | – | – | – | 30.8 | 0.9 | 17.5 | 0.9 | 6.8 | – |
| 23. | *Laurencia papillosa* | 1160.0 | 820.0 | 610.0 | 310.0 | 2400.0 | 1000.0 | 380.0 | 37.8 | 0.5 | 24.0 | 0.5 | 5.5 | – |

### *(1) Gujarat*

| | |
|---|---|
| (A) Alginophytes | (*) 200 Tonnes wet harvested around Sikka. |
| | (*) 500 Tonnes wet algae per Sq. mile harvastable from Jamnagar coast every year. |
| | (*) 50,000 Tonnes of wet algae can be harvested from Okha-Porbandar-Veravai area including islands. |
| (*) and (**) | 10,000 Tonnes in the Gulf of Kutch, |
| (B) Agarophytes (*) and (**) | 205 Tonnes wet in the Gulf of Kutch |

### *(2) Tamil Nadu*

| | |
|---|---|
| (A) Alginophytes | (*) 6,000 Tonnes wet wt harvestable per year from point Calimere. |
| | (*) 75 Tonnes wet wt washed ashore near Rameshwaram. |
| | (*) 100 Tonnes wet wt around Thirunelveli-Tanjor |
| | (**) 1,31,588 Tonnes wet wt in 234.35 Km area Dhanushkhoti Hare Island |
| (B) Agrophytes | (**) 3,300 Tonnes at in 30 mile area Rameshwaram and Kilakarani |
| | (**) 70,454 Tonnes wet wt in 234.25 sq Km area Dhanushkhoti Hare Island |
| | (**) 1,31,588 Tonnes wet wt in 234.25 Km area Dhanushkhoti Hare Island |

### *(3) Maharashtra*

| | |
|---|---|
| (**) | 300 wet wt |

Source : (*) Desai, B.N. 1965. (**) Kaliaperumal, N., V.S.K. Chennubhotla and S. Kalimuthu, 1987.

## Sea weeds in the Gulf of Kutch of Gujarat

Over 200 different species of sea weed have been reported in Gujarat. Sea weeds or marine algae are diverse, attractive, interesting and useful group of aquatic plants which help the ecosystems to thrive. The size of different algae varies from those that are microscopic to those which are 80 to 100 feet along like kelp. (Chhaya, 1997).

**Table 26.4 : Chemical constituents in seaweeds**

| Sl. No. | Seaweed | Protein (%) | Carbo-hydrate(%) | Lipid (%) | Iodine (mg/ 100 g dry weed) | Bromine (%) | Ascrobic acid (mg/ 100 g fresh weed) |
|---|---|---|---|---|---|---|---|
| **Green Algae** | | | | | | | |
| 1. | *Enteromorpha compressa* | 23.8 | 24.8 | 11.4 | – | – | – |
| 2. | *E. intestinalis* | – | – | – | 58.00 | – | – |
| 3. | *E. prolifera* | – | – | – | – | – | 0.22 |
| 4. | *Enteromorpha* sp. | – | – | – | 4.16 | 0.032 | – |
| 5. | *Ulva fasciate* | 25.5 | – | – | 7.40 | – | – |
| 6. | *U. lactuca* | 25.8 | 16.0 | 7.4 | 6.27 | – | 6.10 |

*contd...*

**Table 26.4 *Contd...***

| Sl. No. | Seaweed | Protein (%) | Carbo-hydrate(%) | Lipid (%) | Iodine (mg/ 100 g dry weed) | Bromine (%) | Ascrobic acid (mg/ 100 g fresh weed) |
|---|---|---|---|---|---|---|---|
| 7. | *U. reticulata* | 24.4 | 16.9 | 8.6 | – | – | 5.69 |
| 8. | *U. rigida* | 22.4 | – | – | 4.83 | – | – |
| 9. | *Chaetomorpha aerea* | 10.1 | 31.5 | 8.6 | – | – | – |
| 10. | *C. antennia* | 19.7 | 27.0 | 11.5 | – | 0.105 | – |
| 11. | *C. brachygonia* | – | – | – | – | – | 5.92 |
| 12. | *C. linoides* | 16.7 | 27.0 | 12.1 | 72.00 | – | – |
| 13. | *Cladophora expansa* | – | – | – | 18.06 | – | – |
| 14. | *C. fascicularis* | 16.3 | 49.5 | 15.7 | 64.64 | – | – |
| 15. | *C. fritschii* | – | – | – | – | – | 6.04 |
| 16. | *Clasophora* sp. | 9.2 | 6.6 | 6.5 | 18.83 | 0.024 | – |
| 17. | *Bryopsis plumosa* | 19.2 | 27.0 | 9.0 | – | – | – |
| 18. | *Caulerpa chemnitzia* | 11.8 | 32.6 | 11.4 | – | – | – |
| 19. | *C. cupressoides* | 7.4 | 51.8 | 11.0 | – | – | – |
| 20. | *C. fergusonii* | 7.8 | 23.6 | 7.2 | – | – | – |
| 21. | *C. laetevirens* | 8.8 | 56.3 | 8.8 | – | – | – |
| 22. | *C. peltata* | 24.4 | 45.0 | 11.4 | – | – | – |
| 23. | *C. racemosa* var. *macrophysa* | 24.8 | 33.8 | 10.6 | – | 0.130 | – |
| 24. | *C. scalpelliformis* | 25.2 | 10.7 | 7.6 | – | – | – |
| 25. | *C. sertularioides* | 22.7 | 49.5 | 7.0 | – | 0.027 | – |
| 26. | *C. taxifolia* | 23.6 | 9.7 | 4.1 | – | – | – |
| 27. | *Codium adhaerens* | 7.3 | 40.5 | 7.4 | – | – | – |
| 28. | *C. decorticatum* | 6.9 | 50.6 | 9.0 | – | – | – |
| 29. | *C. dwarkense* | 7.2 | – | – | 5.31 | – | – |
| 30. | *C. tomentosum* | 5.1 | 29.3 | 7.2 | – | – | – |
| 31. | *Halimeda macroloba* | 5.4 | 32.6 | 9.1 | – | – | – |
| 32. | *H. tuna* | – | – | – | 31.30 | – | – |
| 33. | *Udotea indica* | 13.0 | – | – | 215.30 | – | – |
| 34. | *Boodlea composita* | 10.3 | – | – | 29.77 | – | – |
| 35. | *Chaemodoris auriculata* | 13.7 | – | – | 10.43 | – | – |
| 36. | *Ceadophoropsis zollingerii* | 10.3 | 0.3 | 0.5 | – | – | – |
| 37. | *Dictyosphaeria cavernosa* | 6.1 | 42.8 | 10.5 | – | – | – |
| 38. | *Microdictyon agardhianum* | 20.9 | 27.0 | 9.4 | – | – | – |
| 39. | *Boergesenia forbesii* | 7.4 | 21.4 | 11.4 | – | – | – |
| 40. | *Valonipsis pachymema* | 18.8 | 31.5 | 9.1 | – | – | – |
| ***BROWN ALGAE*** | | | | | | | |
| 41. | *Dictyopteris australis* | 8.1 | – | – | 23.48 | 0.039 | – |
| 42. | *Dictyopteris* sp. | – | – | – | 25.81 | – | – |
| 43. | *Dictyota bartayresiana* | – | – | – | – | 0.015 | – |
| 44. | *D. dumosa* | – | – | – | – | 0.022 | – |
| 45. | *Padina australis* | – | – | – | 500.00 | – | 7.86 |
| 46. | *P. gymnospora* | 13.0 | 13.2 | 1.3 | 7.95 | – | – |
| 47. | *P. tetrastromatica* | – | – | – | – | 0.022 | – |
| 48. | *Spatoglossum asperum* | – | – | – | – | 0.055 | – |
| 49. | *S. variabile* | 15.7 | – | – | 16.44 | – | – |

*contd...*

**Table 26.4 Contd...**

| Sl. No. | Seaweed | Protein (%) | Carbo-hydrate(%) | Lipid (%) | Iodine (mg/ 100 g dry weed) | Bromine (%) | Ascrobic acid (mg/ 100 g fresh weed) |
|---|---|---|---|---|---|---|---|
| 50. | *Stoechospermum marginatum* | 14.9 | 15.4 | 3.7 | 5.44 | – | – |
| 51. | *Levringia borgensenii* | – | – | – | 104.50 | – | – |
| 52. | *Colpomenia sinuosa* | 6.6 | – | – | 8.99 | – | – |
| 53. | *Cystoseira* sp. | 11.2 | – | – | 34.19 | – | – |
| 54. | *Hormophysa triquetra* | 16.6 | 3.3 | 0.6 | – | – | – |
| 55. | *Sargasum cinereum* var. *berberifolia* | 9.6 | – | – | 33.20 | – | – |
| 56. | *S. ilicifolium* | 15.1 | 24.0 | 1.1 | – | – | – |
| 57. | *S. johnstonii* | 10.9 | – | – | 39.80 | – | – |
| 58. | *S. myriocystum* | 15.6 | 23.8 | 0.5 | – | – | 66.60 |
| 59. | *S. swartzii* | – | – | – | 28.18 | – | – |
| 60. | *S. tenerrimum* | 12.1 | – | – | 37.21 | 0.040 | – |
| 61. | *S. vulgare* | – | – | – | 29.29 | – | – |
| 62. | *S. wightii* | 16.3 | 24.9 | 1.2 | – | – | – |
| 63. | *Turbinaria conoides* | 15.2 | 14.0 | 3.6 | – | – | – |
| **RED ALGAE** | | | | | | | |
| 64. | *Porphyra* sp. | 16.0 | – | – | – | – | – |
| 65. | *Scinaia indica* | 12.5 | – | – | 5.62 | – | – |
| 66. | *Asparagopsis taxiformis* | 16.2 | – | – | 499.30 | – | – |
| 67. | *Asparagopsis* sp. | – | – | – | 556.70 | – | – |
| 68. | *Gelidiella acerosa* | 8.8 | 57.0 | 3.6 | 54.00 | – | – |
| 69. | *Chondrococcus* sp. | – | – | – | – | 0.054 | – |
| 70. | *Amphiroa anceps* | – | – | – | 5.15 | – | – |
| 71. | *Jania rubens* | 1.5 | 1.8 | 0.4 | – | – | – |
| 72. | *Grateloupia lithophila* | 5.8 | 36.9 | 0.7 | – | – | – |
| 73. | *Halymenia venusta* | – | – | – | 25.00 | – | – |
| 74. | *Gracilaria corticata* | 6.1 | 45.5 | 6.0 | 18.41 | 0.078 | – |
| 75. | *G. crassa* | 4.3 | 30.4 | 0.9 | – | – | – |
| 76. | *G. edulis* | 7.6 | 45.8 | 2.4 | 208.00 | – | 7.25 |
| 77. | *G. foliifera* | – | – | – | 8.07 | – | – |
| 78. | *Agardhiella tenera* | – | – | – | 12.65 | – | – |
| 79. | *Sarconema filiforme* | – | – | – | 107.00 | – | – |
| 80. | *S. furcellatum* | – | – | – | 357.00 | – | – |
| 81. | *Soleria robusta* | – | – | – | 15.54 | – | – |
| 82. | *Hypnea musciformis* | – | – | – | 100.00 | 0.027 | 8.58 |
| 83. | *H. valentiae* | 6.1 | 37.8 | 6.1 | – | – | – |
| 84. | *Hypnea* sp. | 7.5 | – | – | – | – | – |
| 85. | *Centroceras clavulatum* | 20.1 | 4.8 | 3.4 | 20.79 | 0.063 | – |
| 86. | *Acanthophora delile* | – | – | – | 5.78 | – | – |
| 87. | *A. muscoides* | 21.8 | – | – | – | – | – |
| 88. | *A. spicifera* | 4.8 | 29.7 | 0.5 | 90.00 | 0.095 | 4.00 |
| 89. | *Chondria armata* | – | – | – | – | 0.400 | – |
| 90. | *Laurencia papillosa* | 4.3 | 11.6 | 0.6 | 137.00 | – | 5.92 |
| 91. | *Polysiphonia ferulacea* | – | – | – | 39.06 | – | – |
| 92. | *Polysiphonia* sp. | – | – | – | 4.78 | – | – |
| 93. | *Heterosiphonia muelleri* | – | – | – | 10.01 | – | – |
| 94. | *Corallina* sp. | – | – | – | – | 0.020 | – |

**Table 26.5 : Agar and agaroid yielding seaweeds of India**

| Sl. No. | Seaweed | Yield (%) | Gel strength ($g/cm^2$) 1.5% conc. | Gelling temp. (°C) 1.5% conc. | Melting temp. (°C) 1.5% conc. | Places of occurrence |
|---|---|---|---|---|---|---|
| | ***AGROPHYTES*** | | | | | |
| 1. | *Gelidiella acerosa** | 50.8 | 325 | 52 | 99 | Okha, Dwarka, Porbandar, Diu, Veraval, Manapad, Tuticorin, Mandapam, Lakshadweep and Andaman-Nicobar. |
| 2. | *G. indica* | 44.0 | 30 | 52 | 76 | Muttam, Kovalam, Idinthakarai, Manapad and Tiruchendur |
| 3. | *Gracilaria edulis** | 45.0 | 139 | 57 | 99 | Tuticorin, Mandapam, Lakshadweep and Andaman Nicobar |
| 4. | *G. crassa** | 23.0 | 140 | 48 | 84 | Tuticorin, Mandapam and Andaman-Nicobar |
| 5. | *G. arcuata* var. *arcuata* | 48.0 | 67 | 52 | 98 | Kanyakumari, Tiruchendur Tuticorin, Mandapam and Lakshadweep. |
| 6. | *G. corticata* var. *Corticata** | 45.0 | 134 | 49 | 68 | Dwarka, Mumbai, Karwar, Goa, Quilon, Vizhinjam, Muttam, Idinthakarai, Manapad, Tiruchendur, Tuticorin, Mandapam, Mahabalipuram, Chennai, Visakhapatnam and Andaman Nicobar |
| 7. | *G. corticata* var. cylindrica | 43.0 | 15 | 45 | 98 | Idinthakarai, Manapad, Tuticorin, Mandapam and Andaman Nicobar. |
| 8. | *G folifera* | 50.0 | 55 | 41 | 68 | Gopnath, Okha, Mumbai, Muttam, Kovalam, Idinthakarai, Tuticorin and Mandapam. |
| 9. | *G. obtusa* | 35.0 | 19 | 22 | 38 | Muttam, Kovalam, Idinthakarai, Tiruchendur, Manapad, Tuticorin and Mandapam |
| 10. | *G. verrucosa** | 23.0 | 41 | 40 | 55 | Okha, Mumbai, Goa, Tuticorin, Mandapam, Chennai, Pulicat, Chilka and Andaman-Nicobar |
| 11. | *Gelidium pusillum* | 50.0 | 276 | 38 | 86 | Dwarka, Porbandar, Veraval, Mumbai, Karwar, Cannanore, Quilon, Muttam, Kovalam, Idinthakarai, Tiruchendur, Mandapam, Chennai, Visakhapatnam, Lakshadweep and Andaman-Nicobar. |
| 12. | *Pterocladia heteroplatos* | 35.0 | 288 | 38 | 83 | Mumbai, Mandapam, Visakhapatnam, Lakshadweep and Andaman-Nicobar. |
| | ***AGAROIDOPHYTES*** | | | | | |
| 1. | *Hypriea musciformis** | 51.6 | 75 | – | – | Gopnath, Okha, Dwarka, Mumbai, Goa, Karwar, Tuticorin, Mandapam, Visakhapatnam, Lakshadweep and Andaman Nicobar. |
| 2. | *H. valentiae** | 39.0 | 151 | – | – | Mumbai, Muttam, Kovalam, Idinthakarai, Manapad, Tuticorin, Mandapam and Lakshadweep. |

***contd...***

**Table 26.5 *Contd...***

| Sl. No. | Seaweed | Yield (%) | Gel strength ($g/cm^2$) 1.5% conc. | Gelling temp. (°C) 1.5% conc. | Melting temp. (°C) 1.5% conc. | Places of occurrence |
|---|---|---|---|---|---|---|
| 3. | *Acanthophora spicifera** | 120 | – | – | – | Okha, Mumbai, Goa, Karwar, Kovalam, Idinthakarai, Manapad, Tiruchendur, Tuticorin, Mandapam, Porto Novo, Visakhapatnam and Lakshadweep |
| 4. | *Laurencia papillosa* | 19.0 | – | – | – | Okha, Mumbai, Manapad, Tiruchendur, Tuticorin, Mandapam, Lakshadweep and Anadaman Nicobar. |

* Available in exploitable quantity.
*Source* : Chennubhotla *et al.*, 1991.

**Table 26.6 : Algin yielding seaweeds of India**

| Sl. No. | Seaweed | Yield (%) | Places of Occurrence |
|---|---|---|---|
| 1. | *Sargassum wightii** | 31.7 | Mumbai, Goa, Alleppey, Muttam, Kovalam, Idinthakarai, Tiruchendur, Tuticorin, Mandapam, Chennai and Andaman-Nicobar |
| 2. | *S. tenerrimum** | 15.2 | Gulf of Kutch, Okha, Dwarka, Mumbai, Goa, Karwar, Mandapam, Visakhapatnam and Andaman-Nicobar. |
| 3. | *S. myriocystum** | 34.5 | Kovalam, Idinthakarai, Manapad, Tuticorin, Mandapam and Andaman Nicobar |
| 4. | *S. ilicifolium** | 34.9 | Mumbai, Goa, Karwar, Tuticorin, Mandapam, Chennai, Visakhapatnam and Andaman-Nicobar. |
| 5. | *Scinereum* var. *bererifolia** | 29.2 | Gulf of Kutch, Mumbai, Goa, Karwar and Vizhinjam |
| 6. | *S. johnstonii** | 22.3 | Okha |
| 7. | *S. bulgare** | 25.3 | Dwarka, Goa and Visakhapatnam |
| 8. | *S. duplicatum** | 19.1 | Muttam, Lakshadweep and Andaman-Nicobar |
| 9. | *Turbinaria conoides** | 35.6 | Manapad, Tuticorin, Mandapam, Lakshadweep and Andaman-Nicobar |
| 10. | *T. ornata** | 32.2 | Dwarka, Mandapam, Lakshadweep and Andaman-Nicobar |
| 11. | *T. decurrens* | 26.3 | Tuticorin, Mandapam and Andaman-Nicobar |
| 12. | *Cystoseira trinodis* | 30.5 | Okha, Dwarka, Mumbai, Tuticorin and Mandapam |
| 13. | *Hormophysa triquetra* | 25.5 | Okha, Manapad, Tuticorin, Mandapam and Andaman-Nicobar |
| 14. | *Spatoglossum asperum** | 17.1 | Gulf of Kutch, Okha, Dwarka, Porbandar, Veraval, Mumbai, Goa, Karwar, Idinthakar, Manapad, Tuticorin, Mandapam and Visakhapatnam |
| 15. | *Colopomenia sinuosa* | 16.7 | Gulf of Kutch,Okha, Dwarka, Porbandar, Veraval, Mumbai, Goa, Karwar, Idinthakar Manapad, Tuticorin, Mandapam and Visakhapatnam |
| 16. | *Hydroclathrus clathratus* | 14.7 | Dwarka, Tuticorin, Mandapam, Lakshadweep and Andaman-Nicobar. |
| 17. | *Rosenvingea intricata* | 20.5 | Okha, Dwarka, Tuticorin, Mandapam, Chennai, Pulicat and Lakshadweep |
| 18. | *Chnoospora implexa* | 10.6 | Muttam, Kovalam, Tuticroin, Mandapam, Pulicat and Lakshadweep |
| 19. | *Padina boergesenii* | 24.8 | Dwarka, Mumbai, Manapad, Tuticorin, Mandapam, Lakshadweep and Andaman-Nicobar. |
| 20. | *P. tetrastromatica* | 23.3 | Gulf of Kutch, Okha, Dwarka, Mumbai, Goa, Karwar, Vizhinjam, Muttam, Kovalau Tuticorin, Mandapam, Mahabalipuram, Chennai, Pulicat, Visakhapatnam and Andaman-Nicobar. |
| 21. | *Stoechosperumum marginatum* | 23.8 | Okha, Dwarka, Mumbai, Malwan, Goa, Karwar, Cannanore, Idinthakarai, Manap Tiruchendur, Tuticorinand Mandapam |
| 22. | *Dictyota dishotoma* | 21.8 | Okha, Dwarka, Mumbai, Goa, Karwar, Idinthakarai, Tiruchendur, Tuticorin, Mandap Visakhapatnam, Lakshadweep and Andaman-Nicobar. |
| 23. | *D. bartayresiana* | 22.9 | Okha, Dwarka, Mumbai, Goa, Karwar, Kovalam, Tuticorin, Mandapam, Mahabalipuram Chennai, Lakshadweep and Andaman-Nicobar. |
| 24. | *D. dumosa* | 13.3 | Mumbai, Goa and Karwar |

* Available in exploitable quantity.

**Table 26.7 : Algin percentage and viscosity of different *Sargassum* species**

| Name of Seaweed | Algin Percentage | Month | Viscosity 2% | Viscosity 1% | Viscosity 0.5% | Viscosity 0.25% |
|---|---|---|---|---|---|---|
| *Sargassum* | 21.31 | November | | | | |
| *sinereum* | 21.50 | January | 550 | 125 | 104 | 65 |
| *Sargassum* | 15.54 | November | | | | |
| *swartzii* | 12.67 | December | - | 3000 | 500 | 140 |
| *Sargassum* | 15.04 | November | | | | |
| *tenerrium* | 11.83 | January | 884 | 125 | 65 | 50 |
| | 20.13 | | | | | |
| *Sargassum* | 15.34 | November | – | – | – | – |
| *jhonstonii* | 11.97 | January | | | | |

*Source* : Bhandari P.P. 1981-82.

**Table 26.8 : Occurrence of Agar, Carrageen and Algin yielding sea weeds in different parts of Indian Coast.**

| Seaweed | Place of occurrence |
|---|---|
| **Agarophytes** | |
| *Celidiella acerosa* | Okha, Dwarka, Porbandar, Diu, Veraval, Lakshadweep and Andaman-Nicobar Tamil Nadu, Gulf of Mannar. |
| *Gracilaria edulis* | Lakshadweep and Andaman-Nicobar, Gulf of Manar, Tamil Nadu. |
| *G. crassa* | Andaman-Nicobar, Gulf of Mannar, Rameshwaram, Mandapam. |
| *G. corticata* var. *corticata* | Dwarka, Mumbai, Karwar, Goa, Tikkoti, Quilon, Varkala, Vizhinjam, Visakhapatnam, and Andaman-Nicobar, Gulf of Manar, Rameshwaram, Mandapam. |
| *G. foliifera* | Gopnath, Okha, Mumbai, Tikkoti, Andaman-Nicobar, Gulf of Manar, Rameshwaram, Mandapam. |
| *G. verrucosa* | Okha, Mumbai, Goa, Chilka, Andaman-Nicobar, Chennai, Tuticorin. |
| **Alginophytes** | |
| *Sargassum wightii* | Mumbai, Goa, Alepey, Vizhinjam, Andaman-Nicobar, Gulf of Mannar, Mandapam, Rameshwaram |
| *S. teenrrimum* | Gulf of Kutch, Okha, Dwarka, Mumbai, Goa, Karwar, Visakhapatnam and Andaman-Nicobar. |
| *S. ilicifolium* | Andaman-Nicobar, Gulf of Mannar, Rameshwaram Mandapam, Tuticorin, Chennai, Kovalam. |
| *S. cinereum var. berberifolia* | Gulf of Kutch, Mumbai Goa, Karwar and Vizhinjam. |
| *S.johnstonii* | Okha |
| *S. vulgare* | Dwarka, Okha and Visakhapatnam |
| *S. duplicatum* | Lakshadweep Andaman-Nicobar |
| *Tubrinaria conoides* | Lakshadweep, Andaman-Nicobar, Gulf of Mannar, Mandapam, Tuticorin, Ervali, |
| *T. ornate* | Dwarka, Lakshdweep Andaman-Nicobar, Pamban, Mandapam |
| *T. decurrens* | Andaman-Nicobar Pamban, Mandapam. |
| *Cystoseira trinodis* | Okha, Dwarka, and Mumbai. |
| *Hormophysa triquetra* | Okha and Andaman-Nicobar. |
| **Carrageenophytes** | |
| *Hypnea musciformis* | Gopnath, Okha, Dwarka, Mumbai, Goa, Karwar, Visakhapatnam, Lakshadweep, Andaman-Nicobar, Gulf of Mannar, Darga, Ervadi, Tuticorin. |
| *H. valentiae* | Mumbai, Tikkoti, Vizhinjam, Pulicat, Darga, Tuticorin. Kalapakkam, Lakshadweep. |

*Source* : *Mar. Fish. Infor. Serv. T & E. Ser.* No. 119 : January, February 1993.

**Table 26.9 : Industrial Uses of Alginates**

| Industry | Used for | Product |
|---|---|---|
| Food and Bakery | Thickening | Filling Creams, Lamon Curds, Soups |
| Cosmetic | Do | Thoothpaste, Hand creams, Shampoos Lotions |
| Textile | Do | Printing Paste Alginate |
| Rubber | Do | Latex |
| Food and Bakery | Emulsifying Stabilizing | Ice cream, Soft drinks, Baker's emulsions. |
| General | Do | Emulsion paints, Welding electrodes, Insecticides. |
| Pharmaceuticals | Do | Emulsions |
| Food and Bakery | Gelling and bindng | Jellies, Milk jellies, Puddings |
| Pharmaceuticals and Cosmetics | Do | Tablets, Hand and anti-burn lotions |
| Dental | Do | Impression Power |
| Medical | Do | Haemostatics |
| Ceramics | Do | Glazes and engravers |
| General | Do | Sintered products |
| Food | Film and Filament forming | Sausage casings |
| Pharmaceutical | Do | Barrier creams |
| Medical | Do | Absorbent dressings |
| Textile | Do | Soluble yarn |
| Paper | Do | Transparent paper |
| General | Do | Coated papers, washable well paper Antistick and mould release agents, Leather |

## *References*


Anon. 1996. Mineral contents of seaweed *Sargassum* sp. of Gujarat. Department of Bio Sciences, Saurashtra University, Rajkot.

Bhandari, P.P. 1981-82. Mineral seaweeds "Matsya Vikas" No 67-69 pp. 24-31 Commissionerate of Fisheries, Govt. of Gujarat, Ahmedabad.

Chennubhotla, V.S.K., N. Kaliperma and J.R. Ramalingam. 1991. Commercially important seaweeds of India, their occurrence, chemical products and uses. Mar. *Fish. Infor. Serv.* T & V *Serv No 107;* 11-16.

Chhaya, N.D. 1997. Minding our Marine wealth-An appraisal of Gujarat's Coastal Resources. Center of Environmental Education, Nehru Foundation for Development, Ahmedabad.

Desai, B.N. 1965. Seaweed Resources and extraction of alginate and agar. Proceedings of the Seminar of Sea Sakt and plants (Dec. 20-23) CSMCRI, Bhavnagar.

Dholakia, A.D. Y.B. Rava; D.V. Nandasana and K.N. Puruhit. 1998. Collection and supply of seaweed *Sargassum* Sp. plays an important role in Economics of Sikka Fishermen. Current and Emerging Trends in Aquaculture. Edt. M.M. Thomas. published by Daya Publishing House, New Delhi.

Gopalkrishnan, P. 1969. Some marine algae from the Gulf of Kutch, *Phykos* Vol 8 (1&2), 61-69.

Kaliaperumal, N., V.S.K. Chennuhotla and S. Kalimuthu, 1987. Seaweed research and Utilization, CMFRI Bulletin, 41: 51-55.

Kaliaperumal, N. and S. Kalimuthu. 1993. Need for conservation of economically important seaweeds of Tamil Nadu Coast and their time table for their commercial exploitation. *Mar, Fish. Infor. Serv.* T&V *Serv.* No. 119 : 5-12

Sundraraj, V. and B. Srikrishnadhas 2000, Cultivable aquatic organism, Book Published by Narendra Publishing House Delhi.

# 27
# CORALS

## Introduction

Corals form one of the major marines tropical and sub tropical communities of appreciable economic importance. Coral formations are found throughout the tropics and in Indian waters, though restricted to a few places. We have several examples of fringing reefs and atolls. Fringing reefs occur around Andaman and Nicobar Islands, along Palk Bay and Gulf of Mannar at the South eastern coast of India peninsula and in the Gulf of Kutch on the North-West coast. Atolls occur in the Lakhadweep Archipelago where the reefs harbour a rich and varies coral fauna. Corals fall under two groups namely "False corals" and the true "Stony corals". The false corals are again a heterogeneous grouping. (P, Nammalwar. 1096)

In Gujarat coral reefs were found in the Gulf of Kutch on the coasts of Okha, Bet-Shankhodar, Sikka, Jindra, Pirotan, Karumbhar, Munde etc. It was found that these corals were 5000 years old and near Salaya up to 45,000 years old. Of the rich variety of animals that inhabit the tropical seas, the corals, sea anemones and jellyfish are the most abundant. They are grouped together in a single phylum known as Coelenterates or Cnidaria. Corals are marine animals that collectively deposit calcium carbonate to built ornate and some times large colonies. The corals polyps change mineral dissolved in water into lime stone of the reef. The polyps of the most species remain retracted by day, but at night they protrude their tentacles and sweep the sea for passing plankton on which they feed.

## Classifications

Corals are classified in *Coelentrata* group in the same group of jelly fish, Sea animon, etc. They are reverse jelly fish. Where stomach is like empty bag and long and thread like tentacles near mouth is the speciality of this group. In Greak stomach like empty bags are known as "Coelentron" and hence animals of this group are known as *Coelentratas.*

## Taxonomic position and species

Phylum : *Coelenterata*

Class : *Anthoza* (True coral)     *Class : Hydrozoa* (False corals)

| | |
|---|---|
| Order | *Scteretania* |
| *Sub Order* | *(1) Perforata* |
| | *(2) Aporosa* |

*(3) Fungacea*

Families a) Acroporidae

b) Poritidae

c) Favidae

d) Fungidae

The soft body of coral is known as "Polip" when sea water enters in this polip, calcium ions presence in sea water is absorbed by body and calcium carbonate is deposited cylindrical exoskeleton is formed. Which protects polips from enemies.

## Species and Locations

National Institute of Oceanography conducted a study during February 1999 in the Gulf of Kutch. Under water photography and videography was done at 5 M and 10 M depth on east from Pirotan, Karmbhar and Boria and on west up to Bet-Shankhodar. From the photography and videography, it was found that under water hard corals like *Favia, Goniopora, Motipora, Tubastrea, Favites, Sinularia Turbinaria* and soft corals like *Dendronephthya* and *Nephthea* were found. Near Pirotan hard corals were more while on other area soft corals were more in quantity. If you proceed from Pirotan to Bet-Shankhodar live coral reduces from 80% to 50%. This is because of increase in turbidity. This study have changed the prediction that in Gulf of Kutch only 25 to 30% corals are live (Prakruti, 2000, 21:3) If oil spillage is done during transportation of oil from oil tankers even by accident, it may damage coral permanently. (Gujarat Ecological Society)

Collective affords by many tiny marine animal known as polips-smallest part of coral colony-for number of years, large coral rocks were formed. Human should learn it. West coast of Saurashtra is protected by such coral rocks from the damage against cyclone and hence ports and certain industries were developed on the coast. Secondly according to one scientist James Lavlok sea salinity on an average 34 ppt is maintained since many thousands of years. Salt from land is added to sea through rivers. However salinity is maintained that may be because of coral rocks make some reactions and controls salinity. This is how corals also help in maintaining salinity and ecosystem. (Prakruti, 2000, 21:6.) Of the rich variety of animals that inhabit the tropical seas, the corals, sea anemones and jellyfish are the most abundant. They are grouped together in a single phylum known as Coelenterates or Cnidaria. Corals are marine animals that collectively deposit calcium carbonate to built ornate and some times large colonies. The corals polyps change mineral dissolved in water into lime stone of the reef. The polyps of the most species remain retracted by day, but at night they protrude their tentacles and sweep the sea for passing plankton on which they feed.

## Commercially important species of Corals

### True Corals

*1) Acropora indica* *2) Acropora humilis*

*3) Astreopoa Spp.* *4) Monipora Spp.*

*5) Favcia pallida* *6) Porties solida*

*7) Porities porities* *8) Coerulia* (Brain coral)

9) *Corallium* 10) *Fungia* (Solitary form)
11) *Madrepora* 12) *Tubipora* (Red coral)
13) *Gorgonida* (Whip coral) 14) *Fungi* (Mushroom coral)

## False Corals

1) *Pennatula* (Sea pen) 2) *Gorgonia* (Sea fan)
3) *Heliopora* (Blue coral) 4) *Sea lilies*
5) Sea plumes.

There are two methods of breeding (a) Budding (b) sexual.

### *(a) Budding*

Most of these corals are colonial, where number of polyps develop by budding from the parantal of the original individuals, and come to live side by side. They help in budding up the more of the colony by recreating it's own calcareous skeleton.

### *(b) Sexual*

They also reproduce sexually by releasing sperms in surrounding sea water. That enter into another polyps through mouth to fertilize the eggs. After fertilization, the eggs develops an embryo inside the adults body. At certain stage these embryos are released as larval consisting minute cluster of cells called planual. On maturity of larval, their behavior changes and they start exploring sea bed in order to find suitable substratum to settle. Metamorphosis takes place, turning the planual into small juvenile corals polyps, which immediately starts recreating calcareous skeleton.

## Resources

Out of about 225 species of corals reported from India, in Gulf of Kutch 37 species under 24 genera have been reported by the NIO Goa. (Table. 27.1).

Among these 33 species under 20 genera are hermatypes and 4 species under 4 genera are ahermatypes. This low number is mainly due to absence of *Pocilloporidae and Acropora* in Gulf of Kutch fauna. The high turbid condition of the water is the main factor that prevents the growth of *Acropora* and several other small polyped corals here. Among the non scleractinians, *Millepora, Heliopora* and *Tubipora* could not be seen. *Porites,* the most important reef builder has dwindled. The presence of large, dead, semi fossilised intact colonies of marive *Torites* on the raised reef of Okha indicates that the genus was thriving very well in the past. Most of the corals are found partly dead or covered by sediment.

Patel (1976) had assessed qualitatively and quantitatively corals near Poshitra point and he also supported above observation.

Diversity of corals in the Indian Ocean

| Locality | No of species and Sub genera | No of genera |
|---|---|---|
| Gulf of Kutch | 33 | 21 |
| Lakshadweep island | 103 | 37 |
| Phuket (Thiland) | 173 | 66 |
| Andaman and Nicobar island | 185 | 74 |

*Source* : Nagesh, T.S. and T.S. Vishwanath, 2001.

Coral formations are found throughout the tropics and in Indian waters, though restricted to few places.

**Table. 27.1 : List of Coral Species Reported in the Gulf of Kutch**
**Subclass : Hexacorallia Scleratina (Story corals)**

| No. Species | No. Species |
|---|---|
| 1. Psammocora digitata | 20. F.favus |
| 2. Acropora humilis | 21. Favites complanata |
| 3. A.squamosa | 22. F.melicerus |
| 4. Montipora explanata | 23. Goniastrea pectinata |
| 5. M.venosa | 24. Platygyra sinensis |
| 6. M.turgescens | 25. Hydnophora exesa |
| 7. M.hispida | 26. Plesiastrea versipora |
| 8. M.foliosa | 27. Leptastrea purpurea |
| 9. M.monasteriata | 28. Cyphastrea serailia |
| 10. Coscinaraea monile | 29. Symphyllia radiana |
| 11. Siderastrea savignyana | 30. Acanthastrea simplex |
| 12. Pseudosiderastrea tayami | 31. Mycedium elephantotus |
| 13. Goniopora planulata | 32. Paracythus stokesi |
| 14. G.minor | 33. Polycyathus verrilli |
| 15. G.nigra | 34. Tubastraea aurea |
| 16. Porites lutea | 35. Dendrophyllia minuscula |
| 17. P.lichen | 36. Tubinaria crater |
| 18. P.compressa | 37. T.peltata |
| 19. Favia speciosa | |
| Subclass : Octocorallia | |
| Order Alcyonacea (Soft corals) | Family |
| 1 Sclerophytum Polydactylum (Ederuberg) = Alcyonium poly dactylum (Dana) | Aleyonidae |
| 2 Dendronephthya (Spongodes) Dendrophyta | Nephthyidae |
| 3 Dendronephthya (Spongodes) Brevirama | Nephthyidae |
| Order : Axifera (Hornicorals-Seafans) | |
| 4 Astromuricea Stellifera (Seafans) | Muriceidae |
| 5 Echinomuricea uliginosa | Muriceidae |
| 6 Lophogorgia Lulkeni | Gorgonidae |
| 7 Juncella Juncea | Gorgonidae |
| Order : Stelechotokea (Sea pens) | |
| 8 Virgularia rumphil | Virgularidae |

Fringing reefs occur around Andaman and Nicobar Islands, along Palk Bay and Gulf of Mannar on the Southeastern coast of the peninsula and in the Gulf of Kutch on the Northwest coast. The Lakshadweep Archipelago comprises 20 atolls. These reefs harbour a rich and varied coral fauna of appreciable economic importance. The black coral is found to occur in the Gulf of Mannar and in Lakshadweep at depths ranging from 5 to 200 meters. The blue coral is available on large scale in Lakshadweep at moderate depths. (Srivastava et. al. 1991).

## Coral Reefs

Corals reefs are exceptionally diverse and beautiful. They are tropical shallow water Ecosystems that flourish best at temperature between 20-27°C. Extensive reef development is seldom found where ocean temperature fall below 20° c. and hence, coral reefs tend to be restricted to tropical belt between the latitude 30°N to 30°S. The uniqueness of variety of life gives the reef an identity which is not found in any other ecosystem, on the coral reef sponges, anemones shrimps, lobsters, oyster, crabs, snails, sea cucumber, octopus, starfish, sea urchines, tube dwelling worms etc. were found resting on it.

Concerted growth of a variety of corals in a localized habitat gives rise to a coral reef. Reefs are basically classified into three types i.e. (i) Fringing (ii) Barrier and (iii) atoll reefs.

Fringing reefs develop near shore where conditions are ideal for coral growth. Barrier reefs are linear structure developing off shore with well design lagoon between the reef and the shore. Atolls are annular reefs, found both in the deep sea and continental shelf. They are isolated structures, greatly varying in size and generally more or less asymmetrical. Other forms of reefs structure include patch reefs, platform reefs, micro atolls, faros, ridge reef etc. Most of the islands in the Gulf of Kutch support fringing reef, and coral patches are found between the islands.

Coral reef is a work of living organisms. Coral polyps change mineral dissolved contribute building the hard part. Living recreating plants coralline algae repair the wind ward edge of the reef by cementing loose pieces of coral rock into near solid rampants.

Prior to Pleistocene, reef development was more extensive and diverse in the Gulf of Kutch, through the region is tectonically unstable and hydrographically not very conductive for a luxuriant growth of corals. Coral growth is scarce at present, and perhaps be termed more as patch than fringing. Growth in present situation is confined to inter tidal sand stones or wave cut eroded, shallow banks. Mundra and Pirotan are northern limit of reefs or coral patches. Coral reefs in the Gulf of Kutch are particularly vulnerable to pollutants and silt. The reefs in this region are generally restricted to area exposed to strongest tidal current, and living corals are confined to northern and western side of island where the currents are strongest.

Based on this classification the reefs of the Gulf of Kutch are classified as under.

## Coral Reefs types and place in Gulf of Kutch

| | | |
|---|---|---|
| (1). Fringing | Adjacent to the coast or separated by a shallowsmall Channels | Narara, Singach, Sikka Patra. Dhani and Ajad reef. |
| (2). Platform reefs | Forming islands without central lagoon-large size. | Kalubhar, Mundeka Buralchank Pirotan and Jindra reef. |
| (3). Patch | Forming island without central Lagoon. Very small. | Goose and paga reef. |
| (4). Coral Pinnacles | Forming islands without central Lagoon. Very small. | Two small reefs to the south of Goose reef. A small near Paga and four reefs to west of the Bural chank reef. |

*Source* : Marine National Park and Sanctuary Jamnagar.

Area of Coral formations in Gulf of Kutch is given in figure. 27.1.

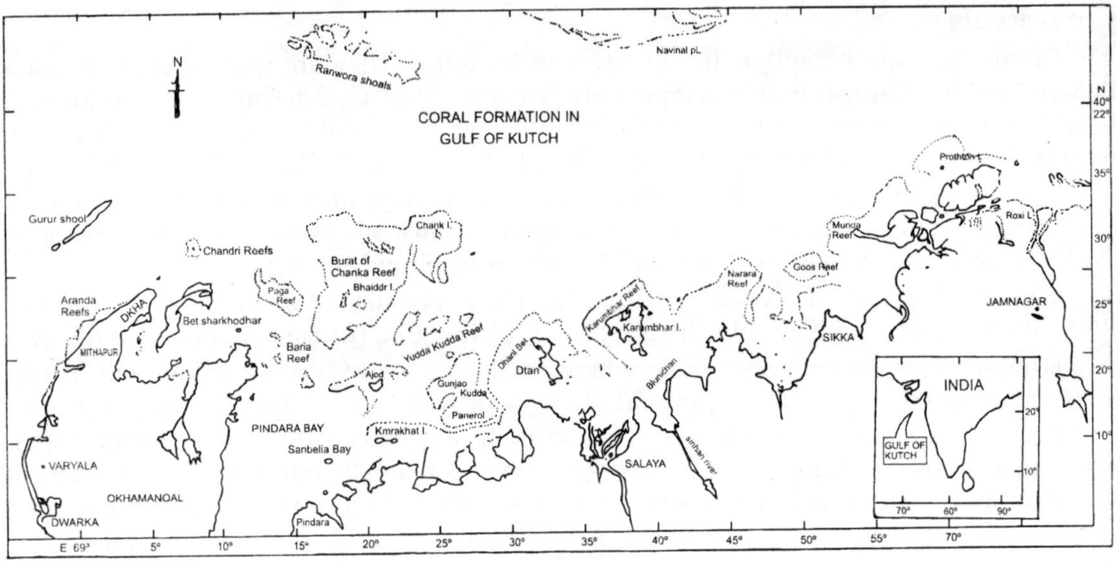


**Fig. 27.1 : Gulf of Kutch showing the areas of coral formation.**

## Ecology of Coral Reefs

### *Physical environment*

A coral reef is not just a conglomeration of colourful calcareous matter. It is a living ecosystem. Its survival, growth and development depend on various physico-chemical parameters such as temperature. Depth, light salinity, nutrients and productivity.

Coral reefs are best developed in areas where the mean temperature is around 23 to 25°C. They cease to grow below 18°C. Depth of the water is another critical factor, which limits the growth of coral reefs. Their growth is restricted to a depth of around 40 m., corals can also occur in relatively deeper waters of about 50-60 m if the water is clear enough to allow light to penetrate to such depths. This indicates that the availability of light is an important factor for their growth and survival. Salinity for coral growth and survival should be between 27-40 ppt. They are often killed under reduced salinity.

### *Biological environment*

The coral reef is also one of the richest ecosystem in terms of boil-diversity. It provides a unique niche for a large variety of organisms including a wide variety of phyto-plankton, zooplankton, benthic algae, sea grasses, sponges, mollusks. Crustaceans and fishes of both food and ornamental value.

Coral reefs are inhabited by several species of sponges. Sponge species like *Cliona and Aka,* cause extensive damage to the reefs due to their habit of boring. Sea urchins and sea cucumber (*Diaelema and Holothruria)* are most abundant echinoderms in this system. Molluscs are represented by gastropods like cone shells, cowries, and conchs and bivalves.

Many crustacean species live in coral reefs, including the commercially important shrimps of the family *Hippolytidae and Paeleomonidae.* Crabs belonging to the family *Haplocarcinidae* and several species belonging to the order *Brachyra* are also found in large numbers. More than 200 species are found in the Indian Ocean.

As many as 603 species of fish have been recorded from the Lakshdweep islands alone. The important food fishes include tuna, sharks, buillfishes, flying fishes and carangids. Tunas alone contribute 72% followed by sharks (22%), Billfishes (5%) and other. Important species of tuna are yellow fin, skipjack and bigeye. From the Andaman and Nicobar Islands, the occurrence of about 600 species has been reported. Lobsters also form significant fisheries also form nearly 1000 mt/year from these islands. Ornamental fishes comprise a great variety of colourful, attractive and often peculiarly shaped fish found around the coral reefs. The total world fish production around coral reefs is about 20 mt./km$^2$/year. However, the sustainable yield is estimated to be around 4-6 mt/km$^2$/year. Indian reefs have a potential yield of 0.2 mt i.e. about 10% of the annual marine fish production. (Nagesh and Vishwanath, 2001).

### *Economic Importance*

Coral reefs have been exploited for their numerous commercial values. Apart from ornamental value they have industrial and medicinal value. Some species like the *Corallum rubrum* is considered a precious, while others like black corals have ornamental value. Corals are used as raw materials in various industries. The most important material is calcium carbonate. Coral blocks and coral sand are used in buildings, construction of roads, and as lime in the cement and carbide industries. Some species of corals like red coral and pipe coral are known to have some medicinal value. They also form, good nursery grounds for the commercially important food and ornamental fishes. Most of the ornamental fishes of export value can be found in this ecosystem. Coral reefs also serve as bird sanctuaries.

### *Social and Economic value of Coral reefs*

Coral reefs rank as the most biologically productive and diverse of all natural ecosystems. Reefs are equivalent to tropical rain forests for their rich biological diversity. A large number of reef building and reef dwelling organisms live on reefs to mankind are both extractive and non-extractive. The non-extractive uses of the tropical reefs centers around recreation, tourism, scientific research, shoreline protection and scuba diving. The extractive uses are very many, only if judicially exploited. Coral reefs are huge mounts of calcium carbonate, which forms the raw material for many lime-based industries such as lime, cement, and calcium carbide. They are also used as building blocks in many parts of the Indo-Pacific. The finfish's fauna of reefs is extremely rich and varied. The estimated fish production from the global reef environments equivalent to about 10% of the marine fish expoited annually. The diversity of fish fauna on reefs varies in space and time. There exist vast resources of reef fishes that are of ornamental value. In addition, reefs harbour lobsters, crabs, mollusk, echinoderms, sponges, coelenterates etc. (Nammalwar, 1996).

## The impact of ecological and human interference on the major ecosystem

### *(a) Siltation*

A high degree of silting in Gulf of Kutch is a major detrimental physical factor that affects the growth of corals as well as many other filter feeding marine animals adversely. In Pirotan many corals are found partly covered by silt and there is clear indication of mass mortality or corals here only due to the deleterious effect of sediments. Many corals with large polyps can combat the bad effect of sediments, while others with small polyps like *Acropora* can never live in highly turbid water. The absence of *Acropora* in gulf of Kutch is mainly due to high degree of silting. Excess waste from Tata chemical, cement industry Thermal power plant etc. have a profound effect on reef communities.

(b) Human interference is a major factor that alters or damages the marine and coastal ecosystems. In Pirotan most of the corals, particularly the massive ones have been quarried and used up for various purposes which has resulted in major destruction.

Dead coral rocks, calcareous sea sand are exploited for production of lime, building blocks, and making calcium carbide. Coral sand is excellent raw material for manufacture of cement M/s Digvijay Cement Co. Sikka had used much more quntity of sea sand from Kalwan, Jindra island, Dhani island, Dara island, Goose island, Pirotan, Narara reef etc. The company was using 1675 tonnes of such raw material per day since 1946-47 to 1985. Company was also extracting material illegally from other islands. This was irreversible damage to coral reef of Gulf.

Destruction of corals during past is alarming in the gulf. The total reef area in the gulf was 217 sq km in 1975 which declined to 180 sq. km. in 1985 and 123 sq km in 1986, with a net loss of 94 sq km of coral cover.

Fishing, walking an coral reefs, collecting of corals just for curiocity, pollution, fishing with poisonous material etc., also cause degradation of the corals.

## Protection and Conservation

Man's interference in the coral reefs ecosystem continues unabated, exploiting this natural resource without bothering about its conservation. Fishing walking on coral reefs, collecting or corals just for curiosity, pollution fishing with poisonous material etc cause degradation of the corals. Corals all over the would are fast dying due to several reasons like dredging, breaking of coral branches during fishing activities, siltation of lagoons, mining of lime and sand stones, sea erosion, activities of many boring organisms. Therefore, proper conservation is the need of today. Location aerial extent, increase/decrease/ damage into total area occupied by the reef, mud deposition, absence and presence of vegetation and management of the reef is necessary. Coral reefs are living bio systems, which have to be conserved and protected. Otherwise we loose not only the colourful corals but also the diverse group of organisms, which thrive together. In India several organizations have been identified for conservation of coral reefs. Marine Parks have been established by some states while the Central Government has identified the Gulf of Kutch as a preserved area or sanctuary.

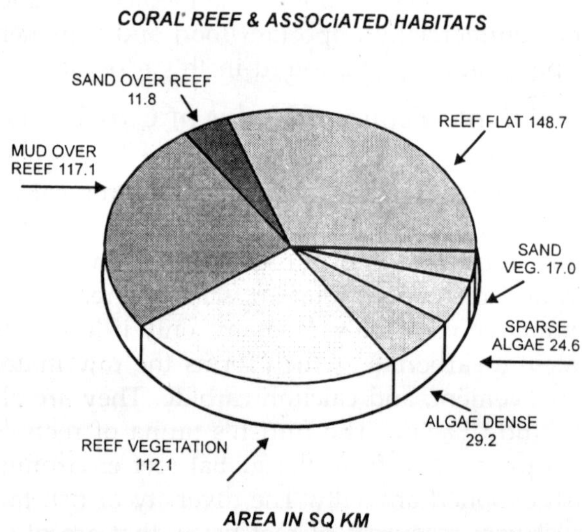


**Fig. 27.2 : Area (in sq. Km.) of Gyulf of Kutch reefs and their Associated. Source : Anjali Bahuguna and Shailesh Nayak., Coral reefs of the Indian coast.**

Fig. 27.3 : Brain coral

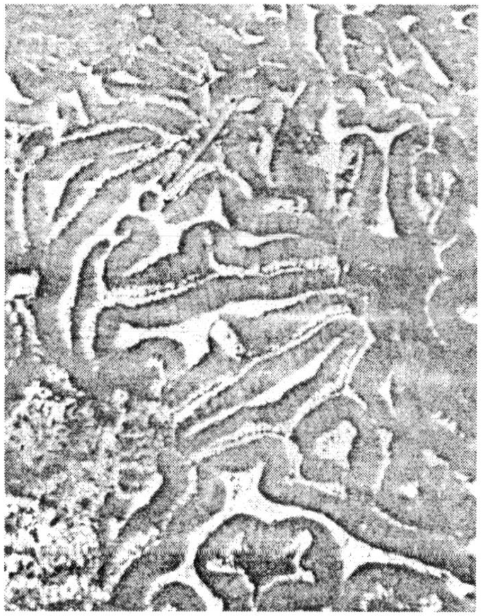

Fig. 27.4 : Mammalian Brain coral

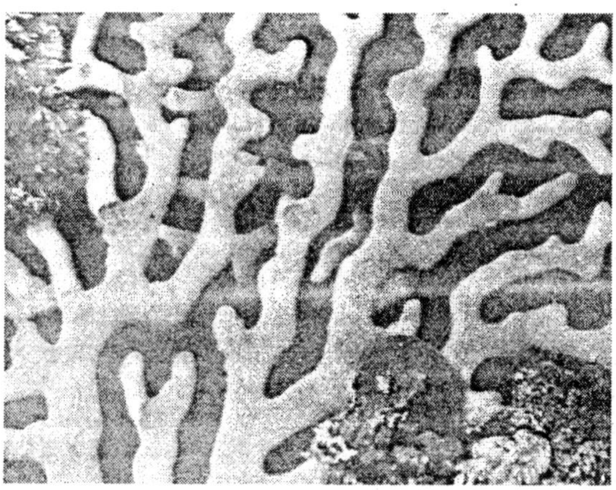

Fig. 27.5 : *Hydroza*

Fig. 27.6 : *Acropora*

Fig. 27.7 : *Fugia acriniformis*

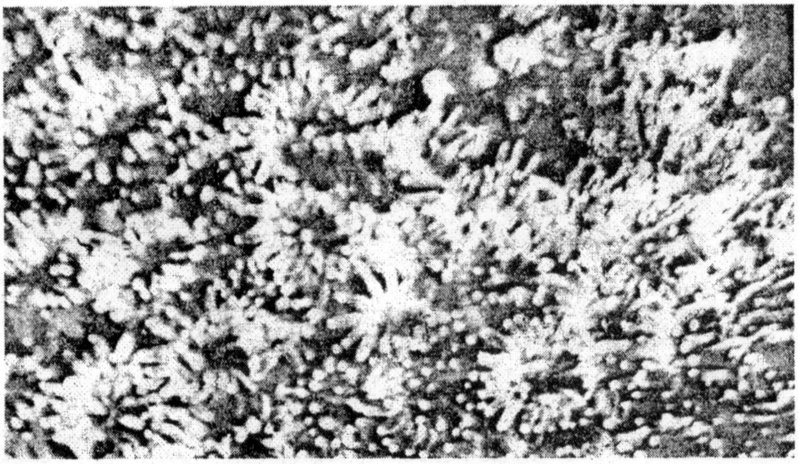

Fig. 27.8 : *Galaxae sp.*

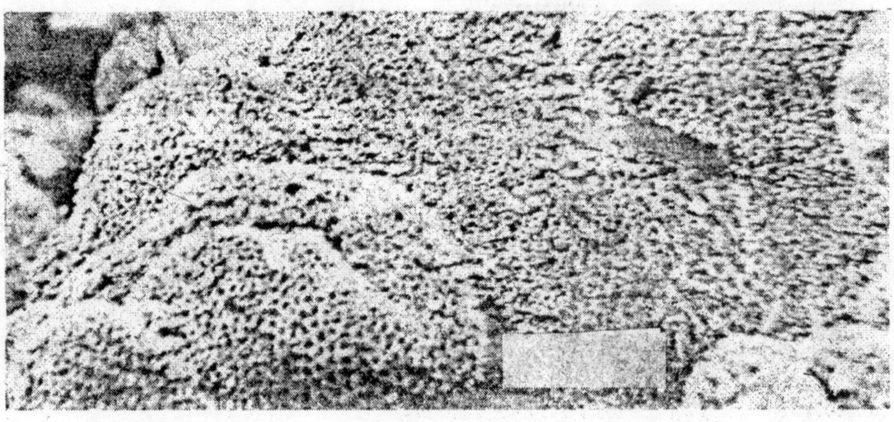

Fig. 27.9 : *Porties*

Fig. 27.10 : *Gonioporam* with its polyps fully extended.

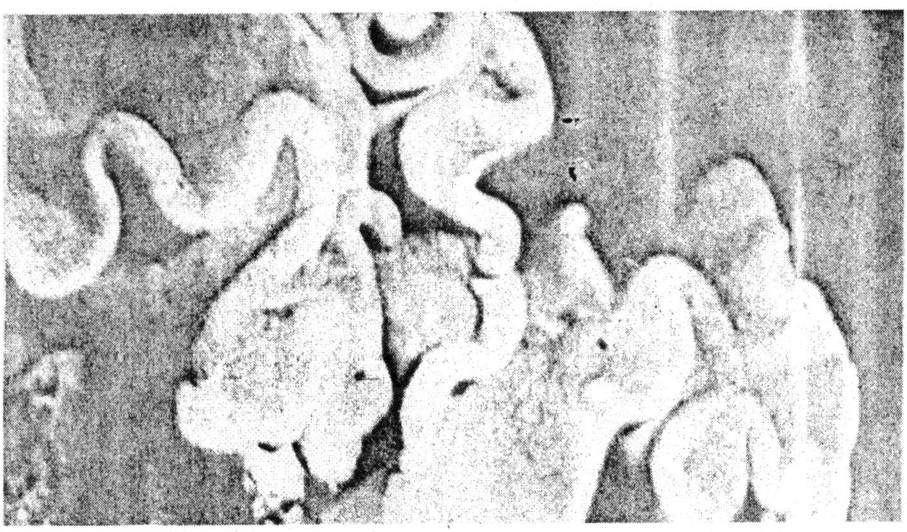

Fig. 27.11 : *Alcyonarian*

Brain coral

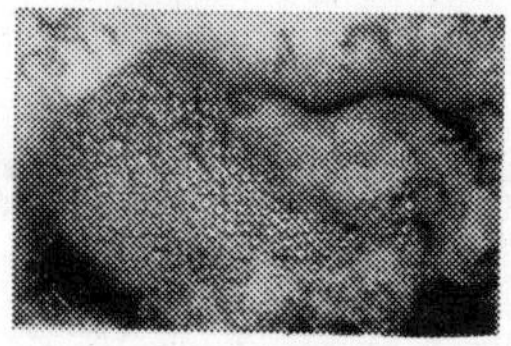

Goniopora myhora

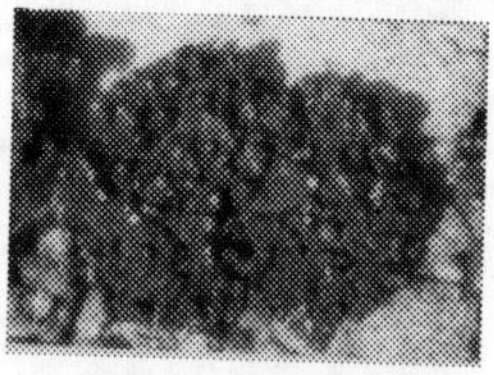

Poraitus lutia

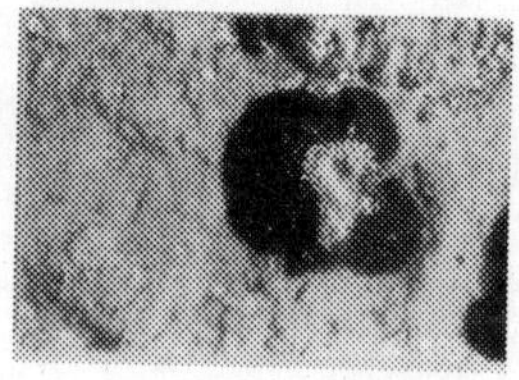

Goniastre picktinata

Goniopora Sp.

Poraitus comprasa.

Simphilia Sp.

Soft coral

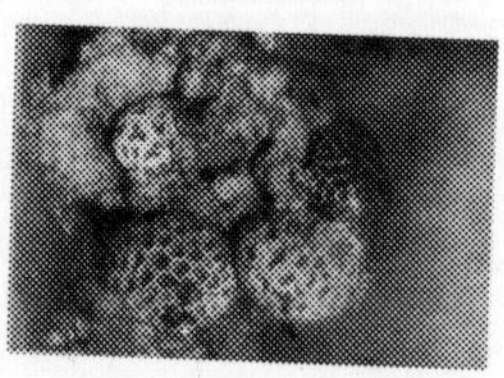

Favia Sp

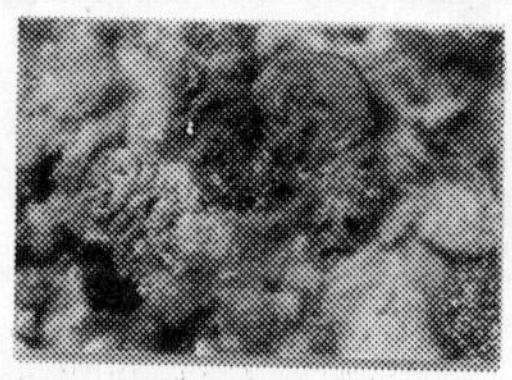

Favia sp.

**Fig. 27.12 : Photographs of different corals**

## *References*

Anjali, Bahuguna and Shailesh Nayak. 1998. Coral reefs of the Indian coast. Scientific note SAC/RSA/RSAG/DOD-COS/SN/16/97, Published by Space Application Centre, (ISRO), Ahmedabad.

Chandrasekar, N. 1996. Sediment transport along the Barrier island and its impact on conservation of Coral reef in the Gulf of Mannar. Presented at Seminar "Aqua meet" on 20-21st September 1996. Tuticorin.

Dholakia A.D. 1997. Coral reefs in the Gulf of Kutch-its importance and conservation (unpublished)

Gopinatha Pillai, C.S., M.S. Ramgopalan, 1979. Preliminary Report on reconnaissance survey of the major coastal and Marine Ecosystems in Gulf of Kutch, *Mar. fish. Infor. Serv.* T&E Serv., 14: 16-21.

Kumaraguru, A.K. 1996. Coral reef in the Gulf of Mannar and the Conservation Strategies Required. Presented at Seminar "Aqua meet" on 20-21"st September 1996 at Tuticorin.

Nammalwar, P., 1996. Conservation and management of Coral Reef in the Gulf of Mannar, Tamil Nadu- A reveiw. Presented at Seminar "Aqua meet" 20-21st September 1996.

Nagesh, T.S. and T.S. Vishwanath, 2001. The coral reef-its ecosystem and fisheries. Info Fish. International; 5: 55-57.

Nayak, S.R., Chauhan, P., H.B., Bahuguna Anjali, and Narendra Nath, A., (1996). ISR-IC Application for coastal zone management. *Current Science.* 70(7): 614-618.

Nayak, S.R., Pandeya Anjali, Gupta, M.C., Trivedi, C.R., Prasad, K.N., and Kadri S.A. (1989). Application of satellite data for monitoring degradation to tidal weland of the Gulf of Kutch. *Western India, Acta, Astronautica,* 20: 171-178.

Pandeya Anjali, Nayak Shailesh and Aggarwal, J.P. (1989). Monitoring of Ecological changes in the Marine National Park using satellit data. Scientific Note, SCA, (ISRO) Ahmedabad, IRS-UP SAC/MCE/SN/24/89.

Patel, M.I., 1976. Corals around Poshitra Point Gulf of Kutch *Asso.* CIFE. Souvenir, 1(6): 11-16.

Singh, H.S. 1994. Management plan for Marine National Park. Jamnagar Published by Chief Conservator of Forest, Wildlife, Gujarat state, pp. 94-97.

Srivastava, U.K., Bakul. H.Dholakia, S. Vathsala, K. Chidambaram, 1991. Fishery Sector of India, Oxford & IBH Publishing Co-Pvt Ltd. New Delhi: 406.

# 28

# MANGROVES

Mangrove is a general term applied to plants, which live in muddy, loose, wet soil in tropical and near tropical waters. They are trees or shrubs that grow between the high water mark of springtides and a little above the mean sea level. The word "mangroves" is derived from a combination of the Portuguese word "manague" for tree and English "grove" for stands trees. A mangrove forest is also called "managal". While the word mangrove is used with reference to specific types of trees.

Mangroves are the salt tolerant forest ecosystems found mainly in tropical and subtropical inter regions of the world. They consist of swamps; forestland and its water spread area. Where conditions are favourable, the mangroves form extensive and productive forests in the sheltered coast lines. They are reservoirs of species of plants and animals, associated to get her over a long evolutionary time and till imperfectly known and not fully understood. Mangrove ecosystem is defined as an economic resource, which has been widely and variously used by coastal people of the tropics for thousands of years.

It is worth noting that the coastal vegetation like seaweeds and the mangroves play a significant part in enriching the coastal sea and stability the shoreline. The important role-played by plants including algae and halophytes with species refer to mangrove forests and swamps in the production of detritus, dissolved organic matter and recycling of nutrients are being increasingly realized. Fish and prawn production in the sea depends on the management of mangroves and other associated marine ecosystems.

Realizing the impotance of this ecosystem, the Government of India has introduced a scheme for conservation and development of mangroves in the country, and National Mangrove Committee consisting of experts has been constituted.

The areas which are rich in mangroves in Gulf of Kutch (Gujarat) area and the authority who is controlling the said area is given in Table. 28.1.

**Table 28.1 : Area and authority controlling Mangroves in Gulf of Kutch**

| Taluka | Area sq km | Authority |
|---|---|---|
| 1) Anjar and Bhachau | 173.0 | Under Kandla Port Trust |
| 2) Jungi Surbari (Bhachau Taluka) | 65.0 | Revenue Department |
| 3) Lakhpat-Abasda Taluka | 529.5 | Mangrove Forest |
| 4) Mundra area | 43.0 | Mundra Port Trust and others |
| 5) Mundra Taluka | 53.3 | Forest Department. |
| 6) Lakhpat Kori Creek area | 80.0 | Revenue Department. |
| Total | 943.8 | |

*Source* : N.D.Chhaya, Ex. Commissioner of Fisheries-Personal discussion.

## Mangroves Resources

As per data furnished by the concerned State Government, total area of tidal forest in India was about 6,740 sq km, constituting about 7% of the world's area under mangroves (Anon, 1987). Another publication of the GOI states that the total area covered by the mangroves in India was estimated about 6,000 sq. km. (Anon 2001). Ground scenario is quite different, as some of the mangroves forests have disappeared due to human activities. Forest Survey of India (FSI), Dehra Dun made independent assessments of mangrove cover at intervals of two years using remote sensing technology. As per the assessment, area under mangrove cover was estimated at 4,871 sq km in 1999, which was higher than its previous estimates. About 380 km of the mainland coast of the country is covered by the mangrove line, which is about 6% of the entire coast of India, whereas 40% of the coast (260 km) of Andaman and Nicobar Islands are lined with mangroves (Singh, 2002)

India harbours some of the best mangroves in the world. These are located in the alluvial deltas of rivers as well as on the Andaman and Nicobar groups of Islands. The Sundarbans of West Bengal represent the largest stretch of mangroves in the country, where tree cover was estimated at 2123 sq. km. Although, Gujarat accounts for an area of 1,031 sq km., majority of its mangroves are scrubby and of low height (FSI, 1999). Area wise, Gujarat stands second to West Bengal with respect to mangroves, but it is considered to be inferior in floristic composition and height.

The Sundarban mangroves are spread over the alluvial flats of the Ganetic delta, which is intersected from north to south by several wide channeled rivers and sluggish, winding creeks interspersed with lagoons. The entire area is subjected to seawater inundation during the spring high tides, whereas extensive flooding of fresh water occurs during the rainy season. Mangroves of the Sundarbans are rich and diverse, which support a variety of animals-amphibians, reptiles, fishes, mammals (like tiger, dolphin and porpoise) and thousands of species of invertebrates. The entire ecological complex of the mangroves is known as an ideal ground for fish, prawn and shrimp. As much as 80% of the fish catch from the lower delta of the Ganga-Brahmaputra comes from the Sundarbans. (Singh, 2002).

The Andaman and Nicobar Island groups have rich marine life. Many of them have mangroves. Quality of forest on the group of islands is outstanding. Height of trees in some areas reaches over 25 m; floral diversity of mangroves in this area is also very high. The mangrove area of Andaman and Nicobar is the third largest in India and constitutes about 11% of the total forest cover. Exploitation (Principally for fuelwood and charcol) has reduced mangrove coverin the island to 966 sq km by 1999. Andhra Pradesh (383 sq. km), Orissa (211 sq km) and Maharashtra (127 sq km) are other important states, which support good mangroves. Ramchandran and Mahanan (1987) reported that Kerala had a large area under mangroves at the mouth of rivers, especially in Kannur district, but now they are restricted to small patches. Eight states and a Union Territory have mangroves and estimated of their areas by the Forest Survey of India, Dehradun have been given in the Table. 28.2.

Untawala (1986), Naskar and Mandal (1999) and Singh (1994) has listed mangrove and mangrove associated flora found in Gujarat, compiled information is given in Table. 28.3.

Puree stands of mangroves in Dhasoidys near Jodia have become scrubby due to heavy pressure from the adjoining salt works. In the exposed areas bordering the Gulf, *Atriples stocksii* and *salicornia branchiate* are dominant species. Due to the high soil salinity and the pressure of camel grazing *Avicennia* stand near Bedi, Navabandar, Singach and Salaya area of scrubby nature. Sikka-Salaya region has extensive growth of Mangroves bordering the mainland as well as several mud banks in the vicinity. Some of the areas had deposits of tarry mater indicating that minor spills occasionally in the Gulf. (NIO-1995).

**Are (in sq. km.) under mangroves as estimated by SAC-1992**

| Area | Status of mangrove | 1975 | 1985 | 1986 | 1992 |
|---|---|---|---|---|---|
| Gulf of Kutch | Dense | 199.6 | 94.2 | 83.6 | 158.8 |
| | Sparse | 76.1 | 128.1 | 135.1 | 763.9 |
| | Total | 275.7 | 222.3 | 218.7 | 922.7 |

*Source* : MNP & S, Jamnagar 1994.

Above figures indicate that there is some improvement in mangrove forests. The improvement in growth of mangrove is because of plantation done by GSFC, Sikka. (Kamdar. 1996.) It has been mentioned that there is a large scale degradation and destruction of dense mangroves in the Park and Sanctuary area around Okha, Pindhara, Bhaidar, Noratapu, Dhani bet, Kalubhar, Narara, Singach, Jindra bet, Dide ke bet and Pirotan. The area under dense mangrove is confined to a few patches is Dide-ka-bet, which earlier occupied the entire bet. There has been improvement in density and area coverage by the mangrove at a few sites during recent past. (MNP & Sanctuary, 1994). List of Mangrove species available in Gujarat is given in Table. 28.4.

## Regeneration of Mangroves

Bhaidar and Pirotan Island have mixed stands of *Rhizophor and Ceriops.* (Fig 28.1 & 28.2) *Bruguiera* Fig. 28.3 and *Aegiceras* have became rare. Pure stand of mangroves near Jodia, Bedi, Navabanfdar, Salaya, Singach became scrubby due to grazing, lopping, ship breaking and salt works.

The vegetation of salt mershes near mangrove forests is prominently of *Aeluropes lagopoides* supported by *Sporobolus* spp. Presence of *Cress cretica* on dry salt and *Cyperus pangorei, Bergia odorata* and *Oldenlandia umbrellata* are probably the indicators of the retrogression due to the detrimental effect of salt works.

**Table 28.2 : State wise mangrove cover (Area in sq. km.)**

| Sr. No | State/Union Territory | Forest Survey of India, Dehradun | | | | |
|---|---|---|---|---|---|---|
| | | 1991 | 1993 | 1995 | 1997 | 1999 |
| 1 | Andaman and Nicobar | 971 | 966 | 966 | 966 | 966 |
| 2 | Andhra Pradesh | 399 | 378 | 383 | 383 | 397 |
| 3 | Goa | 3 | 3 | 3 | 6 | 5 |
| 4 | Gujarat | 397 | 419 | 689 | 991 | 1,031 |
| 5 | Karnataka | - | - | 2 | 5 | 3 |
| 6 | Maharashtra | 113 | 155 | 155 | 124 | 108 |
| 7 | Orissa | 195 | 195 | 195 | 211 | 215 |
| 8 | Tamil Nadu | 47 | 21 | 21 | 21 | 21 |
| 9 | West Bengal | 2119 | 2119 | 2119 | 2123 | 2125 |
| | **Total** | **4244** | **4256** | **4533** | **4830** | **4871** |

*Source* : Forest Survey of India.

**Table 28.3 : Mangrove and Mangrove Associated flora found in Gujarat**

| | | | |
|---|---|---|---|
| **Mangrove Trees** | | | |
| 1) | *Avicennia marina* | 2) | *Avicennia officinalis* |
| 3) | *Avicennia able* | 4) | *Aegicerews cernniculatus* |
| 5) | *Rhizophora mucronata* | 6) | *Rhizophora aplicalata* |
| 7) | *Bruguiera gymnorhiza* | 8) | *Bruguiera cyiidrica* |
| 9) | *Derris heterophylla* | 10) | *Bruguera gymnorrhiza* |
| 11) | *Sonneratiaapetala* | | |
| **Mangroves shrub** | | | |
| 1) | *Ceriops tagal* | 2) | *Ceriops decandra* |
| 3) | *Bruguiiera parviflora* | 4) | *Acanthus ilicifolius* |
| 5) | *Aegiceras corniculatum* | 6) | *Visam orientale* |
| 7) | *Salvadora persica* | | |
| **Associate Flora** | | | |
| 1) | *Aloe vera* | 2) | *Asparagus demosus* |
| 3) | *Aristida Spp.* | 4) | *Aeluropes lagopoides* |
| 5) | *Atri Plex Stocksii* | 6) | *Barassus flabellifera* |
| 7) | *Bergia ordorata* | 8) | *Cyperos conglomerates* |
| 9) | *Capparus Spp* | 10) | *Cressa cretica* |
| 11) | *Eragrostis spp* | 12) | *Eleusine spp* |
| 13) | *Halopyrum mucronatum* | 14) | *Ipomea pes-capra* |
| 15) | *Olenlandia umbellate* | 16) | *Riea spp.* |
| 17) | *Phoemix sylrostris* | 18) | *Suaeda fruticosa* |
| 19) | *Suaeda mudiflora* | 20) | *Salvadora Spp.* |
| 21) | *Sesuvium postulacusrum* | 22) | *Scrius spp.* |
| 23) | *Salicornia branchiate* | 24) | *Tamarix spp.* |
| 25) | *Urochondro setulosa* | | |

*Source* : (1) Untawala (1986); (2) Naskar and Mandal 1999; (3) Shingh. H.S. 1994

Sandy stretches in foreshore is mostly free from vegetation but in sandy area between the rocky shoreline and sandy beach ridge, *Cyperus conglomeratus, Asparagus demosus, Ipomoea pescaprae and Halopysum murcromatum* are observed.

The scrub forest on sandy plains of some of the islands and few patches along the coast consists of *Acacia senegal, A. leucophloea, A.nilotica, Capparis decidua, C.grandies, Balanites aegyptica, Commiphora wighti, Euphorbia nivulia, Foliata, Cadaba fruticosa and Zizyphus numularia, Salvadora oleoides, S.persica, Tamarix spp., Calotropis spp.*

## Mangrove Plantation

Seeds of mangrove species are viviparous and they start germinating on the parent tree, its seed structure is especially adopted for dispersal by seawater, there by helping it to successfully colonize the coast. Particularly in the intertidal zone. The spongy covering of the apricot like fruit makes it buoyant, which facilitate its dispersal by seawater.

**Table 28.4 : List of species of Mangroves available in Gujarat (Gulf of Kutch and M.N.P. area)**

| Sr. No. | Name of Species | Availability |
|---|---|---|
| 1 | *Avicennia marina* | Abundant |
| 2 | *Avicennia alba* | Abundant |
| 3 | *Avicennia officinalis* | Abundant |
| 4 | *Rhizophora mucranata* | Abundant |
| 5 | *Rhizophora apiculata* | Abundant |
| 6 | *Ceriops candolleana* | Abundant |
| 7 | *Bruguiera gymnorrhiza* | Common |
| 8 | *Bruguiera parviflora* | Common |
| 9 | *Aegiceros conniculatum* | Common |
| 10 | *Salicornisa branchiate* | Common |
| 11 | *Sonneratia apetala* | Common |
| 12 | *Suaeda fructicosa* | Common |
| 13 | *Ariples stockssi* | Common |
| 14 | *Aeluropes lagopoides* | Found near salt marsh |
| 15 | *Cyperus conglomerates* | Found on sandy beach |
| 16 | *Asoaragus domosus* | Found on sandy beach |
| 17 | *Ipomoea percaprae* | Found on sandy beach |
| 18 | *Halopysum mucromatum* | Found on sandy beach |
| 19 | *Urochondra setulosa* | Not found in other part of India. Grows only on Pirotan and few other islands. |
| 20 | *Borassus flabellifera* | Once abundant Rare and completely destroyed on the island. |
| 21 | *Phoenix sylrostris* | Once abundant Rare and completely destroyed on the island. |
| 22 | *Eragrostis spp.* | Small grassy patches on island |
| 23 | *Eleusive spp.* | Small grassy patches on island |
| 24 | *Aristida spp.* | Small grassy patches on island |
| 25 | *Cyperacean members* | Small grassy patches on island |

Source : MNP Report.

**Table. 28.5 : Period of flowering and fruiting of Mangroves in some of the species in the Gulf of Kutch**

| Sr. No | Species | Local Name | Flowering Time | Fruit-ripening Time |
|---|---|---|---|---|
| 1 | *Avicennia afficinalis* | Cher | July-August | October |
| 2 | *A.marina* | Bhuradi | May-June | Aug-Sept |
| 3 | *A.alba* | Patcheradi | May-June | Aug-Sept |
| 4 | *Other sp of Avicennia* | Makad | June | September |
| 5 | *Rhizophora Sp.* | Kharod | Jan-Feb | April-May |
| 6 | *Ceriops Sp.* | Kanari | January | April-May |
| 7 | *Aegiceros Sp.* | Chawario | Jan-Feb | May-June |

Source : Singh. H.S. 1994, Status Report on Mangroves in Gujarat State. Gujarat Forest Department, Govt. of Gujarat.

*Acicennia* seeds ripen in theAugust-September period in the Gulf of Kutch. Seeds are abundant on the coastline and on the island in the Gulf. They can be collected directly from the trees and when they get washed up on the sandy shores.

The shape of *Avicennia spp* (Cher) (Fig. 28.4) leaves makes the front bend with the water current and permits the tide water to flow past easily so that even with the strongest of

Fig. 28.1 : *Rizophora mucronata*

Fig. 28.2 : *Ceriops decandra*

Fig. 28.3 : *Brugniera parviflora*

Fig. 28.4 : *Avicennia marina*

tidal flows, the young plants are little affected. Cher seedlings can therefore, colonies those mudflats where the new leaves get exposed to sunlight above the water level for little as couple of hours a day. At this initial stage, the main danger to the seedlings is from stands of algae which are swept in and out with the tides since algal growth gets entangled around the seedlings and smoothers them.

Mangrove nursery is best prepared on the intertidal seabed. The place must receive intertidal waters twice a day. Sparse mangrove are also can be taken up for nursery. There should be 1.0 to 1.5 m of water over the nursery at any time. Scattered beds of 1m x 10m are prepared in the mud. One such bed can provide seedling for planting a 0.4 ha. area. The average distance between two beds should be 60 to 70 M. The nursery bed should be oriented in the direction of the tides to avoid damage due to strong currents. Beds at distance at reduce transport costs.

Polythene bags of size 15cm x 25cm are filled with good quality mud and arrange in row. One-third length of the bag is kept below the surface and mud dugout from the bed is arranged around the bed. Loose mud should not be filled in the bag because it may be washed away by tidal current. Sowing of pregerminated seeds without seed cover provides better results. Resowing of seeds to replace the casualty should done after every 15 days and this should continue till all the bags are full of saplings.

Ripe seeds are reported to germinate within 10 days of sowing and attain 25 to 30 cm, heights in six months. Five to six month old seedlings can be preferably be planted on the site. Plantation is carried out during low tides, when sea water recedes from plantation area. Planting ratio of 2500 seedlings per hectare at a spacing of 2m x 2m is maintained. It takes two years to establish. Once established, the healthy plants seed out aerial roots.

Mangrove plantation is always done during winter. The sea remains disturbed during summer and rainy season and the plantation raised during this period are uprooted.

Post plantation care is very important because there are many adverse factors which operate in the intertidal zone and can destroy seedlings they are

(1) The plants can be washed away by waves.

(2) Can be pulled down by algae.

(3) Crabs and other marine creatures can damage the young plants.

In spite of all obstacles and difficulties, the results of Mangrove regeneration and plantation of the intertidal mud flats have been seen to be very encouraging in Jamnagar District.

The nature of *Rhizophora Spp* (Karod) prevents it from being an effective pioneer species on distant island and more sandy substrates. This species can not colonize very exposed location as effectively as *Avecinnia* does. However, once established, it seems to thrive exceptionally well even where the rise and fall of tide level is considerable. The seed of *Rhizophora Spp* start germinating while the fruit is still attached to the parent tree. The clusters of beam like structure are in fact the germinating radicals, presumably corresponding to the typocatyle in conventional seeds. When it is ready to detach, a collar appearance below the toothed remnants of the clay. When this collar is complete, the slightest touch or jerk as by wind detaches the elongated seeds which drops straight down with its lip getting embedded into the mud like dart. On detachment, the plume is exposed at the point of separation and later unfolds into the shoots. Roots develop from small knobs visible just behind the point of radical. Only those seeds, which get deeply stuck in the mud, grow into seedlings and get established. Those which get washed away will only grow if they get so enmeshed as to permit the roots to get a chance to find a hold on exposed locations. Established trees of this mangrove provide favorite roosting and colonial nesting sites for aquatic birds. It is, therefore, very necessary to preserve the remaining good stands of this species and to raise new ones through artificial planting.

*Ceritops Spp* forms stilt roots when growing on heavy mud but these are not apparent when growing on sandy soil where better aeration is available. The seed resemble those of *Rhizophora* but are shorter and thinner. They too drop down like darts to get embedded in mud. however, since many of the seeds develop too close to the ground because of the short starter of the species, they cannot get enough fall to provide the thrust necessary to get their tips embedded. Such seeds fall sidelong and gernally fail to grow into seedlings.

## Flowering and Fruiting Seasons

Different species of mangroves have different flowering and fruiting season. Details of flowering and fruiting seasons of mangroves of Gujarat is given in Table 28.5. Flowering time for *Rhizophora, Ceriops and Aegiceros* species are January-February and fruit ripening time is April-May. Thereafter for *Avicinnea marine and A.alba* and other *avicinnea Spp.* flowering time is in May-June and fruit ripening time is in August to September. While *A.afficinalis* flowering time is in July August and fruit ripening time is in October.

## Ecological Conditions

On Gulf of Kutch and Gulf of Cambay mangroves have very much scattered distribution, stunted growth and less species diversity. Due to inadequate up stream fresh water supply and very scare monsoon precipitation, these Gujarat coastal mangroves are more saline registrant (Anon. 1987). The highest tides in the Indian sub continent was recorded in Bhavnagar of Gujarat coast and the second high tides are also recorded from the Ganga delta/Hugli-Malta estuaries. Tide water has the greatest impact on the natural growth and regeneration of mangroves, but along with tidal thrust, rain fall, atmospheric humidity, temperature fluctuations and salinity of the deltas are also important. (Naskar and Guha Bakshi, 1987 & 1991).

## Major Factors Governing the Mangrove Ecosystem

### *1) Temperature*

Well-developed mangroves are found only along coastal line where temperature range is between 20°C and 30°C Mangrove species disappear when temperature goes down below 16°C during coldest months. Mangrove species are sensitive to frost though, *A.marina, A.germinans* can tolerate light frost upto 0°C to 4°C, but they do not survive lengthy period of frost. The decreasing temperature and increasing thermic amplitude affect these mangrove species abundance, reduce the forest height and size of the tree.

### *2) Salinety*

Mangroves are halophytes that occupy tidal areas where fresh water plants can not survive. Favorable salinity range is between 10 ppt to 40 ppt. The water and soil salinity may affect the mangroves in three different ways. (1) by osmotic inhibition of water absorption, (2) by specific effects on nutrition and (3) by causing toxicity.

### *3) Soils*

Mangrove grows in muddy, loose wet soils, strong waves and current can uproot seedlings and carry away soft mud. Mangroves are best developed along deltaic coasts or in estuaries and creeks, where soft mud, comprising fine silt and clay and rich in organic matter, is available for growth of seedlings.

### *4) Tidal range*

A wide, horizontal tidal; range is a requisite for extensive growth of mangroves.

### *5) Rainfall*

The rainfall conditions are more decisive for the sequence of mengal distribution of different zones in the tidal region.

Mangroves develop best in the regions experiencing abundant rainfall, evenly distributed throughout the year and when the climate is much regular. 1000 mm to 2000-mm rainfall shows best growth of mangroves.

### 6) *Humid Climate*

In the dry regions like Gulf of Kutch and Pakistan mangroves are dominant with salt tolerant species occupies land zone. On Gujarat coast under semi-arid condition, mangroves are exceptional.

### 7) *Wind Velocity*

The wind flow may have the drying power of the air and its mechanical effect may cause damage to mangrove plants and the ecosystem. The wind may affect of seeds and seedlings dispersal.

## Mangroves on South Gujarat Coast

Coast of South Gujarat is uniform but broken by several small tidal creeks and few major estuaries like Mindhola, the Purna, the Ambica, The Auranga, The Daman Ganga, Mudflats and marsh vegetation are present here. The sandy beach floras are scarce. Mangrove species *Avicennia officinalis* is in patches on the mouth of Kalak estuaries and near the small creeks at Umargoan and the South of Kolak river, a luxurious growth of *Rgizophora spp.* is also present. Scattered mangroves are also reported from the mudflats of the Konai creeks. Apart from these rock coast, coastal dunes, reclaimed mudflats and salt-plans are the other categories of the south Gujarat coasts. (Narkar and Mandal, 1999, pp 149).

Different workers have reported different concentration of mangrove areas varying from 200 sq. km. to 526 sq. km. However, from satellite imagery (Annon, 1992) the mangrove area of Gujarat coast has estimated as 1051.5 sq. km. Mangrove flora of Gujarat are suffering and mangrove areas are reclaiming or denuding continuously due to browsing, for developing harbours, industralization and consequent pollutants discharge of these coastal area; these mangals of Gujarat coast are seen very much degraded. The minimum rain flashing i.e. 60 mm to 400 mm per annum may be the major factors for the stunted and very much degraded mangals.

To under take the National Programme for the development or conservation of Indian Mangroves and mangrove ecosystems, the urgent need is to highlight these Indian mangals by biomonitoring these coastal endangered or most threatened mangrove ecosystem. To fulfil this need Indian Space Application Center (ISRO) Ahmedabad have estimated Indian mangrove areas are 3997 sq. km. The zone wise distributions are shown in Table 28.6.

**Table. 28.6 : Area wise Mangroves of Gujarat**

| Area | Esimated Area in Sq. Km. |
|---|---|
| South Gujarat | 991 |
| Gulf of Kutch | 954 |
| Kutch District | 836 |
| Kori Crick | 80 |
| Jamnagar District | 118 |
| Bhavnagar District | 19 |
| Bharuch District | 12 |
| Surat District | 4 |
| Junagadh District | Negligible |

*Source* : Chhaya, N.D., 1997.

## Mangrove Adaptation

Mangroves have a characteristically low and dense forest type of vegetation with a special root system and viviparous seeds. Most of the mangroves have adjusted to their environment through mechanical adaptation for attachment in soft or loose substrate; formation of respiratory roots and aerating devices; evolution of vivipary; use of specialized means of seed dispersal and development of xerophytic structures.

Usually, mangroves are shallow rooted and like well developed to roots as a result of high salt concentrations and the water saturated organically rich anaerobic substratum. A number of adaptations in root morphology are typical in most mangroves. The roots can be "prop" (Silt like protusions from the lower part of the stem) or "drop" (from branches and upper part of the steam) type that terminated only a few centimeters below ground (e.g. *Rhizophora*). In other cases, surface or horizontal roots (like cables 1 to 5 cm below the surface soil) grow out from the steam base and negative geotropic.

## Ecological Role of Mangroves

The mangrove soil in the area of Jamnagar is sandy loamy to silty loam with pH ranging from 7.8 to 8.6. The organic matter is high. Soils have sodium chloride content form 1.00% to 3.84% and dissolved soils from 1.72% to 3.98%. The calcium carbonte content varies from 7% to 21% and it is as high as 53% where *Avicennia marina* occurs and reaches even 82% with presence of *Urchnodra setulosa.*

Mangroves in association with corals support diverse form of marine life and play following vital ecological role in sustaining the unique and fragile marine ecosystem.

1) Mangroves play a vital role in holding and binding the highly mobile coastal soil by breaking the disintegrating force of tides and storm waves.
2) Ramifying root systems of mangrove act as effective silt causing the decomposition of silt brought down from the mainland by surface runoff. This filtering action prevents the silt from getting deposited on the coral formation.
3) They intercept the prevailing strong salt laden sea winds, thereby encouraging and promoting better and healthier three and shrub growth on their seaward side.
4) Mangrove produce immense quantities of leaf litter which on decomposition forms a rich organic detritus sustaining a host of organism such as Crabs, prawns and fishes. The mangroves are thus, of immense value in the growth and development of marine animals. As much as 90% of marine fish species pass at least one stage of their life cycle in mangrove ecosystems.
5) Mangrove forest offer excellent nesting, roosting and breeding sites for aquatic birds such as herons, egrets, pelicans, ibises, spoonbills, darters, cormorants etc. It also attracts insects and territorial birds. Mangrove forests are excellent Sanctuary for large number of resident as well as migratory birds.
6) They improve the microclimate.
7) Mangroves are valuable sources of fuel, fodder, fruits and fishing material for local people, especially fishermen.
8) Mangroves in the Gulf of Kutch attract thousands of migratory birds during their winter migration. Many of these birds roost or nest on the mangrove trees. (Source: MNP & Sanctuary 1994)

More quantity of snails such as *Vermis sp. Cerithid sp* and *Clypeomorus sp.* were dominant near mangrove plant *Avicennia marina* near Rosy port near Jamnagar (Pandya, J.B. 1993).

## *References*


Anon, 1992. Coastal environment, Space Application Center (SRC) Ahmedabad, RSAM/SAC/COM/SN11/92 ISRO Ahmedabad, p.114.

Anon, 1995. Marine environmental impacts assessment of proposed Refinery Complex at Motikhavdi NIO Report.

Anong, 2001. Project Brief. Andaman and Nicobar Islands: Sustainable Island Biodiversity Managements and Resource development, UNDP/GEF Primary Report, p. 29.

Balasco, F. 1975. The mangroves in India. Institute Francais de Pondicherry, India Shri Arbindo Ashram India, p. 175.

Chapekar, S.B. and S.V. Desmukh, 1987. Mangrove Ecology, 1 (9) : 1-13

Chhaya, N.D. Minding our Marine Wealth, An appraisal of Gujarat coastal Resources; p. 88.

Chhaya, N.D. 1998. West Mangrove Reserve Forest, its marine ecosystem and Fisheries Part II:45-82.

Chhaya, N.D. 1999. Personal Discussion.

Chavan, S.A. 1985. Status of Mangrove Ecosystem in the Gulf of Kutch. Published by Forest Department, Govt. of Gujarat.

F.S.I. 1999. State of Forest Annuarl Report. Forest Survey of India, Dehradun, Ministry of Environment and Forest: p. 113.

Kamdar, G.C. 1996. Mangrove cultivation and Pollution control by Gujarat State Fertilizer and Chemicals Ltd., Report.

Krishman, S. and P.J. Foldu, 1991. Coastal Marine ecosystems and anthropogenic pressure in the Gulf of Kutch. Final Progress Report of WWF, India.

Naskar, K.R. 1987. Mass production of Duck weeds for carp culture. *Proc.* USAID. *training programme on Freshwater aquaculture* CIFA, Bhubaneswar, pp. 56-65.

# 29
# SPONGES OF INDIA

## Introduction

Sponges is an ancient group with evolutionary history of about 570 years and is represented in the extent seas by about 5000 species referable to 790 genera in 80 families. Zoological Survey of India indicates that the total number of sponges species occurring in India is 519 and when compared world's total forms only 10.1%. Another assessment on the level of our qualitative knowledge on the sponges fauna of the Indian seas published by Rao (1990, ICAR Workshop, Kalpakkam) indicates that only 60% of our fauna is documented and hence much more effort has to be put in for getting a complete picture of our sponge fauna.

## Distribution

Sponges are distributed from intertidal to tidal depths and form a major component to the benthic community in some place. Though many species of sponges co-exist in their over all range in distribution, the abundance of each species may vary both in time and space. Easy availability of sufficiently larger specimen is possible only along the areas of their optimum growth.

## Identification

Body form and colour are highly variable even for different specimens of the same species collected from the same area. Hence, general morphology cannot be taken as identifying characters. Spicules, their size, shape, arrangements etc. are the only dependable characters in identifying the species of sponges. Cellular aggregation and totipotent nature of archaeocytes.

## Fertilization

Peculiar mode of fertilization through carries cells, presence of "Protonervous System", symbiotic relationship with bacteria, zooxanthellae, Zoochlorella, higher algae etc. have given an unique position for sponges in animal kingdom.

## Re-generation

Even though sponges are included under multicultural organisms (metazoa), individual cells of sponges can be sieved out using bolting silk or fine meshed cloth and this phenomenon is unique for sponges. Cells thus sieved out when kept in fresh seawater, may aggregate gorming a new individual and this also is specific to sponges.

## Harmfulness

Boring sponges play a decisive role in the destruction of calcium carbonate skeleton of corals, molluscs etc. and hence they pose a serious threat to coral reefs and economically important molluscan beds of the Indian seas. Boring sponges (34 species) are wide spread in the Indian seas and hence a study in this line may be taken up on a long term basis. Sponges etch out minute particles (microchips) from the calcium carbonate substrata whether shell or coral, by enzymatic and physical means but the biochemistry involved in this process is poorly documented with reference to many species endemic to Indian seas.

Fresh water sponges might be used as indicators of pollution. By virtue of restricted physicochemical parameters within which only sponges can normally survive.

## Chemical diversity in sponges

Availability of suitable substratum for attachment is an important factor, which governs the abundance and distribution of marine sedentary organisms. Having settled comfortably amidst several generations of foulers the sponge larva has to grow to its adulthood. While competing for food it has to protect itself from both predators and other animals and in this struggle the various chemicals produced by them been so handy.

Predators are wide spread in the marine environment and they pose a serious threat to the adult sponges which live exposed and are with no mechanism developed for defence. So they resort to overcome this situation by growing in protected areas, growing rapidly, burrowing boring, developing chemical defence or by entering into symbiotic association with animals and plants. For chemical defence the sponges develop a variety of anitibiotic substances, pigments, toxins anti-inflamatory and anti-arthritic compounds. Most of these chemicals are wither ichthotoxic or antimicrobial in nature. Many of the compounds known in the past from sponges have great pharmaceutical potentials as they have respiratory, cardiovascular, gastrointestinal, anti-inflamatory and antibiotic activities. Already three derivatives (Ara-A, Ara-T and Ara-U) of arabinose nucleosides from *Tethya crypta* have been patented as antiviral and anticancer drugs. There is every reason to except that the next few years will see many more patents on compounds of sponge origin. Systematic screening backed up by synthetic chemistry, pharmacology, microbiological testing and clinical trials has only been applied to marine organisms since the early 1970's and is a long and costly road from active crude extract to nontoxic pure synthesized compound.

The different chemical isolated from sponges and their ecological relationships in a few are outlined below:

**Sterols :** Amazing structure diversity is seen in sterols of Demospongiae (Silicious sponges). 14 different sterols are known from *Spirastrella inconstans*, a common sponge in our seas.

1. Cholesterol accounts to 82% of the mixture in *Phyllospongia foliascens*.
2. Portiferasterol is common in many species.
3. Chondrillasterol is common in *Chondrilla spp.*, but present in many other species also.
4. Strongylosterol is common in *Strongylophora durissima*. A common sponge in the Gulf of Mannar.

**Table. 29.1 : Compounds with biological activity found in sponges**

| | | |
|---|---|---|
| **II.** | **Compounds with biological activity.** | |
| | **A. Terpenoids** | |
| 1. | Sesqueterpenes: | |
| | Picrotin, known from *Spirastrella inconstans*<br>Picrotinin known from *S. inconstans* | Ichthyotoxin |
| 2. | Furanosesqueterpenes: | |
| | Hernacin, from *Dysidae heribacea*<br>Variabilin, from *Ircinia variabilis*<br>Latrunculin from *Latrunculia magnifica* | *Ichthyotoxin*<br><br>Cytotoxic to epithelial carcinoma of larynx |
| 3. | Norses Signosceptrellin A,B,C, from *Sigmosceptrella laevis* | |
| 4. | Sesterterpenes. From *Heteronema erecta, Dysidea herbacea, Phyllospongia foliascenes, P.dendyi, P.radiata, Spongia officianalis, Cacospongia scalaris & C. mollior.* | Ichthyotoxin |
| 5. | Diterprenses: Spongia compound 1,5,7, from *Spongia officinalis.* | Antifungal |
| 6. | Meroditerpenoids : Strongylophorin 1,2,3, from *Strongylophora durissima* | Activity not known. |
| **B.** | **Bromophyrole Compounds**<br>1. Oroidin, from *Aagelas oroides* | <br>Activity not known. |
| 2. | Dibromophakellin, from *Phakellia flabellate* | Activity not known. |
| **C.** | **Compound derived from Dibromotyrosine** | |
| | 1. Aerothionin from *verongia aerophoba*<br>2. Aeroplysinin, from *Verongia aerophoba* | <br>Antibiotic. |
| **D.** | **Prenylated Benzoquinones** | |
| | 1. Compound-A from *Ircinia spinulosa* | Confuses the olfactory system in predators. |
| III. | Sponge pigments | |
| | 1. Spongeoporphyrin, from *Cliona spp.* and *Spriastrella spp.* | Royal purple |
| | 2. Uranidine, from *Verongia spp.* | Yellow pigment |
| | 3. Renieratene, from *Verongia spp.* | Brilliant colour. |
| | 4. Isorenieratene, from *Reniera japonic* | Brilliant colour. |
| IV. | Halogen compound | |
| 1. | Bromine and Iodine, from *Chrotella australiensis, Axinella Sp. Spongia officianalis, Dysidea fragilis, Callyspongia fibrosa, Tedania anhelans, Phyllospongia foliascens and Aurova globostellata.* | |
| V. | Nucleosides and Nucleic acids | |
| | 1. Spongouridine, from *Tethya crypta*<br>2. Spongothymidine, from *Tethya crypta* | <br>Antitumour and antiviral |
| | 3. Spongosine, from *Tethya crypta* | |

**Table. 29.2 : Some sponges with known bioactive properrties**

| | | |
|---|---|---|
| *Petrrosia seriata* | Petrosin A,B, | Ichthyotoxin |
| *Haliclona pigmentifera* | | Diruetic activity |
| *Suberites carnosus* | | Spasmogenic |
| *Microciona prolifera* | | Antiviral |
| *Tedania ignis* | 8 fractions | All antibiotic |
| *Halicolna viridis* | | Bacteriostic |
| *Toxadocia violacea*<br>*Halichondria panicea*<br>*Cliona celata* | | Hypotensive and Paralytic |
| *Oligaceras hermprrhages Dysidae etheria* | | Antimicrobial |
| *Haliclona sp., Agelas S., Axinella Sp.,*<br>*Verongia Sp., Homaxinella Sp., Plakortis Sap.,* | | All bacteriostatic |
| *Verongia cauliformis, V.fistularis* | | Antibacterial |
| *Plakortis sp.* | | Antiyeast |
| *Dysidea herbacea* | | Bacteriostatic. |

Source : Thomas, 1996.

**Table. 29.3 : List of Sponges found in the Gulf of Kutch**

| | | | |
|---|---|---|---|
| **2** | | 22. | Reniera perpermollis |
| **(A)** | **Calcareous Sponges** | 23. | Reniera topsenti thiele |
| 1. | Sycon grantioides | 24. | Reniera fibroreticulata |
| 2. | Grantessa hastifera | 25. | Reniera semifibrosa |
| 3. | Heteropia glomerosa | 26. | Halichondria panicea |
| 4. | Leucandra donnani | 27. | Halichondria reticulata |
| 5. | Leucandra sinwaensis | 28. | Siphonochalina crassifibra |
| 6. | Leaucandra dwarkaensis | 29. | Siphonochalina minor |
| **(B)** | **Non Calcareous Sponges** | 30. | Desmacella tubulata |
| 7. | Myriastra haeckeli sollas | 31. | Thrinacophora cervicornis |
| 8. | Jaspis reptans | 32. | Axinella virgultosa |
| 9. | Asteropus simplex | 33. | Phakellia donnani |
| | **Family Geodiidae** | 34. | Auletta lyrata var. glomerta |
| 10. | Geiodia variospiculosa thiele | 35. | Auletta elongata var. |
| | **Family Geodiidae** | 36. | Ciocalypla dichotoma |
| 11. | Donatia seychellensis | 37. | Higginsia sp. |
| 12. | Tuberela aaptos | 38. | Reniera homelli |
| | **Family Chondrosiidae** | | **Family Desmacidonidae** |
| 13. | Choulrilla australiensis carter | 39. | Esperella plumosa |
| 14. | Choulrilla agglutinans | 40. | Desmacidon minor |
| | **Sub Order Sigmatotetraxonida** | 41. | Iotrochota baculifera |
| 15. | Tetilla dactyloidea | 42. | Guitarra indica |
| 16. | Tetilla hirsla | 43. | Psammochela elegans |
| 17. | Tetilla pilula | 44. | Chondropsis kirkil |
| 18. | Tetilla barodensis | 45. | Myxilla arenaria |
| | **Family Haploscleridae** | 46. | Clathria corallitincta |
| 19. | Gellius fibulatus | 47. | Clathia apiculose |
| 20. | Gellius ribleyi | 48. | Echinodilyum gorgonioides |
| 21. | Gellius fibrosa | 49. | Raspailia fruticose |

*Contd.........*

| | | | |
|---|---|---|---|
| 50. | Acarnus tortilis | 60. | Darwinella australiensis |
| 51. | Bubaris radiata | | **Family Spongelidae** |
| | **Family Spirastrellidae** | 61. | Spongelia fragilis |
| 52. | Spirastrella vagabunda | 62. | Spongelia cinerea |
| 53. | Placospongia cariata | 63. | Spongelia elegans |
| | **Family Clionidae** | | **Family Spongidiae** |
| 54. | Clionacoronaria | 64. | Hisppospongia clathrata |
| | Family Seberitidae | | **Nudi branchs** |
| 55. | Suberites carsosus | 1. | Chromodoris petechinalia |
| 56. | Suberites Flabellatus | 2. | Plocamopherus ceylonicus |
| 57. | Suberites cruciatus | 3. | Bornella rangii |
| 58. | Polymastia gemmipara | 4. | Melibe rangii |
| | **Family Aplysillidae** | 5. | Antiopella indica |
| 59. | Megalopastas retiaria | 6. | Pleraelidia semperi |
| | | 7. | Elysia grandifolia |

Source : Singh, H.S. 1994.

**Fig. 29.1 : Sponge.**

## *References*

Shing, H.S. 1994. Management plan for Marine National Park, and Sanctuary, Jamnagar pp. 184-185. Conservator of Forest, Marine National Park, Govt. of Gujarat, Jamnagar.

Thomas, P.A. 1996. An appraisal of the Biological and Biochemical diversity in sponges. *Mar. Fish. Infor. Serv.* T&E *Ser.* No 142 : 1-5.

# 30
# Sea Turtles

## Introduction

Sea turtles are found all along the coast of India including the Lakshdweep and Andaman and Nicobar Islands. There are five species of marine turtle in habitat in the Indian waters. All species are capable of taking long distance migration. The Indian Wildlife (Protection) Act (1972) has placed all species of marine turtles as endangered species in Schedule-I and are thereby completely protected. While for fresh water turtles, in the Act in Schedule-I mentioned, Turtoies (*Testudinidae, Trionychidae)* it will be desirable to specify the species, once more information becomes available.

## Species of Marine Turtle found in India

| Scientific Name | Common Name |
|---|---|
| 1) *Lepidochelys olivacea* | Olive ridley. |
| 2) *Chelonia mydas* | Green Turtle |
| 3) *Eretmochelys imbricata* | Hawksbill Turtle |
| 4) *Carenta carentta* | Loggerhead Turtle |
| 5) *Dmechelys coriacea* | Leatherback Turtle. |

Out of abvoe No 1), 2) and 3) are more common in Gujarat.

## Species of Freshwater Turtles occurring in India

Family : *Emydidae* (Hard shelled Freshwater turtles)

| Scientific name | Common name |
|---|---|
| 1) *Batagur baska* | The common Batagur or Tantong |
| 2) *Cuore amboinensis* | Amboina or Malayan box turtle |
| 3) *Hardella thurgi* | Brahminy river turtle |
| 4) *Morenia petersi* | Turtle |
| 5) *Geoclemys hamiltoni* | Black pond turtle |
| 6) *Kachyga tecta tecta* | India tent turtle |
| 7) *K. tentoria circumdata* | |
| 8) *K. smithi* | |
| 9) *K. tentodia* | |
| 10) *K. dhongoka* | |

| Scientific name | Common name |
|---|---|
| 11) *K. sylhetensis* | |
| 12) *K. kachuga* | |
| 13) *Heosemys silatica* | |
| 14) *Melanochelys tricarinata* | Tricarinata turtle |
| 15) *M. indopensisuloris* | |
| 16) *M. trijuga trijuga* | Common pond turtle |
| 17) *M. t. thermdis* | Common hard shelled terrapin |
| 18) *M. t. cornata* | Malbar pond terrapin |
| 19) *Cyclemys mouhoti* | |
| 20) *C. dentate* | |
| Family : *Testudinidae* (Tortises) | |
| 21) *Geochelone elegans* | Indian star tortoise |
| 22) *G. (monouria) emys* | Burmese brown tortoise |
| 23) *G. elongutu* | Red nosed tortoise |
| 24) *G. travancorica* | The Travancore tortoise. |
| Family : *Trionychidae* (Soft shelled turtle) | |
| 25) *Lissemys punetata punctata* | Indian spotted flat shelled turtle |
| 26) *L. punctata granosa* | Southern soft shelled turtle |
| 27) *Chitra indica* | Narrow headed soft shelled turtle |
| 28) *Trionyx gangeticus* | Ganges soft shelled turtle |
| 29) *T.leithi* | Nagpur soft shelled turtle |
| 30) *T.hurmus* | Peacock soft shelled turtle. |

*Source* : Rajagopalan and Dan 1983.

## A key for Identification of sea Turtle from India

1. a. Skin smooth and without any scutes on head; 7 longitudional narrow ridges on carapace and 5 on plastrom; hrony beak with well defined cusp on each side of upper jaw (break 'W' shaped when viewed from front) and central cusp on lower jaw; flippers without claws ........................................ *Dermochelys coriacea.*

   b. Shell, head and flippers covered with scutes; no longitudinal ridges on carapace or plastron; horny beak not 'W' shaped when viewed from front; flippers with one or two claws ........................................ 2

2. a. Carapace with 4 pairs of lateral scutes ........................................ 3

   b. Carapace with 5 or more pairs of lateral scutes ........................................ 4

3. a. Horny scute imbricated (Overlapping); two pairs of prefrontal scutes; 2 claws on each flipper; carapace is brown with darker markings; skin of neck region pale orange in colour ........................................ *Eretmochelys imbricata.*

   b. Horny scute not imbricated but justaposed; one pair of prefrontal scutes; single claw on each flipper; carapace green with violet markings; skin of neck region yellow to cream in colour ........................................ *Chelonia mydas*

4. a. Plastron with 3 pairs of enlarged inframarginal, scutes without pores; lateral scutes 5; carapace brownish red with light spots and plastron yellow with orange spots .................................................. *Caretta caretta*

b. Plastron with 4 pairs of inframarginal scutes each with pores on hind margin; lateral scuters 6 or more, generally 7; carapace gray and plastrom yellow .................................................. *Lepidochelys olivacea.*

*Source* : Sailas E.G. *et. al.* 1983.

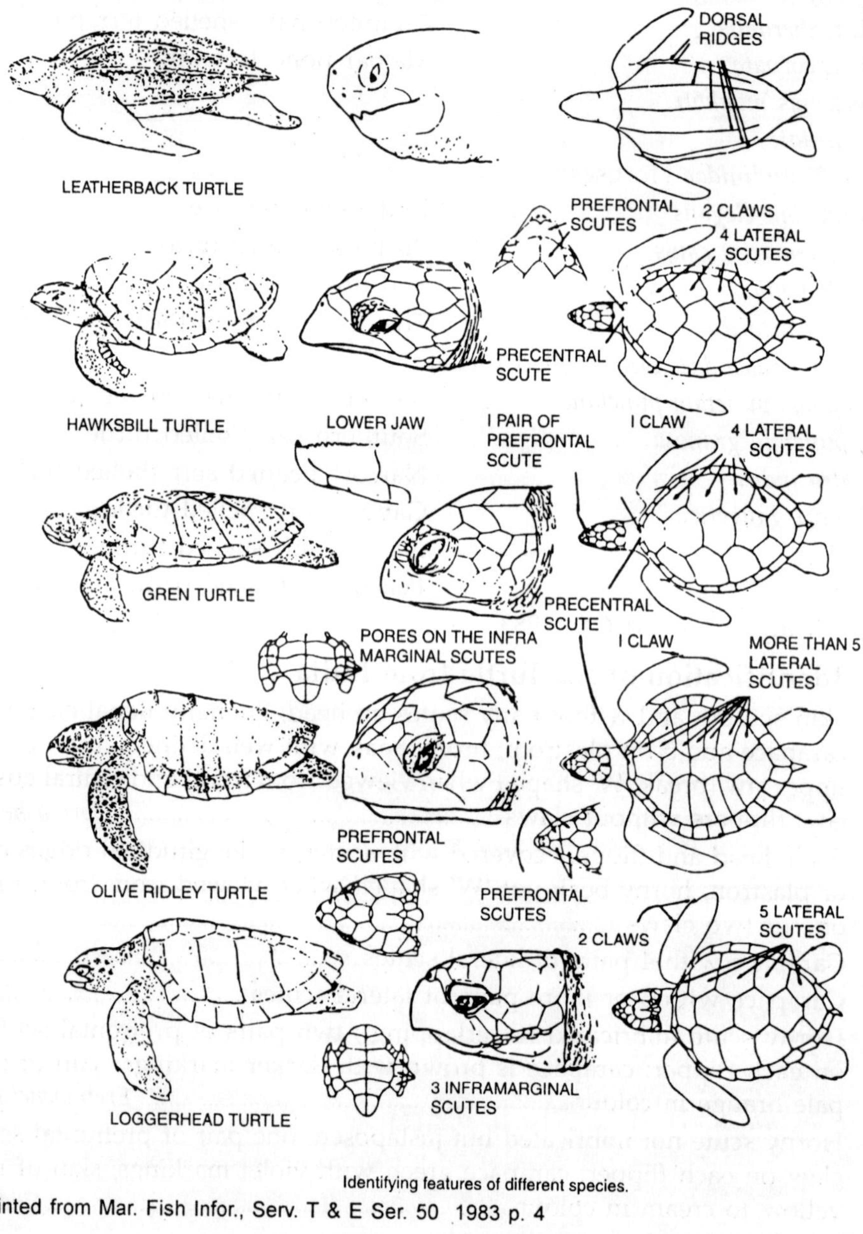


*Source* : Reprinted from Mar. Fish Infor., Serv. T & E Ser. 50 1983 p. 4

**Fig. 30.1 : Identifying feature of different species of Sea Turtle.**

Fig. 30.2 : Dorsal view of *Chelonis mydas*

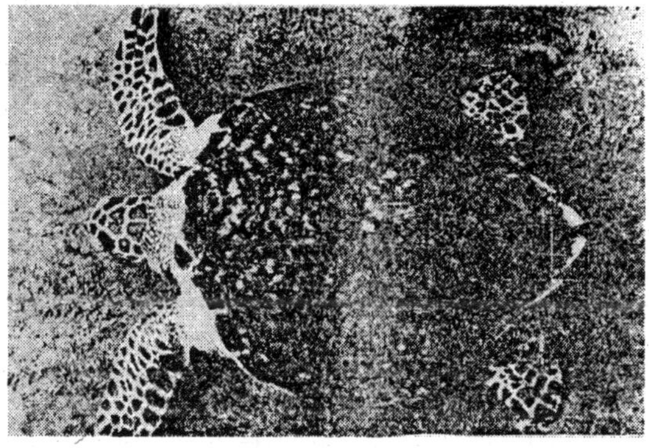

Fig. 30.3 : Dorsal view of *Eretnochelys imbricata*

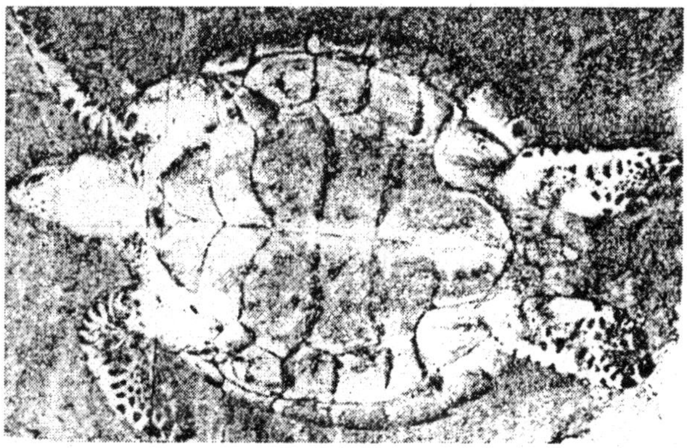

Fig. 30.4 : Ventral view of *E. imbricata*

## Details of species and Nesting Season

### *Lepiochelys olivaca* (Olive Ridely Turtle)

This is the most common sea turtle in Indian waters. Very heavy concentration of this species occurs in Orissa. While it was moderate in Gujarat, Tamil Nadu, Andhra Pradesh, West Bengal and Andaman & Nicobar islands.

Mass nesting occurs in a stretch of 15 km Gahirmaths beach during January-March every year.

### *Chelonis mydas* (Green Turtle)

This is the largest species found in the Indian waters. It occurs on west and east coast of India. Nesting in Gujarat, Andaman & Nicobar and Lakshadweep is moderate. The name Green turtle indicates the green colour of fat. It is predominantly herbivorous and feeds on sea grass.

### *Eretchochelys imbricata* (Hawksbill Turtle)

It is comparatively a small sized turtle and less abundant in the Indian waters. In Gujarat also it is found in rare numbers. It feeds mainly on sponges, crabs, and mollusks. As it frequently feeds on poisonous marine animals, the flesh of this species is often reported to be poisonous.

### *Caretta caretta* (Loggerhad Turtle)

This species is reddish brown in colour and characterized by a large head in relation to the body size. In India recorded only from the Gulf of Mannar. It is carnivorous, feeding on crabs, fish and other benthic animals.

### *Dermochelys coracea* (Leatherback Turtle)

Individuals this species attain a weight of 500 kgs. A thick leathery tissue covers the bones of the shells. They occur in pristine beaches and adjacent waters of Andaman & Nicobar Islands.

### Breeding (Nesting) Season

Nesting season for different marine species of turtle varies in India. Following table gives the details of nesting season of different species in different Maritime states of India.

| Species | Name of State | Nesting season |
|---|---|---|
| *C.mydas* | Gujarat, Maharashtra, Tamil Nadu, | July-January |
| | A&N Islands, | November-January |
| | Lakshadweep | June-September. |
| *L.olivacea* | Gujarat, Maharashtra, Goa, Karnataka and Kerala. | July-September |
| | Tamil Nadu, Andhra Pradesh, Orissa, | December-February |
| | West Bengal, A&N Islands | |
| | Lakshadweep | June-September. |
| *E.imbricata* | A&N Islands | April-January. |
| *D.coriacea* | A&N Islands | December-April |
| *C. caretta* | Not known | Not known |

*Source* : Rajagopalan. M. *et.al.*, 1996.

### Conservation

It was estimated that around 7000 to 7500 turtles were entangled in gill net fishery every year. It is therefore, necessary to create public awareness.

Heavy predation on the eggs of olive ridley especially by jackals, wild boars, hyaena, dogs and other wild animals was noticed. They destroy not only the fresh once but the 5 to 15 days old nests as well. The damage was noticed up to 10 meters inside the mangrove area. Prevention of poaching of turtles by fishermen in the sea is also necessary.

### Need for a National co-ordinated programme for studies on sea turtle

### Area needing priority attention

There is a need for developing a crash programme on sea turtle on scientific lines. This would call for an identification for studies. Concurrent works may have to be undertaken in some of the aspects details below.

1) Survey of the coast line and Bay Islands for identifying nesting beaches and areas to be protected during the nesting seasons. This will also include demarcation of areas to be developed eventually as reserves where human interference will be minimal and thus affording complete protection to turtles during nesting seasons.

2) Developing a national programme where heavy predation by man and animals on nesting turtles and eggs exists. Hatchery programme for incubating and releasing hatchlings into the sea to improve survival at that stage to enhance possible recruitment.
3) More effective implementation of Indian Wild Life (Protection) Act by the State Governments through a programme of training the staff of the concerned Departments in conservation and management of sea turtles and for the implementation of the Wild Life (Protection) Act and regulation.
4) Begin research involving the study of life history, biology and ecology of sea turtles to be carried out with proper dispensation from the concerned authorities.
5) There is a large lacuna in understanding the behavior of all species of turtles from hatchlings to adults, their migratory habits and movements to and from feeding grounds. This is an area, which need important attention.
6) Non-consumptive utilization of turtle resources so that the relationship between man and turtles could change from killing them for its products to utilize them for educational values, captive and display aspects, recreational/tourist potential, ecosystem concept and so on. There is a great need for creating awareness through proper extension and education programmes on sea turtles among the public.
7) Monitoring of turtles incidental catch in fishing operations and finding ways and means of reducing mortality.
8) An impact study on phasing out sbsistence/directed fishery of turtles on the artisanal sector traditionally involved in such activities.
9) Monitoring of the resources to find out whether any of the species could be shifted from Schedule-I to Schedule-II of the Wild Life (Protection) Act, if sufficient scientific data is forthcoming to indicatte the enhancement of resources through proper management measure.

*Source* : Silas E.G. *et. al.* 1983, p. 7.

## *References*

Jayaprakash, A.A., C. Kasinathan and N. Ramamurthy 1993. On the Conservation and Management of Marine Turtles. *Mar. Fish Infor. Serv.* T&E *Ser* 123 : 21-23.

Rajagopan, M. 1983. Leatherbeck Turtle *Dernochelys coriacea* washed ashore at Kovalam, Madras. *Mar. Fish Infor. Serv.* T&E Ser. No 50 : 35.

Rajagopan M.E. Vivekanandan, S Krishna Pillai, M. Srinath and A. Bastian Fernando. 1996. Incidental catch of sea turtle in India. *Mar. Fish Infor. Serv.* T&E Ser 143 : 8-13.

Silas, E.G., M. Rajagopalan, S.S. Dan and Batin Fernando 1985. On the large and Mini arrivals of the Olive redley *Lepidochelys olivacea* at Gahirmatha, Orissa during the 1985 season. *Mar. Fish Infor. Serv.* T&E Ser 64 : 1-16.

Silas, E.G. M. Rajagopalan and A. Bastian Fernando. 1983. Sea turtle of India. Need for a crash programme on conservation and effective management of the Resources. *Mar. Fish. Infor, Serv.* T&E Ser. No 50 : 1-9.

# 31

# MARINE MAMMALS OF INDIA

## Introduction

The marine mammals are one of the important biotic constituents of marine environment. Hershkovitz (1966) has reported 87 species of cetaceans from all the world oceans. The marine mammals of India are comprised of whales, dolphins and dugongs. They are represented by 21 species. As the Indian Ocean is declared as a sanctuary for the whales, the study of whales and dolpihins, some of which are declared endangered species, has attracted international attention. Many countries have intensified their research on the whales and dolphins of the Indian Ocean. The study of sirenians (dugongs) has also received international importance due to their endangered status. (James and Mohan, 1987.)

Though there is no fishery for the dolphins and whales along the India coasts, the smaller cetaceans like dolphins and porpoises are caught in large numbers in the gill nets.

### Classification of Cetaceans

| | |
|---|---|
| Order | Cetacea |
| Sub-order | Mysticeti (Baleen whales) |
| Family | Balaenopteridae |
| | *Balaenoptera musculus* Linnaeus |
| | *Balaenoptera acutorostrata* (Laceoede) |
| | *Balanoptera physalus* Linnaeus |
| | *Balaenoptera borealis*Lesson |
| | *Megaptera novaeangliae* Borowsui |
| Family | Balaenidae |
| | *Eubalaena glacialis* Muller. |
| Sub-order | Odontoceti (Toothed whale) |
| Family | Physeteridae |
| | *Physeter macrocephalus* Linnaeus |
| | *Kogia breviceps* Blainville |
| Family | Hyperodontidae |
| | *Ziphius cavirostris* G.Cuveir |
| Family | Delphinidae (Dolphins and Porpoises) |
| | *Neophocaena phocaenoides* G.Cuvier |
| | *Orcinus orca* Linnaeus |

| | |
|---|---|
| | *Grampus griseus* G.Cuvier |
| | *Pseudorca crassidens* Owen |
| | *Orcella brevirostris* Gray |
| | *Globicephalus macrorhynchus* Gray |
| | *Tursiops truncates aduncus* Ehrenberg |
| | *Sousa chinensis* Osbeck. |
| | *Stenella longirostris* (Gray) |
| | *Delphinus delphis* Linnaeus |
| Family | Susidae |
| | *Platanista gangetica* (Ruxburgh) |
| Order | Sirenia |
| Family | Dugongidae (Dugongs) |
| | *Dugong dugon* (Muller) |

Brief description of the commonly occurring marine mammals along the Indian coasts with their salient features are presented in this chapter to facilitate identification. Fig 31.1 for diagnostic characters of a whale.

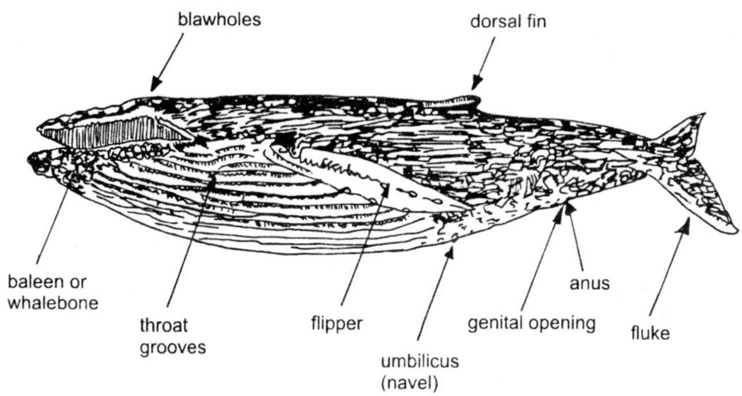


**Fig. 31.1 : Diagnostic characters of a whale.**

## Key to the Identification of Marine Mammals of India

1. 1a Whale bone (baleen plates) absent; teeth present ............ 2
   1b Whale bone (baleen plates) present; teeth absent ............ 16
2. 2a Teeth differentiated in to incisors and molars; lip with developed muzzle, dorsal fin absent ............ *Dugong dugon* (Sea cow)
   2b Teeth uniform, not differentiated (toothed whales and dolphins) ............ 3
3. 3a Teeth on lower jaw only ............ 4
   3b Teeth on both jaws ............ 7
4. 4a Teeth strong ; more than 8 pairs on lower jaw ............ 5
   4b Teeth strong ; less than 8 pairs on lower jaw ............ 6
5. 5a Teeth 9-14 pairs on lower jaw, head 1/6 of body length ; head conical, upper jaw much larger than lower jaw ............ *Kogia breviceps.*

| | | | |
|---|---|---|---|
| | 5b | Teeth 20-30 pairs on lower jaw, head massive, rectangular, 1/3 of body length; lower jaw short and smaller upper jaw projecting; flippers rounded; dorsal fin all defined ........ | *Physeter macrocephalus* |
| 6. | 6a | Teeth 3-7 pairs on lowr jaw, head bulbous, forehead with a vertical crease in center, body with numerous scratches ; flipper tapering ......... | *Grampus griseus* |
| | 6b | Teeth one pair at tip of lower jaw in large males; (Young animals may have more than one air); forehead not distinct; dorsal fin well behind mid point of body; cleft of mouth short; well marked, white bloches in adults ... ........ | *Ziphius cavirostris* |
| 7 | 7a | Eyes rudimentary, not functional; skull with carnial crest; beak well marked; each jaw with 26-30 pairs of small pointed teeth ........ | *Platanista gangetica* |
| | 7b | Eyes well developed, functional; cranial crest absent ........ | 8 |
| 8 | 8a | Teeth spade-like, 15-19 on each jaw; dorsal fin absent ........ | *Neophocaena phocaenoides* |
| | 8b | Teeth conical, pointed, 6-65 on each jaw ........ | 9 |
| 9 | 9a | Teeth 9-15 on each jaw ........ | 10 |
| | 9b | Teeth 20-65 on each jaw ........ | 13 |
| 10 | 10a | Body with large distinct white patches above the eye and base of dorsal fin; teeth strong, 10-12 pairs; dorsal fin erect and larger; flippers paddle shaped ........ | *Orcinus orca.* |
| | 10b | Body without large white patches above the eye and dorsal fin ........ | 11 |
| 11 | 11a | Teeth 7-11 on each jaw; flippers tapering ........ | 12 |
| | 11b | Teeth 12-19 on each jaw; flipper paddles shaped; forehead bulbus; beak absent ........ | *Orcella brevirostris* |
| 12 | 12a | Teeth 7-10 on each jaw, flapper long, 1/5 to 1/7 of body length, forehead bulbous; outer margin of flippers without hump, dorsal fin not pointed. ...... ........ | *Globicephalus macrorhynchus* |
| | 12b | Teeth 8-11 on each jaw; flippers 1/8 to 1/10 of body length, outer margin of flippers with hump ........ | *Pseudorca crassidens* |
| 13 | 13a | Teeth 24-28 on each jaw; beak distinct; adults with gray oval spots ventrally ........ | *Tursiops truncatus aduncus* |
| | 13b | Number of teeth more than 30 in each jaw ........ | 14 |
| 14 | 14a | Number of teeth 34-38 on each jaw; base of dorsal fin extended posteriorly; a dorsal hump present ........ | *Sousa chinensis* |
| | 14b | Number of teeth more than 42 on each jaw ........ | 15 |
| 15 | 15a | Numbr of teeth 45-51 on each jaw, a dark gray stripe from flipper to eye.... ........ | *Stenella longirostris* |
| | 15b | Number of teeth 60-65 on each jaw; palate test; dark grey band from base of flippers to base of lower jaw ........ | *Delphinus delphis* |
| 16 | 16a | Dorsal fin absent; surface of throat not grooved; chin not white; head 1/4 of body; upper border of lower lip much arched; bonnet on head ........ | *Eubalaena glacialis.* |

17 17a Flipper extremely long, 1/3 of body length, lower margin of flippers scalloped ............ *Megaptera novaeangliae*

17b Flippers less than 1/5 of body length, not scalloped below the flippers; baleen plate black, white or partially white ............ 18

18 18a Baleen plates black; throat grooves 80-100 extending to at least umbilicus, length of flippers 1/7 of total length ............ *Balaenoptera musculus*

18b Baleen plates white or partially white; inner side of flippers white ............ 19

19 19a Throat grooves about 50, extending to half way to umbilicus; flipper length 1/8 of total length ............ *Balaenoptera acutorostrata*

19b Baleen plates white or slate coloured; inner side of flippers white ............ 20

20 20a Baleen plates black with frayed white inner edge; flipper length 1/10 to 1/12 of total length; throat grooves 30-60 extending midway between flippers and umbilicus ............ *Balaenoptera borealis*

20b Baleen plates white along anterior third, frayed edge white; flipper length 1/9 of total length; throat grooves 60-90 extending to umbilicus; inner side of flippers white ............ *Balaenoptera physalus*

*Source* : *Mar. Fish. Infor, Serv., T & E Ser.,* 71:1987 ; 11-12.

## 1. *Balaenoptera musculus* Linnaeus (Blue whale)

Throat grooves 80-100, extending to umbilicus; baleen plates jet black; flippers whitish below and gray above.

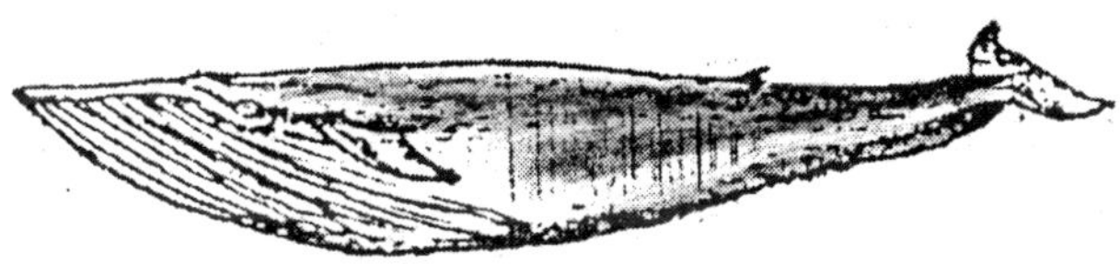

**Fig. 31.2 : *Balaenoptera musculus* Linnaeus (Blue Whale)**

## 2. *Balaenoptera acutorostrata* Lacepede (Minke whale)

Dorsal fin small, markedly falcate. Curved backwards; usually grayish black above and whitish below 'flipper with a prominent white band; baleen plate white with a black streaking.

**Fig. 31.3 : *Balaenoptera acutorostrata* Lacepede (Minke Whale)**

## 3. *Balaenoptera physalus* Linnaeus (Fin whale)

Throat grooves 50 to 90, extending to umbilicus; body grayish black above and white below; flipper uniformly gray dorsally, white ventrally; dorsal fin small, falcate.

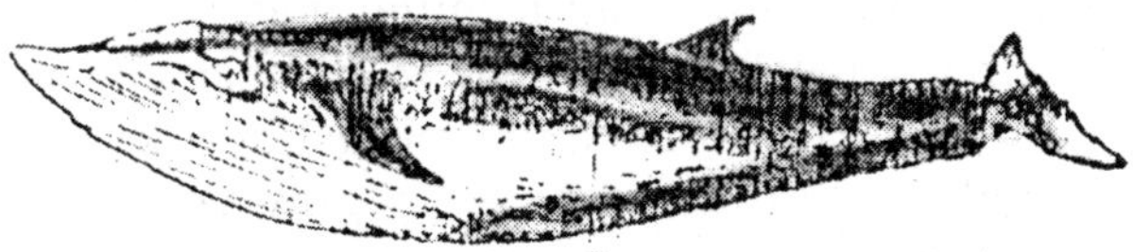

**Fig. 31.4 : *Balaenoptera physalus* Linneus (Fin Whale)**

4. *Balaenoptera borealis* Lesson (Sei whale)

Throat grooves 32-60, distinct and terminating above half way between tip of flippers and umbilicus; baleen plates black but a few partly white; body gray to bluish gray above; dorsal fin deeply concave on the posterior margin; flippers usually small and measures about 1/4 of the total length.

**Fig. 31.5 : *Balaenoptera barealis* Lesson (Sei Whale)**

5. *Megaptera novaengliae* Borowski (Humpback whale)

Throat grooves 19-20, extending from chin to navel; body stockier than the other whales; flipper long and black dorsally and white ventrally and measures 1/4 to 1/3 of the total length; irregular knobs and protuberances on head and flipper; dorsal fin small; whale bone gray-back or white.

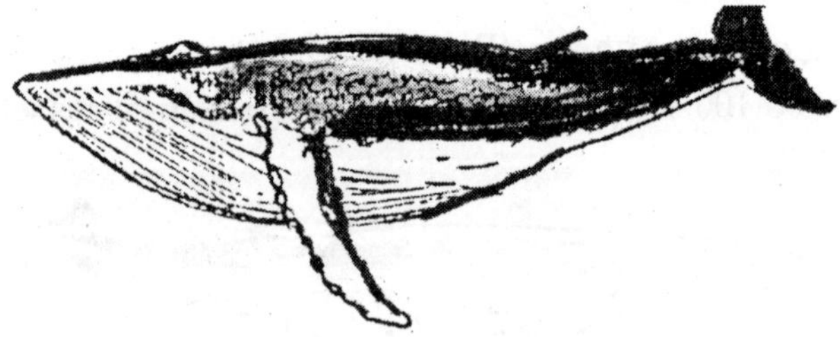

**Fig. 31.6 : *Megaptera novaeanglie* Borowski (Hump back Whale)**

6. *Eubalaena glacialis* Muller (Black right whale)

The baleen plates much longer than in other species; upper jaw with a few horny protuberances; throat grooves and dorsal fin absent.

**Fig. 31.7 : *Eubalaena glacialis* Muller (Black right whale).**

7. *Physeter macrocephalus* Linnaeus (Sperm whale)

Resembles a gigantic tadpole with head forming massive conspicuous part of body; lower jaw weak and disproportionately small with 18-28 large and powerful teeth on each side. No teeth on upper jaw; dorsal fin absent and replaced by a series of low ridges. Flippers broad and rounded. Body black dorsally and silvery gray ventrally.

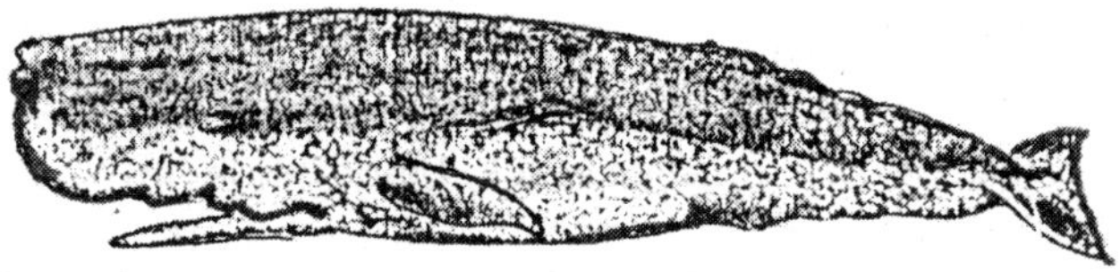

Fig. 31.8 : ***Physeter macrocephalus*** **Linnaeus (Sperm whale)**

## 8. *Kogia breviceps* Blainville (pygmy sperm whale)

Body black above and grayish white below' dorsal fin small and falcate; flippers smoothly curved on forward margin and located well interiorly; body extremely robust and rapidly tapering near the tail; head more or less square in shape; lower jaw narrow and inferior in position, located well behind tip of snout;

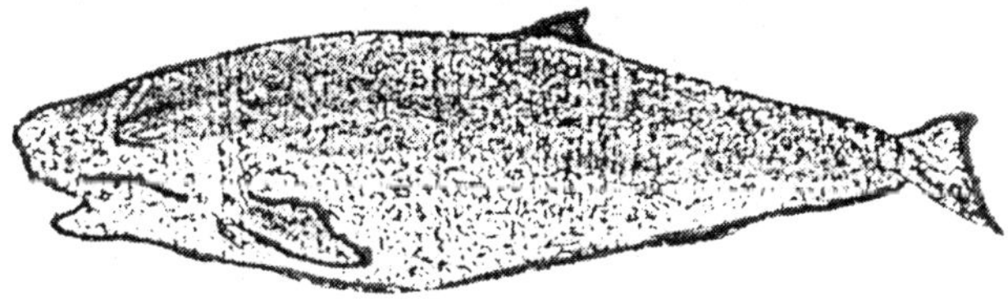

Fig. 31.9 : ***Kogia breviceps*** **Blainville (Pygmy sperm Whale)**

## 9. *Ziphius cavirostris* G. Cuvier (Cuvier's beaked whale)

Beak short; body purplish black above, brown on sides and white below; dorsal surface of head behind beak slopes backwards at an oblique angle from snout; distance from tip of snout to blowhole about 1/10 of total length; teeth not usually visible in females; often 20-30 teeth found in very young animals; characterized by a single pair of teeth on jaw in adult males.

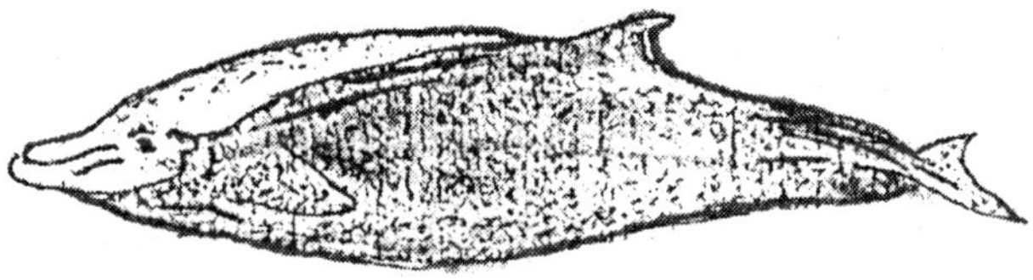

Fig. 31.10 : ***Ziphius cavirostris*** **G.Cuvier (Cuvier's beaked whale).**

## 10. *Neophocaena phocaenoides* G.Vuvier (Black porpoise)

Forehead rounded and almost protuberant in profile; dorsal fin absent; minute tubercles or horny scales present dorsally; body black except for dark gray patches between flippers; teeth conically spade-shaped unlike conically pointed teeth of dolphins; 15-19 teeth on each side of upper and lower jaws.

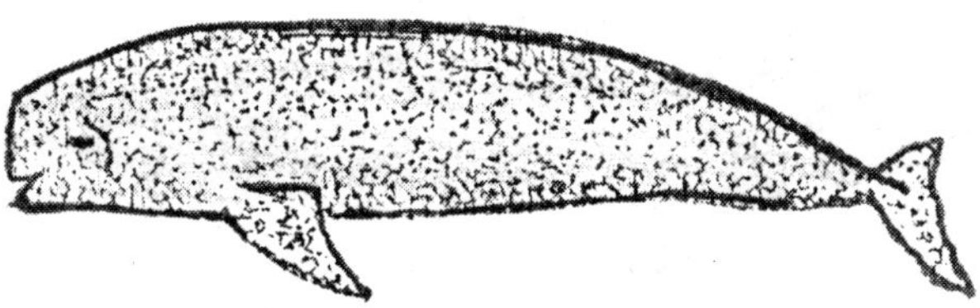

Fig. 31.11 : ***Neophocaena phocaenoides*** **G.Cuvier(Black porpoise).**

### 11. *Orcinus orca* Linnaeus (Killer whale)

Body stream lined; flippers rounded; dorsal fin conspicuous, situated almost midway; flipper length about 1/9 of the body length in young animals and 1/5 in old animals; colour well marked and distinctive; dorsal surface black, belly white; an oval white patch on sides of head just above and behind eye; chin white, lateral white patch a very constant feature; 10 to 12 large powerful conically pointed teeth on each side of upper and lower jaws.

**Fig. 31.12 : *Orcinus orca* Linnacus (Killer whale).**

### 12 *Grampus griseus* G.Cuvier (Risso's dolphin)

Body robust, head bulbous, flipper long pointed. Body light gray to dark, scarred with numerous scratches; not beaked; forehead vertically creased in center; dorsal fin less than 38 cm, erect and distinct; fluke deeply notched. Teeth three to seven pairs on lower jaw only.

**Fig. 31.13 : *Grampus griscus* G. Cuvier (Rissos dolphin).**

### 13. *Pseudorca crassidens* Owen (False killer whale)

The body elongated and slender. Head narrow, gently tapering from blowhole forward. Snout projects a little beyond the tip of lower jaw; dorsal fin small, tip directed backwards, with posterior border concave. Flippers tapering characterised by a broad hump in the front margin near the middle, about 1/10 of the body length. Body entirely black, teeth large, pointed with 8 to 11 pairs in upper and loweer jaws.

**Fig. 31.14 : *Pseudorca crassidens* Owen (False Killer Whale)**

## 14. *Orcella brevirostris* Grey (Irrawady dolphin)

Head convex from blowhole to upper tip; mouth hoizontal. Flipper long, broadly triangular, two times its breadth; center of dorsal fin behind middle of body. Dorsal fin small, falcate with rounded ends. 12 to 19 teeth on each side of upper and lower jaws; small and sharply conical in young animals; flat is older animals.

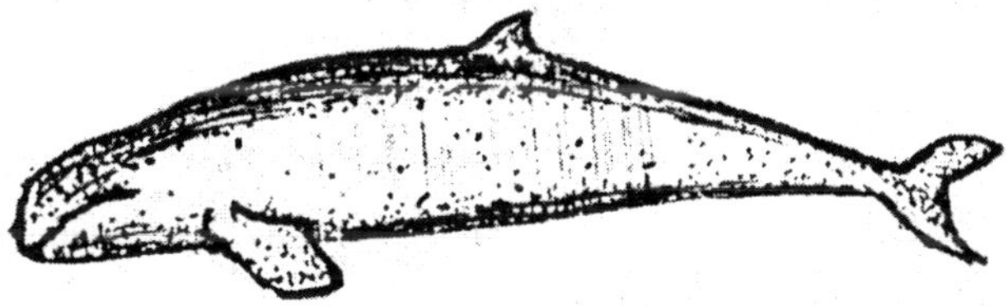

**Fig. 31.15 : *Orcella brevirostris* Gray (Irrawaddy dolphin)**

## 15. *Globicephalus macrorhynchus* Gray (Short-finned pilot whale)

The head somewhat thick and bulbous. In old males the forehead overhangs the mouth to several inches. Flippers 1/16$^{th}$ of the body length or less and sickle shaped. Tail dorso-ventrally thickened in front of fluke; dorsal fin low. Body colours almost entirely black; 7-9 teeth on each side of upper and lower jaws and confined to front portion of the jaws.

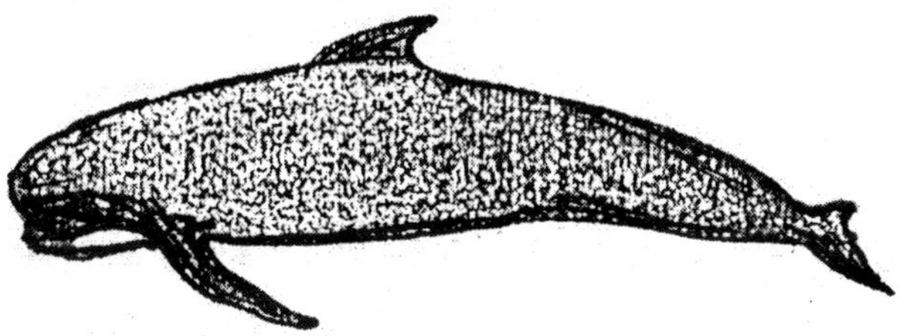

**Fig. 31.16 : *Globicephalus macrorhynchus* Gray (Short finnened pilot Whale)**

## 16. *Tursiops truncates aduncus* Ehrenberg (Bottlenose dolphin)

Characterized by a long well defined snout and a bulbous forehead. Dorsal fin broad at base, tail falcate. Flippers moderate in size, tapering to a point. Teach numbering 24 to 28 on each side of upper and lower jaws. Colouration usually dark gray brown on the back and light gray on the belly. Larger specimen pigmented on the ventral side.

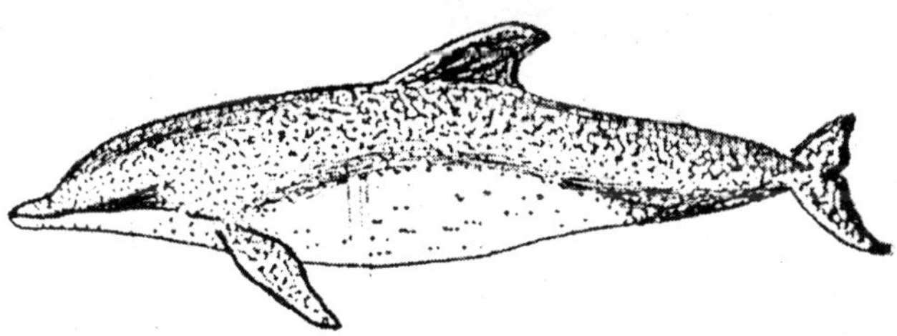

**Fig. 31.17 : *Tursiops truncatus adduncus* Ehrenberg (Bottle nose dolphin)**

## 17. *Sousa chinensis* (Osbeak) (Hymp-back dolphin)

Beak long and forehead bulbous; characterized by a hump on its back; dorsal fin with a little elevation and posterior border feebly indented; flippers recurved and distinctly broad at base. 34 to 37 teeth on each side of upper and loweer jaws. The colour uniformly lead gray; longitudinal blotches on ventral side in larger specimens.

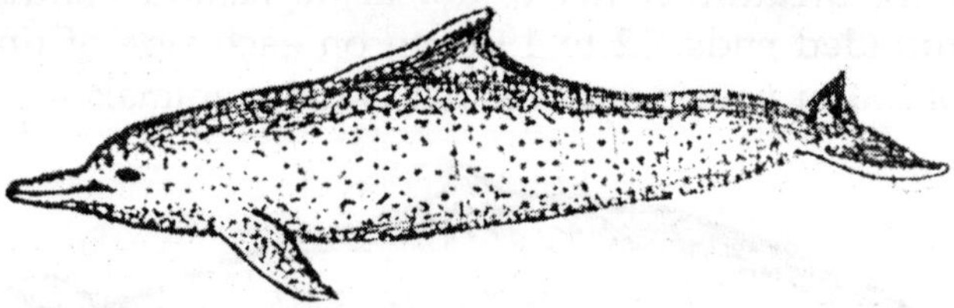

**Fig. 31.18 : *Souse chinensis* (Osbeek) (Hump back dolphin)**

18. *Stenella longirostris* Gray (Spinner dolphin)

Body slender, dorsal fins moderately falcate to traingular and very eret. Tip of snout distinctly black and dark gray to black dorsally; yellowish brown laterally and white ventrally. 44-51 teeth on each side of upper and lower jaws. Teeth small, slightly curved inward, Occurs in herd of several hundred individuals; seen jumping over the waves.

**Fig. 31.19 : *Stenella longirostris* Gray (Spinner dolphin)**

19. *Delphinus delphis* Linnaeus (saddle-back dolphin)

Characterised by a well defined narrow beak and with distinct colourations. Beak acute than *Tursiops aduncus.* Dorsal fin moderately developed and tapers backwards with a concave hind margin. Flippers moderate in size, tapering; body black dorsally, lower surface white, upper and lower jaws with 62-65 teeth on each side.

**Fig. 31.20 : *Delphinus delphis* Linnaeus (Saddle back dolphin)**

20. *Platanista gangetica* Roxburgh (Gangetic dolphin or Susu)

Forehead rises rather steeply to upper surface of head; snout long; dorsal fins almost ridge like. Tail flukes broad; posterior border with a notch in middle; flippers short and broad at their outer end. Eyes very small and animal almost blind. Teeth small about 28-30 on each half of upper and lower jaws. Body lead black and ventral surface pale white.

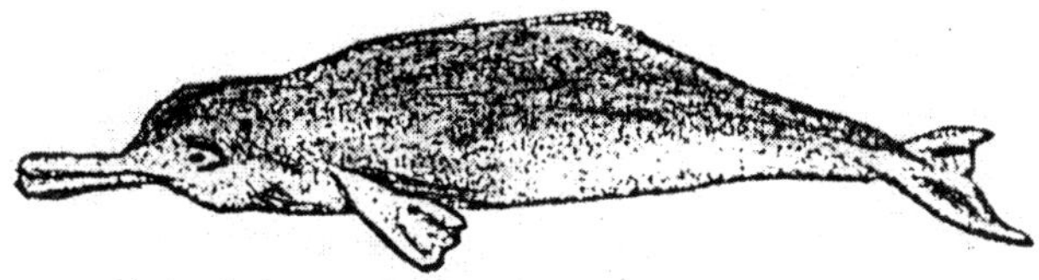

Fig. 31.21 : *Platanista gangetic* Roxburgh (Gangetic dolphin or Susu).

## 21. *Dugong dugon* Muller (Sea cow)

The body spindle shaped, divisible into a head, trunk and tail. Head relatively small, characterized by a muzzle, and a broad flat horse-shoe shaped extension of the upper lip which overlaps sides of mouth; large number of hairs and bristles on the muzzle; eyes small; dorsal fins absent; flippers round towards the end. Dentition with incisors and molars. In male incisors long and pointed. Skin of dugong thick and hairs present on the surface. Tail strengthened by a low ridge mid-dorsally and mid-ventrally and caudal fluke crescentic.

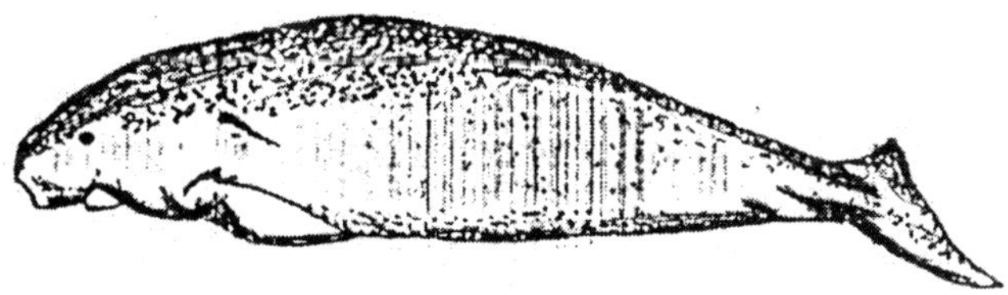

Fig. 31.22 : *Dugong dugon* (Muller) Sea cow

## Food and Feeding

All whales are carnivorous, in the sense that they feed on other animals but while odontocetes have retained a fairly typical pattern of obtaining their food, by catching and eating one at a time, prey items usually between ten or thousand times smaller than themselves. The mysticetes browse or graze through the oceans, staining out, by the thousands, food items which are usually between a hundred million and one million times smaller than the feeding whales.

The common Dolphin have typical dentition where forty-fifty slender, peg-like teeth, about 3 mm in diameter at the gum, are set along each margin of both upper and lower jaw. The spotted dolphin and spinner feed largely on the abundant squid besides numerous fish (FAO 1978)

Details of feeds generally taken by different species are shown in Table 31.1. The average food consumption throughout the year would correspond to about 1.5 to 2% body weight per day. For a big Blue whale 28 m long, weighing about 1,46,400 Kgs, the food requirement would be 4124 Kgs per day. The great weight increase required by the pregnant females. Calculated assimilation efficiencies varied from 76% for a young adult Blue whale to 87% for a mature female Fin whale. These very high values for growth and assimilation efficiency serve to remind as how efficient a system the whales form as harvesters of plankton in the world's ocean.

**Table 31.1 : Showing Food and feeding and Distribution of Marine mammal of India**

| Sr No. | Name of Species | Food and Feeding | Distribution |
|---|---|---|---|
| 1. | *Balaenoptera Musculus* | Euphausiids and shrimps | Along the coast of Bengal, Gujarat, South Karnataka and Tamil Nadu. |
| 2. | *B.acutorostrata* | Copepods, krill and other Zooplankton and anchovies. | Gulf of Mannar, Tuticorin |
| 3. | *B.phylsalus* | Krill and anchovies | Recorded from Bombay and Surat. Occurs in schools of 200-300 in number. |
| 4. | *B.borealis* | Sardines, euphausiids and copepods. | All over the world. |
| 5. | *Megaptera novaeangliae* | Krill and fishes. | A widely distributed in all oceans; often seen along coastal waters. |
| 6. | *Eubalaena glacialis* | Krill and fishes | Abundant before 150 years in Southern hemisphere but later ruthlessly exploited. Now depleted in many areas. |
| 7. | *Physeter macrocephalus* | Squid and fishes | Throughout great oceans through prefers warm waters of tropical and sub-tropical regions. Reported from Madras, Pamban, Karwarcoast and Manglore |
| 8. | *Kogia breviceps* | Cephalopods and fishes | Pacific and Indian Oceans; in temperate and tropical waters. Not a common species along the Indian it was reported once from Waltair coast. |
| 9. | *Ziphius cavirostris* | Squids and deep sea fishes. | Reported from coasts of Europe, America, Africa, Australia, Japan and Lakshadweep islands. |
| 10. | *Neophocaena phocaenoides* | Small squids, shrimps and fishes | It is abundant along the coast of Bombay and tends to frequent rivers and estuaries. In India reported from all along the coasts. |
| 11. | *Orcinus orca* | Fishes, cephalopods, larger marine mammals and turtles | All seas from Arctic to Antarctic; more common in cooler water. In India recorded from Okhamandal in Gujarat coast. |
| 12. | *Grampus griseus* | Squids and fishes. | In India reported near the coast of Sri Lanka. |
| 13. | *Pseudorca crassidens* | Fishes and cephalopods. | In India recorded from Cape Comorinm, Trichendur Madras, Trivendram, Pozhikara and Calicut. |
| 14. | *Orcella brevirostris* | Fishes and crustaceans | In India reported from Waltair coast. |
| 15. | *Globicephalus macrorhynchus* | Mainly on fishes. | Largely in south American coast Mass standing in India reported near Calcutta and Tuticorin. |
| 16. | *Tursiops truncatus aduncus* | Mainly on inshore fishes like sardines and mackerals. | Large number in U.S.A., Japan Small numbers in West Africa, Sri Lanka and India |
| 17. | *Sousa chinensis* | Mainly fishes | China, Occurs along the Indian coast often caught in gill nets in Goa, Manglore and Calicut. |
| 18. | *Pelagic fishes and squids.* | Mainly on pelagic fishes and squids | Pacific and Indian Oceans Reported along the Indian coasts. |
| 19. | *Deohinus delphis* | Fishes and squids | Found with Tuna fishes in Pacific Oceans. In India from Bombay, Goa, Manglore, Calicut and Cochin. |
| 20. | *Platanista gangetica* | Fishes shrimps and molluscs by ploughing its snout in to the mud. | *A highly endangered species and requires urgent protection.* Distributed in Ganges, Brahmaputra and Karnaphuli river systems. |
| 21. | *Dugong dugon* | Fishes and plankton a voracius feeder | Sluggish animals inhabiting the coastal waters; Found in Gulf of Mannar coast Palk Bay Gulf of Kutch and Andaman islands. |

**Table 31.2 : Showing details of Breeding season, Gestation period, Length Maturity etc., of Marine mammals of India**

| Sr. No. | Name of Species | Breeding Season | Gestation Period | Length at the time of birth | Maximum Length | Sexually matured |
|---|---|---|---|---|---|---|
| 1. | *Balaenoptera musculus* | February-March | 9-10 months | N.A. | 26m | N.A. |
| 2. | *B.acutorostrata* | December to May in North Atlantic, January to June in North Pacific | 10 months | 2.8 m | 9 m | N.A. |
| 3. | *B.physalus* | May to August | About a year | 7m | 24m | 3rd year at about 20m |
| 4. | *B.borealis* | Early spring but mating all the year | 10-11 months | N.A. | 19m | N.A. |
| 5. | *Megaptera novaeangliae* | Mating in Winter Migrate to warmer water for breeding | N.A. | 5 m | 16m | N.A. |
| 6. | *Eubalaena glacialis* | N.A. | N.A. | N.A. | 16m | N.A. |
| 7. | *Physeter macrocephalus* | All the year | 10 to 12 months | 4 m | 21m | N.A. |
| 8. | *Kogia breviceps* | N.A. | N.A. | N.A. | 3.5m | N.A. |
| 9. | *Ziphius cavirostris* | N.A. | N.A. | 2 to 3m | 7m | Male 5.4m Female 6.1m |
| 10. | *Neophocaena phpcaenoides* | October | N.A. | N.A. | 1.8m | N.A. |
| 11. | *Orcinus orca* | N.A. | 12 months | 2.5m | 9 m | Male 6.7m Female 5m |
| 12. | *Grampus griseus* | N.A. | N.A. | N.A. | N.A. | Male 3m |
| 13. | *Pseudorca crassidens* | Fairly extensive | N.A. | N.A. | 5.5m | 4.25 to 4.5 m |
| 14. | *Orcella brevirostris* | N.A. | N.A. | 0.8m | 2.4m | N.A. |
| 15. | *Globicephalus macrorhynchus* | Extended breeding and calving season | Calving inteval about three years | 1.4m | 5.3m | Male 3.8m Female 3.0m to 3.2m |
| 16. | *Tursiops truncates* | Extensive | 12 months | 1.0 to 1.2m | 3.7m | Male 2.5 to 2.6m Female 2.2 to 2.4m |
| 17. | *Sousa chinensis* | N.A. | N.A. | N.A. | 3.2m | N.A. |
| 18. | *Stenella longirostris* | N.A. | N.A. | 0.8m | 2.5m | N.A. |
| 19. | *Dolphinus delphis* | N.A. | 10-11 months | 0.8-0.9m | 2.5m | At 4 years |
| 20. | *Platanista gangetica* | April to July | 8 to 9 months | 0.7-0.8m | N.A. | Male 2.1m Female 2.5m |
| 21. | *Dugong dugon* | N.A. | N.A. | 90 Cm | 3 m | Female 2.3m |

*Source : Mar. Fish. Inf. Ser.* T & E *Serv.* 71.

## Maturity

As sexual maturity in females is closely associated with physical size, a young whale that growns faster will become sexually mature earlier. From the observation of about 40 years it was found that mean age of sexual maturity varies from $6^{th}$ to $10^{th}$ year of whale's birth.

Physical maturity is not reached until the male is 20.9 m long and the female 22.5 m in Fin whale. Female sperm whale become sexually mature at about 10 years and physical growth stops of about 11.1 m corresponding to an age of 28-29 years (Best, 1970)

Sexual maturation of some of the whales and dolphin is shown in Table 31.2.

## Reproduction

The male reproductive organs in whales are relatively simple. The penis of the whale resembles in structure that of cattle. The layout of the female reproductive organs in the whale is much more typical of mammalian pattern.

## Courtship

The courtship some times very conspicuous. They roll, smack the water with their enourmous flippers, crash down their flukes on the surface, and even leap right out of the water, to fall back again with tremendous splash. When these activities have reached sufficiently high level of sexual excitement, copulation follows. Usually in side-to-side position, laying at the surface of the water. male Bottlenosed Dlophins swim up from their patterns and mate with their bodies more or less at right angles to each other; mating lsts from two to ten seconds. (Bonner, 1980).

## *References*

Best, P.B. 1970. The Sperm Whale (*Physeter catodon*) off the west coast of South Africa. "Age, growth and Maturity" 246: 1-259. 79; pp. 1-27.

Bonner, W.Nigel 1980. Blandford Mammal Series Whales. Blandford Press, Poole, U.K.

F.A.O. 1978. Mammals in the Sea. Report of the FAO Advisory Committee Marine Resources Research Working on Marine Mammals. F.AO., United Nations, Rome.

Hershkovitz, P 1966. Catlog of living Whales. *Bull. Smith Inst., U.S.Natn. Mus. Washington,* 246: 1-259.

James, P.S.B.R. and R.S. Lal Mohan 1987. The Marine Mammals of India. *Mar. Fish. Infor. Ser.* T & E *Serv., No* 71.

# 32
# PLANKTONS AND THEIR IMPORTANCE IN FISHERY

Planktons are divided in two groups

1) Phytoplankton 2) Zooplankton

## Phytoplankton

Phytoplankton are small plants, mostly microscopic wither are weakly motile or drifted in water subject to the action of waves and currents. Phytoplankton forms the base of the food for many animals and therefore, is extremely important to the ecosystem. Phytoplanktons are the products and belong to first tropical level (autotrophs).

## Zooplankton

Zooplanktons are those organisms, within the aquatic ecosystem. They are intermediatories in the food chain and they frequently graze wither on phytoplankton or on the detritus with its hetrotrophic community. Majority of aquatic animals are represented at some stage within the zooplankton community. Crustaceans are the dominant group among the members, who remain in zooplankton community throughout their life. Most of the fishes are zooplankton feeders. Some of them in their ealy life stages feed on phytoplankton but later, switch to the zooplankton diet. Large quantities of Indian's marine fish catch originate from the estuaries. (Paulinose, *et, al.*, 1998).

Zooplankton characteristics of Gulf of Kutch including major creek system of Nakti, Kandla an Hansthal, for monsoon, post monsoon and pre monsoon period were as under. The area was rich in Zooplankton. Standing stock showing variation of 0.59 to 185.5 ml/ 100 sq m. The maximum biomass was at 112.6 ml/100sq m in monsoon; 109.91 ml/100 sq m in post monsoon and 88.01 ml/100 sq m in pre monsoon. Similarly the population maxima were 942 no/100 $m^3$ during monsoon, 196 no/100 $m^3$ in post monsoon and 80.27 No. / 100 $m^3$ during pre monsoon. *Metapenaeus affinis, M dobsoni, P.indicus, Parapenaeopsis stylefera* and *hardwikii* were the species of penaeid prawns present in the samples. Larvae of *M.affinis* were dominant in the Gulf during monsoon, whereas *Acetes indicus* dominant during pre monsoon and post monsoon. Among the fish larvae, majority of them were in the early stages of development and the size varied from 1 mm to 5 mm. *Boleopthalmus glaucus, Engraulis mystax* and *clupedes* were the most common fish larvae present in this Gulf of Kutch. (Paulinose *et. al.*, 1998)

Most common species found in the Gulf of Kutch were *Harpodon nehereus (21.5%) Coilia (15.6%) Hilsa (8.3%) Clupeid (7.9%), Polynemus tetradacylus (5.5%). Pampus argenteus (4.6%)*

and *Cybium sp. (4.4%)*. The detailed analysis showed that Gulf sustained three times more fishing potential than the inner creel. (Paulinose *et. al.*, 1998).

**Table 32.1 Phytoplanktons and Zooplanktons found in Gulf of Khambhat-Bhavnagar District**

**AT GHOGHA CREEK**

| Sr. No. | Phytoplanktons | Sr. No. | Zooplanktons |
|---|---|---|---|
| 1 | *Casinodiscus Sp.* | 1 | *Copepode* |
| 2 | *Rhizosolania Sp.* | 2 | *Rotifer* |
| 3 | *Chaetocerose Sp.* | 3 | Sponge Spiculs |
| 4 | *Nitzia Sp.* | 4 | Prawn eggs |
| 5 | *Plurosigma Sp.* | 5 | *Lamitods* |
| 6 | *Tricerasium Sp.* | 6 | *Mysids* |
| 7 | *Climacosphinia Sp.* | 7 | *Segitta.* |

**At KATPAR Bhavnagar District**

| Sr. No. | Phytoplanktons | Sr. No. | Zooplanktons |
|---|---|---|---|
| 1 | *Plurosigma Sp.* | 1 | *Copepode* |
| 2 | *Cosinodiscus Sp.* | 2 | *Mysids* |
| 3 | *Tricerasium Sp.* | 3 | Fish larvae |
| 4 | *Rhizosolania Sp.* | 4 | Protozoea stages of prawn |
| 5 | *Climacosphinia Sp.* | 5 | Sponges spiculs. |

**At JAFRABAD**

| Sr. No. | Phytoplanktons | Sr. No. | Zooplanktons |
|---|---|---|---|
| 1. | *Chaetoceros Sp.* | 1. | *Copepods* |
| 2. | *Cosinodiscus Sp.* | 2. | *Mysids* |
| 3. | *Rhizosolania Sp.* | 3. | *Fish larvae* |
| | | 4. | Protozoea stages of prawn |
| | | 5. | Crustacean eggs |
| | | 6. | *Lucifer* |
| | | 7. | *Rotifer* |

*Source* : "Matsyavikas" 1981, Reprinted with the permission from Commissionerate of Fisheries, Gandhinagar.

**Table 32.2 : Major phytoplanktons available in Gulf of Kutch**

| Area | October 1989 GMB Jetty | February 1993 Okha | 93-94 Sikka |
|---|---|---|---|
| Surface | *Rhizosolenia* | *Nitz schia* | *Chaeatoceros* |
| | *Synedra* | *Coscinodiscus* | *Coscinodiscus* |
| | *Chaetorceros* | *Synedra* | *Nitzschia* |
| | *Navicula* | *Navicula* | *Navicula* |
| | *Nitzschia* | *Skeletonema* | *Synedra* |
| | *Pleurosigma* | | *Rhizosolenia* |
| | *Thalassiothrix* | | *Pleurosigma* |
| Bottom | *Rhizosolenia* | *Nitzschia* | *Chaeatoceros* |
| | *Nitzschia* | *Coscinodiscus* | *Nitzschia* |
| | *Biddulphia* | *Thalassiosira* | *Coscinodiscus* |
| | *Chaeatoceros* | *Navicula* | *Navicula* |
| | *Stauroneis* | *Synedra* | *Synedra* |
| | *Pleurosigma* | *Rhizosolenia* | *Pleurosigma* |

**Table 32.3 Major groups of zooplanktons available in Gulf of Kutch**

| February-March 1985-Off Sikka | October 1989-GMB Jetty | February 1993-Okha | 1993-94 Sikka |
|---|---|---|---|
| Copepods | Decapods | Copepods | Decapods |
| Decapods | Medusae | Chaetognaths | Copepods |
| Siphonophores | Copepods | Decapods | Gastropods |
| Medusae | Siphonophores | Polychaetes | Fish eggs |
| Polychaetes | Fish eggs | Amphipods | Fish Larvae |
| Amphipods | Chaetognaths | Gastropods | Siphonophores |
| Ostracods | Lucifer. Sp. | Lucifer. Sp. | Mysids |
| Fish Larvae | Gastropods | Fish Larvae | Chaetognaths |
| Mysids | | Siphonophores | Lucifer.Sp. |

Identification and their scientific name of some of the marine plankton are given in plate no 1 to 4 (Total 127) and attached list. Plate N0 5 shows 37 fresh water plankton found in India. Their scientific name is given in attached list. Planktons found in Gulf of Khambhat-Bhavnagar District of Gujarat is given in Table 32.1. Table 32.2 and Table 32.3 gives the major group of phytoplankton and zooplankton of Gulf of Kutch is given. N.I.O. Council of Scientific and Industrial Research Bombay (1995) have assessed faunal composition of macrobanthos at Sikka during pre monsoon and post monsoon period which is reprinted as Table 32.4. Ramamurthy and Dhavan (1963) had given the copepods in the order of their abundance together with the period of their maxima in the Gulf of Kutch near Kandla. Table. 32.5. Chhaya, (1999) have analyzed plankton range and average during pre monsoon and post monsoon at Khau Thas island and Karo Creek--Gulf of Kutch. Consolidated information is given as Table 32.6.

**Table 32.4 : Faunal Composition (No/m²) of Macrobenthos at Sikka Jetty during premonsoon peiod (1994) and post monsoon period (1993)**

| Faunal Group. | Pre monsoon (1994) | Post monsoon (1993) |
|---|---|---|
| Polychaetes | 4600 | 7438 |
| Copepodes | - | 583 |
| Amphipods | 800 | 692 |
| Isopods | 50 | 29 |
| Octracods | 100 | 88 |
| Decapods | 900 | 175 |
| Tanaidacea | - | 33 |
| Cumacean | 300 | 175 |
| Mollusc | 250 | 1250 |
| Echinoderms | 250 | 17 |
| Nemertines | 100 | - |
| Brachyurans | - | 17 |
| Foraminiferans | - | 4 |

*Source :* Marine environmental impact assessment of proposed refinery complex at Motikhavdi by N.I.O. council and Industrial Research Bombay. 1995.

**Table 32.5 Showing Important Copepods in the order of their Abundance together with the period of their maxima in the Gulf of Kutch near Kandla.**

| Sr. No. | Copepods | Period |
|---|---|---|
| 1 | *Paracalanus spp.* | February & September-December |
| 2 | *Acrocalanus spp.* | March-April and September-November |
| 3 | *Centropage spp.* | May and August-September |
| 4 | *Oithona spp.* | July-September |
| 5 | *Acartia spp.* | July-September |
| 6 | *Eucelanus spp.* | September |
| 7 | *Pseudodiaptomus spp.* | May and September |
| 8 | *Euchaeta spp.* | November-December |
| 9 | *Euterpina spp.* | March and September-October |

*Source* : Ramamurthy and Dhawan 1963.

Table 32.6 : Distribution of plankton range and average (in parenthesis) during premonsoon and post monsoon season 1998 at Khau Thas island and Karo Creek-Gulf of Kutch

| Khau Thar Island | Pre monsoon June 1998 | | |
|---|---|---|---|
| Tide | Level | Count No/$10^4$/L. | Major genera abundance wise |
| Flood | Surface | 25.3 to 31 Av. (28.2) | *Nitzschia, Chaetoceros, Rhizosolenia, Thalassiothix, Thalassionema.* |
| Flood | Bottom | 21.6 to 28.7 Av. (25.2) | *Nitzschia, Thalassiothrix, Thalassionemia, Rhizosolenia, Chaetoceros.* |
| **Post monsoon** | | | **September 1998** |
| Ebb | Surface | 9.3 to 10.00 Av. (9,7) | *Rhizosoienia, Nitzschia, Skeletonema.* |
| Ebb | Bottom | 8.7 to 12.00 Av. (10.44) | *Rhizosolenia, Nitzschia.* |
| **Karo Creek** | **Premonsoon June 1998.** | | |
| Flood | Surface | 29.3 to 35.9 Av. (32.6) | *Nitzschia, Thalassionema, Chaetoceros, Thalassiothrix, Rhizosolenia.* |
| Flood | Bottom | 41.8 to 52.3 Av. (47.1) | *Nitzchia, Thalassiothrix, Navicula, Chaetoceeros, Rhizosolenia.* |
| **Post Monsoon** | | | **September 1998.** |
| Ebb | Surface | 12.3 to 12.7 Av. (12.4) | *Skeletonema,Rhizosolenia, Nitzschia.* |
| Ebb | Bottom | 4.6 to 6.0 Av. (5.4) | *Rhizosolenia, Nitzschia, Skeletonema.* |

*Source* : Consolidated from N.D. Chhaya, 1999. pp. 21-22.

## Identification of various types of plankton

Plate 32.1 :

(1) *Melesera sudeata* (Ehrenberg) KuntzGretille)
(2) *Melosera sudeata* (one cell)
(3) *Stephanopysix palmeriana* (Gretille)
(4) *Stephanopyxi* (Areclae)
(5) *Skeletonema costatum* (Greville)
(6) *Coseinosira polycharda* (gran)
(7) *Thalanicsira suketilies* (osenfeld)
(8) *Coseinodiscus excentricus enhrenberg* (Cell showing arrangement of arcollse)
(9) *Coseinodiscus lineatus ehrenberg*
(10) *C.curvatulus grunow*
(11) *C.radiatus ehrenberg*
(12) *C.granii gough* (Cell in girdle view)
(13) *C.granii gough* (Central rosette)
(14) *C.concinnus* W. Smith (Grdke view)
(15) *C.concinnus* W. smith (Valve view.)
(16) *C.contrails Ehrenberg.*
(17) *C.perfprtus Var. pavillardi* (Forti)
(18) *C.apiculatus Ehrenber* (Central resette)
(19) *C.gigas ehrenberg Var. proetcta hustedt.* (Janish)
(20) *C.gigas ehrenberg* (Margin)

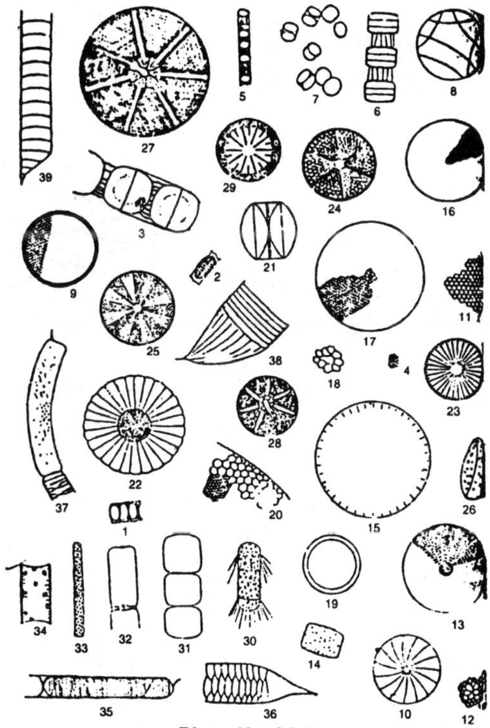


**Plate No 32.1**

(21) *Hyalodiscus steliger Bailey*
(22) *Planktoniella sol.* (Wallich) schutt
(23) *Arachnodiscus chrenbergii* Bailey
(24) *Aetiroptychus undulates* (Bailey) Ralfs,
(25) *A.splendenus* (Shadholt) Ralfs.
(26) *Asteromphalus heptaclis* (Brebisson) Ralfs
(27) *Asterromphalus spp.*
(28) *Actionocyclus chrenbergi Var. ratfsii Hustedt* (Smith)
(29) *A. Spp. Nav*
(30) *Corethron lystrix* terson.
(31) *Landea annulata.* Cleve.
(32) *Schroederella delicatula* (Peragallo)
(33) *Leptocylindrus danicus.* Cleve.
(34) *Curinrdia flaceida* (Castracane) Perigallo.
(35) *Rhizozolenia cylindrus* eleve.
(36) *R.bergonii* Peragallo.
(37) *R.stolter porthii* Paragallo.
(38) *R.robusta* neoman.
(39) *R.imbricata* Bright cell.

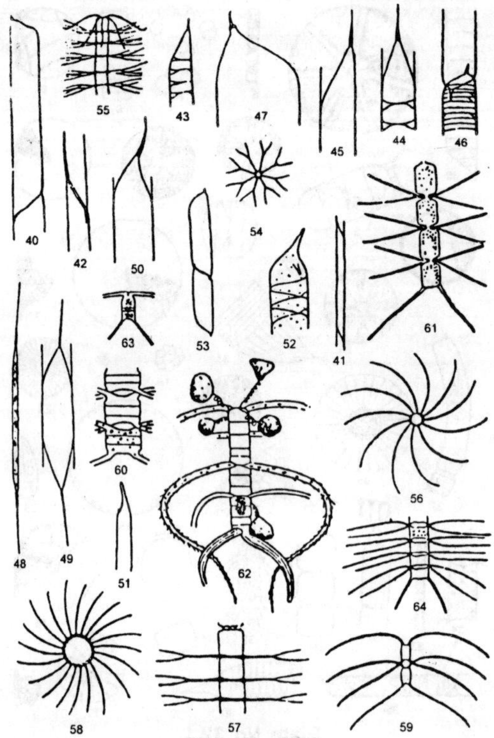


Plate No 32.2

Plate No 32.2

(40) *Rhizosolaria imbricata* (Brite wall)
(41) *R.imbricata Var* Shrubsoleii (Cleve)
(42) *R.slytiformis* Bright well Schulze
.(43) *R.slytiformis* (Showing inter calary)
(44) *R.slytiformis Var. longispina Husteelt.*
(45) *R.slytiformis*
(46) *R.slytiformis var lattissima* Brightwell
(47) *R.slytiformis var lattissima* (Apical process)
(48) *R.setigera* Brightwell.
(49) *R.crassispina Schroder*
(50) *R.calcaravis schulze*
(51) *R.alata* Brightwell F.genuina gran.
(52) *R.alata* Brightwell F.indica.
(53) *R.castracanei peragalla Var. rhimboidea subralamanyan.*
(54) *Bacteriastrum delicatulum* Cleve
(55) *B.hylainum hander*
(56) *B.hyalinum* Var. princeps kari.
(57) *B.vardana* lander.
(58) *B.vardana* (Valve view)

(59) *Chaetoceros atlantenus* Var. heopolitana (Schroder) nustedt.
(60) *C.eibenii* granow
(61) *C.glandazi Mahgin.*
(62) *C.coarctalus lander* (With verticell sp.)
(63) *C.peruvianus* Brightwell
(64) *C.hernzianus* Grunow

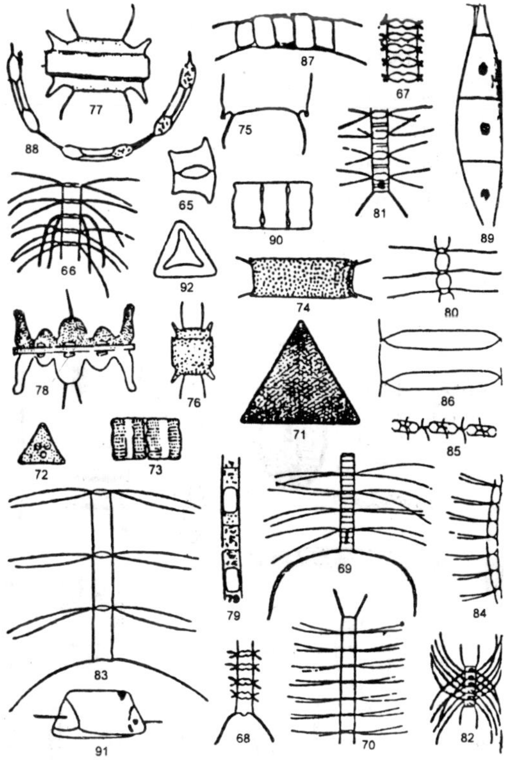


Plate No. 32.3

Plate No. 32.3 :

(65) *Chaetoceros indicus* Subrahmanyan
(66) *C.compressure* Lander.
(67) *C.didymas* Ehrenberg.
(68) *C.diymas Var. protuberanus* (Lander and yend)
(69) *C.affinis* Lander.
(70) *C.affinis Var. weillei* (Gran hustedt)
(71) *Triceratian favus* Ehrenberg.
(72) *T.reticulum* Ehrenberg
(73) *T.alternans* Bailey.
(74) *Brideulphia sinensis* Grey
(75) *B.sinensis* (End of the cell)
(76) *B.mobilnsis* Bailey.
(77) *B.leteroceros* Granow

(78) *B.tuomeyi* Bailey Var. Nov.
(79) *Hemialus lanckii* Grunow.
(80) *Chaetoceros paradoxum* Clove
(81) *C.brevis sehult.*
(82) *C.cliversus Clove Var Nov.*
(83) *C.pscudocerinatue* ostenfed.
(84) *C.curvisetus* Clove
(85) *Encampia zoodiaeus* Ehrenberg
(86) *Climscodium franenfeldianum* Grunow.
(87) *Hemianlus sinensis Greville.*
(88) *H.sinensis* (narrow girdle view)
(89) *Streptotheca catancensis shrubroce.*
(90) *Bellarochca malleus* (Bright well)
(91) *Drithlum bright welli* (West) Grunrck.
(92) *D. sol* Grunow.

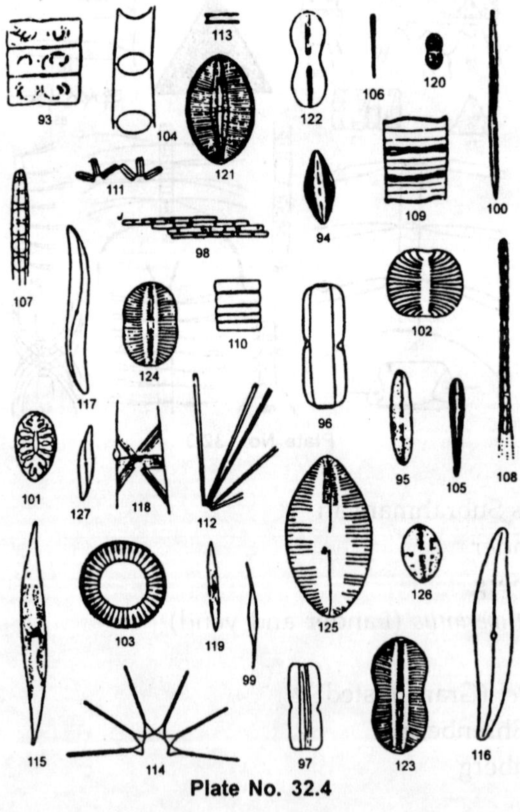

**Plate No. 32.4**

Plate No. 32.4

(93) *Navicula membranance* Clave.
(94) *Tinnularia alpina* W. Smith.
(95) *Trachyneis aspera* (Ehrenberg) Var *genuina cleve.*
(96) *Tropidoneis lepidoptera (Gregory) Cleave*

(97) *Amphiprori gigantean* Grunow Var. *Suleata* (O' Meara) Cleve.
(98) *Bacillaria paradoxa* Gnrelin.
(99) *Nitzehia closterium* (Ehrenberg) W. Smith.
(100) *N.pungens Var,* atlantica Cleve
(101) *Surirella furminensis* Grunow.
(102) *Compylodiseus iyengarii* Subrahmanayan.
(103) *C.horologium* Williams.
(104) *Hemiaulus membranaceous* Cleve.
(105) *Licmophora abbreviate Agerdh.*
(106) *Climacosphenia moniligera* Ehrenberg.
(107) *Climacosphenia moniligera* Ehrenberg.
(108) *Climacosphenia moniligera* Ehrenberg.
(109) *Plagiogramma vanheurekii* Grussow.
(110) *Fragilaria oceanica* Cleave
(111) *Thalassionema nitzchioides* Grunow.
(112) *Thalassiothrix franenfeldii* Grunow.
(113) *Thalassiothrix franenfeldii* (End of the cell)
(114) *Asterionella japonica* Cleve.
(115) *Pleurosigma elongatum* W.Smith
(116) *P.normanii*
(117) *P.angulatum (Queekett) W.Smith Var strigosa*
(118) *P.angulatum Var, strigosa* Structure)
(119) *P.directum* Grunow Var. *membranacea* Subrahmanyan.
(120) *Diploneis weissflogil* (A. Schmidt) Cleve.
(121) *D.smithii* (Brebisson) Cleve.
(122) *D.interrupta* (Kutzing) Cleve.
(123) *D.robustus* Subrahmanayan
(124) *D.robustus* Var.
(125) *Navicula henaleyii*
(126) *N.forciputa* Graville.
(127) *Pleurozisoona actuari*

### *Fresh water Zooplanktons of India.*

Plate No. 32.5

(1) *Polyarthra Spp.*
(2) *Philodina Spp.*
(3) *Barachionus*
(4) *Keratella tropica*
(5) *Paracylops poppei (female)*
(6) *Ostracoda Spp.*
(7) *Ectocyclops phaleratus*
(8) *Cladocera*
(9) *Daphnia*

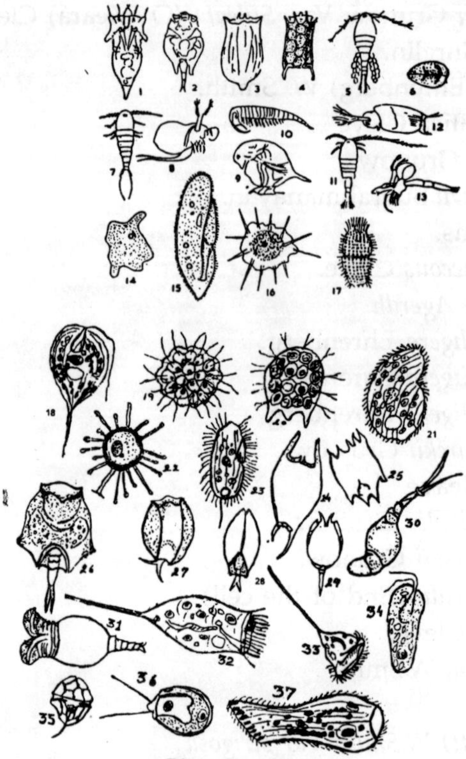

Plate No. 32.5

(10) *Artemia salina*
(11) *Diaptomina*
(12) *Codatenopod*
(13) *Cladocera leptodoria*
(14) *Amoeba proteus*
(15) *Paramecium caudatum*
(16) *Actinophrys*
(17) *Coleps hirus*
(18) *Phacus longicauda*
(19) *Gonium campactum*
(20) *Plantydorina caudate*
(21) *Colpoda steinii*
(22) *Podophrya bengalensis*
(23) *Uronema marina*
(24) *Branchionus diversicornis*
(25) *B.petulus*
(26) *Lepadella bicornis (Ventral view)*
(27) *Lepadella crestata (Dorsal view)*
(28) *Colourela obtuse (Ventral view)*
(29) *Monostyla quadridentata*

(30) *Scaridium longicaudum*
(31) *Philodina citrina*
(32) *Verticell convallaria*
(33) *Vorticella campanula*
(34) *Chilomonas paramecium*
(35) *Glenodinium cintum*
(36) *Chlamydomonas anqulosa*
(37) *Spathidium moniliforma*

## Economic importance of Copepods

Several species of copepods are intermediate hosts to complete the life cycle of many worms, which take heavy toll of human life and other animals. *Cyclops strenuous* and *Diaptomus gracillis* are eaten by perch and it was observed that the larval emerged from the bodies of copepod and pierced the stomach of the fish which acted as a intermediate host. Many small and young fish feed on different species of copepods; and in turn, they constitute a major portion of the diet of many of the larger fish.

Tropical fresh water supports diverse varieties of planktonic flora and fauna depending on a number of seasonal variations in physicochemical parameters including the thermal conditions of such biotopes. The amount and composition of planktons in the same body of water varies at various depths and at various times of the day. The optimal thermal regime for most fresh water diatomes is around 12°C whereas blue-green algae thrives best in a higher temperature ranging between 19 and 23°C (Sehgal, K.L., 1983.)

According to references, total 542 freshwater zooplanktons are identified in India and are grouped in different phylums. Phylum wise numbers are as under :

| Phylum | *Aarthropoda* | No of Sp. | | No of Sp. |
|---|---|---|---|---|
| Class | Copepoda | | | |
| Order | *Calanoida* | 64 | *Cyclopoida* | 47 |
| *Phylum* | *Rotifera* | | | |
| *Order* | *Ploimida* | 204 | *Gnesiotrocha* | 29 |
| | *Philodinidae* | 9 | | |
| *Phylum* | *Ciliophora* | | | |
| *Order* | *Prostomatida* | 23 | *Cryptophorida* | 4 |
| | *Suctoria* | 1 | *Hymenostomatida* | 16 |
| | *Scuticociliatida* | 6 | *Peritrichida* | 18 |
| | *Heterotrichida* | 19 | *Oligotrichida* | 1 |
| | *Hapotrichida* | 5 | *Hypotrichida* | 19 |
| | *Karyorelictida* | 3 | *Haptorida* | 10 |
| | *Pleurostromatida* | 6 | *Colpodida* | 3 |
| | *Nassulida* | 3 | *Cryptophorida* | 4 |
| | *Suctoria* | 10 | *Hymenostomatida* | 16 |
| | *Scuticociliatida* | 6 | *Peritrichida* | 18 |
| | *Heterotrichida* | 19 | *Oligotrichida* | 1 |
| | *Hapotrichida* | 5 | *Hypotrichida* | 19 |
| *Phylum* | *Scrcomastigophora* | | | |
| *Order* | *Dinoflagellida* | 1 | *Euglenidae* | 14 |
| | *Volovocida* | 20 | *Cryptomonadida* | 2 |
| *Sub Phylum* | *Sacrodina* | 9 | | |

*Source* : Battish, S.K. 1992

## Role of Plankton

### *Role of Plankton in the Survival and Growth of Fry*

The main food of the young carp fry is zooplankton, and in the absence of enemies, their survival depends entirely on the production of thick swarms of zooplankton, and the time of stocking the ponds. However, development of thick algal bloom, causes super saturation of dissolved oxygen in water, often resulting in gas trouble to the fry large scale mortality.

Heavy manuring of ponds with cow dung almost invariably results in the production of swarms of zooplankton consisting of rotifers, cladocerans and copepods, in about 10 days. *Volvox, Eudorina, Anabaena* frequently occur in blooms of varying density. Algae also develop and Euglena may form surface scum. The growth and multiplication of algae is very fast, and within a short period of 2-3 weeks, bloom might develop and disappear completely. Among zooplankton, rotifers like *Brachionus, Keratella and Polyarthra* along with copepods, nauplii make their appearance. Zooplankton multiply rapidly and appear is swarms in 4-6 days, and disappear in 7-8 days.

Laboratory experiment have shown that 90 to 100% fry survive is the fry are stocked during swarms of zooplankton and abundant food is available. Zooplankton are preferred by fry as they are easily digested. Phytoplankton is not utilized as food and is difficult to digest. The nutritional value of phytoplankton is also less than the zooplankton.

Application of heavy dose of cowdung in tanks, usually results in early production of zooplankton swarms, and the stocking of the fry has to be adjusted with the time of its maximum production.

## Water pollution and its effect on fisheries

The term 'pollution' broadly refers to any change in the natural quality of environment brought about by chemical physical or biological factors. Environment pollution is unfavourable alteration of our surroundings due to direct or indirect activities of man. Various by-products of man's actions are the pollutants and increase with the rise in population.

Industrialisation and increase in the human population of large cities results in the rivers becoming the drains of waste material. Domestic wastes, sewage and industrial effluents are generally allowed to be added to the rivers without any pre treatment, causing piollution of water. The waste products of some of the industries are extremely poisonous to fish life, and cause depletion of fish population by adverse changes in the physical, chemical and biological properties of fresh water. In addition to the industrial and domestic wastes, a large number of agricultural pesticides and insecticides have further increased the hazards of pollution of water.

## Effects on the Fishes

Various types of pollution mentioned above prove to be harmful to the fishery either directly or indirectly as mentioned below:

1. There is increase in the osmotic pressure.
2. The water becomes acidic and pH lowered below 4.

3. They bring about oxygen depletion.
4. The toxic substance injuries the mucous membrane of the gills and prove to be fatal, by effecting respiration.
5. The suspended matter injures the gill causing mechanical damage to the soft parts.
6. Pollutants reduce the zoo– and phytoplankton and other aquatic organisms, thus indirectly affecting the fish by food depletion.
7. The spawning grounds and nesting material are destroyed, preventing natural spawning of fishes.
8. The fish flesh gets foul odour due to certain pollutants, and becomes unpalatable.
9. Sewage and silt bring about an early ageing of the lakes and ponds.

Experiments conducted during the recent years have shown that besides insecticides, fungicides and fertilisers, heavy metallic ions are also injurious to fresh water fishes. Toxicity of metals to fishes varies greatly depending up on the metal, quality of water and the species tested. Reducing hardness of water markedly increased toxicity of zinc and copper to Puntius conchnius. Marked changes was observed in the various blood parameters of *Mystus vittatus* after exposure to various concentrations of copper and zinc sulphates. There was increase in the hematocrit level, swelling of the erythrocytes and their nuclei after exposure to acute lethal concentrations (15 ppm Cu, 60 ppm Zn). S.S. Khanna, 1992.

## Analysis of Plankton

### *Sampling*

Plankton sampling is done with the help of plankon net, made of extremely fine silk mesh. The size of plankton which can be filtered by nets determined by selecting the size of mesh. The net has a wide mouth (about 11.2 cm) and the other end is fixed with a replaceable bucket.

In flowing and shallow waters, 20 to 25 litre of water is collected through the bucket and filtered through the net. In lakes the net is towed horizontally, vertically or obliquely; In vertical towing the net is lowered to the bottom and then hauled up at a speed of 10 cam/sec. From this the volume of water is calculated as follows: $V = r^2D$

Where V is volume of water sample in litres, "r" is the radius of net orifice in meters and D the depth up to which the net is lowered in meters.

At least 500 ml of water sample should be collected for plankton analysis and in clearer water the quantity may be increased up to 5 litres.

## Preservation

The collected sample is then preserved in 4% formalin solution.

Counting by Sedwick-Rafter cell method–This is most common method used for counting phyto and zooplankton. The Sedwick-Rafter cell is about 50 mm long, 20 mm wide and 1 mm deep and its volume is 1 $cm^3$ or 1 ml. For counting, the sample is first agitated slightly for even distribution of planktons and then 1 ml of it is taken into the cell. The cover slip is put over carefully avoiding the sir bubbles. After some time when the plankton is settled the counting is done under the microscope. Moving the cell horizontally and vetically covering whole area count all the planktons. The method is repeated several times, each time taking 1 ml of preserved and agitated samples till 100 replicates are counted or 100 plankton each of few common species are counted.

$$\text{No. of plankton / ml} = \frac{\text{Number of organisms counted}}{\text{Number of replicates taken}}$$

## Control of Plankton Blooms

Under intensive fertilisation of ponds, the plankton grow rapidly and attain enormous densities, as 20 to 100 x 103 cell/ml of water. Such aggregations of planktons (both phyto and zoo) constitute the bloom.

The blooms are dangerous to fish, because of their respiration causing depletion of oxygen in water and so the serious mortality of fish. Dinoflagellates, phytoflagellates Brackish water species and blue green algae (fresh water species) produce toxins which result in massive killing of fishes.

Using following chemicals does the control of bloom.

1. Ammonia–Aqua ammonia spread on affected water surface at a concentration of 2-3 mg/litre kills *Prymnesium* (a phytoflagellate of brackish water.)
2. Lime manure of phosphate–These kills blue green algae by bringing about a change in pH of water. Their quality is adjusted accordingly.
3. Chrotoxyphos–Chrotoxyphos compound 0 applied at a concentration of 0.1 to 1 mg/lit effectively controls the blooms of zooplanktons.

## *References*

Anon, 1987. Phytoplankton and zooplankton of Bhavnagar District. "Matsyavikas" Commissionerate of Fisheries, Govt. of Gujarat.

Battish, S.K., 1992. Fresh water zooplankton of India, Oxford & IBH Publishing Co. New Delhi.

Chhaya, N.D. 1999. Minding our Marine Weath, An Appraisal of Gujarat Coastal Resources, pp. 21-22.

Dubale, M.S. 1989. Fishery Biology and Fisheries (Gujarati) Gujarat Granth Nirman Board, Govt. of Gujarat, Ahmedabad. pp. 178.

Paulinose, V.T., C.B. Lalithomlia Devi, Vijayalakshmi R. Nair, Neelam Ramaish and S.N. Gajbhaiye, 1998. Zooplankton standing stock and diversity in the Gulf of Kutch with special reference to larval of decapoda and pises. *Indian J. Mar. Sci.* 1998., 27: 340-345.

Ramamurthy and Dhavan 1963. Important Copepodes in the Gulf of Kutch near Kandla.

Sehgal, K.L. 1983. Planktonic copepode of Fresh water Ecosystem Interprint, New Delhi.

# 33
# MARINE RESOURCES

## An Introduction to Fisheries in Gujarat

Gujarat is the Northern most state on the west coast of India. Which lies between 20.1° and 24.7°. North latitude and 68.4° and 74.4° East longitude. Teh present political province of Gujarat came into existence on May 1, 1960, and includes the former United States of Saurashtra, the Kutch State and Gujarat regions of the erstwhile Bombay State. The state is bounded by the Arabian sea (west) Pakistan (north and north-west), Rajasthan (north-west), Madhya Pradesh (east) and Maharashtra (south and south-east). A map of the State is given in Annexure. The area of the State is 195,984 km$^2$.

The State has three distinct geographical regions, viz. The peninsular Saurashtra, desert marshy Kutch and the main land of South, Central and North Gujarat. Two extensive Gulfs-the Gulf of Kutch and the Gulf of Khambhat - are characteristic of Gujarat coast. The State is divided into 19 revenue districts, of which 10 (Kutch, Rajkot, Jamnagar, Junagadh, Amreli, Bhavnagar, Kheda, Bharuch, Surat and Valsad) are coastal. The coast line is 1600 kms. Long and topography of the coastal regions is marked by salt, marshes, sand belts and gravel patches. The southern coast of the Gulf of Kutch is characterised by innumerable coral reefs and islands. The Rann of Kutch is vast expanse of tidal mud flats with saline efflorescence. The natural gradient of the Northern Gujarat plain is north to south, while that of the Southern plain is from east to west. Thus both covering into the Gulf of Khambhat.

The soil pattern is variable. The alluvial plains of the east (in the South Gujarat region) are known for their fertility on account of the deposits of the rivers like Sabarmati, Mahi, Narmada and Tapi. The soil of South Gujarat is deep black, while that of Central and North Gujarat has alluvium. Saurashtra has medium black soil of basaltic origin.

There are about 82 river systems passing through the State. Out of these Narmada (160 km), Tapi (321 km), Mahi (1162 km), Sabarmati (310 km) and Banas (260 km) are the most important. The total river length available to the State is estimated to be 3685 km with an estuarine area of about 21,230 ha. It is estimated that the State has 170,280 ha, of fresh water area in the form of rural ponds, irrigation tanks, reservoirs and lakes.

## Climatic conditions

The climatic conditions of Gujarat are rather extreme. During winter, the temperature falls down to as low as 3 (December-February) in the some regions, while during summer it shoots up as high as 47°C (April-June). The south-west monsoon is the only source of rain fall and the same varier from about 300 mm (Kutch) to about 2000 mm (Valsad).

Drought is frequent phenomenon in the part of the country. Cyclonic winds at a velocity of about 8 to 150 km/hr accompanied by heavy rain fall are not uncommon between June to December.

With onset of South-West monsoon by the middle of May every year, followed by a nutrient replenishing coastal upwelling as well as heavy run off from the river Narmada, Tapti, Sabarmati and others causes an enrichment of the inshore area. Upwelling and heavy run off from the rivers nutriates inshore water. This supports major commercially important fishes such as Pomfrets, Prawns, Bombay duck and Threadfin etc. During monsoon and immediately after monsoon periods, upwelling occurs along the entire coast with variation in intensity. The nutrients brought by the upwelling or turbulence to the surface waters is used by phytoplankton.

Relationship between plankton production and fisheries was investigated by several workers concluded that movements of fish could be related to water movements and plankton bloom and that it is possible that the fishes reach various area as along their route of migration, when on the average feeding conditions were favourable in other words, fish catches coincide with plankton concentration suggesting that fish tends to move in water rich in plankton.

A wider shelf area results invariably in a great abundance of fishery resources, especially demersal groups. In the case of Gujarat, the shelf area is vast and the trawling grounds are extensive. Number of rivers and creeks available in the State also contribute to the nutrients thereby increasing organic production (CMFRI, 1990).

## Potential Resources

The potential yield on north-west coast (Gujarat and Maharashtra) was estimated to be 5,40,000 tons up to 50 m. This consists of Sciaenides, Cat fishes, Pomfrets, Ribbon fishes, Miscellaneous clupeoids and non-penaeid prawns. About 3,40,000 tones of resources are estimated to be available in the area lying beyond. The is composed of Cat fishes, Ribbon fishes, Hourse Makerals adn Oceanic squids are indicated beyond 200 m. Availability of Oceansic squid is assumed to be about 10 times that of all other fishes. The Gujarat coast, particularly off Dwarka and further north is supposed to be richer than southern region (MPEDA 1979). James (1992) have estimated species wise catchable potentials in 0 to 50 m depth as 2,94,000 Tones (Table 33.1).

Species wise major Fishery Resources Potential (in '000 tones) (1996) was estimated for Exclusive Economic Zone (EEZ) of India. (Table 33.4).

According to the Seventh Schedule, Article 246 of the Constitution of India, the fisheries of territorial waters of India are under the jurisdiction of the State Government and fishing and fisheries beyond territorial waters are under the jurisdiction of the Central Government. This creates some confusion, especially if resources are seen from the point of view of their inter-dependence and association. Coastal fishing vessels, registered with respective maritime States, are also moving into the waters beyond the territorial waters as well as moving from the waters of one maritime state to another. Gujarat trawlers fish, for example, beyond the territorial sea and also in the waters of Maharashtra. This clearly highlights the need for greater co-ordination between States and between the State and the Union Government. It may not be possible to the desired legal mechanism to address the needs of conservation

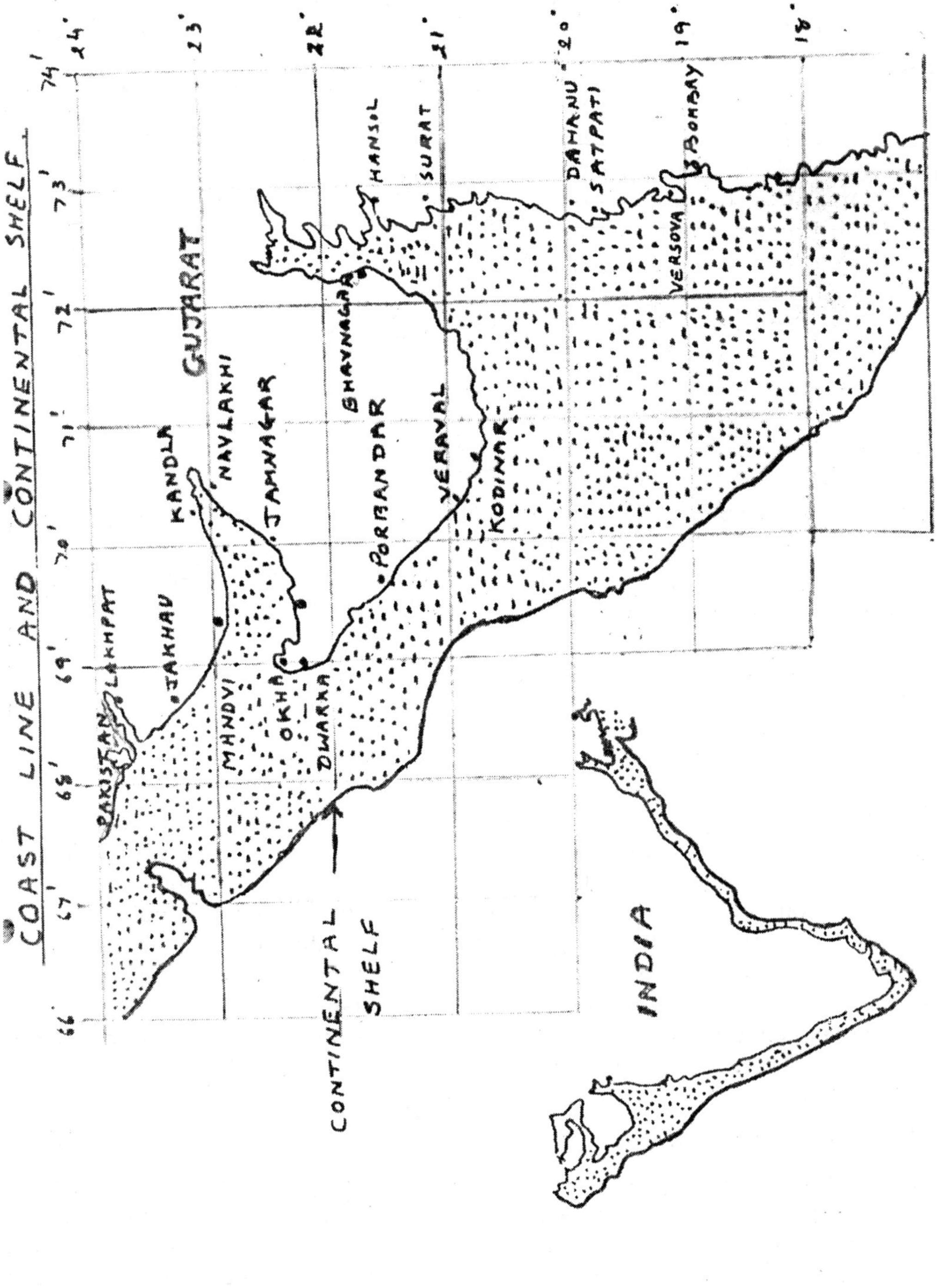


Fig. 33.1 : Details of continental shelf of India and Gujarat.

and management as well as those of allocation between sub-sectors and different States unless marine fisheries move form the State and Union List respectively in to the Concurrent List. This could facilitate the maritime States of India and the Union Government to work in a co-ordinated fashion towards conservation and management. (Sebastia Mathew, 2000).

### *Fisheries Section*

The fisheries sector is of great significance in view of the excellent physical and biological resources. Gujarat has 1600 km long coastline and about 165,000 km$^2$ continental shelf, which respectively form about 30% and 28.53% of the total cost line and continental shelf are available to the country (India) and about 59% of the coastline of the west coast. There has been a tremendous development in the fisheries sector of Gujarat since 1960 and today the State ranks first in the marine fish production. The marine fish production which was to the tune of 79,000 tonnes in 1960 has since increased to about 6,60,068 tonnes (1996-97). Similarly, the inland fisheries which was practically unknown in this region till the formation of the new state of Gujarat, has since resulted in a yield of about 65,278 tonnes of fish (1996-97). The transference of Gujarat into a major fish producing add exporting state has been achieved through various developmental programmes of the States Government.

| District | Length of Coastline (kms) | Percentage in the State's total coastline |
|---|---|---|
| Kutch | 406 | 25.4 |
| Rajkot | 26 | 1.6 |
| Jamnagar | 342 | 21.4 |
| Junagadh | 261 | 16.3 |
| Amreli | 62 | 3.9 |
| Bhavnagar | 152 | 9.5 |
| Kheda | 51 | 3.2 |
| Bharuch | 127 | 7.9 |
| Surat | 83 | 5.2 |
| Valsad | 90 | 5.6 |

District-wise distribution of coastline in ten maritime district of Gujarat is shown as above :

### *The continental shelf*

The continental shelf off Gujarat coast is very wide and providing a shelf area of 164,183 km$^2$ of this 64,810 km$^2$ is within a depth range of 50 m and the reminder i.e. 99373 km$^2$ within the depth range of 50 to 200m. The maximum width of the continental shelf is 191 miles. 270 due west off Umergaon (Valsad dist.) and minimum width is 58 miles between Madhavpur and Miyani (Junagadh dist.). The gradient of the shelf is 1:1.769 at maximum width point and 1 : 537 at the minimum width point. The configuration of the continental shelf off Gujarat coast is shown in Annexure. Though the State has wide continental shelf, by and large, the exploitation is limited to the inshore and offshore waters upto a depth contour of 70 m only. The waters beyond this depth contour have been exploited only very marginally.

## Fisheries Resources of Gujarat

Gujarat is having very large and extended continental shelf, which is shown in Fig. 33.1. Fisheries Survey of India, Government of India, have assessed standing stock of Gujarat using modern mechanized vessel *M.T. Murena*. It has found out division wise catch per hour of Pelagic trawling and plotted on map of Gujarat. (Fig. 33.2) It has also found

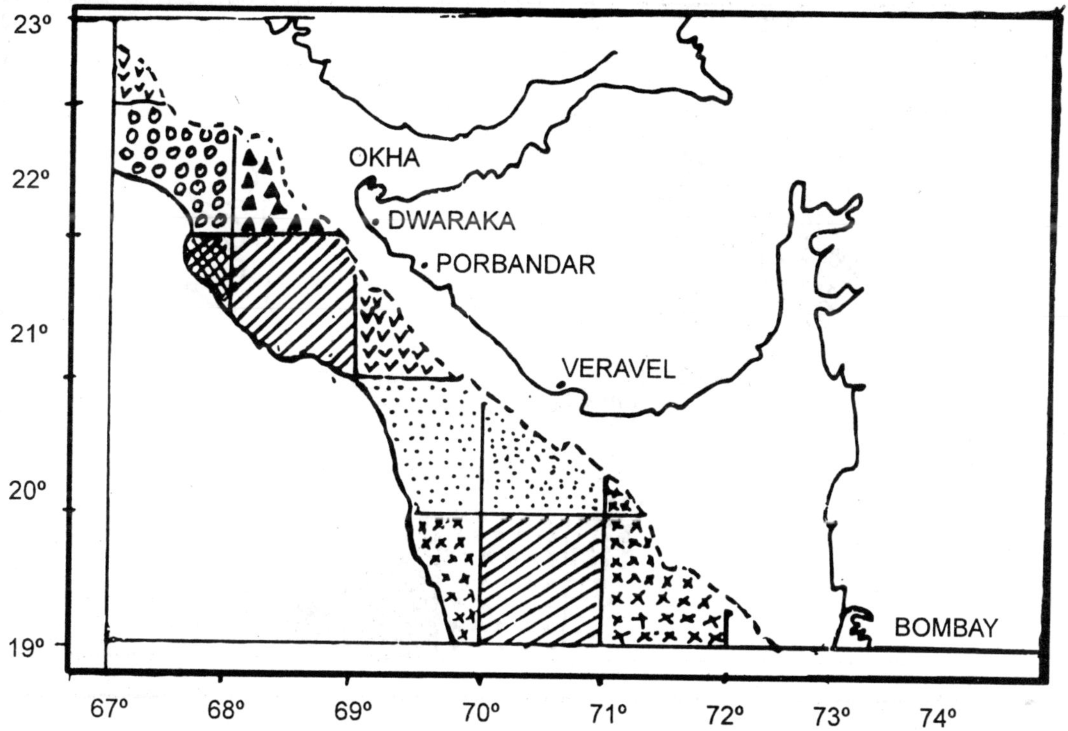


Division-wise catch per hours of Pelagic trawling obtained by M.T. *Murena* during 1st to IVth cruises.

Legend
- 100 Kg /hr
Between 100 & 300 Kg /hr
Between 300 & 500 Kg /hr
Between 500 & 700 Kg /hr
Between 700 & 900 Kg /hr
Between 900 & 1300 Kg /hr
30 Fathoms depth contour
200 Fathoms depth contour

**Fig. 33.2 : Division wise catch per hour of pelagic trawling.**

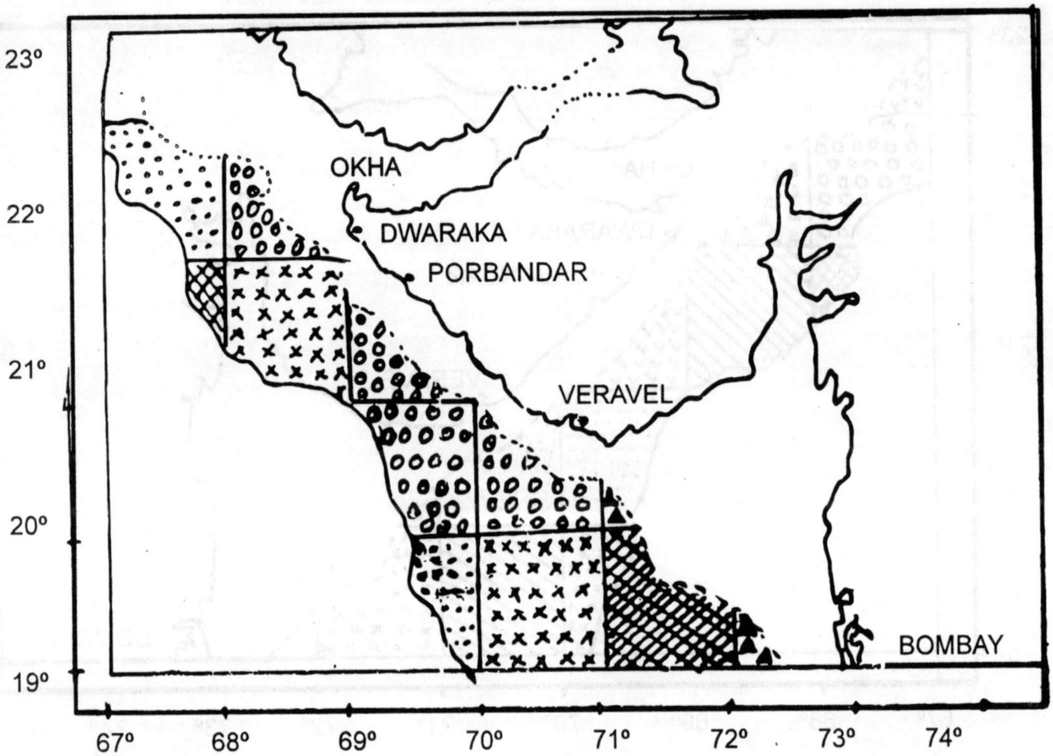


Division catch per hour of bottom trawling by M.T. *Murena* during I[st] to V[th] crusis.
Source :- Ministry of Agriculture and Irrigation, New Delhi.

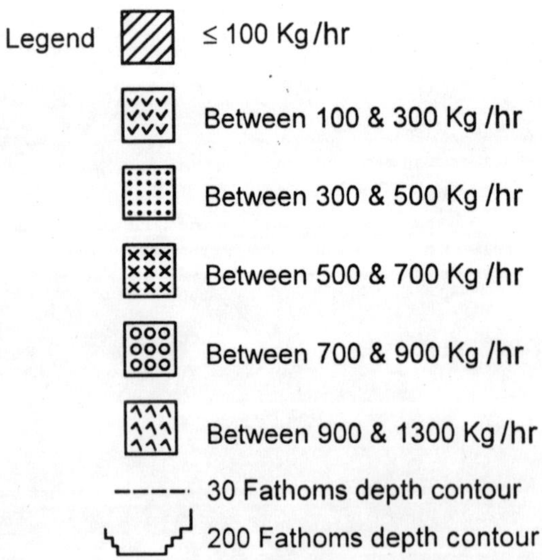


**Fig. 33.3 : Division wise catch per hour of bottom trawling.**

out division wise catch per hour of bottom trawling, the same is shown in Fig. 33.3. Jhingran (1982) has assessed standing stock (in Tones) up to 40 F.M. from different areas of Gujarat. (Table 33.1). Department of Fisheries, Government of Gujarat in 1986-87 had tries to assess stock (catch/hour) of different fishing grounds using different nets. So that fishermen can be advised that in particular fishing ground certain quantity of fish is likely to be obtained. Table 33.2 gives the details of fishing ground, type of fishing nets, months and catch per hour. Table 33.3 gives the details of area and month recommended for profitable fishing near Jakhaw area Dwarka area. Division wise catch/hr. of different species in bottom trawling was done by FSI Govt. of India, in 1972) Table 33.4.

**Table 33.1 : Standing stock Estimates for areas surveyed up to 40 FM depth off Gujarat coast.**

| Sr. No. | Division | Depth range (in FM) | | | | Total | |
|---|---|---|---|---|---|---|---|
| | | 0 - 25 | | 25–40 | | | |
| | | Area (Sq. Km.) | Standing stock (t) | Area (Sq. Km.) | Standing stock (t) | Area (Sq. Km.) | Standing stock (t) |
| 1. | 22-67 | 1,030 | 4,150 | 637 | 2,768 | 1,717 | 6,913 |
| 2. | 22-68 | 7,897 | 49,119 | 2,747 | 11,558 | 10,644 | 60,637 |
| 3. | 22-69 | 5,837 | 1,985 | — | — | 5,837 | 1,985 |
| 4. | 22-70 | 2,232 | 2,678 | — | — | 2,237 | 2,678 |
| 5. | 21-69 | 3,777 | 18,440 | 3,090 | 12,133 | 6,867 | 30,573 |
| 6. | 21-70 | 601 | 5,829 | — | — | 601 | 5,829 |
| 7. | 20-69 | 172 | 835 | 1,717 | 5,959 | 1,889 | 6,794 |
| 8. | 20-70 | 2,404 | 16,924 | 2,404 | 9,227 | 4,808 | 26,151 |
| 9. | 20-71 | 10,129 | 32,919 | 687 | 2,264 | 10,816 | 35,283 |
| 10. | 20-72 | 10,644 | 46,620 | — | — | 10,644 | 46,620 |
| **Subtotal of Gujarat** | | **44,723** | **1,79,499** | **11,332** | **44,005** | **56,055** | **2,23,504** |

*Source* : V.G. Jhingran, Fish and Fisheries of India, 1982 : 563.

**Table 33.2. Details of fish catch survey conducted in different fishing grounds using different nets by Dept. of Fishereis Govt. of Gujarat.**

| Name of Area | Fishing ground | | Month, Year | Type of Net | Catch per hour in Kgs. |
|---|---|---|---|---|---|
| Jakhaw area | 23°-67' | 3 E | 86-87 | Fish Net | 116,116 |
| | 23°-67' | 3 F | 86-87 | Fish Net | 118,364 |
| | 23°-67' | 4 D | 86-87 | Fish Net | 125,870 |
| | 23°-67' | 2 E | 86-87 | Fish Net | 80,750 |
| | 23°-68' | 2A | 86-87 | Fish Net | 59,220 |
| | 23°-68' | 1A | 86-87 | Fish Net | 121,976 |
| | 23°-68' | 1B | 86-87 | Fish Net | 68,038 |
| Dwarka area | 22°-68' | 6 E | 86-87 | Fish Net | 21,400 |
| | 22°-68' | 3 D | 86-87 | Fish Net | 209,750 |
| | 22°-68' | 4 C | 86-87 | Fish Net | 105,260 |
| | 22°-68' | 1 E | 86-87 | Fish Net | 88,100 |
| | 22°-68' | 3 C | 86-87 | Fish Net | 81,648 |
| | 22°-68' | 6 F | 86-87 | Fish Net | 156,633 |

*Table Contd.........*

*Table 33.2 contd.............*

| Name of Area | Fishing ground | | Month, Year | Type of Net | Catch per hour in Kgs. |
|---|---|---|---|---|---|
| | 22°-68' | 6E | 86-87 | Fish Net | 83,000 |
| Jakhaw area | 23°-67' | 3E | October-86 | Trawl Net | 115,750 |
| | 23°-67' | 3E | November-86 | Trawl Net | 116,200 |
| | 23°-67' | 3F | September-86 | Trawl Net | 113,800 |
| | 23°-67' | 3 F | October-86 | Trawl Net | 143,330 |
| | 23°-67' | 3 F | January-86 | Trawl Net | 127,300 |
| | 23°-67' | 4D | October-86 | Trawl Net | 125,870 |
| | 23°-67' | 2E | December-86 | Trawl Net | 80,750 |
| | 23°-67' | 2 A | October-86 | Trawl Net | 81,000 |
| | 23°-67' | 2 A | November-86 | Trawl Net | 56,000 |
| | 23°-67' | 2 A | December-86 | Trawl Net | 77,500 |
| | 23°-67' | 1 A | October-86 | Trawl Net | 128,640 |
| | 23°-67' | 1 A | December-86 | Trawl Net | 60,500 |
| | 23°-68' | 1 B | October-86 | Trawl Net | 72,340 |
| | 23°-68' | 1 B | December-86 | Trawl Net | 62;288 |
| | 23°-68' | 1B | February-86 | Trawl Net | 100,125 |
| | 23°-68' | 2B | September-86 | Trawl Net | 33,000 |
| | 23°-68' | 2B | October-86 | Trawl Net | 46,581 |
| | 23°-68' | 2B | November-86 | Trawl Net | 48,590 |
| | 23°-68' | 2B | December-86 | Trawl Net | 40,120 |
| | 23°-68' | 2B | January-97 | Trawl Net | 9,260 |
| | 23°-68' | 2B | February-97 | Trawl Net | 50,310 |
| | 23°-68' | 2B | March-97 | Trawl Net | 63,740 |
| Dwarka area | 22°-68' | 6E | March-87 | Trawl Net | 2,100 |
| | 22°-68' | 3D | October-88 | Trawl Net | 209,750 |
| | 22°-68' | 4C | May-86 | Trawl Net | 112,000 |
| | 22°-68' | 4C | October-86 | Trawl Net | 181,333 |
| | 22°-68' | 4C | February-87 | Trawl Ne | 43,200 |
| | 22°-68' | 4C | March-87 | Trawl Net | 54,140 |
| | 22°-68' | 1E | March-87 | Trawl Net | 80,100 |
| | 22°-68' | 3C | May-86 | Trawl Net | 130,339 |
| | 22°-68' | 3C | February-86 | Trawl Net | 45,140 |
| | 22°-68' | 3C | March-86 | Trawl Net | 47,650 |
| | 22°-68' | 6F | April-86 | Trawl Net | 167,386 |
| | 22°-68' | 6F | March-87 | Trawl Net | 22,870 |
| | 22°-67' | 6E | April-86 | Trawl Net | 86,000 |

*Source* : With permission of Commissionerate of Fisheries, Govt. of Gujarat. Gandhinagar.

Comparison of fish production (marine and Inland) trend of India and Gujarat from 1950-51 is given in Table 33.5. From this table we can see the growth of fish production. The fish production in Gujarat is from different marine, inland and esturine landing centers. Details of district wise such landing centers is given in Table 33.6. Exact location of such landing centers giving latitude and longitude is given in Table 33.7.

District wise marine fish production of Gujarat State is given in Table 33.8 since 1960-61. Landing figures of 37 years shows that maximum fish landing is obtained from Junagadh district. The fish landing in Gujarat is done by different type of fishing boats. Category wise information of the fishing boats and its growth of Gujarat State since 1980-81 is given Table 33.9.

We know form the table 33.5 that total marine fish landing shows the increasing trends. On checking the landing of miscellaneous fish landing (Table 33.10), it also shows remarkable increase. In the year 1964-65 miscellaneous fish landing was 5,374 M.T. which was 6.04% and in 1969-70 it was 3.83% of total fish landing. As mechanization was increased, landing of miscellaneous marine fish also increased. The landing of miscellaneous marine fish in 1996-97 was 75,471 M.T., which is near to total marine fish landing of the year 1960-61. This may be because fishermen are reducing mesh size day by day. Table 33.11 gives the details of district wise MFV. This also shows that maximum mechanization is in Junagadh district.

**Table 33.3 : Area and month recommended for Profitable fishing near Jakhaw area and Dwarka area. Survey conducted by Dept. of Fisheries Govt. of Gujarat.**

| Sr. No. | Fishing Ground | | Month | Jakhaw area |
|---|---|---|---|---|
| 1 | 23°-67' | 3E | Oct-Nov. | Jakhaw area |
| 2 | 23°-67' | 3F | Sept. Oct. & Jan | Jakhaw area |
| 3 | 23°-67' | 4D | October | Jakhaw area |
| 4 | 23°-68' | 1A | October | Jakhaw area |
| 5 | 23°-68' | 1B | October, December and February | Jakhaw area |
| 6 | 23°-68' | 2B | September | Jakhaw area |
| 7 | 23°-67' | 2E | December | Jakhaw area |
| 8 | 22°-68' | 6E | March | Dwarka area |
| 9 | 22°-68' | 3D | October | Dwarka area |
| 10 | 22°-68' | 4C | May & October | Dwarka area |
| 11 | 22°-68' | 1E | March | Dwarka area |
| 12 | 22°-68' | 3C | May | Dwarka area |
| 13 | 22°-68' | 1F | Nov. & Dec. | Dwarka area |
| 14 | 22°-68' | 1B | November | Dwarka area |
| 15 | 22°-68' | 2A | December | Dwarka area |
| 16 | 21°-67' | 6F | April | Dwarka area |
| 17 | 21°-69' | 6A | Nov., Dec. and Jan. | Dwarka area |
| 18 | 21°-69' | 5B | January | Dwarka area |
| 19 | 21°-69' | 6B | January | Dwarka area |
| 20 | 21°-68' | 1F | December & January | Dwarka area |
| 21 | 21°-68' | 2D | March | Dwarka area |

*Source* : With the permission of Commissioner of Fisheries, Govt. of Gujarat, Gandhinagar.

**Table 33.4 : Division wise catch/hr of different species in bottom trawling by M.T. Murena 1st to Vth cruises.**

| Area Lat./ Long. | Ribbon fish catch/ hr. | Elasmo-branch catch/ hr. | Cat fish catch/ hr. | Dhoma catch/ hr. | Pearch catch/ hr. | Horse Mackeral catch/ hr. | Caranx catch/ hr. | Rani fish catch/ hr. | Pomfret catch/ hr. | Ghol catch/ hr. | Karkara catch/ hr. | Pellona catch/ hr. | Squid catch/ hr. |
|---|---|---|---|---|---|---|---|---|---|---|---|---|---|
| 19/72 | 23 | 39 | 1 | 68 | 1 | 2 | 3 | 4 | – | 17 | 5 | – | 4 |
| 19/71 | 154 | 31 | 8 | 85 | 3 | 1.6 | – | – | – | 21 | 7 | 7 | 3 |
| 19/70 | 25 | 14 | 1 | 48 | 1 | – | 2 | – | – | 3 | 5 | 2 | 0 |
| 19/69 | – | 19 | – | – | 9 | 3 | 1 | – | – | – | – | – | 0 |
| 20/71 | 20 | 82 | 4 | 9 | – | – | – | – | 5 | 20 | 49 | – | – |
| 20/70 | 109 | 21 | 3 | 23 | 1 | 36 | 4 | 6 | 12 | – | 8 | 8 | 4 |
| 20/69 | 200 | 17 | 8 | 6 | 1 | 18 | – | 2 | – | – | – | – | 2 |
| 21/69 | 16 | 92 | 6 | 2 | – | 36 | 1 | 4 | – | 1 | 7 | 1 | – |
| 21/68 | 17 | 66 | – | 14 | 19 | – | 20 | 19 | 4 | 4 | 8 | – | 1 |
| 22/68 | 8 | 27 | 2 | 168 | 13 | – | 1 | – | 7 | 12 | 9 | 10 | – |
| 22/67 | – | – | 8 | – | – | – | – | – | 4 | 1 | – | – | 1 |

**Table 33.5 : Fish Production Trend of India and Gujarat from 1950-51 to 1996-97 (Production in Lakhs M.T.)**

| Year | Marine | | | Inland | | | Total | | |
|---|---|---|---|---|---|---|---|---|---|
| | India | Gujarat | Percent share of Gujarat | India | Gujarat | Percent share of Gujarat | India | Gujarat | Percent share of Gujarat |
| 1950-51 | 5.34 | 0.00 | 0.00 | 2.18 | 0.00 | 0.00 | 7.52 | 0.00 | 0.00 |
| 1960-61 | 8.80 | 0.70 | 8.98 | 2.80 | 0.00 | 0.00 | 11.60 | 0.79 | 6.81 |
| 1970-71 | 10.86 | 1.51 | 13.90 | 6.70 | 0.00 | 0.00 | 17.56 | 1.51 | 8.60 |
| 1980-81 | 15.55 | 2.19 | 14.08 | 8.87 | 0.17 | 1.52 | 24.42 | 2.36 | 9.66 |
| 1981-82 | 14.45 | 2.21 | 15.29 | 9.99 | 0.19 | 1.00 | 24.44 | 2.40 | 9.82 |
| 1982-83 | 14.27 | 1.93 | 13.52 | 9.40 | 0.19 | 2.02 | 23.67 | 2.12 | 8.96 |
| 1983-84 | 15.29 | 2.23 | 14.68 | 9.87 | 0.21 | 2.13 | 25.06 | 2.44 | 9.74 |
| 1984-85 | 16.98 | 2.91 | 17.14 | 11.03 | 0.24 | 2.18 | 28.01 | 3.15 | 11.25 |
| 1985-86 | 17.16 | 3.06 | 17.83 | 11.60 | 0.24 | 2.07 | 28.76 | 3.30 | 11.47 |
| 1986-87 | 17.13 | 3.16 | 18.45 | 12.59 | 0.24 | 1.95 | 29.42 | 3.40 | 11.56 |
| 1987-88 | 16.58 | 3.27 | 19.72 | 18.01 | 0.23 | 1.77 | 29.69 | 3.50 | 11.83 |
| 1988-89 | 18.17 | 4.14 | 22.78 | 13.33 | 0.22 | 1.65 | 30.32 | 4.36 | 13.83 |
| 1989-90 | 22.75 | 4.32 | 18.99 | 14.02 | 0.27 | 1.93 | 36.77 | 4.59 | 12.48 |
| 1990-91 | 23.00 | 5.00 | 21.74 | 15.36 | 0.46 | 2.99 | 38.36 | 5.46 | 12.23 |
| 1991-92 | 24.47 | 5.30 | 21.66 | 17.10 | 0.40 | 2.34 | 41.57 | 5.70 | 13.78 |
| 1992-93 | 25.76 | 6.09 | 23.64 | 17.89 | 0.51 | 2.80 | 43.65 | 6.60 | 15.15 |
| 1993-94 | 26.49 | 6.20 | 23.41 | 19.95 | 0.65 | 3.26 | 46.44 | 6.85 | 14.75 |
| 1994-95 | 26.92 | 6.45 | 23.96 | 20.94 | 0.70 | 3.34 | 47.86 | 7.15 | 14.95 |
| 1995-96 | 28.25 | 5.98 | 21.17 | 21.25 | 0.60 | 2.82 | 49.50 | 6.58 | 13.29 |
| 1996-97 | 28.57 | 6.60 | 23.10 | 22.83 | 0.65 | 2.84 | 51.40 | 7.25 | 14.10 |

*Source* : Reprinted with the permission of Commissionarate of Fisheries Govt.of Gujarat. Gandhinagar

**Table 33.6 : Information of Marine Inland and Estuarine landing center (District wise) of Gujarat state**

| Sr. No. | Name of District | Marine | Inland | Estuarine | Total |
|---|---|---|---|---|---|
| 1 | Jamnagar | 22 | 8 | | 30 |
| 2 | Rajkot | 1 | 5 | 9 | 15 |
| 3 | Surendranagar | - | 5 | 10 | 15 |
| 4 | Bhavnagar | 11 | 5 | - | 16 |
| 5 | Amreli | 10 | - | - | 10 |
| 6 | Junagadh | 25 | - | - | 25 |
| 7 | Kutch | 68 | 5 | - | 73 |
| 8 | Banaskantha | - | 10 | - | 10 |
| 9 | Sabarkantha | - | 15 | - | 15 |
| 10 | Ahmedabad | - | 9 | 4 | 13 |
| 11 | Kheda | 1 | 34 | 5 | 40 |
| 12 | Panchmahal | - | 147 | - | 147 |
| 13 | Vadodara | - | 67 | 9 | 76 |
| 14 | Bharuch | 18 | 55 | 22 | 95 |
| 15 | Surat | 13 | 146 | 14 | 173 |
| 16 | Valsad | 44 | 44 | - | 88 |
| 17 | Mehsana | - | 9 | - | 9 |
| 18 | Dang | - | 4 | - | 4 |
| | **Total** | **213** | **568** | **73** | **854** |

*Source* : Reprinted with the permission of Commissionarate of Fisheries Govt. of Gujarat, Gandhinagar.

Some fishermen of Veraval area have realized and understood this and from the year 2002 they have started using bigger square type mesh size in code end. This has resulted in reduction of trash material, fish juveniles and smaller size fish in their catch. Handling and spoilage because of unwanted trash material was reduced. This has increased quality of prime fish and got more prices. Big size fish fetches good prices hence it was compensated and they were happy. It is necessary that all fishermen understand this and maintain future stock for teh survival of our next generation and industries.

Dholakia (1986) have recorded main available marine fishes of Gujarat with scientific, English and Gujarati names (Table 33.12). Dholakia (1998) have also recorded marine water molluscans of Gujarat from Saurashtra coast. (Table 33.13) However, much more work is to be done in field. Sarvaiya and Chhaya (1983) have listed different molluscans found in Gulf of Kutch and Northern-Western region of Saurashtra (Table 33.14). They have also analyzed and nutritive value of some of the molluscans was recorded. (Table 33.15).

Marine National Park, Department of Forest, Government of Gujarat have declared list of Rare, Endangered, Threatened and Vulnerable species in Gulf of Kutch, Gujarat (Table 33.16).

On checking of Table 33.18 it can be seen that though total catch of almost all species has shown increasing trend. However, catch per boat is reduced. This is because during 1980-81 number of mechanized boats were 3459, while after 20 years number were increased to 11372 mechanized boat registered with Department of Fisheries, Governmet of Gujarat.

**Table 33.7 Latitude and Longitudes of Fish Landing Centres and Ports of Gujarat.**

| Sl. No. | Name of Fish landing Centre & Port | Latitude | Longitudes |
|---|---|---|---|
| 1. | Kandla | 22°58' N | 70°13' E |
| 2. | Okha | 22°29' N | 69°05' E |
| 3. | Bhavnagar | 21°46' N | 72°09' E |
| 4. | Bedi | 22°33' N | 70°02' E |
| 5. | Porbandar | 21°38' N | 69°36' E |
| 6. | Navlakhi | 22°58' N | 70°27' E |
| 7. | Veraval | 20°54' N | 70°22' E |
| 8. | Mandvi | 22°50' N | 69°21' E |
| 9. | Sikka | 22°31' N | 69°49' E |
| 10. | Jakhau | 23°14' N | 68°36' E |
| 11. | Mundra | 25°45' N | 69°43' E |
| 12. | Salaya | 22°19' N | 69°36' E |
| 13. | Mangrol | 21°07' N | 70°06' E |
| 14. | Nawabandar | 22°44' N | 71°05' E |
| 15. | Jafrabad | 20°52' N | 71°23' E |
| 16. | Albert Victor | 20°57' N | 71°32' E |
| 17. | Bheri Rajula | 20°57' N | 71°31.5' E |
| 18. | Broach | 21°42' N | 73°02' E |
| 19. | Lakhpat | 23°48' N | 68°52' E |
| 20. | Jodia | 22°43' N | 70°17' E |
| 21. | Beyt | 22°30' N | 69°20' E |
| 22. | Dwarka | 22°15' N | 69°03' E |
| 23. | Lamba | 21°53' N | 69°19' E |
| 24. | Madhavpur | 21°15' N | 69°57' E |
| 25. | Sutrapada | 20°50.5' N | 70°29' E |
| 26. | Kodinar | 20°48' N | 70°50' E |
| 27. | Kotda | 20°45' N | 70°30' E |
| 28. | Madhwada | 20°45' N | 71°15' E |
| 29. | Ghogha | 21°41' N | 72°16' E |
| 30. | Talaja | 21°18' N | 72°11' E |
| 31. | Cambay | 22°19' N | 72°37' E |
| 32. | Onjal | 20°45' N | 72°55' E |
| 33. | Navsari | 20°50' N | 72°15' E |
| 34. | Bilimora | 20°48' N | 72°57' E |
| 35. | Umarsadi | 20°45' N | 72°50' E |
| 36. | Umaragoan | 20°07' N | 72°44' E |
| 37. | Katpar | 21°09' N | 71°46' E |
| 38. | Chorwad | 21°04' N | 70°15' E |

**Table 33.8 : District wise Marine Fish production of Gujarat state (Production In MT)**

| Sr. No. | Year | Valsad | Surat | Bharuch | Kheda | Bhav-nagar | Amreli | Junagadh | Jamnagar | Rajkot | Kutch | Total |
|---|---|---|---|---|---|---|---|---|---|---|---|---|
| 1 | 1960-61 | 27000 | – | – | – | – | 23874 | 25475 | 1997 | – | 1066 | 79412 |
| 2 | 1961-62 | 17800 | – | – | – | –- | 37170 | 25354 | 2786 | – | 2044 | 85154 |
| 3 | 1962-63 | 17642 | – | – | – | – | 29453 | 31421 | 2308 | – | 1857 | 82681 |
| 4 | 1963-64 | 18000 | – | – | – | – | 34827 | 28141 | 2306 | – | 1726 | 85000 |
| 5 | 1964-65 | 24850 | – | – | – | – | 34640 | 22696 | 4064 | – | 2767 | 89017 |
| 6 | 1965-66 | 30000 | – | – | – | – | 38909 | 34382 | 4191 | – | 2425 | 109907 |
| 7 | 1966-67 | 43528 | – | – | – | – | 31149 | 34790 | 2752 | – | 2956 | 115175 |
| 8 | 1967-68 | 45919 | – | – | – | – | 43908 | 28023 | 3778 | – | 3274 | 124882 |
| 9 | 1968-69 | 55000 | – | – | – | – | 42600 | 29043 | 2836 | – | 2269 | 131748 |
| 10 | 1969-70 | 55491 | – | – | – | – | 46693 | 32249 | 2891 | – | 2671 | 139995 |
| 11 | 1970-71 | 60188 | – | – | – | – | 50214 | 32219 | 4657 | – | 3912 | 151190 |
| 12 | 1971-72 | 56157 | – | – | – | – | 46802 | 34472 | 5457 | – | 4135 | 147023 |
| 13 | 1972-73 | 45836 | 2742 | 3914 | 1285 | – | 43466 | 44587 | 4562 | – | 4811 | 151203 |
| 14 | 1973-74 | 42350 | 1969 | 4691 | 954 | – | 49760 | 64280 | 6760 | – | 6883 | 177647 |
| 15 | 1974-75 | 20958 | 1521 | 487 | 74 | 697 | 21741 | 104387 | 2982 | 281 | 4302 | 157430 |
| 16 | 1975-76 | 12002 | 3089 | – | – | 744 | 31787 | 149612 | 5065 | 2159 | 3842 | 208300 |
| 17 | 1976-77 | 8769 | 784 | 1026 | 234 | 517 | 30978 | 154300 | 21679 | 1292 | 5400 | 225379 |
| 18 | 1977-78 | 5660 | 280 | 1100 | 297 | 281 | 24162 | 123893 | 14463 | 426 | 6333 | 176895 |
| 19 | 1978-79 | 2382 | 844 | 1695 | 280 | 424 | 33349 | 160223 | 21797 | 2680 | 6298 | 229971 |
| 20 | 1979-80 | 4355 | 797 | 1158 | 251 | 822 | 21529 | 146921 | 24755 | 455 | 5706 | 206749 |
| 21 | 1980-81 | 5506 | 1212 | 1701 | 343 | 1249 | 17178 | 157040 | 2305 | 1731 | 4405 | 218872 |
| 22 | 1981-82 | 7695 | 1334 | 1293 | 366 | 760 | 37985 | 128695 | 34167 | 1978 | 6334 | 220607 |
| 23 | 1982-83 | 7847 | 1127 | 1105 | 1108 | 1155 | 21116 | 115241 | 29687 | 498 | 13785 | 192669 |
| 24 | 1983-84 | 9542 | 1255 | 2593 | 1824 | 1834 | 29868 | 124315 | 27443 | 1302 | 23315 | 223291 |
| 25 | 1984-85 | 14570 | 625 | 1575 | 1586 | 873 | 54047 | 150915 | 31521 | 701 | 34295 | 290708 |
| 26 | 1985-86 | 12831 | 858 | 712 | 1280 | 585 | 33887 | 194095 | 25238 | 1674 | 35417 | 306577 |
| 27 | 1986-87 | 11446 | 1859 | 1427 | 994 | 737 | 22792 | 216871 | 27998 | 504 | 31314 | 315942 |
| 28 | 1987-88 | 20556 | 1009 | 1879 | 843 | 1203 | 27872 | 203761 | 40232 | 463 | 29743 | 327560 |
| 29 | 1988-89 | 18525 | 1531 | 1912 | 2315 | 1249 | 30890 | 263769 | 44161 | 2825 | 46896 | 414073 |
| 30 | 1989-90 | 30072 | 2057 | 2114 | 1516 | 2454 | 49564 | 247215 | 45354 | 2452 | 49566 | 432364 |
| 31 | 1990-91 | 35125 | 4484 | 2111 | 988 | 5904 | 49488 | 280840 | 54299 | 1845 | 65378 | 500462 |
| 32 | 1991-92 | 45050 | 4399 | 2319 | 774 | 4106 | 35019 | 310387 | 63452 | 2681 | 61830 | 530017 |
| 33 | 1992-93 | 50229 | 6415 | 2428 | 881 | 5101 | 78715 | 335100 | 66202 | 1058 | 62974 | 609103 |
| 34 | 1993-94 | 55559 | 3996 | 3068 | 976 | 1958 | 72040 | 358629 | 58887 | 1489 | 63234 | 619836 |
| 35 | 1994-95 | 58760 | 7765 | 3397 | 1018 | 4693 | 33610 | 398961 | 58912 | 1318 | 76827 | 645261 |
| 36 | 1995-96 | 70853 | 8158 | 3882 | 1013 | 4153 | 36658 | 331915 | 68088 | 1014 | 72617 | 598351 |
| 37 | 1996-97 | 84984 | 8418 | 4257 | 1315 | 2045 | 48927 | 356438 | 76157 | 855 | 76662 | 600000 |

*Source* : Reprinted with the permission of Commissionerate of Fisheries, Govt. of Gujarat, Gandhinagar.

**Table 33.9 : Category Wise Information of the Fishing Boats of Gujarat State**

| Year | Trawler | Gill netter | FRP | OBM | Other boats | Total Mech. boats | Total Non-Mech. boats | Total boats |
|---|---|---|---|---|---|---|---|---|
| 1980-81 | 1781 | 622 | – | 843 | 213 | 3459 | 6023 | 9482 |
| 1981-82 | 1780 | 672 | – | 908 | 247 | 3607 | 6195 | 9802 |
| 1982-83 | 1705 | 721 | – | 1343 | 247 | 4016 | 6998 | 11014 |
| 1983-84 | 1757 | 774 | – | 1458 | 256 | 4245 | 7529 | 11774 |
| 1984-85 | 1885 | 867 | – | 1566 | 289 | 4557 | 7749 | 13306 |
| 1985-86 | 1919 | 956 | – | 1673 | 310 | 4858 | 8018 | 12876 |
| 1986-87 | 2062 | 1050 | – | 1854 | 347 | 5313 | 8498 | 13811 |
| 1987-88 | 2230 | 1124 | 226 | 1776 | 359 | 5715 | 8965 | 14680 |
| 1988-89 | 2522 | 1756 | 492 | 1806 | 427 | 7003 | 8966 | 15969 |
| 1989-90 | 2712 | 1770 | 859 | 1822 | 485 | 7650 | 8735 | 16384 |
| 1990-91 | 2814 | 1946 | 1044 | 1838 | 489 | 8140 | 8677 | 16817 |
| 1991-92 | 3055 | 2211 | 1386 | 1862 | 510 | 9024 | 8825 | 17849 |
| 1992-93 | 3456 | 2315 | 1618 | 1957 | 524 | 9870 | 8745 | 18615 |
| 1993-94 | 3941 | 3058 | 2161 | 1813 | 530 | 11503 | 8323 | 19826 |
| 1994-95 | 4634 | 3110 | 2545 | 1814 | 545 | 12648 | 8370 | 21018 |
| 1995-96 | 5685 | 3143 | 2810 | 1817 | 562 | 14017 | 8646 | 22663 |
| 1996-97 | 6027 | 3205 | 3012 | 1827 | 600 | 14671 | 8851 | 23522 |
| 1997-98 | 6390 | 3275 | 3551 | 1854 | 628 | 15693 | 8918 | 24616 |
| 1998-99 | 6747 | 3482 | 3998 | 1899 | 635 | 16763 | 9222 | 25985 |

*Source* : Commissionerate of Fisheries, Government of Gujarat. (with permission)

**Table 33.10 : Share of Miscellaneous Fish landing in total marine fish landing in Gujarat**

| Year | Total fish landing of Gujarat | Miscellaneous fish landing | Per cent share |
|---|---|---|---|
| 1964-65 | 89,017 | 5,374 | 6.04 |
| 1969-70 | 1,21,995 | 4,655 | 3.82 |
| 1976-77 | 2,25,279 | 26,605 | 11.80 |
| 1980-81 | 2,18,872 | 28,847 | 13.18 |
| 1981-82 | 2,20,607 | 24,923 | 11.30 |
| 1982-83 | 1,92,669 | 24,545 | 12.74 |
| 1983-84 | 2,23,291 | 26,784 | 12.00 |
| 1984-85 | 2,90,708 | 34,404 | 11.83 |
| 1985-86 | 3,06,577 | 39,313 | 12.82 |
| 1986-87 | 3,15,942 | 39,757 | 12.58 |
| 1987-88 | 3,27,560 | 41,467 | 12.65 |
| 1988-89 | 4,14,075 | 55,202 | 13.33 |
| 1989-90 | 4,32,384 | 40,063 | 9.26 |
| 1990-91 | 5,00,462 | 53,198 | 10.63 |
| 1991-92 | 5,30,017 | 55,935 | 10.55 |
| 1992-93 | 6,09,103 | 60,538 | 9.39 |
| 1993-94 | 6,19,836 | 68,182 | 11.00 |
| 1994-95 | 6,45,261 | 65,383 | 10.13 |
| 1995-96 | 5,98,351 | 67,488 | 11.28 |
| 1996-97 | 6,60,068 | 75,471 | 11.43 |

On analysing above data of more than 30 years, it can be seen that miscellaneous fish landing is ranging between 11% of the total marine fish landing of Gujarat.

## Important Fauna

Important species of fauna found in the Marine waters of Gujarat at Marine National Park (MNP) and Sanctuary are as under :

### *Phyllum porifera*

About 70 species of sponges (Soft and Calcarious types) have been recorded in the region. They occur in fantastic array of colours, viz. pink, red, brown, blue, light blue, yellow, orange and what not.

### *Phyllum celenterata*

This group is mainly represented by Jelly fishes, sea anemones and corals. Animals of this group are most widely distributed and most abundant ones. Most important animals in this group are corals. More than 37 species of corals are recorded in this area.

**Table No. 33.11 : District wise number of Marine Fish Landing Centers and MFV in Gujarat.**

| Name of District | No of landing centers | No of MFV on 31-3-99 Mech-anised | Non-Mech | Total |
|---|---|---|---|---|
| Valsad | 36 | 3294 | 573 | 3867 |
| Surat | 13 | 11 | 43 | 54 |
| Kheda | 1 | 0 | 0 | 0 |
| Bharuch | 18 | 11 | 13 | 24 |
| Bhavnagar | 11 | 117 | 16 | 133 |
| Amerali | 11 | 1253 | 174 | 1427 |
| Junagadh | 25 | 9836 | 417 | 10,253 |
| Jamnagar | 23 | 1041 | 103 | 1344 |
| Rajkot | 1 | 85 | 81 | 166 |
| Kutch | 51 | 974 | 224 | 1171 |
| Total of Gujarat | 188 | 16622 | 1844 | 18,439 |

*Source* : Commissioner ate of Fisheries Govt. of Gujarat

**Table 33.12 : List of Important Fishes available in Gujarat with their Scientific, Common English and Local Names.**

**Marine Water Fishes**

| Sr. No. | Scientific Name | English Name | Local Name |
|---|---|---|---|
| **SHARKS-SKATES AND RAYS:** *Class : Elasmobranchii* | | | |
| 1. | *Stegostoma vrius* | Zebara, shark, nurse shark | Lalo, Magru |
| 2. | *Carcharias melnopteruts* | Black shark, man-eating shark, black finned shark | Patari, Magra |
| 3. | *Galeocerda arcticus* | Tiger shark | Patari |
| 4. | *Scoliodon Sorrokowah* | Yellow dog shark, sharp nosed shark | Sandho, Musi |
| 5. | *Scoliodon palasorrah* | Gray-dog-shark | Musi |
| 6. | *Carcharhinus manisorrah* | Black shark | Musi |
| 7. | *Carcharhinus limbatus* | Gray-shark | Magra, Musi |
| 8. | *Carcharhinus dussumieri* | White checked shark | Magra |
| 9. | *Carcharhinus ellioti* | Elliot's grey shark | Patari |
| 10. | *Sphyrna blochii* | Hammer headed shark, arrow headed hammer head shark | Kannar, Kanmusi |
| 11. | *Sphyrna tubes* | Squat headed hammer head shark | Kannar, Kanmusi |
| 12. | *Rhincodon typus* | Whale shark | Barel |
| 13. | *Rhizoprionodon typus* | Gray-dog-shark | Pishori |
| 14. | *Pristis microdon* | Small toothed saw fish | Churio |
| 15. | *Pritis cuspidatus* | Pointed saw fish | Churio |
| 16. | *Rhynchobatus diiddensis* | White spotted shovelnose ray | Dhose, Bhuthar |
| 17. | *Rhynobatus granulatus* | Granulated shovenose ray | Dhose, Bhuthar |
| 18. | *Acetobatus narinari* | Spotted eagle ray | Vagaliu |

*Contd.........*

*Table 33.12 – Contd...*

| Sr. No. | Scientific Name | English Name | Local Name |
|---|---|---|---|
| 19. | *Acetomylaeus maculatuls* | Bat ray, eagle ray | Kagdi, Vagaliu |
| 20. | *Himantura uarnak* | Marbled string ray | Tarabla, Bur |
| 21. | *Mobula diabonus* | Devil ray | Varala, Karaj |
| 22. | *Trygon zugei* | Pale edged string ray | Buvar |
| 23. | *Narcline timlei* | Spotted electric ray | Manduko, Jangaru |
| 24. | *Trygon Sephen* | Cow tail ray | Buvar, Bur |
| 25. | *Raja mamillidens* | Skate | Tarbla |
| 26. | *Nebrius ferruginoum* | Rusty shark, Nurse shark | Bhura magra |
| 27. | *Alopias vulpinus* | Man-eating shark, Fox shark, Makorel shark | Magru |
| 28. | *Chiloscyllium indicum* | Ridge-lack cat shark, Nurse shark | Magra |
| 29. | *Atelomycterus marmoratum* | Cat shark, Marbled cat shark | Lala |
| 30. | *Rhina ancylostoma* | Bow-mouthed angle fish, mud shark, guiter fish | Dhons |
| 31. | *Amphotistius imbricata* | Scaly siting ray, sting ray | Patara |
| 32. | *Amphotistius zugeri* | Pale-edged sting ray, sting ray | Buvari, Bavra |
| 33. | *Acetobatus flagellum* | Spotted eagle ray, eagle ray bullary | Kodio, Karaj |
| 34. | *Rhinoptera javanica* | Javanese cow ray, cow ray | Boor |
| 35. | *Dasyatis sephen* | Cow tell ray, whip ray | Boovari, Boor |
| 36. | *Gymnura poecilura* | Long tailed butterfly ray | Patari, Boor |
| 37. | *Nacine brunnea* | Brown electric ray, electric ray | Patara |
| 38. | *Narcine indica* | Spotted electric ray, electric ray | Manduko, Jangara |
| 39. | *Galeocorda articus* | Tiger shark, requiem shark | Bhuvar |
| 40. | *Myrmillo manazo* | Gummy shark, requiem shark | Musiya |
| 41. | *Hypoprian macloti* | Maclot's shark | – |
| | **BONNY FISHES EEL** | | |
| 42. | *Muraenesox talabonoides* | Indian pink eel | Vam |
| 43. | *Muraenesox cinereus* | Silver conger eel | Vam |
| 44. | *Gymnotharax undulatus* | Leopard moray | Naro |
| 45. | *Gymnotharax favagineus* | Black-lined moray, Pointed eel | Hingchar |
| | **CAT FISHES** | | |
| 46. | *Batrachocephalus mino* | Frog headed cat fish | Khaga, Khagdi |
| 47. | *Trachysurus caelatus* | Engraved cat fish | Khaga |
| 48. | *Tachysurus sona* | Dusky cat fish | Khaga, Singda |
| 49. | *Tachysurus jella* | Small eye cat fish | Dharva |
| 50. | *Tachysurus dussumieri* | Marine cat fish | Shingati |
| | **CLUPEOIDS** | | |
| 51. | *Chirocentrus Dorab* | Silver bar, wolf herring | Dai, lapdi |
| 52. | *Sardinella longiceps* | Indian oil sardine | Tarli |
| 53. | *Sardinella fimbriata* | Fringe-scale sardine | Pedvi |
| 54. | *Hilso ilisha* | Indian shad | Chaksi, Palla |

*Contd.........*

*Table 33.12 – Contd...*

| Sr. No. | Scientific Name | English Name | Local Name |
|---|---|---|---|
| 55. | *Hilso toil* | Giant herring | Palva, Modar |
| 56. | *Coilia dussumieri* | Golded anchovy | Mandli |
| 57. | *Coilia reynaldi* | Anchovy | Mandli, Pafat |
| 58. | *Stolephorus batavienris* | White bait, Batavian anchovy | Fansi |
| 59. | *Thryssa mystax* | Mustached anchovy | Palli |
| 60. | *Thryssa malbarica* | Malbar anchovy | Pallia |
| 61. | *Nematalosa nasus* | Long ray bony beam | Palli |
| 62. | *Escualosa thoracata* | White sardine | Bhanjo, Morva |
| 63. | *Ilisha eleganta* | Slender shad | Phansi, Palvi |
| 64. | *Ilisha megaloptera* | Jewelled shad | Kati |
| 65. | *Opisthopterus tardoore* | Long finnel herring, razor edge | Kati |
| 66. | *Pellona ditchela* | Indian pellona | Kati |
| 67. | *Elops machnata* | Giant herring | Rajputani, Kalur |
| 68. | *Harpodon nehereus* | Bombay duck | Datniya, Bumla |
| 69. | *Saurida tumbil* | Greater lizard fish | Chor bumla |
| 70. | *Elops saurus* | Giant herring, Tarpan | Rajputani, Kalru |
| 71. | *Thryssa purava* | Anchovy | Palli |
| 72. | *Thryssa hamiltoni* | Hamilton's anchovy | Palli |
| 73. | *Thryssa dussumieri* | Dussumieri's anchovy | Palli |
| 74. | *Kowala coval* | White sardine | Motawa, Bhangi |
| | **FLYING FISH** | | |
| 75. | *Cypselurus cosmatus* | Flying fish | Jira |
| 76. | *Exocoetus volitans* | Two winged flying fish | Jira, Kunga |
| | **HALF BEAKS & FULL BEAK** | | |
| 77. | *Hemirhamphus marginatus* | Bared half beak | Kunga |
| 78. | *Hemirhamphus georgii* | Long billed half beak | Kunga |
| 79. | *Strongylura strongylure* | Full beak gar fish | Kagda, Gira |
| 80. | *Strongylura crocodila* | Fork tail alligator gah fish | Gira, Kunga |
| | **PERCHES** | | |
| 81. | *Epinephelus tanuina* | Greasy grouper rock perch | Vekhu, Vekhali, Salai |
| 82. | *Epinephelus fasciatus* | Banded reef cod | Vekhi |
| 83. | *Epinephelus fario* | Trout reef cod | Vekhru |
| 84. | *Epinephelus merra* | Spotted rock cod, sea bass | Salai |
| 85. | *Promicrops lanceolatus* | Greater grouper | Vekhu, Salai |
| 86. | *Lutianus johni* | Snapper, Moses Perch | Gurka, Tamb |
| 87. | *Lutianus argentimaculatus* | Red snapper | Gurka Tamb, Ratdo |
| 88. | *Lutianus fulviflamma* | One spotted golden snapper | Tusha |
| 89. | *Lethrinus frenatus* | Bridled pig-face bream | Chuncha, Dhamil |
| 90. | *Lethrinus ramak* | Yellow banded pig face bream | Chuncha |
| 91. | *Ganthodentex aurolineatus* | Gold lined larged eyed bream | Chuncha |
| 92. | *Nemipterus japonicus* | Japanese thread fin bream | Rani |

*Contd........*

***Table 33.12 – Contd...***

| Sr. No. | Scientific Name | English Name | Local Name |
|---|---|---|---|
| 93. | *Chrysophrys cuvieri* | Sea bream | Chayo |
| 94. | *Chrysophrys datnia* | Sea bream | Timra |
| 95. | *Chrysophrys berda* | Black bream | Timro |
| 96. | *Argyrop spinifer* | Long spined red bream | Kisi, Ratdo |
| 97. | *Drepane punctata* | Moon fish | Chand, Fanas |
| 98. | *Ephippus orbis* | Spade fish | Vada, fani |
| 99. | *Lates calcarifer* | Cock-up, giant sea perch | Bekti, Dangri |
| 100. | *Labotes surinamensis* | Triple tail | Katkola |
| 101. | *Mylio berda* | Black sea bream | Kaedikisi, Kharpal |
| 102. | *Pomadasys maculatus* | Spotted grunter | Karkara, Kokara |
| 103. | *Pomadasys hasta* | Lined silver grunter | Karkara, Kokara |
| 104. | *Pomadasys argyreus* | Silver grunter | Ghothar |
| 105. | *Rhonciscus furcatus* | Baned grunter | Dhamil |
| 106. | *Scatophagus argus* | Spotted butterfish | Kaski, Koyee |
| 107. | *Therapon jarbua* | Crescent perch, Tiger-perch | Naida, Grangta |
| 108. | *Gaterin cincutus* | Three banded sweetlip | Chini |
| 109. | *Variola louti* | Painted coral trout | Vakhali |
| 110. | *Callyodon harid* | Red lined parrot fish | Popat |
| 111. | *Platyglossus dussumieri* | Rainbow fish | Babat |
| 112. | *Platyglossus marginatus* | Speckled rainbow fish | Babat |
| 113. | *Epinephelus lanecolatus* | Grouper rock-cod, Sea bass | Salai |
| 114. | *Epinephelus maculatus* | Trout reef-cod, Sea bass | Vekhru |
| 115. | *Gaterin schotaf* | Grey sweetlip | Cheni |
| 116. | *Gaterin picoides* | Sweetlip | Cheni |
| 117. | *Acanthopagrus berda* | Black bream, black rock cod | Timro |
| 118. | *Cronidens forskalii* | Sea bream, Pargy | Faradi |
| 119. | *Cronidens indicus* | Sea bream, Pargy | Faradi |
| 120. | *Sargus noct* | Sea bream | Kisi, Faradi |
| 121. | *Epinephelus sexfasciatus* | Trout reef cod, rock pourdh, sea bass | Salai |
| 122. | *Rhabdosargus sarba* | Sea bream | Khrod |
| 123. | *Lethrinus cinereus* | Bridled Pigface bream, rudder fish | Chuncha |
| 124. | *Lethrinus miniata* | Long nossed pigfaced beam | Chuncha |
| 125. | *Psammonperca waigiensis* | Waigen sea perch | Bakti, Dhangri |
| 126. | *Gerres Filaomtosus* | Long rayed silver biddy, silver perch | Farali, Keto, Koto |
| 127. | *Gerres oblongus* | Oblong silver biddy, Majarra | Keta, Koto |
| | **GOAT FINS** | | |
| 128. | *Upeneus sulphureus* | Yellow goat fish | Chiri |
| 129. | *Upeneus vittatus* | Yellow stripped goat fish | Chiri |
| | **THREAD FINS** | | |
| 130. | *Eleutheronema tetradactylum* | Indian salmon | Ravas |
| 131. | *Polynemus indicus* | Monk fish, giant thread fin | Dara |

*Contd.........*

**Table 33.12 – Contd...**

| Sr. No. | Scientific Name | English Name | Local Name |
|---|---|---|---|
| 132. | *Polynemus heptadactylus* | Seven thread tassel fish | Shiri |
| 133. | *Polynemus plebeius* | Common tassel fish | Kund |
| 134. | *Johnieops sina* | Drab jew fish | Dhoma |
| 135. | *Johnieops macrorhynus* | Jew fish | Dhoma |
| 136. | *Johnius dussumieri* | Croaker | Dhoma |
| 137. | *Johnius vogleri* | Croaker | Dhoma |
| 138. | *Otolithus ruber* | Rosy jew fish | Dhoma |
| 139. | *Otolithus argentens* | Silver banded jew fish | Dhoma |
| 140. | *Otolithus biauritus* | Dori | Koth |
| 141. | *Protolibea diacanthus* | Two spinned jew fish | Ghol |
| 142. | *Otolithoides brunneus* | Rosy jew fish, croaker | Koth |
| 143. | *Johnius glaucus* | Silver jew fish, drum croaker | Dhoma |
| 144. | *Nibea semiluctosus* | Drab jew fish | Dhoma |
| 145. | *Otolithus maculatus* | Silver banded jew fish | Dhoma |
| 146. | *Nemipterus japonicus* | Japanese thread bream | Dhoma |
| 147. | *Polynemus tetradactylus* | Four thread tassel fish Indian salman | Ravas |
| | **RIBBON FISHES** | | |
| 148. | *Lepturacanthus savala* | Silver ribbon fish | Patti |
| 149. | *Trichiurus maticus* | Cutlass fish, ribbon fish | Lapdi |
| 150. | *Trichiurus haumela* | Large headed ribbon fish | Patti, Lapdi |
| 151. | *Eupleurogramnus muticus* | Ribbon fish, Cutlass fish | Lapdi |
| 152. | *Trichiurus lepturus* | Large headed ribbon fish | Lapdi, Patti |
| | **CARANGIDS** | | |
| 153. | *Megalapsis Cordyla* | Horse Macherel | Bangda, Khadvo |
| 154. | *Decapterus russelli* | Russll's Scad | Pira bangda |
| 155. | *Scomberoides lysan* | Port hole fish, leather jacket | Shag |
| 156. | *Alepes djeddaba* | Dijeddaba scad | Khadvo |
| 157. | *Alepes kalla* | Golden scad | Lal bangda |
| 158. | *Atropus atropus* | Horse mackerel | Bangda, Khadvo |
| 159. | *Megalaspis cordyla* | Torpedo trevelly | Chapri |
| 160. | *Carangoides malabaricus* | Malbar trevelly | Chapri, Bangda |
| 161. | *Rachycentron canadus* | Black king fish | Modasa |
| 162. | *Selar crumenopthalmus* | Big eyed scad | Karba bangda |
| 163. | *Citula malabaricus* | Malbar trevelly, Horse mackeral | Chapri, Bangda |
| 164. | *Caranx sexfasiatus* | Six handed trevelly | Chapri, Khadro |
| 165. | *Scomberoides tol* | Slender queen fish, leather jacket | Sag, Chalia |
| 166. | *Scomberoides Santipetri* | Saint Peter's leather's skin, leather jacket | Ala, Chalia, Sag |
| 167. | *Scataphgaus argus* | Spotted butter fish, Scat | Kui, Kaski, Vada, Sugeli |
| 168. | *Singanus aramin* | White spotted spine foot rabit spine foot fish | Kasar |
| 169. | *Trachinotus blochii* | Bloch's dart | Naryela |

*Contd........*

*Table 33.12 – Contd...*

| Sr. No. | Scientific Name | English Name | Local Name |
|---|---|---|---|
| 170. | *Monocanthus choiracephalus* | Pig faced leather jacket, trigger fish | Chalia |
| 171. | *Stephanelepis diaspros* | Reticulated leather jacket, file fish | Chalia |
| | **SILVER BELLIES** | | |
| 172. | *Leiognathus spendens* | Splendid pony fish | Katli |
| 173. | *Leiognathus fasciatus* | Silver belly | Katri |
| 174. | *Leiognathus lactarius* | White high jaw'jumper | Dhari, Dhagri |
| 175. | *Leiognathus equala* | Greater pony fish | Katri, Chandalia |
| | **BIG JAWED JUMPER** | | |
| 176. | *Lactarius lactarius* | Big-jaw jumper, white fish | Dangri |
| | **POMFRETS** | | |
| 177. | *Parastromateus niger* | Black pomfret | Havala, Addiyo |
| 178. | *Pampus argenteus* | Silver pomfret | Vichuda, Paplet |
| 179. | *Pampus chinensis* | Chinese pomfret, white pomfret | Pathu, Kafri, Paplet |
| | **MACKERELS** | | |
| 180. | *Rastrelliger kanagurata* | Indian mackerel | Bangda |
| | **SEER FISHES** | | |
| 181. | *Scomberomorus commerson* | Barred seer fish | Surmai, Chapra |
| 182. | *Indocybium guttatum* | Spotted spanish mackerel | Surmai, Bhangda |
| 183. | *Scomberomorus lineolatus* | Streaked seer fish | Surmai, Chapri |
| | **TUNNIES** | | |
| 184. | *Euthynnus affinis* | Little tuna | Gedra |
| 185. | *Auxis thasard* | Frigate mackerel | Gedra |
| 186. | *Katsuwonus pelamis* | Shipjack tuna/striped tuna | Gedra |
| 187. | *Thunnus albacares* | Yellow-fin tuna | Gedra |
| | **BARRACUDA** | | |
| 188. | *Sphyraena jello* | Banded barracuda | Bhungar |
| 189. | *Sphryaena obtusta* | Blunt jewed sea pike barracnda | Bhungar, Kunga |
| | **MULLETS** | | |
| 190. | *Mugilio Cephalus* | Flathead gray mullet | Boi, Bodaki |
| 191. | *Valamugli speigleri* | Speigleri's gray mullet | Makan |
| 192. | *Liza stronglycephalus* | Yellow-tail mullet, round head mullet | Ghandhiyo |
| 193. | *Mugil seheli* | Blue-spot mullet | Mankan |
| 194. | *Mugil dussumieri* | Gray mullet | Boi |
| 195. | *Mugil cunnesius* | Yellow tail mullet | Gandhio |
| 196. | *Mugil tade* | Green back mullet | Gandhio |
| 197. | *Mugil parsia* | Gold spot mullet | Gandhio |
| 198. | *Mugil vaigiensis* | Dimond-scale mullet | Gandhio |
| 199. | *Mugil macroiepis* | Borneo mullet | Gandhio |
| | **UNICORN COD** | | |
| 200. | *Bregmaceros macclellndii* | Indian-cod | Chiri |

*Contd.........*

**Table 33.12 – Contd...**

| Sr. No. | Scientific Name | English Name | Local Name |
|---|---|---|---|
| | **FLAT FISHES** | | |
| 201. | *Psettodes erumei* | Indian Halibut | Hario, Moti Jeeb |
| 202. | *Solea elongata* | Elongate sole | Khetar |
| 203. | *Brachirus orientalis* | Oriental sole | Datro |
| 204. | *Pseudorhombus arsius* | Left hand flounder, large tooth flounder | Khetar, Jipti |
| 205. | *Cynoglossus lingua* | Long tongue sole | Jeeb |
| 206. | *Cynoglossus macrostamus* | Malbar tongue sole | Jeeb |
| 207. | *Cynoglossus lida* | Shoulder-spot, tongue sole | Khetar, Jeeb |
| 208. | *Zebrias quagga* | Zebra sole | Khetar, Dataro |
| 209. | *Platycephalus indicas* | Indian flat head, flat fish | Khukhari |
| 210. | *Parapercis pulchella* | Rosy grab fish | Vekerdungi |
| | **GOBIES AND MUDSKIPPERS** | | |
| 211. | *Thypauchen vagina* | Red goby | Lal levta |
| 212. | *Periophthalmus koelreuteri* | Mudskipper | Levta |
| 213. | *Boleophthalmus dotatus* | Goby | Andha gulla |
| 214. | *Boleophthalmus dussumieri* | Goby | Levti |
| 215. | *Gobius striatus* | Goby | Andergola |
| 216. | *Scortelaos tenuis* | Goby | Andergola |
| | **OTHER SMALL GROUPS** | | |
| 217. | *Batrachus gangene* | Frong fish | Gunga |
| 218. | *Hippocampus kuda* | Sea horse | Jalghodo |
| 219. | *Diodon hystrix* | Spotted porcupine | Govido, Shahudi |
| 220. | *Cyrinodon dispar* | Killi fish | Pipuda |
| 221. | *Chanos chanos* | Milk fish | Sani, Sanu |
| 222. | *Histiphorus gladius* | Banana sail fish, sward-fish | Ghodo |
| 223. | *Triacanthus brevirostris* | Short-nosed tripod-fish | Trikandi |
| 224. | *Mene maculata* | Moon fish | Chand, Vanyani |
| 225. | *Linophora vagabunda* | Butter fish | Jubir |
| 226. | *Chaeotodontops collaris* | Butterfly fish, Cord fish | Chatardi |
| 227. | *Batrachus grunniens* | Frog fish | Gunga |
| 228. | *Istiophorus gladius* | Indian sail fish, Peacock fish | Ghodo |
| 229. | *Makaira indica* | Indian spear fish, bill fish | Ghodo |
| 230. | *Parapegasus natans* | Floating dragon fish | – |
| 231. | *Trachyrhamphus serratus* | Crested pipe fish | – |
| 232. | *Syngathus cyanospilus* | Pipe fish | – |
| 233. | *Fistularia petimba* | Smooth flute mouth, flute fish, cornet fish | Wansari, Vansri |
| 234. | *Choridactylus multibarbis* | Orange-banded goblin fish | Sheri |
| 235. | *Minous monodactylus* | Gray goblin fish | Lore |
| 236. | *Amphanius dispar* | Killifish, toothed carp minnow | Pipuda |
| 237. | *Gerres setifer* | Black tipped silver biddy | Morli |

*Contd........*

***Table 33.12 – Contd...***

| Sr. No. | Scientific Name | English Name | Local Name |
|---|---|---|---|
| 238. | *Sillago sihama* | Silver whiting, sand whiting | Sing, Vekarkunga |
| 239. | *Pterois volitans* | Winged fire fish, fire fish scorpion fish | Gofanio |
| 240. | *Pterois russell* | Scorpion fish, flying dragon | – |
| 241. | *Amphiprion polymnus* | Anemone fish, demsel fish | – |
| 242. | *Abudefluf bengalensis* | Demoiselle | – |
| 243. | *Pomacentrus sindersis* | Domoiselle | – |
| 244. | *Coryphaena hippurus* | Dolphin | Imero, Daria Gai |
| 245. | *Platycephalus scaber* | Rough flat head, Flat head | Khukhari |
| 246. | *Echeneis naucrates* | Slender sucher fish, sherk sucker | Lutchik |
| 247. | *Echeneis remora* | Remora, Common remora | Meghar |
| | **ANGEL FISH/CORAL FISH** | | |
| 248. | *Drepane punctata* | Spotted scat; spotted bat fish; moon fish; sickle fish | Chand, Fanas, Vaniani |
| 249. | *Ephippus orbis* | Spade fish; scat | Vada, Fani, Fanas |
| 250. | *Pomacanthodes semicirculatus* | Bue angel fish; angel fish | Chitri |
| 251. | *Pomacanthodes annularis* | Ringed angle fish | – |
| 252. | *Chaetodon vagabundus* | Vagabond coral fish; coral fish | Jubir |
| | **PUFFER FISH** | | |
| 253. | *Gastrophysus lunaris* | Green rough-backed flow fish | Kaku |
| 254. | *Arothron loepardus* | Banded leopard blow fish | Kaku |
| 255. | *Arothron immaculatus* | Immaculate blow fish | Popcha |
| 256. | *Arothron stellatus* | Starry blow fish | Kaku |
| 257. | *Ostracion gibbosus* | Thron-backed box fish | Kaku |
| | **CRUSTACIANS** | | |
| | **(A) Penaeid prawn** | | |
| 258. | *Penaeus indicus* | Indian white prawn | Jumbo |
| 259. | *P. monodon* | Tiger prawn | Tiger |
| 260. | *P. merguiensis* | Banana prawn | White Jumbo. |
| 261. | *P. semisulcatus* | Flower | Flower. |
| 262. | *P. japonicus* | Kurma prawn | Kurma |
| 263. | *P. penicillatus* | Yellow prawn | Hite zinga |
| 264. | *Metapenaeus monoceros* | Indian prawn | Kapsi |
| 265. | *M. kutchensis* | Kutch prawn | Kutchi zinga |
| 266. | *M. affinis* | Indian prawn | Medium sumadi. |
| 267. | *M. bravicornis* | Yellow prawn | Medium |
| 268. | *Parapenaeopsis stylifera* | Kiddi prawn | Karkari |
| 269. | *P. sculpilis* | Rainbow prawn | Kolmi |
| 270. | *P. hardwicki* | Spear prawn | Kolmi |
| 271. | *Solenocera crassicornis* | Coastal mud prawn | Goiner. |
| | **(B) Non Penaeid Prawn** | | |
| 272. | *Acetes indicus* | Paste prawn | Jawla |
| 273. | *Acetes arethaeus* | Paste prawn | Jawla |

*Contd........*

**Table 33.12 – Contd...**

| Sr. No. | Scientific Name | English Name | Local Name |
|---|---|---|---|
| 274. | *Hippolysmata ensirostris* | Pink prawn | Dhobi |
| 275. | *Hippolysmata acirostos* | Hunter shrimp | Dhobi |
| 276. | *Palaemon tenuiper* | Shrimp | Karadi, Ambdi |
| 277. | *Plaemon styliferus* | Shrimp | Karadi |
| | **(C) Lobsters** | | |
| 278. | *Penulirus polyphagus* | Rock lobster | Titan |
| 279. | *Penulirus penicillatus* | Spiny lobster | Lobster, Titan. |
| 280. | *Penulirus ornatus* | Spiny lobster | Lobster, Titan. |
| 281. | *Thenus orientalis* | Sand lobster | Kako. |
| | **(D) Crabs** | | |
| 282. | *Matuta lunaris* | Sand crab | Reti-kurchala. |
| 283. | *M. planipes* | Sand crab | Reti-kurchala |
| 284. | *Scylla serrata* | Stone crab, mud crab | Dhebo |
| 285. | *Portunus pelagicus* | Raticulata crab | Karchalo |
| 286. | *P. sanguinolentus* | Spotted crab | Karchalo |
| 287. | *Charybdis cruciata* | Cross crab | Samudario |
| 288. | *C. lifera* | Blue crab | Karchalo |
| 289. | *C. callianassa* | Crab | Kurchal |
| 290. | *C. crientalis* | Crab | Kurchal |
| 291. | *C. natator* | Crab | Kurchal. |
| 292. | *Thalamita pynma* | Crab | Karachalo |
| 293. | *T. crenata* | Crab | Karchalo |

Dholakia 1986, Chhaya-1990

**Table 33.13 : Details of Identified Marin Water Molluscans of Gujarat : Saurashtra**

| S. No. | Scientific Name | S. No. | Scientific Name |
|---|---|---|---|
| SIKKA | | 15. | *Bursa spinosa* |
| 1. | *Conus araneosus* | 16. | *Paphia-ala-papilines* (Roding) 1798 |
| 2. | *Conus figulinus* | 17. | *Potomides fascum* (Horn shell) |
| 3. | *Cyprathium rebeculum* | 18. | *Donax variabilies* |
| 4. | *Cypraea annulus* (Linne) 1758 | 19. | *Poromya granulata* |
| 5. | *Cypraea onyx* (Linne) 1791 | 20. | *Oliva reticularis* |
| 6. | *Bursa margaritala* | 21. | *Bursa granularis* |
| 7. | *Cardium setasum* (Redfurn) 1846 | 22. | *Pinctada modrasensis* |
| 8. | *Murex trapa* (Roding) 1798 | 23. | *Pinctada fucata* |
| 9. | *Murex virgineus* (Roding) 1798 | 24. | *Bulla occibentall* A. Adams |
| 10. | *Tibia curta* | 25. | *Chino fluctifroga* (Sowerby) |
| 11. | *Turbo argyrostomus* | 26. | *Diodora dysoni* (Reeve) |
| 12. | *Turbo margaritaceus* | 27. | *Nerita fulgurans* |
| 13. | *Durapa nodulusa* C.B. Adams | 28. | *Conus muriculatus* |
| 14. | *Lathyrus samaragdulus* | 29. | *Thais bufo* (Muncinella) (Lamark) 1845 |

*Contd........*

Table 33.13 – Contd...

| S. No. | Scientific Name |
|---|---|
| 30. | *Mactra turgida* (Gmelin) 1790 |
| 31. | *Telescopium telescopium* |
| 32. | *Protothaca staminea* (Conard) |
| 33. | *Cerithium scabridum* (Phip) 1848 |
| 34. | *Pectan diegensis* |
| 35. | *Tapes aurens* |
| 36. | *Trachycardium egmonfianum* |
| 37. | *Tellina Ala* (Hanley) 1844 |
| 38. | *Tellina mevopsis* |
| 39. | *Babylonia spirata* |
| 40. | *Murex adustus* (Lamarck) |
| 41. | *Scapharca inaequivalvis* |
| 42. | *Pyrene vulpecula* (Sowerby) 1847 |
| 43. | *Turritella cooperi* |
| 44. | *Vasum latirifome* |
| 45. | *Turritella communis* |
| 46. | *Tellina bodegensis* |
| 47. | *Buda striata* (Bruguiere) |
| 48. | *Erosaria lamaroki* (Cowrie) |
| 49. | *Pecten crassicostatus* |
| **GOOSE** | |
| 1. | *Venus chemnitzi* (Hanely) 1844 |
| 2. | *Venus reticulata* (Linne) 1758 |
| 3. | *Cerithium morus* (Lamarck) 1843 |
| 4. | *Nassarius fossatus* (Dumeri) 1806 |
| 5. | *Nassarius frivittatus* |
| 6. | *Turritella acropora* (Dall) |
| 7. | *Rapana bulbosa* (Dilwyn) 1817 |
| 8. | *Nerita dombeyi* (Reculz) 1841 |
| 9 | *Nerita rumphii* (Reculz) 1841 |
| 10. | *Pinna atropurpura* (Sowerby) 1825 |
| 11. | *Voluta musica* (Linne) 1758 |
| 12. | *Mactra stultorum* |
| 13. | *Cerithiopsis subaluta* (Montagu) |
| 14. | *Turbo castaneous* (Gmelin) |
| 15. | *Nassarius reticulus* |
| 16. | *Turritella acropora* (Dall) |
| 17. | *Lioconcha lorenziana* (Dilluyn) |
| 18. | *Drupa heptgonalis* (Reeve) 1845 |
| 19. | *Lithophaga anbillarum* |
| 20. | *Tagolus californianus* |
| 21. | *Cerithiopsis subuluta* |
| **POSHITRA** | |
| 1. | *Barbatia domingensis* |
| 2. | *Spiroglyphus ıtuellus* (Morch) 1800 |
| 3. | *Cerithium charneum* (Bruguiere) |
| 4. | *Venus mercenaria* (Linne) |
| 5. | *Nassarius obsoletus* (Say) |
| 6. | *Natica tigrina* (Roding) 1798 |
| 7. | *Nassa hepatica* |
| 8. | *Cerithium scabridum* (Phip) 1848 |
| 9. | *Stombina* spp. |
| 10. | *Ceritobelisais bruguiera* |
| 11. | *Tellina squalide* |
| 12. | *Littorina undulata* (Grey) |
| 13. | *Cypraea pallida* (Grey) 1863 |
| 14. | *Crassostrea cristagalli* (Linnaeus) |
| 15. | *Donex cuneatus* |
| 16. | *Gibbula magans* |
| 17. | *Pinctada vulgaris* (Schumacher) |
| 18. | *Drupa margariticola* (Broderip) |
| 19. | *Stellaria solaris* |
| **PIROTAN** | |
| 1. | *Cerithium citrinum* (red furn) 1846 |
| 2. | *Semele striata* (Ruppell) 1798 |
| 3. | *Semele flavescens* |
| 4. | *Umbonium vesbianum* (Linne) 1806 |
| 5. | *Solariella obscura* (couthouy) 1842 |
| 6. | *Pinna atropurpura* (Swerby) 1825 |
| 7. | *Acmuea pelta* (Eschscholtz) |
| 8. | *Cerithium obeliscus* (bruguiere) 1792 |
| 9. | *Cancellaria rebianlata* |
| 10. | *Acropogia gausta* |
| 11. | *Angaria plicata* (Kiener) |
| 12. | *Patella cernica* |
| 13. | *Barbatia obliquata* |
| 14. | *Cardium assimile* |

*Contd.........*

*Table 33.13 – Contd...*

| S. No. | Scientific Name |
|---|---|
| 15. | *Glycimeris taylori* (Angar) |
| 16. | *Arca fusca* (bruguiere) |
| **NARARA** | |
| 1. | *Venus verrucosa* |
| 2. | *Irus lamellifera* |
| 3. | *Carita bicolor* (Lamarck) 1819 |
| 4. | *Chama macerophylla* |
| 5. | *Astarte arctica* |
| 6. | *Vexillum cornense* (holetype) |
| 7. | *Barbatia candide* |
| 8. | *Dentalium octangulatium* (Donovan) 1804 |
| 9. | *Dentalium aprinum* (Linne) 1766 |
| 10. | *Dentalium elpis* (Winckworth) 1927 |
| 11. | *Pirula ficus* |
| 12. | *Trochus tentorium* (Gonelin) 1791 |
| 13. | *Ocenebia circumtexta* (Fleming) 1828 |
| 14. | *Mitromorpha olivoidea* |
| **OKHA** | |
| 1. | *Cypraea ocelata* (Linne) 1767 |
| 2. | *Cyparea figris* (Linne) 1758 |
| 3. | *Cyparea caurica* (LInne) 1758 |
| 4. | *Tellina idae* |
| 5. | *Gafrarium dispar* (Chemnitz) 1795 |
| 6. | *Architectonica nobilis* (Roding) |
| 7. | *Pinctada chemnitgi* |
| 8. | *Trochus radiatus* (Gmelin) 1791 |
| 9. | *Natica lineata* (Lamarck) 1838 |
| 10. | *Natica didyma* (Roding) 1798 |
| 11. | *Bembicium eauratum* |
| 12. | *Nerita undata* |
| 13. | *Succinea godivariana* |
| 14. | *Turbo coronatus* |
| 15. | *Acro inaequivalvis* |
| 16. | *Diodora funiculata* |
| 17. | *Pyrene zebra* |
| 18. | *Solen sicarius* (Gould) |
| 19. | *Gafrarium divaricata* (Chemnit) 1782 |

| S. No. | Scientific Name |
|---|---|
| **HARSHAD** | |
| 1. | *Crossostrea madrasensis* |
| 2. | *Crossostrea follium* |
| 3. | *Echinics nodulasus* |
| 4. | *Sinum cuvierianum* |
| 5. | *Astrea semisostata* (kiener) |
| 6. | *Licconcha lorenziana* |
| 7. | *Erosaria lamarchi* |
| 8. | *Lithopahaga* spp. |
| 9. | *Dinocardium robustum* |
| 10. | *Mitrolumnu olivoidea* |
| **BHOGAT** | |
| 1. | *Conus verrucosus* |
| 2. | *Mauritia eglantina* |
| 3. | *Pila gbbosa* |
| 4. | *Plectopylis bensoni* |
| 5. | *Mercenaria mercenaria* |
| 6. | *Oncomeluma* |
| **MANJAR ISLAND** | |
| 1. | *Cerithium variabile* (G.B. Adums) |
| 2. | *Sacostrea cuculata* |
| 3. | *Ostrea semicostata* |
| 4. | *Barbatic tenera* |
| **CHAKHADI ISLAND** | |
| 1. | *Acirsa boreabis* |
| 2. | *Neoropun tuberculata* |
| 3. | *Cassidula mustelina* |
| **SHIRVAL ISLAND** | |
| 1. | *Pseudanachis duclosiana* |
| 2. | *Nerita maura recluz* (1841) |
| **HANUMAN DANDI** | |
| 1. | *Turbo coronatus* |
| 2. | *Cancellaria reticulata* |
| 3. | *Arca gnbernaculum* |
| **PORBANDAR** | |
| 1. | *Tellina radiata* Linne. |
| 2. | *Arene gemma.* (Tuom and Holm) |

*Source* : Dholakia, A.D. 1998. Annual Report, Fisheries Research Station, Guj. Agri. University, Sikka.

There cannot be any diversity of colour and shape in life on the coral reefs. Reef itself is the work of living animals. Coral polyps change minerals dissolved in the sea water into the limestone of the reef. Clams and tubeworms contribute by building the hard part. Lime secreting plants (Coralline algae) repair the windward edge of the reef by cementing loose pieces of coral rock into near solid rocks. It is surprising that the coral polyps constitute only about 10% of the living matter on the reef. Rest of the 90% consisting of sponges, anemons, shrimp, lobster, crabs, cownle, sea cucumber, octopus, sea urchin, tube dwelling worms etc.

**Table : 33.14 : List of Marine Edible Molluscs of the Gulf of Kutch and Northern-Western region of Saurashtra**

| Scientific Name | Common Name | Collection area |
|---|---|---|
| CLASS : AMPHINEURA | | |
| *Ischnochiton* sp. | Coast of mail shell | Okha, Chushapir. |
| CLASS : GASTROPODA | FAMILY : PETELLIDAE | |
| *Cellana radiata* (Born) | Limpets | Okha, Hanuman-dandi. |
| *Trochus stellatus* Gemilin | Top shells or button shells | Okha, Adatra Poshetra, Hanuman dandi, Chushapir, Baide, Nora. |
| *T. radiatus* Gemilin | Top shells or button shells | Okha, Adatra, Poshetra, Hanuman dandi, Chushapir, Baide, Nora. |
| *Umbornium vestiarium* (Linn) | Top shells or button shells | Okha, Adatra, Poshetra, Hanuman dandi, Chushapir, Baide, Nora |
| FAMILY : TURBINIDAE | | |
| *Turbo intercostalis* Menke | Turban shell | Balapur Okha, Adatra |
| *T. coronatus* Gmeiin | Turban shell | Poshitra, Dwarka, Boria, Pashu, Nora, Chusnapır, Baida, Hanuman dandi. |
| *Astroa semicostata* (*Kiener*) | Turban shell | Okha, Poshitra, Pashu, Hanuman dandi. |
| FAMILY : NERITIDAE | | |
| *Nerita albicilla* Linn. | Nerites | Okha, Adatra, Poshitra, Hanuman dandi. |
| *N. dombeyi* Recluz | Nerites | Okha, Adatra, Poshitra, Hanuman dandi. |
| FAMILY : POTAMIDIDAE | | |
| *Telescopium telescopium* Linn | Horn shell | Sikka, Baide, Balapur Bay. |
| FAMILY : STROMBIDAE | | |
| *Strombus urecus* Linn. | Wind shell | Hanuman dandi. |
| FAMILY : NATICIDAE | | |
| *Natica tigrina* (Roding) | "Shankula" Gujarati name | Okha, Hanuman dandi, Baida. |
| *N. lamarckii* Chenu | "Shankula" Gujarati name | Hanuman dandi. |
| FAMILY : MURICIDAE | | |
| *Murex virgineus* (Roding) | Murices | Okha, Adatra, Poshetra, Boria. |
| *M. adustus* Lamarck | Murices | Baida, Nora, Pashu, Hanuman dandi. |
| *Thais rugosa* (Born) | Purples | Okha, Poshitra. |
| *T. rudolphi* (Lamarck) | Purples | Off Pindhara |
| *T. Carinifora* Lamarck | Purples | Nora, Hanuman dandi. |

*Contd.........*

*Table 33.14 – Contd...*

| Scientific Name | Common Name | Collection area |
|---|---|---|
| FAMILY : VASIDAE | | |
| *Xancus pyrum* (Linn) | Sacred chank | Okha, Arafdmda, Baidda, Pashu, Nora. |
| FAMILY : OOLIVIDAE | | |
| *Oliva gibbosa* (Born) | Olives | Okha, Hanuman dandi |
| *O. nebulosa* Lamarck | Olives | Okha, Hanuman dandi |
| FAMILY : BUCCINIDAE | | |
| *Babylonia spirata* (Linn) | Whelks | Hanuman dandi, Balapur Bay, Off Dwarka fishing ground (22-68/2E) |
| CLASS : PELECYPODA | | |
| FAMILY : ISTREIDAE | | |
| *Crassostrea gryphoides* (Newton & Smith) | Edible oyster | Dwarka, Armda, Hanuman dandi, Muddy creecks of Kutch. |
| *C. discoides* (Gould) | Edible oyster | Armda Creek, Off Poshitra, Port Okha, Dwarka, Porbandar. |
| *Ostrea madrasensis* Preston | Edible oyster | Pirotan Island. |
| *O. folium* Gmelin | Edible oyster | Pirotan Island. |
| *C. cucullata* (Born) | Rocky oyster | Aramda, Dwarka, Adatra,Hanuman Dandi. |
| FAMILY : PTERIDAE | | |
| *Pinctada fucata* (Gould) | Pearl oyster | Pirotan, Movada, Ajad, Chankh. |
| FAMILY : ANOMIIDAE | | |
| *Placenta placenta* (Linne) | Window panel oyster | Balapur Bay, Pindhara Bay |
| FAMILY : PINNIDAE | | |
| *Pinna vexillum* Born | Pen shell or Feather shell | Poshitra, Sikka |
| *P. bicolor* Gmelin | Pen shell or Feather shell. Atroparpurea sowerby | Pen shell or Feather shell |
| *P. atrooarpurea* Sowerby | Pen shell or Feather shell | Pirotan Island |
| *P. nigra* (Dil) | Pen shell or Feather shell | Pirotan Island |
| FAMILY : VENERIDAE | | |
| *Paphia textile* (Gmelin) | Clam | Baida, Poshitra |
| *P.malabarica* (Chemnitz) | Clam | Pirotan Island |
| *Meretrix meretrix* (Linne) | Clam | Pirotan Island |
| *M. casta* (Chemnitz) | Clam | Veraval |
| *Gafrarium tumidum* Roding | Clam | Pirotan Island. |
| FAMILY : MESODESMATIDAE | | |
| *Mesodesma glabratum* (Lamiarck) | Small clam | Pirotan |
| FAMILY : DONACIDAE | | |
| *Donax cuneatus* Linne. | Wedge shell | Hanuman dandi |
| FAMILY : SOLENIDAE | | |
| *Solen lamarckii* Desh | Rozer shell | Balapur Bay. |
| *S. truncatus* Wood | Rozer shell | Balapur Bay. |

*Table Contd........*

*Table 33.14 – Contd...*

| Scientific Name | Common Name | Collection area |
|---|---|---|
| FAMILY : ARCIDAE | | |
| *Area granosa* Lamarck | Ark shell | Pirotan Island |
| FAMILY : MYTILIDAE | | |
| *Mytilus viridis* Linnce | Green mussel | Pirotan Island. |
| *Modiolus metcalfei* (Henley) | Weaving mussel | Pirotan Island |
| CLASS : CEPHALOPDAE | | |
| FAMILY : LOLIGINIDAE | | |
| *Loligo indica* Pfeffer | Squid | Off Dwarka fishing ground |
| FAMILY : SEPIIDAE | | |
| *Spiella intermis* (F'erussco & d'Orbigny) | Cuttle fish | Off Dwarka fishing grounds. |
| *Sepis acnleats* (F'erussco & d'Orbigny) | Cuttle fish | Balapur Bay. |
| FAMILY OCTOPODIDAE | | |
| *Octopus honkogensis* Hoyle | Octopus | Okha, Poshitra, Hanuman dandi, Baida. |

*Source*: Sarvaiya and Chhaya: 1983; Saurashtra Molluscus.

Most of the edible molluscs are distributed in three classes namely gastropoda, pelecypoda and cephalopoda. Amongst marine products, molluscs stand as a next edible group to fish and crustaceans. Some of the molluscs which are edible and have a good food value, yet they are not utilized as food in Gujarat as well as in India.

**Table 33.15 : Nutritive value of some of the molluscs found on the coast of Saurashtra–Gulf of Kutch**

| S. No. | Species | Meat % | Shell % | Moisture % | Protein % on dry wt. base | Ash on dry wt. base | Salt % on dry wt. base | Fat % |
|---|---|---|---|---|---|---|---|---|
| 1. | *Sepilla intermis* | 65.7 | – | 83.5 | 77.5 | 5.18 | 2.33 | 4 |
| 2. | *Octopus Sp.* | 84.31 | – | 83.5 | 76 | 7.28 | 5.32 | 4.75 |
| 3. | *Loligo Sp.* | 79.4 | – | 86 | 77 | 5.31 | 2.03 | 5.80 |
| 4. | *Solen truncatus* | 41.3 | 26.6 | 81.49 | 84 | 10.80 | 4.6 | – |
| 5. | *Placenta placenta* | 13.2 | 79.5 | 82.97 | 56.20 | 12.86 | 6.26 | 9.80 |
| 6. | *Pinna vexillum* | 16.1 | 75.00 | 77.74 | 70.00 | 25.0 | 3.17 | – |
| 7. | *P.bicolor* | 18.4 | 74.9 | 81.56 | 55.13 | 18.00 | 3.87 | – |
| 8. | *P.atropura* | 16.0 | 79.5 | 78.72 | 48.13 | 26.0 | 3.58 | – |
| 9. | *Pitra erycina* | 15.6 | 59.3 | 83.11 | 54.0 | 13.4 | 6.84 | 8.21 |
| 10. | *Crassostrea gryphoides* | 7.7 | 78.6 | 76.57 | 50.32 | 9.15 | 3.51 | 13.25 |
| 11. | *Crassostrea cuculata* | 7.6 | 79.9 | 81.96 | 49.8 | 10.15 | 2.75 | 14.50 |
| 12. | *Murex virgineus* | 7.8 | 82.5 | 73.13 | 72.21 | 7.57 | 2.91 | 1.0 |
| 13. | *Thias rugosa* | 8.6 | 80.1 | 72.72 | 68.64 | 9.97 | 2.66 | 3.80 |

*Source*: Sarvaiya and Chhaya, 1983; Saurashtra Molluscs. Published by P.R.Sarvaiya.

## *Phyllum echiuroidea*

The group is represented in the MNP mainly by species of *Bonellia*. The *Bonallia* has become well known among zoologists because of the means, by which its future sex is determined. In most animals the decision as to an individual will become male and female is made at thee time of fertilisation. In *Bonallia,* this decision remains completely open even at the time that the creature is a free swimming larvae. If this larvae comes into contact with a female, due to certain chemicals secreated by female, it becomes male and lives attached to the female. If it does not come in the contact of female, it burrows in the rocks and develops into a female. Thus, sex determination in *Bonallia* is by chemicals which is unique feature in the species.

**Table 33.16 : List of Rare, Endangered, Threatened and in Vulnerable Species in Gulf of Kutch (Present Status)**

| No. | Species | Common name | Status |
|---|---|---|---|
| (A) | Phyllum Vertebrata | | |
| (1) | Sea turtle | | |
| | *Dermochelys Coriacea* | Leathen back turtle | Endangered |
| | *Chelonia mydas* | Green turtle | Endangered |
| | *Lepidochelys olivacea* | Olive Ridley | Endangered |
| (2) | Fish | | |
| | *Hippocampus kuda* | Sea Horses | Endangered |
| | *Ambhiriom polymnous* | Fish | Endangered |
| | *Narcinetilen* | Electric rayfish | Vulnerable |
| | *Fistularea petimba* | Flying fish | Vulnerable |
| | *Exocoetus volitans* | Flying fish | Vulnerable |
| | *Parapegasus natans* | Flying fish | Vulnerable |
| | *Tetradom spp* | Puffer fish | Vulnerable |
| | *Drepane punctata* | Chamba fish | Vulnerable |
| | *Pterois volentans* | Scorpion fish | Vulnerable |
| | *Dyodon hystrix* | Porcupine fish | Vulnerable |
| | *Linophra vagaduna* | Fish | Vulnerable |
| | *Mene maculata* | Fish | Vulnerable |
| (B) | *Dolphin delphis* | Dolphin (mallar) | Threatened |
| | *Dugong dugong* | Dugong (lulli) | Rare and endangered |
| | Whales | Whales | Threatened |
| (C) | Phylum Coeleterata | | |
| | *Stoichacts gigentrum* | Giant Sea anemone | Vulnerable |
| | *Cerianthus* (with phoronis) | | Vulnerable |
| | All species of Corals | | Threatened |
| | *Acropora* spp | | Endangered and probably exinct |
| | *Alcyonium polydactylym,* Dana | Soft corals | Threatened |
| | *Paracyanthus indicus* | Soft corals | Endangered |
| | *Virgularia* spp | | Threatened |
| | *Astromuricea stellifera* | Red sea fan | Endangered |
| | *Dendrontha* spp | Sec cabbage | Vulnerable |
| (D) | *Phylum echiuroidea* | | |
| | *Rubricetatus pirotansis* | Bonnelia | Rare and Endangered |
| | *Ikedella mistakiensis* | Bonnelia | Rare and Endangered |

*Contd........*

**Table 33.16 – Contd...**

| No. | Species | Common name | Status |
|---|---|---|---|
| (E) | *Phylum echinodermata* | | |
| | *Salmacis bioolor* | Sea urchin | Threatened |
| | *Planeria* spp | Flat worm | Vulnerable |
| | *Polynoe* spp | Sea worm | Vulnerable |
| (F) | Nudibranchs | | |
| | Majority of available types | | Vulnerable |
| (G) | Phylum mollusca | | |
| | *Strombus gigas* | Conch | Endangered |
| | *Xuncus phyrum* | Sacred Conch | Threatened |
| | *Lambia* spp | Spider Conch | Threatened |
| | *Murex* spp | Murex oyster | Vulnerable |
| | *Pinctada fucata* | Pearl oyster | Threatened |
| | *Solen trunchetus* | Razer shell | Vulnerable |
| | *Pinna bicolar* | Ear shell | Vulnerable |
| | *Telescopium* spp | Telescopium shell | Vulnerable |
| (H) | Phylum crustacea | | |
| | Lobsters | | Vulnerable |
| | *Atergatis inermis* | Rani crab | Vulnerable |
| | *Saron marmoratus* | Prawn | Vulnerable |
| | *Trachypenaeus pescadoreensis* | Prawn | Vulnerable |
| | *Solenocera crasicornis* | Prawn | Vulnerable |
| | *Metapenaeopsis Stridulans* | Prawn | Vulnerable |
| | *Palaemon semmelinkii* | Prawn | Vulnerable |

*Source*: Singh, H.S. Marine National Park, and Sanctuary-Jamnagar.

**Table 33.17 : List of Recorded Live Species of Flora at Narara Reef**

Phyllum: Porifera

1. Cillona spp.

**Phyllum : Coelenterata**

1. *Zooanthus* spp.
2. *Goniostrea pectinata* (Ehrenbery)
3. *Montipora explanta* (Brueggenmann)
4. *Favities complanta* (Ehrebery)
5. *Favia favus* (Forskal)
6. *Turbinaria peltata* (Esper)
7. *Acanthastrea* spp.
8. *Goniopora planuluta* (Ehrenbery)
9. *Porites lutea* (Med-H)
10. *Lobophyllia* spp.
11. *Stoichactis gignanthum*

**Phyllum : Echinodermata**

1. *Ophiothrix* spp.
2. *Holothuria* spp.

**Phyllum : Arthropoda**

1. *Balunus* spp.
2. *Scylla* spp.
3. *Lophoxanthus* spp.
4. *Atergatis integerrinous*
5. *Squilklla* spp.

**Phyllum : Mollusca**

1. *Murex adustus* (Lamarek)
2. *Murex trapa* (Roding)
3. *Pinktada fucata* (Gould)
4. *Thais* spp.
5. *Umbonium verstiarium* (Linn)
6. *Pinna bicotor* (Gmelin)
7. *Trochus radiatus* (Gmelin)
8. *Trochus stellatus* (Gmelin)
9. *Turbo intercosalis* (Menke)
10. *Telescopium telescopium* (Linne)

*Contd.........*

**Table 33.17 – Contd...**

11. *Nerita albicilla* (Linne)
12. *Pyrene* spp.
13. *Hemifufusus conchlibium* (Linne)
14. *Cypraea onyx*
15. *Cypraea caurica*
16. *Cerethium morus* (Lamarck)
17. *Cerethium citrinum* (Sowerby)
18. *Aplvsia bendicti* (Eliot)
19. *Dris* spp.
20. *Cantharus undosus*
21. *Onchiadium* spp.
22. *Octopus* spp.
23. *Turritella coperii*

**Phyllum : Chlorophyta**

1. *Caularapa reemosa*
2. *Caularapa corynephora* (Montagene)
3. *Helimeda tuna* (Ell and Sol) Lamour
4. *Ulva reficulata*
5. *Copanenia sessile*
6. *Hydroclatharus* spp.
7. *Bryopsis plumosa*
8. *Hydroclathrus clatharus*
9. *Ulva lactuca*
10. *Codium* spp.
11. *Enteromorpha* spp.
12. *Codium iyengarii* (Boergsrn)

**Phyllum : Phaeophyta**

1. *Sargassum swartzii* (Turn)
2. *Sargassum wightii* grev.
3. *Padina* spp.
4. *Cystosira* spp.

**Phyllum : Rhodophyta**

1. *Halymenia porphyroides*
2. *Solieria robusta*
3. *Scinaia furcellata*
4. *Halymenia Ceylanica*
5. *Gellidium* spp.
6. *Gracilaria* spp. *Acanthophora* spp.

**Table 33.18 : Comparison of catch per boat of 1980-81 and 1999-2000 for Gujarat State**

| Name of fish | Fish landing 1980 81 in MT. | Catch per boat | Fish landing 1999-2000 | Catch per boat |
|---|---|---|---|---|
| No of Mechanized boat | 3459 | | 11372 | |
| White Pomfret | 9539 | 2757 | 11133 | 978 |
| Black pomfret | 3248 | 939 | 3489 | 307 |
| Bombay Duck | 36270 | 10486 | 88166 | 7751 |
| Hilsa | 4740 | 1370 | 1710 | 150 |
| Shark | 6306 | 1823 | 16588 | 1458 |
| CatFish | 6625 | 1915 | 24526 | 2156 |
| Ribbon fish | 4104 | 1187 | 59616 | 5241 |
| Shrimp (Small) | 8192 | 2368 | 48032 | 4223 |
| Shrimp (Medium) | 6745 | 1950 | 7284 | 640 |
| Shrimp (Jumbo) | 1805 | 522 | 1126 | 99 |
| Lobster | 238 | 69 | 511 | 45 |
| Squid/Cuttle fish | 5182 | 1498 | 23423 | 2059 |
| Scianeids | 59999 | 17346 | 224417 | 2959 |
| Eel | 5305 | 1537 | 3347 | 294 |
| Crabs | 921 | 266 | 1657 | 146 |
| Carangids | 11086 | 3140 | 2206 | 194 |

**Table 33.19 : Estimated annual catchable potentials in 0 to 50 m depth**

| Name of Fish | Potentials (In Tonnes) |
|---|---|
| Elasmobranch | 12254 |
| Eels | 2794 |
| Cat fishes | 1349 |
| Oil Sardine | 166 |
| Other Sardine | 261 |
| Anchovies | 1 |
| Other Clupeids | 35420 |
| Bombay duck | 57108 |
| Lizard fishes | 698 |
| Perches | 8528 |
| Scienieds | 33820 |
| Ribbon fishes | 23859 |
| Carangids | 4241 |
| Silver bellies | 660 |
| Pomfrets | 11655 |
| Mackerals | 70 |
| Seer fishes | 6299 |
| Tunnies | 4189 |
| Flat fishes | 2742 |
| Penaeid Prawns | 16436 |
| Non-Penaeid prawns | 14385 |
| Cephalopods | 6339 |
| Others | 40655 |
| Total | 293929 |

*Source* : P.S.B.R. James, Present status, Problems and strategies for Development of Coastal and off shore Fisheries pp 25 in National Workshop of Development of Marine Fisheries for higher productivity and export 9-10 June 1992 at Cochin.

**Table 33.20 : Potential Marine Fisheries Resources of EEZ**

| Species | Fishery resources Potential (in '000 T) |
|---|---|
| Elasmobranch | 168 |
| Eels | 7 |
| Aanchovies | 53 |
| Lesser Sardins | 306 |
| Bombay Duck | 104 |
| Lizards fish | 48 |
| Perches | 239 |
| Cat fishes | 123 |
| Sciaenids | 142 |
| Pomfrets | 54 |
| Ribbon fishes | 311 |
| Carangeds | 447 |
| Mackerai | 224 |
| Sel fish | 42 |
| Coastal Tuna | 279 |
| Flat fish | 38 |
| Penaeid Prawns | 54 |
| Cephalopods | 71 |
| Others | 443 |

*Source*: Report of the working group on Revalidation of the Potential Marine Fisheries Resources of the Exclusive Economic Zone of India, 1996.

## *Phyllum annelids*

This group is mainly represented in the MNP and Sanctuary by following species.

1. *Sabella* spp. (Fan worm)
2. *Neris versicolor*
3. *Tubicolor* (Polychaetes)
4. *Eurythoe complanta.*

## *Phyllum crustacea*

This animal group is mainly represented by Barnacles, Lobsters and crabs. Among these, Barnacles and crabs are most abundant. About 30 different species of crabs are found in area. About 27 species of prawns are recorded in this area.

### *Phyllum molluscs*

This group of invertebrates is very well represented in the Marine National Park and Sanctuary by Snails, Mussels, Oysters, Shells, Octopus, Sea-horse, Chiton, Sepia etc. More than 200 species of this group are recorded in shallow and deeper water and the intertidal zone. Important species of the group are as follows:

1. *Pattalla spp.*
2. *Trochus spp.*
3. *Turbo intercostalis*
4. *Nerita spp*
5. *Natio spp*
6. *Turitalla spp.*
7. *Telescopium*
8. *Carithidea spp*
9. *Murex spp*
10. *Octopuss hogkingensis.*

**Table : 33.21 : Scope of exploitation in depth beyond 50 m.**

| Species | Estimated potentials of India (In '000 T.) |
|---|---|
| Tinas | 450 |
| Carangids | 300.5 |
| Ribbon fishes | 216 |
| Thread fin breams | 110 |
| Elasmobranchs | 71 |
| Cat fishes | 63 |
| Bull's eye | 55 |
| Crockers | 22 |
| Lizard fishes | 21 |
| Cephalopods | 21 |
| Perches | 15 |

*Source* : CMFRI Annual Report 1996-97. p 144

### *Phyllum echinodermata*

This spiny-skinned marine animals are mostly bottom dwelling. Examples of the animals found in MNP and sanctuary area are as follow:

1. Lamprometra palamata.
2. Astropecten polyacanthus.
3. Pateria spp.
4. Sealilly.
5. Feather star.
6. Star fishes.
7. Brittle star
8. Sea urchin.
9. Sand dollar etc.

### *References*

Anon, 1979. A compendium of Indian Fisheries and sea food export industry. M.P.E.D.A. on third Indian Seafood Trade Fair 9-11 Feb : 8

Anon, 1963. Techno Economic survey of Gujarat by National Council Applied Economic Research, New Delhi.

Anon, 1996 (1). C.M.F.R.I. special publication No 38 : 47

Anon 1996 (2). Annual Report of CMFRI, p. 1482

Anon 1997. Annual Report of CMFRI, p. 144

Anon 1998. Report of the working group on revalidation of the potential Marine Fisheries Resources of the Exclusive Economic Zone of India. Government of India.

Chhaya, N.D. 1999. Observations on Environmental Impact Assessment of Coastal Marine Ecosystem in Kutch Dist. Forum for Planned Industrialization of Kutch.

Deveraj, M.S. Srirama Chandra Murty, R. Santhiadhas and K.K. Joshi. 1998. The New Economic Policy and Prospective for Marine Fisheries Research and Development in India. *Fishing Chimes,* 18(8) : 18-29

Dholakia, A.D. 1998. Station wise Details of Identified water molluscans of Gujarat. Annual Report, 1998-99, Fisheries Research Station, G.A.U. Sikka.

James, P.S.B.R. 1992. Estimated Annual Catchable Potentials in 0 to 50 m depth. National Workshop of Development of Marine Fisheries of for higher Productivity, p. 25

Murty, V.S. 1997. Vision – 2020 – CMFRI perspective plan. CMFRI Cochin, p. 70.

Solanky H.U. Mini Raman, Bena Kumari, R.M. Diwivedi and A. Narain. 1998. Seasonal trends in Fishery Resources off Gujarat: Salient observation using satellite NOAA AYHRP. *Indian. J. Mar. Sci.* 27 : 438-442.

Unnikrishnan, N. and Jecob Thomas. 1997. Areawise Fishing Effort of Fisheries Survey of India, Govt. of Inda. *Resour. Infor. Ser.* Vol 3(4).

# 34
# Inland Fishery Resources

## Inland Resources of India and Gujarat

Inalnd Fishery Resources in India are very rich. About 30 per cent of the total fish the inland capture fisheries alone contribute production. Inland fishery can be conveniently divided in to two categories; *Fresh water* and *Brackish waters.* The fresh water fishery comprises the great river system, fresh water lakes, irrigation canals, tanks, ponds and reservoir, and the brackish water includes the estuaries (river mouths), lagoons, backwaters and brackish water lakes. The managements of capture fisheries are provided by the rivers (riverine estuaries and the lakes but the culture fisheries are provided by tanks, ponds and jheels etc).

### *Fresh Water Fisheries*

Besides extensive river system there is a variety of inalnd water ranging from icy-cold water tanks to tropical fresh water ponds. There are various species of fishes suitable for cultivation in ponds and confined waters. The Major fresh water inland water bodies are the *Ganges system,* the *Brahmaputra system* and the *Indus system* in the northern part of India. The *Ganga system* is one for the largest river system in the world. The river Ganges takes its origin from Gangotri in the Himalayas and finally joins the Bay of Bengal. The Ganga sytem has a total length of about 8000 km. The river Ganges comprises the range Himalayas in the beginning and then covers the states of Haryana, Uttar Pradesh, Bihar, Bengal and also a part of Rajasthan and Madhya Pradesh. The river Yamuna is a major component of the Ganga system and it borders the states of Pubjab and Haryana and also flows through the territory of Delhi. Others principle tributaries of the Ganga system are the river Ramganga. Gomati, Ghaghara, Kosi, Gandak, Tons and Sone. The important tributaries of the river Yamuna are the rivers Chambal, Betwa and Ken.

The main constituents of the East Coast system are the Mahanadi, The Krishna, The Godavari and the Cauvery, having a length of nearly 6430 km. The river Mahanadi is the largest river of Orissa. Others less important rivers of Orissa are the Brahmani, Daya, Bhargavi and Subernrekha. Tributaries of the river Cauvery from the Nilgiri have the main tributaries of the river Godavari. Some of the minor tributaries are the Purna, Maner and Sabari. Paingunga and Wardha are the secondary tributaries of Wainguanga. The main tributaries of Krishna are the Bhima and the Tungbhadra.

The West coast river system has a length of 3380 km and includes the basins of the Narmada and Tapti rivers mainly. Other short rivers are just like the terrestrial streams.

The river Narmada has as many as 18 tributaries. The river Tapi flows through the states of Madhya Pradesh, Maharashtra and Gujarat before joining the Arabian sea. The river Narmada originates from the Amarkantak hill in M.P. covers the States of Madhya Pradesh and Gujarat, finally joping the Gulf of Khambhat. (Yadav, B.N., 2002)

The riverine fishery of the plain supports the fisheries of the following fishes:

(1) Major carps: *Catla catla, Labeo rohita, L.bata, L. fimbratus, L. calbasu, Cirrhinus mrigala, Schizothorax spp.* and also *Tor* and *Puntius* spp etc.

(2) Catfishes: *Mystus seenghala, Mystus aor, Wallago attu, Bagarius bagarius, Pangasius pangasius, Silonia silondia, Ompok bimaculatus, Rita rita, Heteropneustes fossilis, Clarias batrachus, Snabas testudineus, Channa gachua, C. punctatus, Channa strietus, Eutropiichthys vacha* etc.

(3) Cludeids: *Hilsa ilisha, Notopterus, notoperus, N. chitala, Setipinnapha, Gadusi chapra* etc.

Commissioner of Fisheries have recorded that total length of river in India is 1,73,289 km, while in Gujarat it is 3,865 km, which is 2.23% of total rive length of India. Similarly area of reservoirs of India is 20.90 lakh hectare, while in Gujarat it is only 2.43 lakh hectare, which is about 11.62%. Area of pond and tanks of India covers 22.54 lakh hectare, while of Gujarat it is only 0.71 lakh hectare, which is only 3.15% of total ponds and tanks of India.

## Inland Resources of India and Gujarat

Commissioner of Fisheries have recorded that total length of river in India is 173289 km while in Gujarat it is 3865 km, which is 2.23% of total river length of India. Similarly area of Reservoirs of India is 20.90 lakh hectare, while in Gujarat it is only 2.43 lakh hectare which is about 11.62%. Area of ponds and tanks of India covers 22.54 lakh hectare while in Gujarat it is only 0.71 lakh hectare which is only 3.15% of total ponds and tanks of India. State wise inland water resources of India is shown in Table 34.1

Details of district wise reservoirs of Gujarat with capacity is given in Table 34.2. Details of small, medium and large reservoirs of Gujarat is shown in Table 34.3. District wise Fresh water Dams of Saurashtra with their capacity is given in Table 34.4. Dholakia (1986) and Chhaya (1990) have recorded about 97 fresh water fishes with their scientific, English and local (Gujarati) names. Consolidate information is given in Table 34.5. Key to the identification of some fresh water culturable fishes was given by Kaur and Dhavan (1997) (Table 34.6). A list of about 68 fresh water molluscans with their scientific name is given in Table 34.7.

Commissionerate of Fis ries, Govt. of Gujarat, recorded inland fish production from all fresh water resources. The production is tabulated from 1971-72 to 1998-99 (Table 34.8) From the table it can be seen that inland production was almost steady during 1981-82 to 1988-89. However, thereafter production was increased rapidly.

Commissionerate of Fisheries, Govt. of Gujarat has consolidated sector wise inalnd fish production from 1996-97 to 2000-01 Gujarat, is reprinted as Table 34.9. Species wise inland fish production of Gujarat state is recorded by Commissionerate of Fisheries, Govt. of Gujarat from 1996-97 to 2000-01 as Table 34.10.

State wise inalnd fish production of India from 1995-96 to 1999-2000 is shown in Table 34.11. Fish production trend of India and Gujarat from 1950-51 to 2000-01 is reported as Table 34.12.

**Table 34.1 : Inland Water Resources in India**

| Sr. No. | State/ UTS | Length of Rivers & Canals (km) | Area of Reservoirs (Lakh ha.) | Area under Tanks & Ponds (Lakh ha.) | Beels. Oxblow & Direlict Waters (Lakh ha.) | Brackish water (Lakh Hact) |
|---|---|---|---|---|---|---|
| 1. | Andhra Pradesh | 11514 | 2.34 | 5.17 | 0.00 | 0.64 |
| 2. | Assam | 4820 | 0.02 | 0.23 | 1.10 | 0.00 |
| 3. | Bihar | 3200 | 0.60 | 0.95 | 0.05 | 0.00 |
| 4. | Goa | 250 | 0.03 | 0.03 | 0.00 | 0.00 |
| 5. | Gujarat | 3865 | 2.43 | 0.71 | 0.12 | 3.76 |
| 6. | Haryana | 5000 | 0.00 | 0.10 | 0.10 | 0.00 |
| 7. | Himachal Pradesh | 3000 | 0.42 | 0.01 | 0.00 | 0.00 |
| 8. | Jammu & Kashmir | 27781 | 0.07 | 0.17 | 0.06 | 0.00 |
| 9. | Karnataka | 9000 | 2.20 | 4.14 | 0.00 | 0.08 |
| 10. | Kerala | 3092 | 0.30 | 0.30 | 2.43 | 2.43 |
| 11. | Madhya Pradesh | 20661 | 2.94 | 1.19 | 0.00 | 0.00 |
| 12. | Maharashtra | 1600 | 2.79 | 0.50 | 0.00 | 0.10 |
| 13. | Manipur | 3360 | 0.01 | 0.05 | 0.40 | 0.00 |
| 14. | Meghalaya | 5600 | 0.08 | 0.02 | 0.00 | 0.00 |
| 15. | Nagaland | 1600 | 0.17 | 0.50 | 0.00 | 0.00 |
| 16. | Orissa | 4500 | 2.56 | 1.14 | 1.80 | 4.17 |
| 17. | Punjab | 15270 | 0.00 | 0.07 | 0.00 | 0.00 |
| 18. | Rajasthan | 0 | 1.20 | 1.80 | 0.00 | 0.00 |
| 19. | Sikkim | 900 | 0.00 | 0.00 | 0.05 | 0.00 |
| 20. | Tamil Nadu | 7420 | 0.52 | 6.91 | 0.00 | 0.56 |
| 21. | Tripura | 1200 | 0.05 | 0.12 | 0.00 | 0.00 |
| 22. | Uttar Pradesh | 31200 | 1.50 | 1.62 | 1.33 | 0.00 |
| 23. | West Bengal (P) | 2526 | 0.17 | 2.76 | 0.42 | 2.10 |
| 24. | Arunachal Pradesh | 2000 | 0.00 | 0.00 | 0.00 | 0.00 |
| 25. | Mizoram | 1395 | 0.00 | 0.02 | 0.00 | 0.00 |
| 26. | Andaman & Nicobar | 115 | 0.01 | 0.03 | 0.00 | 0.00 |
| 27. | Chandigarh | 2 | 0.00 | 0.00 | 0.00 | 0.00 |
| 28. | Delhi | 150 | 0.04 | 0.00 | 0.00 | 0.00 |
| 29. | Lakshadweep | 0 | 0.00 | 0.00 | 0.00 | 0.00 |
| 30. | Pondichery | 247 | 0.00 | 0.00 | 0.01 | 0.01 |
| 31. | Dadra & Nagar Haveli | 54 | 0.05 | 0.00 | 0.00 | 0.00 |
| 32. | Daman & Diu | 12 | 0.00 | 0.00 | 0.00 | 0.00 |
| | **Total** | **171334** | **20.50** | **28.54** | **7.87** | **14.22** |

*Sources*: Fisheries Statistics of Gujarat 2000-01 (Reprinted with permission)

## Fisheries Resources of Surat Region of Gujarat

Except Dangs and some of the pockets in Vansda, Dharampur and Umargaon Talukas, most of the area in the region has ample scope to exploit untapped fishery wealth from rivers, canals, steams, food plain wet lands, reservoirs, village ponds, brackish water (khar) lands, estuaries etc.

Table 34.2 : District Wise Reservoirs of Gujarat

| S No. | District | Less than 20 ha. | | 20 ha to 500 ha | | 500 ha to 1000 ha | | More than 1000 ha | | Total. | |
|---|---|---|---|---|---|---|---|---|---|---|---|
| | | No of Reservoir | Area at FRL Ha. | No. of Reservoir | Area at FRL Ha. | Nof of Reservoir | Area at FRL Ha. | No. of Reservoir | Area at FRL Ha. | No. of Reservoir | Area at FRL Ha. |
| 1. | Ahmedabad | 0 | 0 | 1 | 296 | 9 | 9 | | 9 | 1 | 296 |
| 2. | Amreli | 0 | 0 | 2 | 651 | 2 | 1210 | 1 | 3285 | 5 | 5146 |
| 3. | Banas Kanths | 0 | 0 | 0 | 0 | 1 | 578 | 2 | 6615 | 3 | 7193 |
| 4. | Bharuch | 0 | 0 | 4 | 634 | 0 | 0 | 1 | 3677 | 5 | 4311 |
| 5. | Dang | 2 | 17 | 0 | 0 | 0 | 0 | 0 | 0 | 20 | 17 |
| 6. | Jamnagar | 0 | 0 | 10 | 2989 | 0 | 0 | 3 | 5018 | 13 | 8007 |
| 7. | Bhavnagar | 1 | 0 | 9 | 2622 | 4 | 3188 | 3 | 8941 | 17 | 14751 |
| 8. | Junagadh | 4 | 0 | 13 | 3907 | 3 | 1895 | 1 | 1581 | 21 | 7383 |
| 9. | Kheda | 0 | 0 | 0 | 0 | 0 | 0 | 1 | 2086 | 1 | 2086 |
| 10. | Kutch | 0 | 0 | 14 | 3678 | 4 | 2936 | 1 | 1145 | 19 | 7759 |
| 11. | Mehsana | 0 | 0 | 1 | 277 | 0 | 0 | 1 | 10700 | 2 | 10977 |
| 12. | Punch Mahal | 0 | 0 | 4 | 950 | 5 | 3965 | 3 | 27249 | 12 | 32164 |
| 13. | Rajkot | 1 | 0 | 11 | 2721 | 4 | 2445 | 8 | 13798 | 24 | 18964 |
| 14. | Sabar Kantha | 1 | 0 | 3 | 936 | 0 | 0 | 5 | 11850 | 9 | 12786 |
| 15. | Surat | 1 | 16 | 2 | 272 | 1 | 581 | 2 | 96200 | 6 | 97069 |
| 16. | Surendra Nagar | 0 | 0 | 0 | 0 | 3 | 1963 | 1 | 1914 | 4 | 3877 |
| 17. | Vadodra | 0 | 0 | 1 | 130 | 1 | 567 | 1 | 2904 | 3 | 3601 |
| 18. | Valsad | 0 | 0 | 2 | 674 | 0 | 0 | 1 | 5144 | 3 | 5818 |
| Total | | 10 | 33 | 77 | 20737 | 28 | 19328 | 35 | 202107 | 150 | 242205 |

*Source*: Fisheries Statistic of Gujarat, 2000-01

Table 34.3 : Details of Large and Medium Reservoiurs of Gujarat (As on 1998-99)

| Name of Reservoir | District | Area at FRL (HA) | Name of Reservoir | District | Area at FRL (HA) |
|---|---|---|---|---|---|
| LARGE RESERVOIRS | | | | | |
| 1. Kadana | Panch Mahal | 16600 | 2. Panam | Panch Mahal | 8980 |
| 3. Kakarapar | Surat | 44200 | 4. Ukai | Surat | 52000 |
| 5. Setrunji | Bhavnagar | 6734 | 6. Damanganga | Valsad | 5144 |
| 7. Dharoi | Mehana | 10700 | | | |
| MEDIUM RESERVOIRS | | | | | |
| 1. Dantiwada | Banaskantha | 4047 | 2. Sipu | Banaskantha | 2568 |
| 3. Karjan | Banaskantha | 3677 | 4. Vanaakbori | Kheda | 2086 |
| 5. Deo | Panch Mahal | 1668 | 6. Hathamati | Sabarkantha | 3236 |
| 7. Meshwo | Sabarkantha | 1116 | 8. Watrak | Sabarkantha | 4475 |
| 9. Guhati | Sabarkantha | 1713 | 10. Mazum | Sabarkantha | 1310 |
| 11. Sukhi | Baroda | 2904 | 12. Sankroli | Aamreli | 3285 |
| 13. Ranghola | Bhavnagar | 1162 | 14. Sukhbhadar | Bhavnagar | 1045 |
| 15. Sasoyi | Jamnagar | 1468 | 16. Uni | Jamnagar | 1730 |
| 17. Sani | Jamnagar | 1820 | 18. Aamipur | Junagadh | 1581 |
| 19. Bhadar | Rajkot | 4400 | 20. Moj | Rajkot | 1203 |
| 21. Phophal | Rajkot | 1207 | 22. Machhu-1 | Rajkot | 1200 |
| 23. Machhu-2 | Rajkot | 1396 | 24. Aji II | Rajkot | 1439 |
| 25. Demi II | Rajkot | 1023 | 26. Nyari | Rajkot | 1930 |
| 27. Brahmi | Surendranagar | 1914 | 28. Tapar | Kutch | 1145 |

*Source:* Reprinted with permission of Commissionerate of Fisheries, Govt. of Gujarat.

During the last two decades the amrine fisheries sector has progressed from artisanal traditional to major economic activity in Surat region. The details are as under.

| | Detail | 1977 | 1997 |
|---|---|---|---|
| (1) | Mechanised Vessels | 516 | 3257 |
| (2) | Non mechanised Vessels | 1002 | 548 |
| (3) | Marine Fish production | 9553 MT | 84567 MT |
| (4) | Catch per Unit/annum | 5.125 T | 15.09 T |

## *Riverine Fisheries*

The major river systems of Surat region are (1) Tapi (2) Kim (3) Mindhole (4) Purna (5) Ambica (6) Kaveri (7) Kharera (8) Auranga (9) Par (10) Kolan (11) Damanganga. The length of river system in Surat region is estimated to be 1100 km.

The species available in fresh and estuarine water of the Surat region are:

| | | |
|---|---|---|
| (1) | Indian Major Carps | *Catla catla, Labeo rophila C. mrigala, L. calbasu, L. fimbriatus* |
| (2) | Mahsheer | *Tor tor* |
| (3) | Major Carps | *L. bata, L. gonius, L. reba* |
| (4) | Cat fishes | *Clarius batrachus, Wallago attu, Mystus singala, M. aor, M. cavasois. Ompock bimaculetus* |
| (5) | Feather backs | *Notopterus notopterus, N. chitala.* |
| (6) | Murrels | *Channa merulius. C. striatus. C. puctatus. C. gachua* |
| (7) | Prawns | *Machrobranchium rosenbergii, M. idea, M. rude, Palaemon tenupes, Penaeus monodon, P. indicus. P. merguiensis, P. semisulcatus Metapenaeus monoceros, M. affinis, M. bravicornis, M. dobsoni, Parapenaeopsis spp.* |

The catch trends of Uklai reservoir in respect of Catla. Rohu, Mrigal and Mahsheer indicate decline, whereas the production of catfishes has increased. Hilsa, which is an acclimatized marine species in the reservoir, is endangered and Mahsheer is noticed to be on the verge of extinction. The decline of such precious species is alarming.

**Table 34.4 : Fresh Waters Fishes**

| Sr. No. | Scientific Name | English Name | Local Name |
|---|---|---|---|
| 1. | *Mastocembelus pancalus* | Spiny eel; mastacembelid eel | Bam, Chepli |
| 2. | *Mastocembelus armatus* | Spiny eel; eel; throny backed ell; mastacenbeild ell | Bam, Vam |
| 3. | *Macrognathus aculeatus* | Lesser spiny eel; mastacembelid | Vam |
| 4. | *Wallago attu* | Fresh water shark; sheat Fish; eurasian cat fish; wallago | Padiyo Padon, Malli Valago, Pahadi |
| 5. | *Ompok macrophthalmus* | Fresh water cat fish | Gungavri |
| 6. | *Ompok bimaculatus* | Eurasian cat-fish; butter cat-fish; sheat fish | Gunbavri, Pabda |
| 7. | *Ompok pabda* | Fresh water fish | Padba, Tamolana |
| 8. | *Rita rita* | Fresh water cat fish | – |
| 9. | *Rita pavimentata* | Fresh water cat fish | – |
| 10. | *Plotossus anguillaris* | Plotoside cat fish; striped cat fish; eel cat fish; cat fish eel | – |
| 11. | *Heteropneustes fossills* | Airsac cat fish; scorpion fish; stinging cat-fish. | Shingi |
| 12. | *Clupisoma garua* | Butchwa | Gangavari |
| 13. | *Mystus seengala* | Long-nose cat fish; giant river cat fish; cat fish; bagrid cat fish | Shingada, Katiya, Tergani |
| 14. | *Mystus punctatus* | Bagrid cat fish | Khagi |
| 15. | *Mystus vittatus vittatus* | Striped dwarf cat-fish; bagrid cat fish | – |
| 16. | *Bagarius bagarius* | Cat fish | – |
| 17. | *Mystus gulio* | Long whiskers cat-fish; bagrid cat-fish | Khaga |

*Contd.........*

*Table 34.4 – Contd...*

| Sr. No. | Scientific Name | English Name | Local Name |
|---|---|---|---|
| 18. | *Mystus bleekeri* | Freh water cat-fish; bagrid cat-fish | Katiya |
| 19. | *Mystus aor* | Fresh water cat-fish; bagrid cat-fish | Aaor |
| 20. | *Mystus cavasius* | Fresh water cat-fish; bagrid cat-fish | Katiya |
| 21. | *Xonentodon cancila* | Fresh water gar fish; needle fish | Kuthavra, Sarvati, Kangda |
| 22. | *Notopterus chitala* | Feather back | Chital |
| 23. | *Lepidocephallchthys thermalis* | Lesser loach | Chipi |
| 24. | *Lopipocephalichthys guntea* | Loach | Chipi |
| 25. | *Aplocheilus panchax* | Toothed carp minnow; lesser top minnow; killi fish | Dindiya |
| 26. | *Aplocheilus lineatum* | Striped top minow; killi fish; toothed carp fish | – |
| 27. | *Notopterus notopterus* | Barber's knife; razor fish; feather back | Patra, Pholi Panare, Potola |
| 28. | *Hypophthalmichthys molitrix* | Silver Carp; white big head; chinese carp | Silver carp |
| 29. | *Oxygastyer bacaila* | Minnow | Chal |
| 30. | *Oxygastor clupeoides* | Minnow | Chal, Chili, Chaliya |
| 31. | *Oxygastor phulo* | Minnow | Chal, Chili, Chaliya |
| 32. | *Chela laubuca* | Rasbora; winged rasbora | Chal, Dindiya, Munaya |
| 33. | *Osteobrama cotio cotio* | Silver fish | Gurda, Moyala |
| 34. | *Osteobrama bakeri* | Rohtee | Rotali |
| 35. | *Danio malabaricus* | Giant danio | Ranjan |
| 36. | *Danio (Brachydanio) rerio* | Zebra fish | Ranjan |
| 37. | *Barilius barna* | Minnow | Jurva |
| 38. | *Danio devario* | Minnow | Melva |
| 39. | *Schismatorhychtus (Nukta) nukta* | Carp | – |
| 40. | *Labeo angra* | Carp | Fatar chatu, Golchi Goheri |
| 41. | *Labeo dyocheilus* | Carp | Goheri |
| 42. | *Labeo potail* | Carp | Fatar chatu, Golchi Goheri |
| 43. | *Labeo boggut* | Carp | Loi (Kutch), Gauri |
| 44. | *Labeo bata* | Minor carp; carp; bata | Raeya, Chiri, (In Kutch) |
| 45. | *Labeo ariza* | Carp | Bhanga |
| 46. | *Labeo kontius* | Cauvery carp; pig mouthed carp | Karchi |
| 47. | *Labeo calbasu* | The kanoshi; major carp; orangefein labeo; kalbasu | Kansi, Kalavat, Kalot, Dai (Kutch) Kavchi, Kalbsu |
| 48. | *Labeo fimbriatus* | -do- fringed lipped carp; carp | Belji, Belzi |
| 49. | *Labeo dussumeiri* | Common labeo; carp | Kursa, Gauri, Loi |
| 50. | *Labeo gonius* | Major carp; carp | Karsi, Kursa, Khursi |
| 51. | *Labeo porcellus* | Orenge-fin Labeo; carp | Malia |
| 52. | *Labeo rohita* | Major carp; Rohu; carp | Rohu |
| 53. | *Garra lamta* | Stone ophiocephalus | – |
| 54. | *Noemacheilus botia aureus* | Striped loach; loach | Songujari |
| 55. | *Cirrhinus mrigala* | Major carp; mrigal; Hamilton's carp | Nagri, Nagari Marahaki (Kutch) |

*Contd........*

*Table 34.4 – Contd...*

| Sr. No. | Scientific Name | English Name | Local Name |
|---|---|---|---|
| 56. | *Cirrhinus latia* | Carp | Bhagna |
| 57. | *Cirrhinus reba* | Minor carp; reba; mirror carp | Reba, Rahia, Reiya, Murkhi |
| 58. | *Puntius chola* | Green bard; bitter carp; bard | Dhebri |
| 59. | *Puntius parrah* | Barb | Dhebri |
| 60. | *Puntius anphlbia* | Scarlet banded barb; barb | Dhebri |
| 61. | *Puntius melanostigma* | Barb | Dhebri |
| 62. | *Puntius arenatus* | Barb | Dhebri |
| 63. | *Puntius pleurotaenia* | Barb side-striped barb | Dhebri |
| 64. | *Puntius puckolli* | Barb | Dhebri |
| 65. | *Puntius sarana* | Indian trout; olive barb; olive carp barb | Darai, Sarna, Puthi |
| 67. | *Cyprinus carpio* | European carp; German carp; common carp | Common carp, Carp |
| 68. | *Puntius chrysopoma* | Barb | Dhebri |
| 69. | *Tor tor* | Mahseer; Mahsler; tor; mahaseer; barb; deep-bodied mahaseer | Kudna, Mahaseer, Bhingad |
| 70. | *Tor mussullah* | Barb | – |
| 71. | *Puntius hexastichus* | Barb | Dhebri |
| 72. | *Puntius bovanicus* | Barb | Dhebri |
| 73. | *Puntius sophore* | Barb | Karva (Kutch), Dhebri |
| 74. | *Tor khudree* | Barb; deccan mahseer | Bhasera, Kudna |
| 75. | *Catla catla* | Major carp; catla | Katla, Bavas, Bavus |
| 76. | *Ctenopharyngodon idella* | Grass carp; chinese carp; white ammur | Ghash machhi, Grass carp |
| 77. | *Puntius chrysopterus* | Bard | Dhebri |
| 78. | *Puntius stigma* | Barb | Dhebri |
| 79. | *Amblypharyngodon mola* | Silver fish | Melva |
| 80. | *Punctius vittatus* | Barb | Dhebri |
| 81. | *Puntius conchonius* | Barb | Dhebri |
| 82. | *Puntius ticto* | Fire fin barb; barb | Dhebri |
| 83. | *Esomus danricus* | Flying barb | Ranjan, Rajna |
| 84. | *Rasbora vaterifloris* | Rasbora; vateria flower rasbera | Rajni |
| 85. | *Rasbora rasbora* | Rasbor | Ranjan Ranjna |
| 86. | *Rasbora daniconlus* | Common rasbra | Derai, Ranjan, Jonir (Kutch) Munnava |
| 87. | *Barilius bendelisis* | Minnow | Jurva, Murkhi |
| 88. | *Rhinomugil corsula* | Mullet; grey mullet; fresh water grey mullet | Manga, Mankan |
| 89. | *Glossogobius georis* | Goby | Lepta, Modu |
| 90. | *Channa marulins* | Snakehead; murrel; giant snake head; giant murrel, large murrel | Saap, Saval, Marel, Larje marel |
| 91. | *Channa striatus* | Murrel; striped snake-head; live fish; common murrol; snake head | Saval, Marel, Saap Kadva, Daku |

*Contd........*

*Table 34.4 – Contd...*

| Sr. No. | Scientific Name | English Name | Local Name |
|---|---|---|---|
| 92. | *Channa punctatus* | Murrel; green snake head; live fish; snake head | Kadva, Daku, Saap marel |
| 93. | *Chanda nama* | Glassy perchlet | Chandbijva, Glassfish, Kachki, Chandva, Kathaya |
| 94. | *Chanda ranga* | Glassy perchlet; Indian glass fish | Chandbijva, Kachki, Glassfish |
| 95. | *Aplocheilus blochii* | Toothed carp minnow; killi fish | Dindiya |
| 96. | *Tilapim mossambic* | Tilapia | Tilapia |
| 97. | *Etrplus suratensis* | Banded etroplus | Atroplus |

*Source* : (1) Dholakia, A.D., 1986. List of important fishes available in Gujarat with Scientific, English and Local names. pp.
(2) Chhaya, N.D. 1990. List of fishes of Gujarat in Fisheries Management and Science, published by Gujarat Granth Nirman Board, Govt. of Guj. Ahmedabad.

**Table 34.5 : List of District wise Fresh water Dams of Saurashtra with their Capacity.**

**Bhavnagar District**

| Sr. No. | Name of Dam | Height in foot | Sr. No. | Name of Dam | Height in foot |
|---|---|---|---|---|---|
| 1. | Shetrunji dam | 34.00 | 2. | Randhola | 30.00 |
| 3. | Malan | 34.00 | 4. | Rojaki | 32.50 |
| 5. | Surajwadi | 35.00 | 6. | Goma | 20.00 |
| 7. | Ghelo Somanth | 25.75 | 8. | Ghelo Etaraio | 25.00 |
| 9. | Bhimdad | 28.00 | 10. | Sukhbhadar | 23.67 |
| 11. | Bagad | 16.56 | 12. | Kharo | 21.70 |
| 13. | Rajaval | 25.10 | 14. | Hamirpara | 17.10 |
| 15. | Lakhanka | 30.30 | 16. | Kalubhar | 21.80 |
| 17. | Malpara | 16.73 | 18. | Malgadh | 18.70 |
| | | | **Amreli District** | | |
| 1. | Khodiar | 75.0 | 2. | Munjiyar | 24.5 |
| 3. | Shingoda | 61.66 | 4. | Dhatarwadi | 34.14 |
| 5. | Vadia | 16.56 | 6. | Raydi | 28.37 |
| 7. | Thabi | 20.00 | 8. | Sankroli | 25.00 |
| | | | **Junagadh District** | | |
| 1. | Madhuvanti | 51.0 | 2. | Zanzeshri | 32.14 |
| 3. | Ambajal | 34.44 | 4. | Hiran-1 | 42.50 |
| 5. | Hiran-2 | 29.00 | 6. | Shigoda | 61.68 |
| 7. | Amirpur | 7.5 | 8. | Kalindri | 20.96 |
| 9. | Dhrafad | 21.32 | 10. | Machundri | 32.81 |
| 11. | Raval | 62.35 | 12. | Hasnapur | 34.00 |
| 13. | Khambhala | 34.00 | 14. | Fodara | 37.00 |
| 15. | Vrijmi | 30.84 | 16. | Adwana | 12.13 |
| | | | **Rajkot District** | | |
| 1. | Bhadar | 34.00 | 2. | Moj | 44.00 |
| 3. | Fofal | 25.50 | 4. | Venu-2 | 19.68 |
| 5. | Aaji-1 | 29.00 | 6. | Aaji-2 | 30.04 |
| 7. | Aaji-3 | 26.74 | 8. | Bangavadi | 16.33 |
| 9. | Dami-1 | 20.00 | 10. | Lalpari | 15.00 |
| 11. | Machhu-1 | 49.01 | 12. | Machhu-2 | 33.01 |
| 13. | Dami-2 | 19.64 | 14. | Ghodadhoi | 16.99 |
| 15. | Chaparvadi-1 | 15.12 | 16. | Chaparvadi-2 | 25.00 |
| 17. | Ishwaria | 19.03 | 18. | Karmal | 13.12 |

*Contd.......*

*Table 34.5 – Contd...*

| Sr. No. | Name of Dam | Height in foot | Sr. No. | Name of Dam | Height in foot |
|---|---|---|---|---|---|
| 19. | Gondli | 30.15 | 20. | Vchpari | 18.00 |
| 21. | Nyari-1 | 21.91 | 22. | Nyari-2 | 20.67 |
| 23. | Motisar | 14.73 | 24. | Faldangbeti | 11.75 |
| 25. | Khodapipar | 7.87 | 26. | Veri | 9.33 |
| | | | **Jamnagar District** | | |
| 1. | Sasoi | 22.00 | 2. | Panna | 15.00 |
| 3. | Ghee | 20.00 | 4. | Fulzar-1 | 22.00 |
| 5. | Fulzar-2 | 15.00 | 6. | Sapada | 18.50 |
| 7. | Vijarkhi | 27.50 | 8. | Rangmati | 15.37 |
| 9. | Vartu-1 | 23.00 | 10. | Daiminsar | 12.80 |
| 11. | Sonmati | 21.28 | 12. | Und-1 | 25.92 |
| 13. | Kankavati | 13.12 | 14. | Sani | 11.81 |
| 15. | Vartu-2 | 26.89 | 16. | Und-2 | 12.30 |
| 17. | Sindhani | 10.99 | 18. | Shetha bhadthri | 14.76 |
| 19. | Veradi | 16.90 | 20. | Vadisang | 12.47 |
| 21. | Fulzar (koba) | 24.44 | | | |
| | | | **Surendranagar District** | | |
| 1. | Bhrahmani | 27.00 | 2. | Limbdi-Bogavo-1 | 20.00 |
| 3. | Nayka | 19.00 | 4. | Dholidhaja | 20.00 |
| 5. | Vansal | 8.86 | 6. | Dhari | 15.00 |
| 7. | Falku | 16.24 | 8. | Morsal | 12.80 |
| 9. | Saburi | 19.02 | 10. | Nibhni | 15.09 |
| 11. | Limdi-Bogavo-2 | 12.96 | 12. | Thrivenithaga | 17.06 |
| | | | **Porbandar District** | | |
| 1. | Sorthi | 16.89 | | | |

**Table 34.6 : Key to the Identification of Some Freshwater Culturable Fishes.**

| | | |
|---|---|---|
| 1. | Body is covered with scales and jaws are without teeth | 2 |
| | Body is devoid of scales and jaws are with teeth | 10 |
| 2. | Scales are samll, silvery in colour | 3 |
| | Scales are large | 5 |
| 3. | Anal fin long and continuous with caudal fin | 4 |
| | Anal fin short and not continuous with caudal fin | *Hypophthalmichthys molitrix* |
| 4. | Dorsal profile convex | *Notopterus chitala* |
| | Dorsal profile not convex | *Notopterus notopterus* |
| 5. | Head is covered with large plate like seales | *Channa Spp.* |
| | Head is not covered with Large scales | 6 |
| 6. | Mouth wide upturned and superior with prominent lower lip | 7 |
| | Mouth small and inferior or terminal | 8 |
| 7. | Barbels absent | *Catla catla* |
| | One pair of barbels present | *Cyprinus carpio* |
| 8. | Lips thick and fringed | *Labio rohita* |
| | Lips thin | 9 |
| 9. | One pair of small barbels present | *Cirrhina mrigala* |
| | Barbels absent | *Ctenopharyngodon idella* |
| 10. | Dorsal fin with spine, adipose dorsal may or may not present | *Mystus Sp.* |
| | Dorsal fin spineless, adipose dorsal absent | 11 |
| 11. | 2 pair of barbels present and dorsal fin with 4-5 rays | 12 |
| | 4 pair of barbels present and dorsal fin with 6-8 rays | *Hetropneustes fossilis* |
| 12. | Gape of mouth small, ends before the anterior margin of eye | *Ompok sp.* |
| | Gape of mouth large, ends beyond the posterior margin of eye | *Wallago attu* |

*Source* : Kaur and Dhadwan, 1997.

**Table 34.7 : List of Fresh Water Molluscans**

| | | | |
|---|---|---|---|
| 1. | *Cyclophorus* (Litosstylus) involvulus (Muler) | 2. | *Cyclophorus aurantiacus* (Schumbacher) |
| 3. | *Vivipara bengalensis* (Lamarch) | 4. | *Vivipara dissimillies* (Muller) |
| 5. | *Pila globosa* (Swainson) | 6. | *Bithynia stenothyroides* (Dohrn) |
| 7. | *Sulcospira hugeli* | 8. | *Faunus ater* (Linne) |
| 9. | *Pauludomus* (*Stomatodon*) stomatodon (Benson) | 10. | *Melania* (Plotia) *scabra* (Muller) |
| 11. | *Melania tuberculata* (Muller) | 12. | *Lymnaea luteola* (Lamarck) |
| 13. | *Lymnaea acuminata* (Lamarck) | 14. | *Lymnaea pignguis* (Dhorn) |
| 15. | *Lymnaea auricularia* (Drapaarnaud) | 16. | *Planorbis* (*Indoplanorbis*) exustus (Deshayes) |
| 17. | *Anisus* (Gyraulus) *convexiusculus* (Hutton) | 18. | *Opeas gracile* (Hutton) |
| 19. | *Zootecus insularis* (Ehrenberg) | 20. | *Macrochlamys infausta* (Blanfford) |
| 21. | *Zootecus chion* (Pfeiffer) | 22. | *Macrochlamys pedina* (Benson) |
| 23. | *Cryptozona* (Xestina) *belangeri var. bombayana* (Pfeiffer) | 24. | *Ariphanta laevipes* (Muller) |
| 25. | *Ariphanta bajadera* (Pfeiffer) | 26. | *Planispira proxima* (Ferussac) |
| 27. | *Parreysia corrugata var. negpoorensis* (Lea) | 28. | *Lamellidens marginalis* (Lamarck) |
| 29. | *Lamellidens corrianus* (Lea) | 30. | *Corbicula regularis* (Prime) |
| 31. | *Cyclophorus* (Glossostylus) *indicus* (*Deshayes*) | 32. | *Cyclotopsis semistriata* (Sowerby) |
| 33. | *Digoniostoma pulchella* (Benson) | 34. | *Alocinma orcula* (Benson) *var. producta (Nevill)* |
| 35. | *Paludomus obesa* (Philippi) | 36. | *Rachis punctatus* (Anton) |
| 37. | *Glessula notigena* (Benson) | 38. | *Glessula ceylanica* (Pfeiffer) |
| 39. | *Glessula dilkarngense* Godwin-Austen. | 40. | *Sitala denselirata* Preston |
| 41. | *Kaliella bullula* (Hutton) | 42. | *Macrochlamys tenuicola* (Adam) |
| 43. | *Macrochlamys* (Eurychlamys) *platychlamys* | 44. | *Xesta* (*Fretum*) *semirugata* (Beck) |
| 45. | *Ariophanta bistrialis* (Beck) | 46. | *Planispira* (*Trachia*) *crassisctata* (Benson) |
| 47. | *Eulota scalpturita* (Benson) | 48. | *Indonaia coerulea* (Lea) |
| 49. | *Lamellidens consobrinus* (Lea) | 50. | *Corbicula striatella* Deshayes |
| 51. | *Corbicula peninsularis* Bergh | 52. | *Peltodoris rubescenss* Bergh |
| 53. | *Platydoris pulchra* Elliot | 54. | *Achatina fulica* Bowdich |
| 55. | *Viviparus bengalensis* (Lamarck) | 56. | *Viviparus variatus* (Frauenfelf) |
| 57. | *Digoniostoma pulchella* (Benson) | 58. | *Melanoides tuberculatus* (Muller) |
| 59. | *Melanoides lineatuss* (Gray) | 60. | *Melanoides scabra* (Muller) |
| 61. | *Indoplanorbis exustus* (Deshayes) | 62. | *Gyraulus convexiusculus* (Hutton) |
| 63. | *Lymnaea acuminata* Lamarck. | 64. | *Lymnaea iuteola* Lamarck |
| 65. | *Planispira asperella* (Pfeiffer) | 66. | *Eurychlamys platychlamys* (Blanford) |
| 67. | *Indonaia caerulea* (Lea) | 68. | *Corbicula strriatella* (Deshayes) |

**Table 34.8 : Inland Fish Production Trends from 1971-72 to 1998-99**

| Sr. No. | Year | Inland Fish Production in Tone | Total Value in Lakh Rs. | Value in Thousand Rs/Tone |
|---|---|---|---|---|
| 1. | 1971-72 | 14160 | 141.67 | 1.005 |
| 2. | 1972-73 | 14560 | 146.60 | 1.023 |
| 3. | 1973-74 | 16400 | 213.20 | 1.300 |
| 4. | 1974-75 | 7455 | 124.91 | 1.675 |
| 5. | 1975-76 | 12695 | 276.08 | 2.175 |
| 6. | 1976-77 | 15261 | 460.47 | 3.017 |
| 7. | 1977-78 | 13135 | 417.78 | 3.181 |
| 8. | 1978-79 | 15651 | 857.10 | 5.476 |

*Contd........*

***Table 34.8 – Contd...***

| Sr. No. | Year | Inland Fish Production in Tone | Total Value in Lakh Rs. | Value in Thousand Rs/Tone |
|---|---|---|---|---|
| 9. | 1979-80 | 16343 | 837.69 | 5.129 |
| 10. | 1980-81 | 17331 | 1086.57 | 6.270 |
| 11. | 1981-82 | 19323 | 1330.52 | 6.886 |
| 12. | 1982-83 | 19750 | 1818.97 | 9.210 |
| 13. | 1983-84 | 21108 | 2484.32 | 11.769 |
| 14. | 1984-85 | 23614 | 2971.23 | 12.477 |
| 15. | 1985-86 | 24172 | 3323.72 | 13.751 |
| 16. | 1986-87 | 24451 | 3539.00 | 14.474 |
| 17. | 1987-88 | 22551 | 3626.81 | 14.474 |
| 18. | 1988-89 | 22315 | 3826.94 | 17.150 |
| 19. | 1989-90 | 27146 | 5231.74 | 19.273 |
| 20. | 1990-91 | 45687 | 9940.61 | 21.758 |
| 21. | 1991-92 | 39870 | 9112.10 | 22.855 |
| 22. | 1992-93 | 51154 | 12321.61 | 24.087 |
| 23. | 1993-94 | 65019 | 22492.07 | 34.593 |
| 24. | 1994-95 | 70100 | 25029.12 | 35.705 |
| 25. | 1995-96 | 60158 | 18421.20 | 30.621 |
| 26. | 1996-97 | 65278 | 21044.01 | 32.238 |
| 27. | 1997-98 | 70450 | 24025.00 | 34.102 |
| 28. | 1998-99 | 80068 | 28505.00 | 35.601 |

Source : Based on Fisheries Statistics of Gujarat 2000-01 (With permission)

From the table 38.4 it can be seen that inland fish production from 1971-72 to 1998-99 (i.e. in 28 years) was increased to five times, while value realization was increased to 35 times.

**Table 34.9 : Sector Wise Inland Fish Production of Gujarat**

| Sector | 1996-97 | 1997-98 | 1998-99 | 1999-00 | 2000-01 |
|---|---|---|---|---|---|
| Reservoir | 8017 | 12290 | 22772 | 13212 | 7812 |
| Brackish Water | 240 | 173 | 176 | 140 | 212 |
| Riverine | 9637 | 7372 | 9769 | 8484 | 3877 |
| Esturine | 16410 | 20912 | 18607 | 10118 | 10647 |
| FFDA | 5606 | 4011 | 5316 | 5817 | 2492 |
| Other | 25368 | 25692 | 22268 | 31725 | 13645 |
| Nal Sarovar | 0 | 0 | 1160 | 836 | 1910 |
| Total | 65478 | 70450 | 80068 | 70332 | 40595 |
| Growth Rate | 8.51% | 7.92% | 13.65% | –12.16 | –42.28 |

*Source* : Fisheries Statistics of Gujarat, 2000-01 (Reprinted with permission)

**Table 34.10 : Species Wise Inland Production of Gujarat State (Production in M.T.)**

| Sr. No. | Name of Fish | 1996-97 | 1997-98 | 1998-99 | 1999-00 | 2000-01 |
|---|---|---|---|---|---|---|
| 1. | Catla | 7653 | 8329 | 13114 | 11091 | 4608 |
| 2. | Rohu | 9351 | 100048 | 12707 | 9021 | 4738 |
| 3. | Mrigal | 5883 | 6322 | 8236 | 7396 | 3685 |
| 4. | Kalbasu | 133 | 140 | 667 | 607 | 412 |
| 5. | Minor Carp | 147 | 161 | 140 | 104 | 4 |
| 6. | Wallago-Attu | 2167 | 1754 | 2930 | 2560 | 1020 |
| 7. | Scorpion | 165 | 175 | 167 | 248 | 46 |
| 8. | Murrel | 1646 | 2330 | 2051 | 3039 | 1061 |
| 9. | Catfish | 5162 | 5577 | 5628 | 5632 | 3265 |
| 10. | Bombay Duck | 582 | 625 | 567 | 450 | 324 |
| 11. | Clupids/Hilsa | 8849 | 9514 | 9673 | 7053 | 3960 |
| 12. | Mullets | 2626 | 2811 | 3143 | 2612 | 1462 |
| 13. | Eel | 445 | 447 | 271 | 480 | 402 |
| 14. | Prawn (Shrimp) | 6373 | 6848 | 6043 | 3595 | 5434 |
| 15. | Prawn (Medium) | 2391 | 2568 | 1753 | 3072 | 2556 |
| 16. | Prawn (Jumbo) | 475 | 512 | 764 | 1051 | 357 |
| 17. | Crab | 32 | 56 | 104 | 77 | 91 |
| 18. | Mahasheer | 53 | 645 | 10 | 150 | 25 |
| 19. | Bekti | 203 | 35 | 30 | 43 | 3 |
| 20. | Levta | 603 | 208 | 507 | 724 | 438 |
| 21. | Mic. | 10339 | 11335 | 11653 | 11323 | 6699 |
| | **Total** | **65278** | **70450** | **80068** | **70328** | **40590** |

*Source*: Fisheries Statistics of Gujarat, 2000-01 (Reprinted with permission)

**Table 34.11 : Inland Fish Production Over States (Production in M.T.)**

| Sr. No. | State/UTs | 1995-96 | 1996-97 | 1997-98 | 1998-99 | 1999-2000 |
|---|---|---|---|---|---|---|
| 1. | Andhra Pradesh | 203.97 | 207.31 | 6.00 | 260.83 | 380.58 |
| 2. | Arunachal Pradesh | 1.85 | 1.85 | 7.00 | 2.30 | 2.40 |
| 3. | Assam | 155.06 | 154.61 | 155.13 | 155.71 | 159.77 |
| 4. | Bihar | 239.58 | 249.78 | 208.54 | 202.29 | 254.74 |
| 5. | Goa | 3.61 | 3.30 | 3.24 | 3.18 | 3.51 |
| 6. | Gujarat | 60.00 | 65.28 | 70.80 | 80.00 | 70.33 |
| 7. | Harayana | 28.01 | 30.00 | 32.05 | 32.52 | 30.00 |
| 8. | Himachal Pradesh | 5.94 | 6.27 | 6.69 | 6.79 | 7.00 |
| 9. | Jammu & Kashmir | 16.52 | 17.80 | 18.53 | 18.85 | 19.01 |
| 10. | Karnataka | 87.35 | 101.65 | 950.28 | 95.00 | 126.65 |
| 11. | Kerala | 49.59 | 52.10 | 57.51 | 65.86 | 73.90 |
| 12. | Madhya Pradesh | 91.28 | 110.53 | 115.16 | 119.59 | 127.43 |
| 13. | Maharashtra | 77.00 | 109.00 | 127.00 | 125.50 | 135.39 |
| 14. | Manipur | 12.50 | 12.71 | 13.70 | 15.31 | 15.50 |
| 15. | Meghalaya | 3.58 | 3.58 | 3.09 | 1.98 | 4.68 |
| 16. | Mizoram | 2.50 | 2.55 | 2.70 | 2.78 | 2.89 |
| 17. | Nagaland | 3.00 | 4.00 | 3.70 | 4.50 | 5.00 |
| 18. | Orissa | 134.85 | 143.50 | 153.43 | 159.90 | 135.30 |
| 19. | Punjab | 26.00 | 32.00 | 36.00 | 44.50 | 47.18 |
| 20. | Rajsthan | 12.40 | 14.30 | 15.10 | 12.00 | 12.97 |
| 21. | Sikkim | 0.15 | 0.15 | 0.14 | 0.14 | 0.14 |
| 22. | Tamil Nadu | 108.00 | 109.00 | 109.50 | 110.20 | 112.00 |
| 23. | Tripura | 25.71 | 27.47 | 27.90 | 28.41 | 29.34 |
| 24. | Uttar Pradesh | 145.70 | 149.43 | 160.02 | 183.03 | 192.72 |
| 25. | West Bengal | 740.00 | 765.00 | 786.02 | 823.50 | 865.70 |
| | **Union Territories** | | | | | |
| 26. | Andaman & Nicobar | 0.05 | 0.04 | 0.04 | 0.04 | 0.06 |
| 27. | Chandigadh | 0.08 | 0.08 | 0.00 | 0.00 | 0.00 |
| 28. | Dadra & Nagar Haveli | 0.04 | 4.00 | 0.02 | 0.02 | 0.02 |
| 29. | Daman & Diu | 0.00 | 0.00 | 0.00 | 0.00 | 0.00 |
| 30. | Delhi | 4.00 | 4.10 | 4.20 | 4.41 | 4.30 |
| 31. | Lakshadweep | 0.00 | 0.00 | 0.00 | 0.00 | 0.00 |
| 32. | Pondichery | 4.00 | 0.04 | 4.10 | 4.10 | 4.20 |
| | **Total** | **2242.32** | **2381.43** | **2222.59** | **2563.24** | **2822.70** |
| | Average Growth | | | 8.73% | 5.14% | 10.12% |
| | Gujarat Production | | | 70 | 80 | 70 |
| | Average Growth | | | 16.67% | 14.29% | –12.50% |
| | | | | 2.87% | 3.12% | 2.48% |

*Source*: Fisheries Statistics of Gujarat 2000-01 (Reprinted with permission)

**Table 34.12 : Fish Production Trend of India and Gujarat from 1950-51 to 2000-01 (Production in lakh M.T.)**

| Year | Marine | | | | | Inland | | | | | Total | | | | |
|---|---|---|---|---|---|---|---|---|---|---|---|---|---|---|---|
| | India | Growth Rate | Gujarat | Growth Rate | Percent Share of Gujarat | India | Growth Rate | Gujarat | Growth Rate | Percent Share of Gujarat | India | Growth Rate | Gujarat | Growth Rate | Percent Share of Gujarat |
| 1950-51 | 5.34 | 0.00 | 0.00 | 0.00 | 0.00 | 2.18 | 0.00 | 0.00 | 0.00 | 0.00 | 7.52 | 0.00 | 0.00 | 0.00 | 0.00 |
| 1960-61 | 8.80 | 6.48 | 0.79 | 0.00 | 8.98 | 2.80 | 2.84 | 0.00 | 0.00 | 0.00 | 11.60 | 5.43 | 0.79 | 0.00 | 6.81 |
| 1970-71 | 10.86 | 2.34 | 1.51 | 9.11 | 13.90 | 6.70 | 13.93 | 0.00 | 0.00 | 0.00 | 17.56 | 5.14 | 1.51 | 9.11 | 8.60 |
| 1980-81 | 15.55 | 4.32 | 2.19 | 4.50 | 14.08 | 8.87 | 32.39 | 0.17 | 0.00 | 1.92 | 24.42 | 3.91 | 2.36 | 5.63 | 9.66 |
| 1990-91 | 23.00 | 4.79 | 5.00 | 12.83 | 21.74 | 15.36 | 7.32 | 0.46 | 17.06 | 2.99 | 38.36 | 5.71 | 5.46 | 13.14 | 14.23 |
| 1996-97 | 28.57 | 2.42 | 6.60 | 3.20 | 23.10 | 22.83 | 4.86 | 0.65 | 4.13 | 2.85 | 51.40 | 3.40 | 7.25 | 3.28 | 14.11 |
| 1997-98 | 29.50 | 3.26 | 7.02 | 6.36 | 23.80 | 24.38 | 6.79 | 0.70 | 7.69 | 2.87 | 53.88 | 0.48 | 7.72 | 648 | 14.33 |
| 1998-99 | 26.96 | -8.51 | 5.52 | -21.37 | 20.47 | 25.63 | 5.13 | 0.80 | 14.29 | 3.12 | 52.59 | -0.24 | 6.32 | -18.13 | 12.02 |
| 1999-00 | 28.34 | 5.12 | 6.70 | 21.38 | 23.64 | 28.23 | 10.14 | 0.70 | -12.50 | 2.48 | 56.57 | 0.76 | 7.40 | 17.09 | 13.08 |
| 2000-01 | – | – | 6.20 | –7.46 | – | – | – | 0.40 | -42.86 | – | – | – | 6.60 | -10.81 | 0.00 |

*Source*: (1) CMFRI upto 1950-2001; (2) State Govt./UTS Fisheries Statistics of Gujarat, 2000-01, p. 65 (reprinted with permission)

## *References*

Anon, 2000. Annual Reports of CMFRI, 1980 to 2000, Central Marine Fisheries Research Institute, Cochin.

Anon, 2002. Annual publications of Fisheries statistics of Gujarat from 1980 to 2001-02, Commissioner of Fisheries, Govt. of Gujarat, Gandhinagar.

Chhaya, N.D. 1990. List of Fisheries of Gujarati Fisheries Management and Science. Gujarat Granth Nirman Board, Govt. of Gujarat. Ahmedabad.

Dholakia, A.D. 1986. List of important fishes available in Gujarat with Scientific, English and Local names in Fish Processing and Preservation and Quality control, pp. 134-144. Dholakia D.P. Ahmedabad.

Yadav, B.N. 2002. Fish and Fisheries, Daya Publishing House, New Delhi, pp. 187-188.

35

# Coastal Aquaculture in India

## Introduction

In India, although by record, many millions hectares of potential areas for prawn cultivation are identified as suitable for prawn culture, nearly 90% of the total estimated area is still not under the administrative control of respective departments of Fisheries of the maritime States. Thus, difficulty in availability of these lands for beneficiaries is a major constraint.

Though Gujarat State ranks second in having potential backwater area next only to West Bengal, the culture practice in this part of the country is very much infancy. However, in light of the total awareness that has been created all over the country in prawn farming, suitable policy of allotting land for aquaculture has been created by Gujarat. But real explosion in coastal aquaculture will start only, when the administrative control of such areas is passed on to concerned agencies.

It is expected that landing from capture fisheries will remain stagnant more or less and the increase in demand can only be met by increasing production from culture sources. (Varghese and Varghese, 1986) in India, capture prawn production has shown declining trend (Rao, R.M. 1983, Santhana Krishnan, 1988) The industry and Government both have expressed that it may not be possible to meet the increasing demand from natural stock. (Dwivedi, 1987).

If India wants to maintain her position in prawn exports, there is an urgent need for increasing production from culture on priority, so as to meet the growing demand of export industry for its raw material to maintain and improve the socio-economic status of processing industry and that of fishermen. (Dholakia, A.D. 1994).

Aquaculture has emerged as a reliable source and has made rapid strides to augment fish/prawn production in the world. It not only creates employment opportunities to the skilled and unskilled rural poor but also earns foreign exchange for the country so one of the Government of India's policy objective is to encourage brackish and inland fisheries through development programmes mainly benefiting the poorest of the poor in the fishing communities. India is endowed with rich natural resources in the coastal zone in the form of brackish water/estuaries for taking up shrimp farming. India has about 1.19 million ha brackish water area of which only 83,000 ha are now under shrimp farming mostly by adopting traditional practice (50,000 ha) and partly by extensive method of culture (R Paul Raj. CMFRI, 1995).

Most of the maritime states are now taking brackish water aquaculture as an important activity for benefiting the rural communities. However, the ecological importance of the coastal area should be taken in to consideration before using it for aquaculture or any other development acitivity. This is because; the fast expanding prawn farming along the coasts may pose a serious threat to the coastal ecosystem. It may lead to solidification of the underground aquifers and damage the underground water resources as well as the fertile coastal agricultural land. This is more serious in case of intensive and semi intensive shrimp farming where water quality is maintained by pumping, stocking rates are high and a proportionately high amount of feces, molted shells and other waste material, excessive lime organic wastes, pesticides and chemicals flush in to neighboring mangroves and coastal wastes and there by affect them adversely. For sustainable aquaculture, it is better to have a buffer zone around aqua cultured zone as suggested by Prof. Algarswami, Director, Central Brackish Water Institute, Madras, (Business India, 1993). More over cutting of mangroves, exchanging out the water in the same canal, farming on the agricultural land etc., should be avoided.

Space Application Centre (SAC), Ahmedabad has taken a project "Coastal Land use Mapping for Brackish water Aquaculture Site Selection". Under this project SAC has prepared coastal land use maps on 1,50,000 scale for all maritime states using remote sensing satellite data. Using these maps database have been created for selected sites of West Bengal, Orissa, Gujarat and Maharashtra to prioritize the sites for brackish water aquaculture development. Geographical Information System (GIS) is a powerful tool for collecting, Storing, organizing, retrieving transforming and displaying data from the real world for a practical purpose in this study this system is used. (Anon, ISRO 1995).

## Policy for Coastal Aquaculture

Ministry of Agriculture, Govt. of India is the nodal ministry, which looks after fisheries development through the respective state governments. As most of the coastal areas suitable for farming belong to the state governments, Ministry of Agriculture, Govt. of India has issued guidelines for classification, use and lease of brackish water lands for shrimp/fish farming. Under these guideliness the brackish water areas are to be classified and recorded separately in the state records and should be earmarked for aquaculture only. It has suggested that 60% of the brackish water area involving low capital investment on development may give to fishermen and small and marginal farmers with financial and technology support through central and state schemes, 20% of brackish water area which requires medium size capital investment may be leased to medium entrepreneurs and balance of the land involving high investment may leased to the progressive entrepreneurs who may have financial backing and management capacity for introducing advanced technology. Based on the guidelines, states like Orissa, Andhra Pradesh, Maharashtra, Gujarat, Tamil Nadu and Union Territory of Pondichery have brought out land leasing categories of people. Ministry of Agriculture is also assisting the maritime states to establish brackish water Fish Farmer's Agencies (BFDA) to promote prawn farming in the states by meeting 50% cost of the projects. The BFSA's assist small prawn farmers by giving technical and financial assistance. Many state governments have already established BFDA's. Ministry of Agriculture is also haveing a proposal to establish 5 prawn hatcheries in different states with UNDO assistance. One such hatchery is coming up in Goa, which is likely to go in to operation shortly. Other projects are under different stages of progress. The IVAR institutes, agricultural universities and fisheries colleges are also in the process of refining the technology on various aspects of shrimp farming. All the state governments have also schemes to promote shrimp farming.

## Financial Institutions

Most of the nationalized and scheduled banks, state co-operative and agriculture development banks promote shrimp farming by giving credit which is normally refinanced by National Bank for Agriculture and Rural Development (NBARD). Assistance from other financing organizations like Industrial Credit Investment Corporation of India (ICICI) and Shipping Credit and Investment Company of India (SCICI) is also available to integrated export oriented shrimp farming units. Shrimp farming activities have also been covered under insurance scheme.

## Employment Opportunity

Shrimp farming development provides immense employment potential for the local people. It provides direct employment to at least two persons for managing and operating one-hectare culture area. It also provides indirect employment to extent of 1100 man-days per ha during construction, input supplies, handling of harvest materials etc. In other words, 1 lakh ha is developed for shrimp farming, it will provide direct employment to 2 lakh people and generate indirect employment of 110 million man-days which will revolutionize the economic conditions of the local people. It will also provide opportunity to establish 200 hatcheries, 100 feed mills, which will create other employment opportunities.

As indicated in the foregoing paragraphs, India is having vast potential for coastal aquaculture development. The state governments have started alloting government land on long-term lease. Technology for farming, production of seed and feed for extensive farming is also available. Import of quality feed is allowed with duty concession. Institutional finance is available from various banks. Several promotional agencies extend both technical and financial assistance. Establishment of 100% EOU aquaculture projects are also encouraged. There is a continuous and increasing demand for shrimp in the international markets offering high price.

## Quality standards of water for Prawn Culture Farm

The most important factors of the water, which need to be considered, are salinity, turbidity, pH, and possible source of pollutants from industrial, agricultural and domestic origin. Checking of the Ammonia, dissolved iron and hydrogen sulphide content are also necessary. Distributions of rainfall in the area and salinity and temperature fluctuations have to be taken into account to decide the period of culture. The favourable water quality should be as follows :

| | *Parameters* | *Range* |
|---|---|---|
| 1) | Temperature (°C.) | 26-33 |
| 2) | Salinity (ppt) | 15-35 |
| 3) | Transparency (CM) | 25-60 |
| 4) | Dissolved oxygen (ppm) | 3-12 |

| | | |
|---|---|---|
| 5) | pH | 7.5 to 8.7 |
| 6) | Total Ammonia (ppm) | less than 1.0 |
| 7) | Free Ammonia (ppm) | less than 0.25 |
| 8) | Nitrite (ppm) | less than 0.25 |
| 9) | Hydrogen sulphide (ppm) | less than 0.25 |
| | **Heavy metals** | |
| 1) | Mercury (ppm) | less than 0.0025 |
| 2) | Copper (ppm) | less than 0.1 |
| 3) | Iron (ppm) | less than 0.01 |
| 4) | Zinc (ppm) | less than 0.25 |
| 5) | Cadminium (ppm) | less than 0.15 |
| | **Pesticides** | |
| 1) | Malathion (ppb) | less than 0.0004 |
| 2) | Parathion (ppb) | less than 0.001 |
| 3) | Arzodin (ppb) | less than 0.01 |
| 4) | Paraquet (ppb) | less than 0.01 |
| 5) | Endosulfan (ppb) | less than 0.01 |
| 6) | Butachion (ppb) | less than 10.00 |

Note : ppt = parts per thousands, ppm = parts per million, ppb = parts per billion

*Source* : Hand Book MPEDA, 1991.

## Resources available for coastal aquaculture in India

Resources such as suitable water areas, variety of cultivable species, their seed and large number of fish farmers for immediate starting of coastal aquaculture are available in all maritime states of India. Table 35.1 provides certain basic data pertaining to the potentials available in the sector for general information. (The numbers given in the table refer to)

### *Fishes*

1. *Chanos chanos* (Milk fish)
2. *Mugil spp* (Mullets)
3. *Etroplus suratensis* (Pearl spot)
4. *Lates clacarifer* (Giant brackish water perch)
5. *Sillago sihama* (Lady fish or sand whiting)
6. *Polynemus indicus* (Thread fin.)
7. *Eleutheronema tetradactylus* (Thread fin.)
8. *Elops sp*
9. *Hilsa ilisha*

### *Prawn*

10. *Penaeus monodon* (Tiger prawn)
11. *P. indicus* (White prawn)
12. *P.merguiensis* (Banana prawn)

Table 35.1 : Information relating to potential area, available potential species and season of availability of seed for coastal aquaculture in different Maritime States of India.

| State/Union Territories | Potential inshore area 0-18 m available for open sea Farming Million Ha. | Estuarine and Brackish water area Mill. Ha. | Species of fishes and shellfishes that could be used for intensive culture | Main seasons of availability of seeds | | | | | |
|---|---|---|---|---|---|---|---|---|---|
| | | | | Fishes | Prawns | Mussels | Molluscs Edible Oyster | Pearl Oyster | Clams and cockles |
| Gujarat | 4,752 | 0.376 | 1,2,3,4,11,12,18,19, 20, 23,24,25,28, 29,30,31,32, | All months | Feb.-April & Sept. | N.A. | Apr.-Mar., July-Sept. | March-June Oct.-Feb. | Sept.-May |
| Maharashtra | 0.593 | 0.081 | 1,2,3,4,6,7,12,14 16,17,20,22,23,25, 26,27,28,30,31,32, | Nov.-June and Sept. | Oct.-Dec. | N.A. | Sept.-May | N.A. | Sept.-May |
| Goa | 0.119 | 0.019 | 1,2,3,4,10,11,12,14, 15,16,20,22,23,28,29,30, | Oct.-May | Oct.-Dec. & Feb.-May | N.A. | N.A. | N.A. | N.A. |
| Karnataka | 0.259 | 0.008 | 1,2,3,4,10,11,14,15, 16,20,22,23,25, 26,27,28,30,32. | All months | Oct.-April. | March-Oct. | All months | N.A. | All months. |
| Kerala | 0.259 | 0.243 | 1,2,3,4,10,11,14,15, 16,19,20,21,22,23, 24,25,27,28,30,31,32. | Oct.-Dec. and April-Sept. | Nov.-Dec. and Feb.-May. | Oct.-Nov. and June-Aug. | Oct.-Dec. and March-June. | April-May and Sept.-Oct. | All months |
| Tamil Nadu | 1,606 | 0.080 | 1,2,3,4,10,11,13 14,20,21,22,23, 24,25,26,28,29,30 | April-June and Oct.-Feb. | Feb.-May and Aug.-Dec. | March-June-Aug. and Oct.-Dec. | Oct.-Dec. and March-April | April-May and Sept.-Oct. | All months. |
| Pondicherry | 0.067 | 0.0084 | 1,2,3,10, 11,14,20,22,23 | N.A. | Nov.-Dec. | N.A. | N.A. | N.A. | N.A. |
| Andhra Pradesh | 0.414 | 0.200 | 1,2,3,4,10,11,12 14,17,19,20,22,23, 25,26,27,29, | June-Aug. and Oct.-Feb. | Sept.-April | N.A. | All months | N.A. | All months |
| Orissa | 0.768 | 0.299 | 1,2,3,4,10,11,14, 16,19,20,22,23, 25,28. | May-June and Dec-Feb. | Feb.-May and Aug-Sept. | N.A. | All months | N.A. | All months |
| West Bengal | 0.078 | 0.405 | 1,2,3,4,9,10,11, 17,19,20,23. | March-July and Dec.-Jan. | Feb.-May and Aug.-Sept. | N.A. | N.A. | N.A. | N.A. |

*Source* : *Mar. Fish. Infor. Serv.* T & E *Ser. No 13, 1978, : 12-13.*

13. *P.semisulcatus* (green tiger prawn or pink flower prawn)
14. *Metapenaeus monoceros* (Indian prawn)
15. *M.dobsoni* (Flower-tail prawn)
16. *M.affinis* (Indian prawn)
17. *M.brevicornis* (Yellow prawn)
18. *M.kutchensis* (Kutch prawn)
19. *Macrobranchium rosenbergii* (Giant fresh water prawn)

### *Crabs*

20. *Scylla serrata* (Green crab)

### *Lobster*

21. *Panulirus homarus* (Indian spiny Lobster)

### *Molluscs*

22. *Perna spp* (Brown & green mussel)
23. *Crassostrea spp.* (Edible oyster)
24. *Pinctada fucata* (Pearl oyster)
25. *Meretrix spp.* (Backwater clam)
26. *Anadara granosa* (Blood clam)
27. *Ketelysia* (Backwater clam)

### *Seaweeds*

28. *Gracilaria sp.* (Red alga)
29. *Gelediella sp.* (Red alga)
30. *Sargassum sp.* (Brown alga)
31. *Enteromorpha sp.* (Green alga)
32. *Ulva sp.* (Green alga)

**Table 35.2 : District wise details of brackish water cultivable land, allotment and prawn culture Farms in Gujarat (Up to 1994-95)**

| Name of District | Potential cultivable land (Ha) | Land allotted for shrimp Farming | No of Farms |
|---|---|---|---|
| Valsad | 20235 | 1428 | 129 |
| Surat | 24300 | 30 | 6 |
| Bharuch | 47500 | 505 | 63 |
| Amreli | 80095 | 25 | 5 |
| Junagadh | 19020 | 62 | 2 |
| Jamnagar | 22300 | 50 | 10 |
| Kutch | 68400 | 425 | 26 |
| Rajkot | 2000 | 20 | 4 |
| Bhavnagar | 5200 | 46 | 4 |
| Kheda | 20735 | N.A. | N.A. |
| Ahmedabad | 49780 | N.A. | N.A. |

*Source* : M.P.E.D.A. 1997.

Marine Product Export Development Authority (MPEDA), Ministry of Commerce, Govt. of India, is helping to start Aqua farm. They are giving guidance in following topics :

1. Aquaculture Engineering and Water Quality Management
2. Aquaculture Feed
3. Fresh water Fishes
4. Fishery potential of Andamans
5. Trout-Eel-Fresh water Prawn-Gray Fish
6. Identifying feature of cultivable species
7. Live feed
8. Molluscs
9. Ornamental Fishes
10. Shrimp Hatchery
11. Directory of Aquaculture

MPEDA have established Prawn Farming Offices in all Maritime States.

Addresses of all offices are as under :

1. MPEDA Prawn Farming Project comples, Vallarpadam. Cochin-682031.
2. MPEDA Regional Centre (P.F.), XXXVIO/563, 1st Floor, T.D. Road, Cochin-682011.
3. MPEDA Regional Centre (P.F.) Jayadev Nagar, Lewis Road -Bhubaneshwar-751002.
4. MPEDA Regional Centre (P.F.) 322-210, Parajasakthi Nagar, Vijaywada-520010.
5. MPEDA Regional Centre (P.F.) Plot No 19, Nirmala Nagar, Vallam No Road, Thanjavur_613007, Tamil Nadu.
6. MPEDA Regional Centre (P.F.) P.B. No 130, Halar Road, Valsad-369001 Gujarat.
7. MPEDA Sub-Regional Centre (P.F.), Rossanna, 1st Floor-High Church Street, Karwar-581301, Kakinada.
8. MPEDA Sub-Regional Centre (P.F.), 913/A, Kamat Lane, Alibag-402201, Maharashtra.
9. MPEDA sub-Regional Centre (P.F.) 3rd Floor, Ultadanga-VIP Road, Calcutta-700054, West Bengal.
10. Orissa Shrimp Seed Production Supply & Research Centre (OSPARC), Station Road, Ganjam, OSS Railway, Ganjam Berhampur-760005.
11. The Andhra Pradesh Shrimp Seed Production, Supply & Research Centre (TASPARC) 48-79, Rama Talkies Road, 1st Floor, Srinagar Visakhapatnam-530016, Andhra Pradesh.
12. West Bengal Scampi Seed Production Supply & Research Centre (WESPARC), P-161/1, 3rd Floor, VIP Road, Ultadanagar, Calcutta-740054, West Bengal.

## *References*

Anon. 1978. Resources available for coastal aquaculture in India. *Mar. Fish. Infor. Serv.* S & E *Serv.* No 3:12-13.

Anon. 1991. Hand book on Shrimp Farming. M.P.E.D.A. Cochin.

Anon. 1995. Brackish water Aquaculture Site Selection using Techniques of Geographical Information System (GIS). Marine and water Research Division, SAC, ISRO, Ahmedabad pp. 1-5.

Anon. 1997. Booklet-Aquaculture Farms of Gujarat Prawn culture Office, MPEDA, Valsad.

Dholakia, A.D. 1994. Marine prawn fishery and its culture in Saurashtra with special reference to *Penaeus Merguiensis* (De Man) Ph.D. Thesis, Saurashtra University.

Dwivedi, 1987. Frontier areas for research and utilization of marine living resources. *National symposium on utilization of living resources of Indian seas.* 19-20 Dec. Bombay.

Meera Shenoy, 1993. The price of Aquaculture, *Business India,* July.

P. Paul Raj. 1995. Employment opportunities in Aquaculture. Conference on Sustainable Aquaculture, Anna University. Madras. April 5-6.

Rao, R.M. 1983. Shrimp and fish resources of India. *Proceeding National Symposium on Shrimp Seed Production and Hatchery Management* : 21-22 May: 49.

Santhna Krishnan, 1988. Fisheries regulations with specific reference to shrimp farming Training course on shrimp culture and hatchery management: 185-2002.

Varghese and Varghese 1986. Breeding and propagation of marine prawns. *National symposium on utilization of Living resources of the Indian Seas.* Ab. 59.

# 36

# Marine and Fresh Water Aquatic Diseases

## Aquatic Diseases, Symptoms, Control and Treatments

The usual way of treating the disease is by adding prophylactic or therapeutic level of microbialagents to the sort of control/treatment methods are the requirement of large amount of expensive drugs and their discharge to environment, creating risk for animal and human health. Hence a method of disease treatment through the food chain using much smaller quantities of drugs would be more economical and safer for environment. (Anil *et. al.*, 1998.)

*Artemia* and rotifers are filter feeders and have successfully been applied as biological carriers for transfering essential nutrient to predator larvae. Similarly they can be used as tool for transfering therapeutics through the food chain.

Therapy tested on *Vibrio anguillarum* challenged sea bass larvae *Dicentratchus labrax* indicated highly significant improvement in survival larvae when administered with medicated *Artemia* with TMP Tri Metho Prim SMX (Sulpha Metho Xaxole) in 1:5 part either preceding or following the bacterial challenge, compared to the challenged no treated fish larvae Touraki *et. al.*, 1996.

Viral diseases are a major threat to the aquatic industry all over the world. Presently there are about 12 known virus affecting shrimp. Out of which some like the White Spot Baculo Virus (WSBV), Infectious Hypodermal and Hematopoietic Necrosis Virus (IHHNV) and Taura syndrome virus have caused serious mortalities. Other viral agents such as Monodon Baculo Virus (MBV) are a hazard in the presence of environmental stress factors whereas the significance of some shrimp viruses such as hepatopanvceatic parvo virus is not clearly understood.

Viruses are extremely small micro-organisms which cannot be seen even with the help of ordinary microscope. Expensive electronic microscope is required to see viruses. However, when viruses enter host cells and multiply there (either in the nucleus or in the cytoplasm), they leave their imprint there. Viral exclusions in host cells help us to detect viral infections and that can be seen in histopathological sections using an ordinary microscope. Thus histopathological studies are the mainstay of diagnosis of viral diseases of shrimp. (Karunasagar and Karunasagar, 1997).

However, above methods suffer from following limitations:

(1) Routine histopathology often is generally unable to provide identification in a meaningful time frame, if one considers the time taken for diagnosis and rapid rate at which disease spreads in the culture system.

(2) Detection of a symptomatic and early stages of infection is not possible.

(3) The cost of setting up and maintaining the isolation laboratories for enhancement and bio-assay method is prohibitive.

The unique molecular tool helps in amplifying a specific sequence of DNA. Polymerase Chain Reaction (PCR) base diagnostic system can be detected extremely low amount of viral nucleic acid. In principle, presence of single viron in a sample can be detected using PCR methos.

(Panchaythapani and Corpron 1997).

A Systemic baculovirus has been proved beyond any doubt as the primary causative factor for White Spot Syndrome (WSS). The prominent clinical signs are sudden reduction in feed consumption, loose cuticle, reddening antennae cut and characteristic circular white spot on the inner side of the carapace.

This new disease is attributed to at least four viruses reported by different names.

(1) Buculoviral Hypodermal and Haematopoietic Necrosis Virus (BHHNV).

(2) Rad-shaped Virus of *Penaeus japonicus* (RV-PJ).

(3) Systemic Ectodermal and Mesodermal Baculovirus (SEMBV).

(4) White Spot Baculovirus (WSBV).

(Mohan and Shanker. 1997).

Control and prevention of infectious diseases in aquaculture is a function of management. Many diseases of fish are secondary to environmental input and can often be prevented through proper management or environmental manipulation. Maintenance of high quality environment is importnat by insuring an adequate water supply that is free of pesticides, industrial and municipal contaminants and other pollutant to insure adequate dissolved oxygen, avoid elevated ammonia and carbon dioxide concentrations, prevent excessive accumulation of organic matter and avoid temperature shock.

Another approach to disease control is the use of prophylactic treatments when fish are handled. A prophylactic treatment of salt, potassium permanganate or other mucous stimulators will help injured areas to heal and inhibit "secondary infections." Diseases will occur in aquaculture regardless of the quality or intensity of management. It is essential to treat the fish with appropriate chemicals, drugs or antibiotics when these occur. Parasites and some external bacterial diseases can be treated by adding chemotherapeutics to the water. It is necessary to know four important points when treating fish with chemicals, because if these are not known, the treatment may not be successful or the treatment may kill fish. These are:

(1) Know the chemicals.

(2) Know the fish.

(3) Know the disease.

(4) Know the water.

### *Know the Chemicals*

The toxicity of the chemical or drug to be used must be known for the species of fish to be treated. Some species are more sensitive to some chemicals than others. Also, some chemicals are neutralized by sunlight, high pH, and high temperatures and are affected by total alkalinity. Some chemicals are toxic to plants and contribute to oxygen depletion or result in chemical removal of oxygen from the water.

### *Know the Fish*

Different species or ages of fish are affected differently by some chemicals. Some species are more sensitive to certain chemicals, while within a species there may be variation of toxicity to some chemicals of different age groups. If certain drug or chemical has not been used in a given water with a given fish species or age group of that species, a bioassay should be done. This involves treating a few fish in a small vessel containing the water to be treated and observing the effects of the chemical on the fish.

### *Know the Disease*

The disease to be treated must be known. A given chemical is most effective against specific parasitic or bacterial agents, with some exceptions, thus the cause of the disease must be known. Also multiple infections of two or more types of pathogens often cause disease at the same time. In this case, a sequence of treatments would be prescribed. The cause of a disease can be determined only through a proper necrosy because some parasites and bacteria produce similar clinical signs. This will dictate the proper treatments. Selection of the incorrect treatment could have disastrous consequences. If drugs or chemicals are used indiscriminately more fish may die because of improper treatment than would have died without any treatment.

### *Know the Water*

Before applying a treatment, the volume of water to be treated should be accurately known and not determined by guesswork. If the volume is not known, an overtreatment could kill the fish while an undertreatment would be ineffective. Water chemistry parameters such as total alkalinity, pH, and temperature should be known because these factors will influence the efficiency and toxicity of some chemicals to fish.

## Method of Treatment

A variety of methods of drugs and chemical applications has been used and there are no specific procedures or methods that are bette than others. The most appropriate (effective and economical) method of application must be determined for each situation requiring treatment. There are six basic methods of therapeutic application that be used to treat an active infection or can be used prophylactically. These are dip, flush, prolonged, indefinite, feeding and injection.

### *Dip*

A strong solution of the chemical to be used is made and the fish are immersed (dipped) in the solution for 15 to 60 seconds. Extreme caution must be used here because the difference between an effective dip treatment and toxic concentration may be very small. It is best to treat a few fish to determine their reaction before treating the entire population. The dip method is usually used with number of easily confined fish with external parasite or bacterial infections.

### *Flush*

The flush method involves adding a specific amount of chemical to tank and allowing it to flush through, without interruption of water flow. The flush treatment last from a few minutes up to an hour. It is usually used for fungus infections on eggs or ectoparasite and bacterial infections on fish.

### *Prolonged*

The prolonged treatment exposes the fish to a concentration of chemical or drug is for comparatively short-term bath for a specified period. The chemical/drug is added to tank at a specific concentration and left for a predetermined time with water flow interrupted. The fish should be observed continuously during the prolonged treatment and if they show any sign of stress of discomfort, such as gaspiring, flshing or loss of equilibrium, the chemical should be removed immediately.

### *Feeding*

Treating systemic bacterial infections and intestinal parasites requires incorporation of antimicrobial or anthelminthic drugs in the feed. The amount of antimicrobials and anthelminthics is applied based on fish body weight to be treated; therefore standard units of treatment are given of active ingredient per 45 kg of fish per day and for a predetermined number of days.

### *Injection*

Valuable broodfish, potential broodfish caught from wild populations, or small number of other valuable fishes can be treated for some diseases by injections of antimicrobials. The injection can either be given intraperitoneally (IP) or intramuscularly (IM), IP injections are given in the posterior area of the peritoneal cavity, care being taken not to puncture an internal organ. If the needle is inserted at a 45 angle at the base of the pelvic fin, no internal injury will occur. IM injections are given in the thick dorsal masculature near the dorsal fin.

## Diseases in Marine water Fish and Shell Fish

Generally fishes have great resistance to disease when compared to prawns so long as they weakened by unfavorable food, lack of Oxygen, extremely high temperature and salinity or other ecological influences.

Many biological factors poses hazards to the successful rearing of healthy marine animals. Some pathogens cause of debilitation and mortality include dietary deficiencies, wounds, poisons and environmental factors such as temperature and salinity.

Major disease in Marine water shell fishes, specially in prawns are as under :

## Prawn Diseases

1) Larval mycosis.
2) Milk or Cotton disease.
3) Ciliate disease
4) Fungal disease
5) Cramped tail
6) Attack of pathogen
7) Gill disease
8) Nacrosis of the muscle
9) Luminous Bacterial Disease

## Prawn Diseases

### *(1) Larval Mycosis*

Cause of Infection: This disease is caused by fungus *Legenidiam callinectys*

*Symptoms*

In most cases the host issue is completely distracted and reptured by fungal mycelia. Mortality have been reported to be rapid in hatchery tanks usually in protozoeal stages. It has also been observed on eggs, nauplier and mysis stages. Mortality rate is 100%. This disease develops more fast when there are dead eggs and dead larvae in the tanks.

*Treatment/Prevention*

Dead egg and larvae should be removed immediately. Disinfection of contaminated tank and sterilization of incoming water have been reported as effective control measures. (Chandramohan, C, *et. al.*, 1980).

### *(2) Milk or Cotton Disease*

Cause of infection is *Microsporideans.*

*Symptoms*

They mainly infect the tail and abdomen of the hosts, making them chalky white appearance (Similar to cooked prawns) and cottony in texture. This parasite invades even the digestive tract, gonads and heart. Microscopic examination of smears from infected muscle will show characteristic micropordian spores. From nature and aquaculture rarely exceed 15%. This disease has reported to be higher in monsoon months.

*Treatment/Prevention*

Adequate water quality must be maintained.

Destroy infected individuals and avoiding contacts of infected brood stock with off spring.

### *(3) Cliates Disease*

Cause of infection :

Stalked peritrichoes ciliates, *zoothamnium penae* and *Epistylis Spp.*

*Symptoms*

Have fuzzy appearing mat on the appendages and occasionally on eyes and carapace. Infected prawn become restless and frequently come to the surface or crowd the edge of ponds.

*Treatment/Prevention*

This may occur because of low oxygen level (below 5 ppm.). Hence either exchange water or start paddle aerator. Filtration and sterilization of culture water and removal of extraneous organic material in the tank.

### *(4) Fungle Disease*

Cause of infection

Attack of fungus *Fusarium solani* or *Pythium afertile.*

*Symptoms*

When suspected tissues are scraped and observed under light microscope presence of boat shaped microcornidia confirms the pathogen *Fusariam solani.* Reported from invade tissues mainly through wounds. Results in progressive paralysis and followed by rapid death. If observed under light microscope hyphal strands of invading fungus *Puthium afertile* can be clearly seen.

*Treatment/prevention*

Rough handling of the animal and cannibalism should be avoided. Strict measures of disaffection of culture tank may prove successful.

### *(5) Cramped Tails*

*Symptoms*

The flexure of the tails which is rigid and can not straightened. Sometimes muscle bulges out between 3rd and 4th abdominal segments, practically cramped animal swim about with humped abdomen. While in advance stage stops movement and lie on sides and hence cannibalism takes place. Other prawn will kill that prawn.

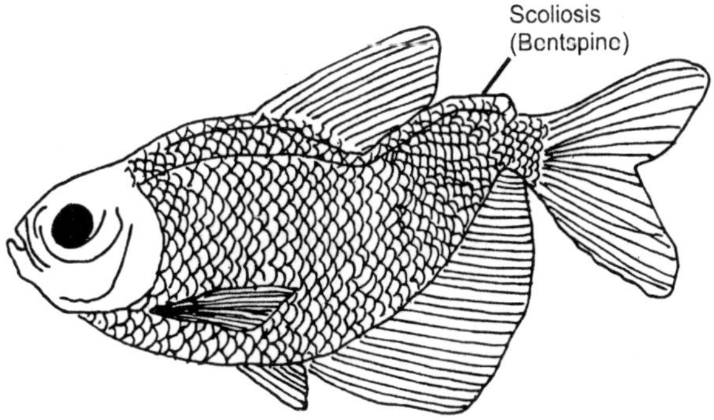


**Fig. 36.1 : Cramped tail**

*Treatment/prevention*

Sudden change of salinity to extremes, very low temperature and handling stress have been recorded as possible cause. Try to minimize.

### *(6) Attack of Pathogenic Bacteria*

### *(a) Vibrio Sp*

*Symptoms*

In the initial stage of this disease, some parts of the larvae in the tank are observed to have yellow vermilion and red colour along the nervous system throughout the body. It proceeded by incomplete moulting in the mysis and post larval stages.

The larvae lose some parts of their appendages and eventually becomes handicapped and die in the following one or two day.

### *(B) White turbid liver*

This disease is often seen. The mid gut gland of the larval becomes white turbid in general. As the disease progresses the white turbidity becomes more apparent and well defined. The heavily infected post larvae, mostly 6 mm to 0 mm in length can easily be distinguished as they float inactively on the surface of the water and exhibit white intestinal lines. It also observed in rare cases that blotches of brownish black pigment emerge in the mid gut. This is because of pathogenic bacteria which was reponsible for mass killing of the shrimp.

*Treatment/prevention: For (A) and (B)*

After retition of oral application of varied concentration of Sulfisozole, Nifurstryrie Acid and Chloramphenicol compounded in formula feed found that these drugs were effective in saving the pathogen inoculate shrimp. (Kunihiko Shiguenow, 1987).

### (7) Gill Disease

This disease is caused by bacteria. In course of growing at high stocking density in a culture tank, the shrimp happen to encounter this infections of gill disease, especially in spring season. In the initial stages of this disease, the colour of gill turns dull orange-yellow or light brown.

*Treatment*

Bathing in furazolidone at a concentration of 2 ppm-3 ppm is recommended. The bathing is undertaken for two to four consecutive nights, depending on the conditions. (Kunihiko Shigueno 1987).

### (8) Necrosis of the Muscle

*Symptoms*

Prawns showed discolouration spread focally or irregularly on the abdominal segments, the discolouration often extending marginally on each segment. Resulting in the partial degeneration of the tissue, and the shell is peeled off.

Prawns with extensive involvement of the muscle have a marked diminution of body movement and may die within eighteen to twenty-four hours.

*Treatment*

Use certain antibiotics such as furacin, furanace, chloramphenicol and oxytetracycline of $KMnO_4$ could be used in treating in initial stage.

### (9) Luminous Bacterial Diseases

Pathogenic luminous bacteria infect the eggs and tissues of early larvae, postlarvae and juveniles of Penaeid shrimps. *Vibrio harveyi and V.splendidus* are the predominant pathogens for luminescent vibriosis in shrimps. In infected individuals, the pathogens multiply in the haemolymph. The mid-gut content of the spawner was found to be main source of *V.harveyi* infection.

The bacterium *V.harveyi* at $10^{-8}$ CFU $ml^{-1}$ caused 70-100% mortality of *P.merguiensis.* Significant mortality occurred in larve of *P.monodon* within 48 hours when exposed to $10^2$ CFU $ml^{-1}$ of bacteria. While 100% mortality was found at $10^6$ CFU/shrimp. (Jayabalan, *et. al.*, 1996)

*Symptoms*

The larvae of *P.monodon* infected by *V.horveyi* emit greenish luminescence in dark. The thoracic region turns opaque white in weak and moribund larve having feeble swimming movements which finally settle at the bottom to form a luminescent mat. Microscopic examination of the haemocoel and gut would reveal the presence of densely packed bacteria.

*Treatment*

Sodium infustyrenate at a bactericidal dose of 10 mg for 24 hours is found as promising drug as it does not cause morphological deformities in nauplii. However, it causes deformities in older stages of larvae. An exposure of infected *P.monodon* nauplii to 10 ppm pure furazolidone for 6 hours and mysis for 12 hours for 4 to 5 consecutive days have been recommended.

## *Fresh Water Fish Diseases*

Fish mortality often occurs due to environmental stress followed by Parasitic infection and bacterial viral, fungal and Protozoan infection etc. The occurrence and magnitude of infection are closely related to general health of the fishes themselves.

Following are some of the main disease and treatment for recovery.

### *(1) Fungus Infection*

The most common species of fungus affecting fish is *Saprolegnia parasitica.* Live major carps which get mechanically injured during transportation or rough handling are liable to fungal attack unless properly disinfected. Recent expreriments have indicated that probably fungus is a secondary infection and bacterial is the primary cause.

*Symptoms*

Fish becomes weak and lethargic and gradually dies after ulceration or exfoliation of the skin followed by hemorrhage; exposure of the jaw bones; blindness: congestion at the base of pectoral and anal fins; inflammation of the liver or intestine etc.

*Treatment*

Dip treatment in 3% common salt solution or in 1:2000 solution of copper sulfate, or in 1:1000 solution of Potassium permanganate for 5 to 10 minute or until fish shows distress (Hora and Pillay, 1962).

### *(2) Gill Rot*

Gill rot is another disease caused by the filamentous fungus *Branchiomyces sanguinis* which obstructs the veins in the gill filaments. The disease occurs is the hottest time of the year and is associated with excessive organic matter such as putrefying vegetation and shortage of water.

*Symptoms*

Red flecking on the gill filaments which at later stage become grayish-white and may finally drop off altogether leaving the cartilaginous support exposed. Such fish die of suffocation and may be seen gasping for air the surface.

*Treatment*

Prevention of pollution, creasing artificial feeding, letting fresh water into the pond, addition of quick lime (51-102 kg/hr) Jhingran 1982). In case of limited infections 5 to 10 minutes' bath in 3 to 5% salt solution or in 5 ppm $KMnO_4$ solution (Alikunhi, 1957).

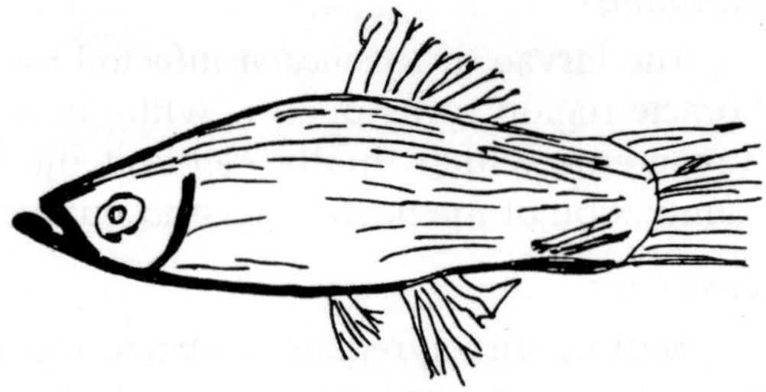

Fig. 36.2 : Finrot

### (3) Fin and Tail Rot

This disease affect to both adult and young. The infection, during its early stage, appears as a white line on the margin of the fin, spreading and imparting frayed appearance to the appendage which eventually putrefies and disintegrated. The disease is contagious and may cause considerable damage.

*Treatment*

Bath for 1-2 minute in 1:2000 solution of Copper sulphate till the fish shows distress. (Hora and Pillay, 1962).

### (4) Dropsy

Infections dropsy is serious epidemical disease affecting major carps. The pathogenic bacterium responsible for the disease is *Aeromonas spp* (Jhingran, 1982). Heavy organic pollution has created pathological features and resolute in fish mortality, (Bali 1983).

The Pathological feautres were as under :

(1) Exophthalmic condition of eyes.
(2) Soft bulging abdomen.
(3) Abdominal scale protruded and gaping
(4) Multiple body lacerations and frayed fins with secondary external infection of the fungus *Saprolegnia*.
(5) Inflammed intestines.
(6) Reddish fluid filing the coelomic space.

This condition expect external fungal infection indicated "Infections dropsy" caused by pathogenic bacteria *Aeromonas spp*.

*Sympotoms*

Accumulation of a fluid inside the body cavity, scale protrusion, exophthalmic condition and inflammation of intestine. (Gopalakrishnan, 1961).

*Treatment*

Following treatment can be given per day :

(1) Five minutes dip in 5 ppm common salt solution.
(2) Local application of concentrated aqueous solution of accriflavin on frayed fins and lacerated surface.

(3) Intro peritoneal injection of 1 ml of 500 ppm aqueous solution of chloramphenicol for the dropsy pathogen.

(4) Five minute bath in 5 ppm aqueous solution of accriflavin as general profilactic measure. Recovery started after 24 hours of the treatment and dropsy condition disappeared within 72 hours. Secondary infection frayed fins and lacerated surface completely healed in 7 days.

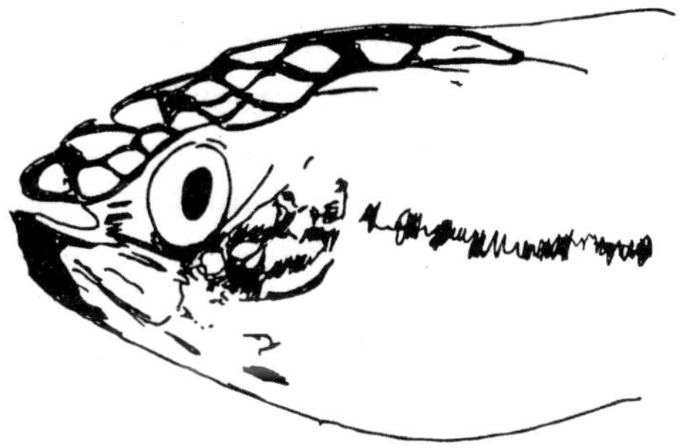

Fig. 36.3 : White Spot

## 5. *White Spot or Ichthyophthiriasis*

This disease is caused in major carps by the protozoan ciliate, *Ichthyophthirius multifilis* which infects different regions of the body externally and causes simple hyperplasia of the site of infection causing formation of pustules. (Jhingran 1982).

### *Symptoms*

Presence of small whitish cysts (1 mm in diameter) on the skin, gills and fins.

### *Treatment*

Dip fish in 1:5000 formalin solution for 7 to 10 days, one hour daily, or in 2% common salt solution for more than 7 days.

Other ciliate infections observed in India are caused by *Trichodina* infecting *indica* different fishes *Scyphidia pyriformis* affecting rohu, catla, mrigal and reba, (Tripathi, 1954) Tripathi (1954). He also reported that *Trichoding indica* and *Scyphidia pyriformis* are killed in 5 to 10 minute by 2-3% common salt solution, 1:1000 acetic acid solution and 1:5000 to 1:6000 formalin solution.

## 6. *Costiasis*

Infection caused by *Costia necatrix* is the most common mastigophoran infection observed in Indian carps. Indications of costiasis are the presence of bluish coating on the skin of the fish and presence of large amount of mucus. The parasite causes irritation and disturbs respiration. Other mastigophoran parasites recorded are *Bodomonas rebae* in mrigal, rohu and catla. (Tripathi 1954).

### *Treatment*

Give bath of 10 minute duration in 3% common salt solution. (Gopalakrishnana, 1964). Tripathi (1954) found that *Bodamonas rebae* was killed in 5 to 10 minutes by 2-3% common salt solution.

## 7. *Worm Disease by Trematodes*

### *(a) Monogenetic Trematodes*

Of various monogenetic trematodes reported in fishes, *Gyrodactyus* and *Dactylogyrus* cause serious infections (Jhingran, 1982) *Gyrodactylus* infects skin and gills where as *Dactylogyrus* affects only the gills.

*Symptoms of Gyrodactylosis*

Fading of colors, drooping of scales, occurrence of excessive mucus on caudal pedancle and fins, peeling of skin etc. When the infection is in gills, it is likely that death results from hyper-secretion of mucus on gill surface, brought about by irritation. Infected fishes may be rubbing their body against hard surface in order to get rid of the parasites (Jhingran, 1982).

*Treatment*

Alternate bath in 1:2000 acetic acid and sodium chloride solutions have been found effective for major carps; dip treatment of 5 minutes duration in 5% common salt or of 5 to 10 minutes duration on in 1:5000 formalin solution. (Gopalakrishnan, 1964).

### 8. Worm Disease by Cestodes

Fishes are very prone to infection by tape worms, through the latter apparently seem to do little harm.

*Symptoms*

Infected fish appeared dull, sickly, and with parts of the alimentary canal swollen or completely choked by cestode cysts. Gall bladder is also affected in some cases.

*Treatment*

Dip in 1:1000 solution of glacial acetic acid and disinfection with 1:10,000 $KMnO_4$. A single treatment with 5 ppm Gammexane was observed to cure most of the major carps infected with leechesand in a similar second treatment removed all parasites. (Jhingran 1982).

### 9. Crustacean Disease

The parasite copepods *Argulus* and *Lernaea* attach themselves to the fish with their body buried in the scale pockets. *Argulus* attaches itself to the body of the fish by means of suckers and hooks, but it can also swim freely in waters several instances of argulosis associated with fish mortality have been reported from Indian ponds.

*Symptoms*

The infected fish become very weak and emaciated. Stunted growth, loss of scale and appearance of red spots at sites of infection.

*Treatment*

To find out *Argulus* infected fish bamboo pales are fixed inside the water so that the affected fish may rub their bodies against them to get rid of their ecto-parasites. A bath in 1:1000 glacial acetic acid for 5 minutes followed by 1% salt solution bath for one hour, cures the fish.

### 10. Attack by Carp Lice Lernaea

*Lernaea* is a minute rod-like external parasite which attaches itself to the host fish any where on the body by means of anchor-like appendages present on the head of the parasite. The anterior portion of the parasite is buried deep into the skin of the fish, and only the posterior region may be visible externally. The majority of carps are subject to infection particularly during the summer months.

*Treatment*

Give bath in weak $KMnO_4$ for 2-3 minutes. 1.0ppm gammexane treatment also gives good results.

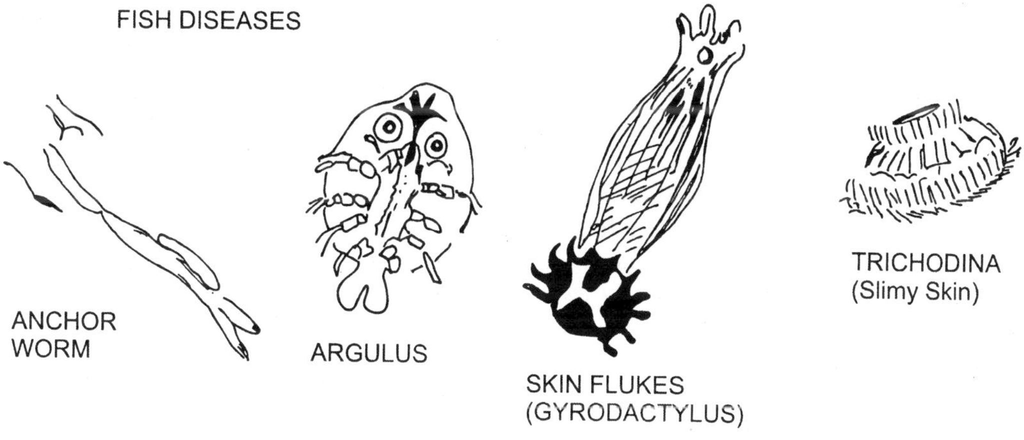


**Fig. 36.4 : (a) Anchor worm (b) Argulus (c) Skin fluckes (d) Trichodina.**

## *11. Furunculosis*

*Symptoms*

Boils and/or ulcers may be present; skin lesion tinged with blood having or not having pus, anorexia, pale gills, with or without lesion, sluggish movements, fraying of the fins, inflamed intestine, myocardial nicrosis, liquefation of the kidney, speen and petechial hemorrahage of the liver may occur.

The aetiological agent of this disease is *Aerpmones salmonicidas*. It is a gram negative.

*Control*

This disease can be prevented or controlled by adopting suitable prophylactic measures, environmental manipulation, breeding resistant varieties, oral vaccination, maintaining optimum stocking, balanced diet, regular disease check up, strict hygiene and proper sanitation are important prophylactic measures.

## Disease in Fresh Water Prawn *M.rosenbergii*

As per information collected from the literature, among the major disease problem in India, parasitic infections account for 70%, while the bacterial and fungle disease problems accounts for 27% and 2.5% respectively. It is necessary to know the disease and their treatment.

Though prawns are less susceptible to diseases than shrimps, several diseases have been identified in the hatcheries and grow-out pond systems. Most of these diseases occur due to poor water quality, nutritionally deficit diet and poor hygienic conditions.

The major diseases which cause heavy mortalities are:

1) Modicycle larvae Diseases (MCD)
2) Exuvia Entrapment Disease (EED)
3) Bacterial necrosis and Microbial Epibiont Disease.

*Treatment*

Some of the antibiotics like Erythromysin, Oxytetracycline, Chloramphnicol etc., have been in use in some of the hatcheries, but the beneficial effects of these drugs have not been established so far. The common epibiont protozoans i.e., *Vorticella Spp., Zoothamnium Spp.* can be easily controlled with 20 ppm. Formaline for 24 hours in a static bath.

In the recent years, Ozone is gaining increased popularity for water purification and sterilization due to its low cost. The Ozone has been in use in various aquaculture systems in different parts of the world. The Central Institute of Fisheries Education (CIFE), Bombay has initiated the use of Ozone as prophylactic and therapeutic agent in *M.rosenbergii* hatchery. Ozone effectively controls the filamentous bacteria, bacterial necrosis and fungus at various larval stages of *M. rosenbergii.* (Reddy, 1997).

*Macrobranchium rosenbergii,* maintained in "grow out" ponds for a long period appears to have more fauling problems and shell diseases.

## Shell Disease

This disease is also known as brown spot, black spot or burn spot. The cause of this disease is considered to be multifactoral, with epicuticulor and cuticular damage or abnormality due to mechanical, nutritional, chemical or other factors followed by secondary bacterial and/or fungal infections. (Kamonporn Tonguthai, 1992).

## Black Gill Disease

This disease is caused by precipitating chemical and nitrogenous waste products, which are involved in melanization of the gills. (Johnson, 1982).

**Table 36.1 : Curently used Chaemotheraoeutic Agents with dosage to control Viricides, Bactercides, Fungicidies, Algicides, and Parasitic Diseases.**

| Chemicals/Drugs | Dose | Treatment for Pathogens |
|---|---|---|
| | VIRICIDES | |
| Betadirie | 100 ml/1 painted on | Virus |
| Benzal Konium Chloride | 0.02g/1 painted on | Virus |
| Volan a adhesive ointment it is applied. | Mixed with cortisone | Lymoicystosis |
| Sodium hypochloride | 10 ppm for 24 hours | Virus. |
| | BACTERICIDES | |
| Nitrofurans | 7-10 g/100 kg of fish/day for 10 days as feed supplement | Bacterial disease. |
| Sulphadiazine+Trimethoprim (Bactrisol) along with Chlorotetracycline (Aurofac-20) | 5g/100kg fish to 7g/100 kg fish for 7 days | Surface ulcerative and systemic type *(A.hydrophila)* and unidentified disease in gills. |
| Oxytetracycline (TM-50) or Chlorotetracycline (Aurofac-20) or Nitrofurans (Bifuran) | 7-10g/100 kgs fish for 10 days. | Columnaris disease. |
| Acriflavin, neutral | 10ppm | Columnaris disease |
| Alcohl iodated Coryne | 100 ppm bath for 10 minutes | Acinetobacter. Bacterial kidney disease. |
| Ampicillinm Sodium | Undermined | Gram+ve & Gram-Ve bacteria. |
| Aquarol | 80 ppm | Bacterial fin rot. |
| Acrycil | Intrapertoneal injection of 1 c.c. of a 1% solution, then 5% solution, three times. | Bacteria (Carp pod) |

*Contd.......*

*Table 36.1 – Contd...*

| Chemicals/Drugs | Dose | Treatment for Pathogens |
|---|---|---|
| Aureomycin | 13 ppm | Flexibacter-Columnaris Vibrio. |
| Bacitracin | 250-300 mg/gal of water. | Gram positive bacterial |
| Betadine | 10 min bath at 10-100 ppm | Acinetobacter, Furunculosis. |
| Benzakonium chloride | 0.5 ppm | Bacterial gill disease, fin rot. Myxobacterial disease. |
| Chloramphenicol | 80 ppm for long duration bath or 50-75mg/kg of body wt. per day for 5-10 days. | Gramve bacterial, Bacteria aemorrhagic septicaemia (BHS) Fin rot, Furunculosis, pseudomonas. |
| Colistin sulphate | Undetermined | Gram positive & pseudomonus. |
| Copper sulphate | 1-2 ppm | Bacterial fin rot |
| Erythomycin | 10 mg/kg of fish/day in fed for 21 days | *Corynbacterium* & several gram+ve & Gram-ve bacteria |
| Ethionamide | As per the condition | Fish Tuberculosis |
| Fumagillin | 0.5 ppm | Bacterial gill disease (BGD) |
| Furanence | 1ppm or 0.4 to 0.9 mg/kg of fish per day for 1 month. | BGD, BHS, Cold water disease, fin rot, pseudomonas, Vibrio anguilarum. |
| Gentamycin sulphate | 20 mg/gal of salt water | Aeromonas, pseudomonas. |
| Griseefulvin | 10 ppm | Gram-ve bacteria |
| Indophores | 100 ppm for 10-15 min bath | Pseudomonas, BHS |
| Kanamycin | 0.02-0.05 g/l of water | Aeromonas, Pseudomonas, mycobacterium, Vibrio. |
| Malachite green | 1g/10 $m^2$ of tank surface 2-3 times alternate day. | Columnaris. |
| Methiolate | Swab of fish ulcer | Fin rot, Ulcer. |
| Neomycin | 250 mg/gal saltwater | Gram+ve & gram-ve bacteria, pseudomonas. |
| Nitrofurantion sodium | As per prevailing condition | Do |
| Nitrofurazone | 10 ppm | Vibrio enguillarum |
| Novobiocin | As per the prevailing condition | Gram+ve & Gram-ve bacteria |
| Oxalinic acid | Orally 3 mg/kg of fishonce, daily for 5 days. | Aeromonas & Columaris |
| Oxytetracycline | 10 ppm in water | As furance |
| Ozone | 26-65 ppm | All types of bacteria |
| Penicillin, potassium | 1 ppm daily | Bacteria (gill infection). |
| Polymycin, B sulphate | 50 mg/kg body wt. In feed. | Gram-ve bacteria. |
| Potassium permanganate | 20 ppm for one hour | Bacteria on skin. |
| Quatenary ammonium | 2-4 ppm active ingredients 1 hr. bath | BGD, Cold water compounds disease columinaris. |
| Streptomycin | As pe Chloramphenicol | Gram+ve & Gram-ve bacteria, BHS, fish tuberculosis, pseudomonas. |
| Sulphamethazin | 200 mg/kg of body weight per day with food for 14 dyas | Columnaris, Furunculosis BKD, fish tuberculosis, pseudomonas |
| Sulfanilamide | 100-250 mg/l doe long duration bath. | All bacterial diseases. |

*Contd........*

*Table 36.1 – Contd...*

| Chemicals/Drugs | Dose | Treatment for Pathogens |
|---|---|---|
| | **FUNGICIDES** | |
| Acrcriflavin, neutral | 3-1 ppm as prolonged bath | *Saprolegnia* |
| Aureomucin | 10-2-ppm in water | *Saprolegnia* |
| Bacitracin | 250-300 mg/gal of water | Marine fungus |
| Calcium cyanamide | 200 g/m$^2$ distributed on the bottom and banks of drained but still we ponds | *Branchiomyces Sp.* |
| Chloramin-T | 10 ppm | *Saprolegnia* |
| Copper sulphate | 0.2 to 0.5 ppm in 2-3 instalments at 3-4 days interval. | *Saprolegnia, Branchiomysis Sanguinis.* |
| Firmalin (40%) | 100 ppm doe foor 3 hrs or 40 ppm for 24 hrs | *Achyla, Aphanomyces, B.demigrans, B.sanguinis* |
| Furance | 0.05-0.2 ppm added to ponds | *Phythium, Saprolegnia* |
| Gention violet | 0.3 pm | *Saprolegnia* |
| Iodoform | 2 ppm | *Saprolegnia* |
| Malachite | 5 ppm as a 1 hr flush daily | *Achyla, Aphanomyces Ichthyophpnus hofen, Saproleginia* |
| Methylene blue | 3 cc of 1% solution in 101 of water as long duration bath (3-5 days) | Antifungal agents on fish eggs, *Saprolegnia* |
| Oxytetracycline | 12.5 ppm dialy | *Saprolegnia* |
| Ozone | As mentioned in Bacteriacides | *I.hoften,* All types of fungus |
| Phenpxethol | 100-200 ppm for 12 hrs | *Saprolegnia* |
| Potasium dichromate | 20 ppm | *Saprolegnia, Achylasis* |
| Potassium permanganate | 5-10 ppm for 2 hrs | *Saprolegnia* |
| Silver nitrate | 10,000 ppm applied topically | *Saprolegnia* |
| Sulfanilamide | As mentioned in bacterial disease. | All types of fungal disease. |
| Sodium chloride | 10-15 g 1 of water as a short bath of 20 minutes. | *Saprolegnia* |
| | **ALGICIDES** | |
| Acrolein | 3 ppm in water | Herbicides |
| Ammonium sulphate | 100 ppm | *Pyrmnesium* |
| Copper sulphate | 2-3 ppm | Blue green algae, *pyrmnesium* |
| Cupravil | 1-2 kg/ha | All types of algae Acetic acid |
| Diquat | 2-4 ppm of the active ingredients for 1 hr. | Blue green algae |
| Ethylmercuric | 0.5 ppm | *Pyrmnesium Phosphate* |
| Pridyle mercuric acetate | 0.2 ppm | *Pyrmnesium phosphate* |
| Paraquat, dichloride | 0.5 ppm | Blue green algae |
| | **PARASITIC** | |
| Sodium chloride | 50 kg/ha, applied in 2-3 installments at 3-4 days intervals | *Myxobolus Spp. Epistylis Spp. Zoothminum Spp.* |

*Contd.......*

*Table 36.1 – Contd...*

| Chemicals/Drugs | Dose | Treatment for Pathogens |
|---|---|---|
| Oxytetracycline or Chloratetracyline | 5g/100kg of fish to prevent secondary bacterial infection | *Myxobolus Spp.* |
| Malathion | 0.2 ppm level in pond applied 2-3 times at 4 days interval | *Dactylogyrus Spp.* *Gyrodactylus Spp.* |
| Malathion or Dichlorovos | 0.15 to 0.20 ppm | *Argulus Spp.* |
| Malathion or Dichlorovos | 0.05 to 0.1 ppm | *Lernaea Spp.* |
| Malathion or Dichlorovos | 0.05 to 0.1 ppm applied in 2-4 installments at 4-7 days interval. | *Ergasilus Spp.* |

**Table 36.2 : Diseases in shrimp.**

**Disease in the breeding larvae**

| S. No. | Bacteria/Attach | Observation | Treatment |
|---|---|---|---|
| 1. | Attach of Vibrio Sp. | Yellowish vermilion and red colour along the nervous system and throughout body. | Oral application of varied concentration of Sulfisozole, Nifurstyrie acid. Chloramphanicol. |
| 2. | White-turbid liver | The midgut gland of the larval becomes white-turbid in general. Infected larvae float inactively on the surface of water. | Oral application of varied Concentration of Sulfisozole, Nifurstyrie acid chloramphanicol. |
| 3. | Saprolegnia and Fusarium solani | Black gill imperfect fungi shrimp dies within two weeks. | No treatment suggested. |
| 4. | Bacterial gill disease | In the initial stage the colour of gill turns dull orange yellow or light brown. | Bathing with Furazolidon at concentration of 2 ppm to 3 ppm. The bathing is under taken for 2 to 4 consecutive nights. |
| 5. | Bacterial disease | Prawn showed white discolouration spread focally or irregularly on the abdominal segment. Discolouration often extending marginally on each segment. Resulting in to degeneration of the tissue. | Fduracin, Furanance chloramphicol and Oxytetracycline either by direct addition or in the feed. |
| 6. | Fungal disease | Fungal growth is seen on the body. | Malachite Green Oxylate (0.006 ppm) or Trifluralin (0.01 ppm) or Firance antibiotics Gallymycin and fungicide, Trifluralin. |
| 7. | *Aeromenes salmonicidas* | Boils and/or ulcer skin lesson tinged with blood having or not having pus, anorexia, Pale gills, sluggish movement | Oral vaccination maintaining optimum stocking, balanced diet, strict hygiene and proper sanitation. Terramycin @ 5-10 mg per 100 mg body weight daily for 7 to 21 days. Intera-periotoned Injection of the antibiotic at a level of 0.02 mg per gm body weight. |

*Source* : Journal of the *Indian Fish Asso,* 18, 1988: 289-298.

## *References*


Anon, 1988. Diseases in shrimp larvae. *J. of Indian. Fish.* Asso. 18, 1988: 289-298.

Anil, T.M., N. Jayabalab, K. Ravi, K. Abokare 1998. Bio-encapsulated Drug Delivery Technique for Treatment of Diseases in Fish and shell fish larval culture. *Fishing Chimes,* 18 (5): 9-10.

Bali, V.K. 1983. Infectious Dropsy and its control in *Cyprinius carpiia. Fishing Chimes* : 3(2): 31.

Basanta Kumar Das, Subas Chandra Mukerjee and Kausalya Kumari Nayak. 1994. Chemotherapeutics in fish diseases. *Fishing chimes,* 14(7): 40-42.

Chandramohan, D.,P. Lakshmanaperulnala Swamy and K. Devedran. 1980. Prawn Diseases. *Seafood Export Journal,* XII (7): 9-19.

Gopalakrishnan, V. 1964. Recent developments in the prevention and control of parasites of fishes cultured in Indian waters. *Proc. Zool. Soc. Calcutta* 17: 95-100.

Gopal Rao, K.C.V. Mohan and D. Seenappa. 1990. The use of chemotherapeutic agents in fish culture in India. In Diseases in Asian Aquaculture-I, pp. 510-511.

Hora, S.L. and Pillai, T.V.R. 1962. Handbook on fish culture in the Indo-pacific region. FAO Fish. *Biol. Tech. Pap.* (14): 204.

Jayabalan, N., Seema Bala and K. Ravi. 1996. Luminous Bacterial Disease in Shrimps. *Fishing chimes,* 15(11): 32-33.

Jhingran, V.G. 1982. Fish and Fisheries of India. Book published by Hindustan publishing Corporation (I): 486-491.

Karunasagar, I. and Indrani Karunasagar 1997. Polymerase chain Reaction (PCR) based methods for diagnosis of viral diseases of shrimps. *Fishing chimes,* 17 (7): 13-14.

Kunihiko Shigneno, 1987. Shrimp farming in Japan pp. 111-118. Book published by Association for International Technical Promotion.

Mohan, C.V. and K.M. Shankar, 1997. White Spot viral disease management in shrimps: Needs for scientific approach, *Fishing chimes,* 17 (2): 41-43.

Nammalwar, P., and T Thangaraj. 1980. Problems of diseases on Marine Fish and Prawn Farming in India. *Seafood Export Journal,* XII (12) : 9-12.

Panchyunthapani, D and Kenneth E.Corpron. 1997. Rapid diagnostic methods for viral diseases of shrimp. *Fishing chimes,* 17 (2) 31.

Pillai, Thankappan, C. 1980. Fish and shell fish diseases in culture system. I Concept of diseasse and pathogen. Seafood Export journal, XII (2) : 13-14.

Pillai, Thankappan. C. 1980. Fish and shell disease n culture system. II Heterophic bacteria and kinds of infection *Seafood Export Journal,* XII (9) : 21-23.

Pillai, .C. 1980. Fish and shell fish diseaases in culture system III. Defense against infection.
*Seafood Export Journal,* XII (10) : 19-21.

Pillai, T.C. 1983. Fish and shell fish diseases in culuture system XI Furunculosis. *Seafood Export Journal* XV (2) ; 13-14.

Touraki, M. Mourelatos, S. Karamalidou, G., Kalattzopoulous, and Kastritsis, C. 1996. Bio-encapsulation of chemotherapeutics in *Artemia* as a means of prevention and treatment of infectious diseases of marine fish fry. *Aquaculture Engineering,* 15(2) : 133-147.

# 37
# CRAFT AND GEAR

## Fishing Boats

Depending upon the development and modernization, size, type of fishing boat and design of boat changes. Some time according to suitability of local fishing ground, type of availability of fish and type of net being used for certain fishery, choice and liking of local fishermen, size and design of fishing boats are adopted. In this chapter we will study area wise different fishing boat used in past and presently, with latest adoption.

Since ancient fishing boats were prepared from wood of different varieties and in different shapes as per requirement. Now a day Fiberglass and Ferocement fishing boats are also introduced. However, basic parts of boats remain more or less unchanged. They are as under (Fig. 37.I.1)

| | | |
|---|---|---|
| 1) Keel | 2) Hog | 3) Fore dead wood |
| 4) Fore foot knee | 5) Lines of planking | 6) Aft dead wood |
| 7) Shaft log | 8) Stern post | 9) Horn Timber |
| 10) Transom post | 11) Transom | 12) Apron |
| 13) Stem | | |

Name of other parts are as under:

| | |
|---|---|
| 1) Hull | Main body of boat. |
| 2) Deck | Upper surface of the boat |
| 3) Gunwale | Border of the deck, which gives extra height to boat. |
| 4) Mast | Pole in the center of deck where lighting facililties are provided |
| 5) Fish hold | Space to preserve fish. |
| 6) Engine room | In mechanized boat separate engine room is provided. |
| 7) Star board | Right side of boat |
| 8) Left board | Left side of boat |
| 9) Fore (Bow) | Front part |
| 10) Aft | Back side. |

The diagram, design and other specifications of different boats being used in Gujarat are given from Fig. 37.I.2 to 37.I.6. Different types of boats are used in different districts of Gujarat. Table 37.I.2 gives the details of such boats. From the table we can see that now a day trawler boat, and fiberglass boats are common in all the districts of Gujarat, which shows the trend of modernization. Total number of fishing boat in different district as in

Classification of fishing Gear material

- Natural Fibers
  - Vegetable Fibers
    - Seed Fiber
      1. Cotton
      2. Kapok
    - Stalk Fiber
      1. Palx
      2. Hemp
      3. Jate
      4. Kennaf
      5. Remie
      6. Linen
    - Leaf
      1. Sisal
      2. Manila
      3. Yucca
      4. Hemp
    - Fruit coir
  - Animal Fiber
    1. Wool
    2. Hair
    3. Silk
- Natural Polymer
  - Cellulose
    - Pegenerated Cellulose
      1. Viscose
      2. Cupronium
    - Cellulose Derivatives
      1. Hatate
      2. Tricetate
  - Protein Base
    - Vegetable Base
      1. Algenate
      2. Vicara
      3. Ardil
      4. Silkool
      5. Azion
    - Animal Base
      1. Lamital
      2. Fibrolon
      3. Casenka
      4. Casien
      5. Merilova
      6. Wipolan
- Chemical Fiber
  - Synthetic polymers
    - Polycondensation
      - Polyester
        1. Terylene
        2. Dacron
        3. Teirtal
        4. Terygal
        5. Terilenka
        6. Diolen
        7. Trevira
        8. Lanón
      - Polyamide
        1. Nylon Perlon
        2. Grilon
        3. Enkalon
        4. Lilon caprolan
        5. Delifion
        6. Sleelon
      - Mixed Polymers
        1. Dynol
        2. Saran
        3. Vinyyon
        4. Harlon
        5. Veral
    - Polymeric Compound
      - Polyvinyl Chloride
        1. PCU
        2. Pece
        3. Rhonyl
        4. Thermovy
        5. Vinyon H.H.
        6. Pecea
      - Polyvinyl Alcohol
        1. Vinylon
        2. Kuralon
        3. Kannebion
        4. Croomona
        5. Mewlon
        6. Synthofil
        7. Woodlon
      - Polyacryl Nitrate
        1. Orlon
        2. Acrilan
        3. Pan
        4. Dralon
        5. Reelon
        6. Courtelle
        7. Crylor
        8. Dolan
        9. Prelana
        10. Wolerylen
      - Polyethylene
        1. Polythene
        2. Courlene
        3. Reevon
        4. Wynene
        5. Teflon
    - Polyddicative compound
      - Polyethylene
        1. Perlon
        2. U-Fiber-32
- Mineral Source
  1. Iron
  2. Zinc
  3. Aluminum
  4. Lead
- Inorganic
  1. Glass
  2. Quarts
  3. Basalt
  4. Asbestos

**Table : 37.I.1 : Classification of fishing Gear material**

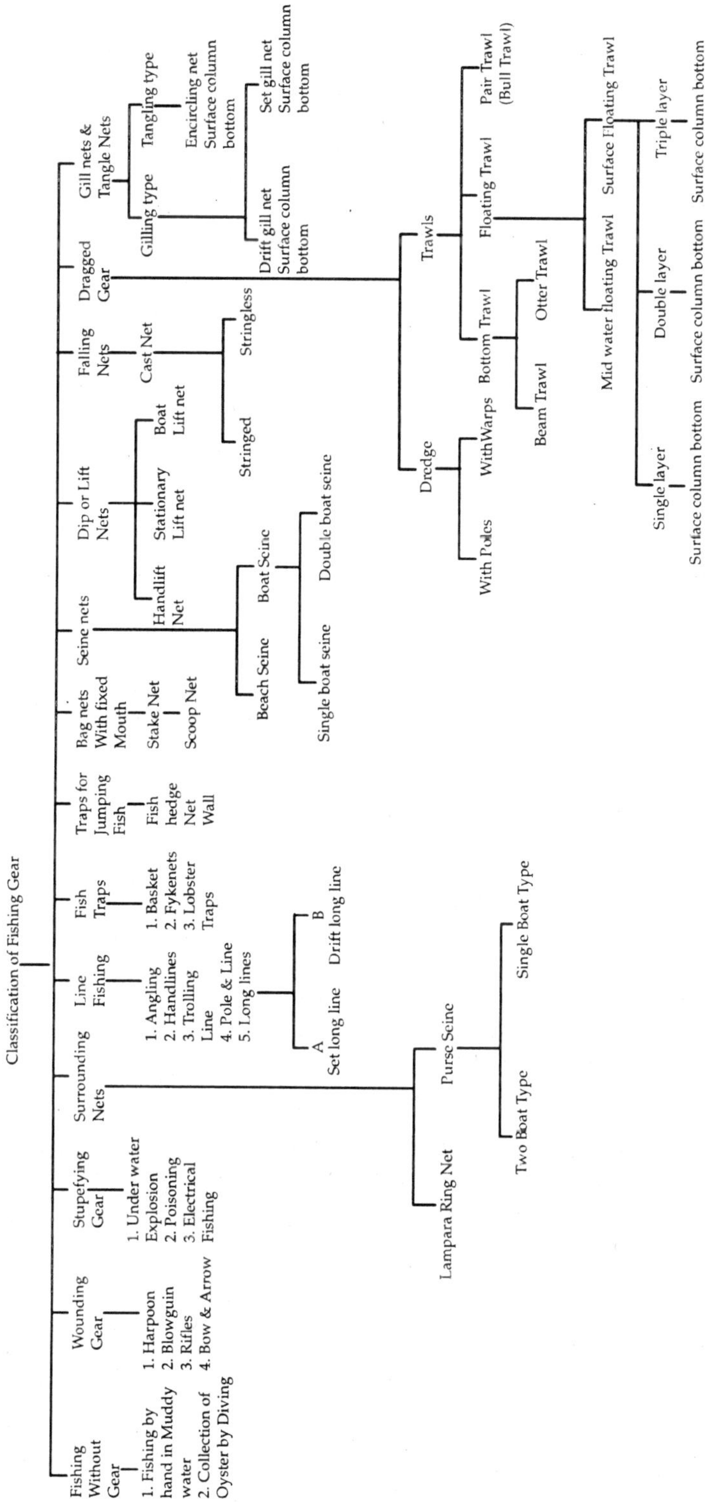


**Table : 37.I.2 : Calssification of Fishing Gear**

1996-97 are given in table 37.I.3. While table 37.I.4 gives the category wise information of the fishing boat of Gujarat State since 1990. From this table it can be seen that the growth of mechanization is fast. Table 37.I.5 gives the details of district wise fishing boats of Gujarat as on 31.3.98. Growth of mechanized and non-mechanized boats in Gujarat since 1960 is given in table 37.I.6. From table it can be seen that growth of mechanization is 2080% compared to 1960 whille for non-mechanized boat it only 174%.

**Table 37.I.3 : Details of Mechanized boats being used in Gujarat**

| S. No. | Name of Boat | Size of boat | Type of Engine | Speed knot/hr. |
|---|---|---|---|---|
| 1 | Trawler | (i) 48' x 12' x 6'<br>(ii) 52' x 14' x 7' | (i) 88 H.P.<br>(ii) 165 H.P. | |
| 2 | Fiber glass | – | – | – |
| 3 | Kutch Hoda | 37'10" x 8'7" x 4'5" | 20 HP | 7 Knot |
| 4 | Kutch Machwa | 21'5" x 6'6" x 3'0" | 10 HP | 6 Knot |
| 5 | Porbandar Machwa | 28'1" x 6'5" x 3'1.5" | 6 HP | 6 Knot |
| 6 | Veraval Lodhia | 33'9" x 6'3" x 2'9" | 6 to 9 HP | 6 Knot |
| 7 | Veraval-Vanakbara Type boat | 40'9" x 9'6" x 4'6" | 30 HP | 7.5 knot. |
| 8 | Jafrabad Machwa | 36'10" x 8'10" x 3'10" | 27 HP | 7.25 Knot |
| 9 | Jafrabad Hoda | 34' x 8'4" x 3' | 15 HP | 6.75 Knot |
| 10 | Safal Lodhia | 35' x 8' x 3'9" | 20 HP | 7 Knot |
| 11 | Safal Machwa | 31' x 8'6" x 4' | 20 HP | 7 Knot |
| 12 | Gujarat Vahan | 40' x 11' x 4' | 35 HP | 7 Knot |
| 13 | Salaya Carrier boat | 63' x 12'6" x 6'2" | 75 to 100 HP | 8.5 Knot |
| 14 | Halar Machwa | 23'10" x 7' x 3'10" | 10 HP | 6 Knot |

**Table 37.I.4 : Details of Local Boat used in different district of Gujarat**

| | Name of District | Type of boat being used |
|---|---|---|
| 1. | Amrali | Jafrabad Machwa, Vanakbori Machwa, Trawler boat. |
| 2. | Junagadh | Veraval Lodhia, Porbander Machwa, Dugout cannos, fiber glass boat. |
| 3. | Jamnagar | Halar Machwa, Fiber glass boat, Trawler boat, Dingi boat. |
| 4. | Rajkot | Navlakhi Hoda, Flat bottom Malia Hoda, Trawler boat, Fiber glass boat. |
| 5. | Kutch | Kutchhi Machwa, Kutchhi Hoda, Flat bottom boat, Surajbari boat, Trawlers and Fiber glass boat. |
| 6. | Valsad and Surat | Gujarat Vahan, Trawler boat, Fiber glass boat. |

**Table 37.I.5. : Category wise information of the fishing boats of Gujarat state**

| Year | Trawler | Gill netters | FRP | OBM | Other | Total mech. boats | Total non-mech. boats | Total boats |
|---|---|---|---|---|---|---|---|---|
| 1990-91 | 2814 | 1946 | 1044 | 1838 | 498 | 8140 | 8677 | 16817 |
| 1991-92 | 3855 | 2211 | 1386 | 1862 | 510 | 9024 | 8825 | 17849 |
| 1992-93 | 3456 | 2315 | 1618 | 1957 | 524 | 9870 | 8745 | 18615 |
| 1993-94 | 3941 | 3058 | 2161 | 1813 | 530 | 11503 | 8323 | 19826 |
| 1994-95 | 4634 | 3110 | 2545 | 1814 | 545 | 12648 | 8370 | 21018 |
| 1995-96 | 5685 | 3143 | 2810 | 1817 | 562 | 14017 | 8646 | 22663 |
| 1996-97 | 6027 | 3205 | 3012 | 1827 | 600 | 14677 | 8851 | 23522 |

*Source* . Reprinted with the permission of Commissionerate of Fisheries Govt. of Gujarat Gandhinagar.

Table 37.I.6 : Growth of Fishing Boats in Gujarat since 1960

| Details | 1960-70 | 1970-80 | 1980-90 | 1990-97 |
|---|---|---|---|---|
| Total mechanized boats Nos | 314 to 1276 | 1276 to 3459 | 3459 to 8140 | 8240 to 14671 |
| No of boat increased | 962 | 2183 | 4681 | 6531 |
| Growth percentage in decade | 306.36 | 171.08 | 135.33 | 80.23 |
| Growth % compared to 1960 | 306.36 | 695.22 | 1490.76 | 2080 |
| Total non-mechanized boat nos | 3217 to 3691 | 3691 to 6023 | 6023 to 8677 | 8677 to 8851 |
| No of boat increased | 474 | 2332 | 2654 | 174 |
| Growth % in decade | 14.73 | 63.18 | 44.06 | 2.0 |
| Growth compared to 1960. | 14.73 | 72.49 | 43.83 | 5.4 |

*Source* : Dept. of Fisheries, Govt of Gujarat. Annual Reports.

From above Table 37.I.7. it can be seen that growth of mechanization is 2080% compared 1960 while for non-mechanized boat is only 174%.

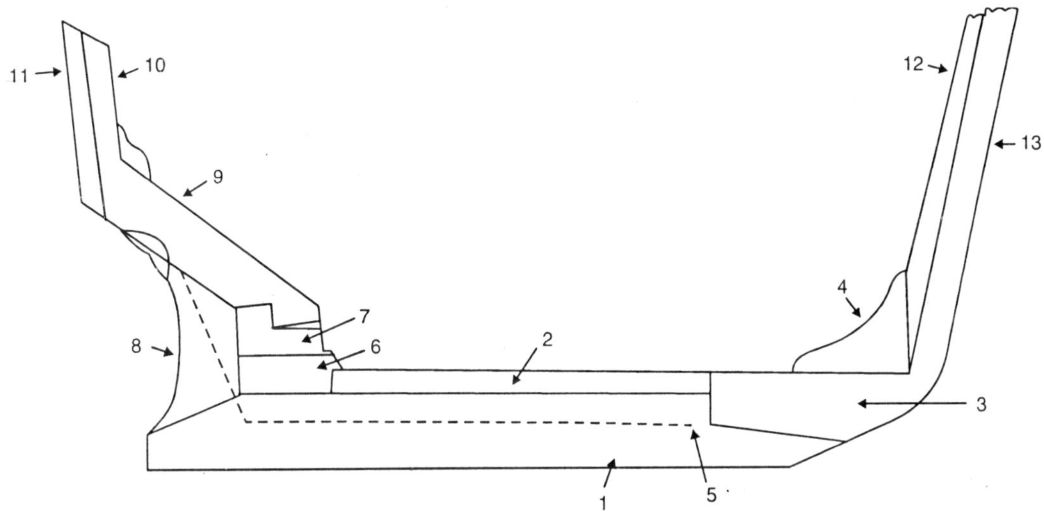


(1) Keel (2) Hog (3) Fore dead wood (4) Fore foot knee (5) Lines of planking
(6) Aft dead wood (7) Shaft log (8) Stern post (9) Horn Timber (10) Transom post
(11) Transom (12) Apron (13) Stem

Fig. 37.I.1 : Terminology used in Boat Building

Name of boat, characteristics of the boat and general size of boat varies in different States. Similarly in each State different types of boats are being used. Table 37.I.7 gives the details of fishing crafts used in different States of India with their local name, characteristics of the crafts, length etc. (Pravin *et al.*, 1998)

Now details of different type of boats used in Gujarat are given hereunder.

Table 37.I.7 : Statement showing details of District wise fishing boats of Gujarat as on 31-3-98

| | | Mechanised | | | | | | | Non-mechanised. | | | | | | | |
|---|---|---|---|---|---|---|---|---|---|---|---|---|---|---|---|---|
| Sr. No. | District | Trawler | Gill net | FRP/ BM | FRP/ OBM | Wooden/ OBM | Other | Total | Vahar | Flat bottom | Dug out | Tarapa | Tin boat | Other | Total | Total of 9 and 16 |
| 1 | 2 | 3 | 4 | 5 | 6 | 7 | 8 | 9 | 10 | 11 | 12 | 13 | 14 | 15 | 16 | 17 |
| 1 | Junagadh | 4871 | 246 | 26 | 2501 | 1443 | 298 | 9345 | - | 130 | 112 | - | - | 175 | 417 | 9762 |
| 2 | Jamnagar | 154 | 480 | - | 279 | 61 | - | 974 | - | - | 3 | - | - | 298 | 301 | 1275 |
| 3 | Bhuj (Kutch) | 18 | 471 | 122 | 1 | 57 | - | 669 | 232 | 136 | 1 | - | - | 154 | 523 | 1192 |
| 4 | Rajkot | - | 40 | 41 | - | - | - | 81 | 73 | 528 | - | - | - | 109 | 710 | 791 |
| 5 | Amerai | 207 | 13 | 32 | 342 | 293 | 330 | 1217 | - | - | 14 | - | - | 160 | 174 | 1891 |
| 6 | Bhavnagar | - | 16 | 31 | - | - | - | 47 | - | - | - | - | - | 74 | 74 | 121 |
| 7 | Surendranagar | - | - | - | - | - | - | - | - | - | - | - | - | 324 | 324 | 324 |
| 8 | Banaskncha | - | - | - | - | - | - | - | - | - | - | - | 55 | - | 55 | 55 |
| 9 | Kheda | - | - | - | - | - | - | - | 9 | 13 | 43 | - | - | - | 136 | 136 |
| 10 | Vadodra | - | - | 1 | 1 | - | - | 2 | - | 137 | 327 | 128 | 100 | - | 692 | 694 |
| 11 | Panchmahal | - | - | - | - | - | - | - | - | - | - | - | 519 | 521 | 1040 | 1040 |
| 12 | Bharuch | - | - | 25 | 21 | - | - | 46 | 7 | 1384 | 5 | - | - | - | 1396 | 1442 |
| 13 | Surat | - | 23 | 1 | 27 | - | - | 51 | 207 | 689 | 1 | - | - | 902 | 1799 | 1850 |
| 14 | Valsad | 1180 | 1986 | 100 | 0 | 0 | 0 | 3266 | 390 | 96 | 90 | - | - | 6 | 582 | 3848 |
| 15 | Sabarkantha | - | - | - | - | - | - | - | - | - | - | - | 243 | - | 243 | 243 |
| 16 | Ahmedabad | - | - | - | - | - | - | - | - | 327 | - | - | - | - | 327 | 327 |
| 17 | Mehasana | - | - | - | - | - | - | - | - | - | - | - | 125 | - | 125 | 125 |
| | Total | 6390 | 3275 | 379 | 3172 | 1854 | 628 | 15698 | 918 | 3440 | 596 | 128 | 1042 | 2794 | 9818 | 24616 |

**Table 37.1.8 : Details of fishing crafts used in different states of India**

| State | Name of the Boat | Characterstics of the Crafts | Length |
|---|---|---|---|
| Gujarat coast | Kotia, Machwa, coaster nauris & Dhangies | Great fore reaching bow, a deep foot and racking stem, Latin, sail plank built, square | 31' * 8.5' * 4' |
| | Lodhia | Plank built square | 35' * 4' * 3'9" |
| | Maliya | Premitive double ended flat bottomed, carvel planked | 18' * 22' * 4.42.2' – |
| | Gujarat machwa | Plank built | |
| Maharashtra coast | Large Machwa or Bombay Battela padar | Carvel planked | 47' |
| | Pattamar | Great sheer foxe and aft, large curved overhand at bow and great beam in quarters, stem rounded, two masted | |
| | Konkan fishing boat | | 30' to 45' |
| | Satpati Hody | Double ended, carvel shape, out rigger used | 22' * 40' * 2.5' * 8' |
| | Ratnagiri Machwa out riggercanoe | Single masted, Lateen rigged dug out canoe | |
| Kondan & Kerala coast | Beppa dug out odam, Odam dug out | Dug our, Plank, built canoe, Steering being effected by a big paddle, used for propallsion as well as control, no rudder | |
| | Malabar dug out smaller | | 12' to 14' |
| Madras, Andhra, | Cattamaram | 4 to 5 logs fied together, Raft type | 20' to 40' |
| Orissa | Boat cattamaram | Improved from cattamaram | 23' * 3' |
| | Boat canoe | Like Malabar dug-out spread by wedged and heghiened by faired 9" to 10" wash-strake. | 22.5' * 2.5' |
| Madras | Vallam | Long, narrow and keel less, boat canoes side rounded | Largest 40' Smallest 27' |
| | Dhoni Tuticorin | Plank built and Both and sharp | 29' |
| | Dug-out canoe out-rigger | Carvel built, long narrow, used for pearl fishing | |
| Coromandal coast | Vala Vathai | Narrow, carvel-built boats or canoe. | 43' * 4.5' * 2.5' |
| | Palagai, Kallu, Vattai | Plank buit canoe, similar to dug-out | |
| | Kalla dhoni | Like a thief it runs fast | |
| | Cottamaram | The developed form is large Pariyamaram used in fishing with thuri, edu and pamthe valai nets | |
| | Teppas | Another type of Cattamaram | 22' * 4.6' |
| | Masula | Planks stiched | 21' to 2.5' * 5' to 5'.10" |
| | Navas | Narrow, round bottom boat without a keel heavily framed and carvel planked with a short fore deck and long aft deck | 31'.2" * 5'.2" * 2'.3" |
| West Bengal | Bat chari | Stem as high as bow | 28' to 60' |
| | Chot boats | Carvel plank | 34' |
| | Diamond harbour boat | The boat has a fore deck and art deck | 20' |
| Andaman | Dug-out out rigger canoe | Smaller hulls, fashioned on identical outrigger, frame is ingle and booms are slended poles | |
| Nicobar | Out rigger canoe | Radicaly different from that of the Andamans. These are better built and more elegant in form and never possess more than two booms | |

*Source* : Pravin *et. al.*, (1998).

**Table 37.I.9 : Main varieties of wood being used for the construction of fishing boats**

| Sr. No. | Common English Name | Local Name | Scientific Name | Class |
|---|---|---|---|---|
| 1. | Teak, Indian Oak, ship tree | Sag, Sagun, Sagwan | *Tectona gradis* | Verbenaceae |
| 2. | Benteak, Nana wood | Bili Nandi, Nana, Nandi, Vanda, Benteak | *Lagerstroemia lanceolata* | Lathraceae |
| 3. | Anjeli wood | Aainee, Anjeli, Patfanas, Chanfans | *Artocarpus hiresuta* | Moraceae |
| 4. | Sajjad | Sajjad | *Terminalia tomentosa* | Combretaceae |
| 5. | Mango | Amab, Ambi | *Mangifera indica.* | Anacardiaceae |

They have calculated composition of motorized and mechanized crafts in Gujarat. According to them trawlers 41%, Gill netters are 22%, FRP (OBM) are 19%, wooden (OBM) are contributing 13%, Dol netters are 4%, FRP (IBM) boats was being introducing and contributing only 1%, Different varieties of wook like Teak, Indian Oak, Beteek, Nana wood, Anjali wood, Sajjad, Mango etc., are being used for the construction of fishing boats. Their English name, local name, Scientific name and class are given in Table 37.I.8.

To conserve forest, to save trees and environment, after 1980's along with wooden fishing boats different size of fiberglass boats were introduced. Fiberglass Reinforced Plastic (FRP) is being used as raw material for marking trawlers, canoes, rescue boats etc. FRP boats are robust, elegant, seaworthy, economic and lightweight. Specification of newly introduced FRP boats are given in table 37.I.9. Photographs of some of the FRP boat are given as Fig 37.I.7 to 37.I.12.

## Dug-Out Boat : (Anon 1962)

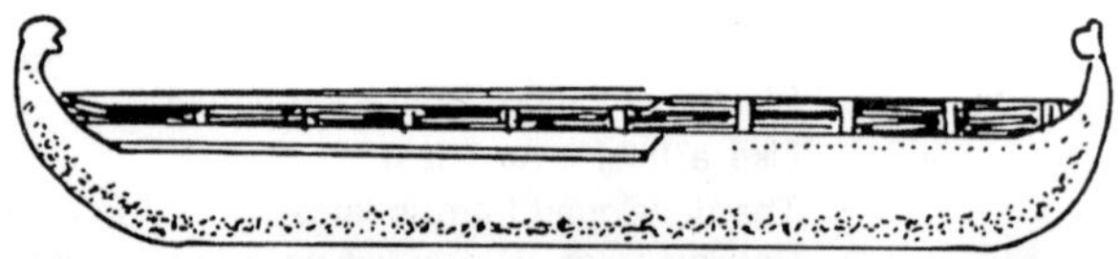

Fig : 37.I.2 Dug-out Canoe

These types of boats are scooped from one single wooden log. Backside (Aapt) of this boat is curved, bunt and thick. While front side (Fore) is pointed and little bit elevated. As there is no keel in this boat beaching on sandy shore is very easy. These are mainly used on Kerala and Kanara coast, and also between Coalachel and Kathiawar. The large dugout (*Vanchi, odams*) form the main fishing crafts of the Malbar coast, operating a variety of nets. This boat is non-mechanized boat and sales are used.

Specifications of this boat are as under:

| | |
|---|---|
| OAL | 24′ to 26′ |
| Beam | 3.25′ to 3.75′ |
| Depth | 2.0′ to 2.25′ |
| Draft | 1.25′ to 1.5′ |

Because of shortage of wood, now a days fiber glass is used as replacement of wood and accordingly it is fabricated.

**Table 37.I.10 : Specification of some newly introduced FRP boats**

| Sr. No. | Name of Boat | OAL | Beam | Depth |
|---|---|---|---|---|
| 1. | 49 ft. Trawler | 49' | 12.75' | 7.0' |
| 2. | 46 ft. Trawler | 46' | 12.75' | 7.0' |
| 3. | 37 ft. Trawler cum gillnetter | 37' | 11'.03" | 5'6" |
| 4. | 20 ft. Chine Cano type boat | 20' | 06' | 02' |
| 5. | 10 ft. Dinghy | 10' | 4'10" | 1'6" |
| 6. | 8 ft. Dinghy U bottom | 8' | 3'10" | 1'10" |
| 7. | 36'6" Fishing canoe | 36'6" | 5'4" | 2'9" |
| 8. | 32'6" Fishing Canoe | 32'6" | 5' | 2'9" |
| 9. | 30' Fishing Canoe | 30' | 5'3" | 2'9" |
| 10. | 21' Fishing Canoe | 21' | 4' | 2'3" |
| 11. | 16' Fishing Canoe | 16' | 2' | 11" |

*Source* : M/S Elite Shipyard Pvt. Ltd. Veraval (Personal communication)

## Flat Bottom "Malia" Boat

Specifications of the boat:

OAL 18′ to 21′
Beam 4′ to 4.5′
Depth 1′ to 1.25′
Draft 1′ to 1.25′

Malia boats are used in Gulf of Kutch in Malia area hence they are called "Malia boat". Where sea water are shallow. It is mainly used for prawn fishing. Propulsion is with sales. This boat is classified in non-mechanized boats. Boat is made with wooden planks. In the boat total three left to right wooden planks are provided for sitting arrangements. (Chhaya 1990).

**Fig. 37.I.3 : Malia flat bottom boat**

## Plank Built Boat

Specifications of the boat:

OAL 6 m to 14 m.
With 0.9 m to 3.3 m
Depth 0.6 m to 1.0 m

These are dugouts which are further enlarged with planks on the sides Propulsion is with the help of sails. This boat is also classified in non-mechanized boats.

(Anon 1962).

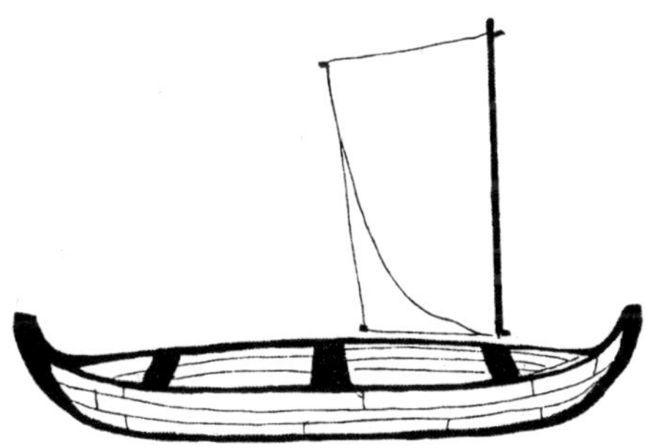

**Fig. 37.I.4(a) : Plank-built boat**

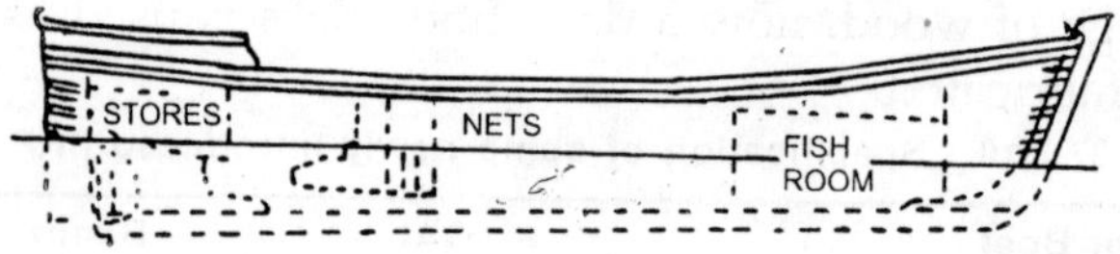


Fig. 37.I.4(b) : Second variety of Machwa – A bult-up boat of Saurashtra

## Halar Machwa

Specifications of the boat:

| | |
|---|---|
| OAL | 18′ to 32′ |
| Beam | 5′ to 9.5′ |
| Depth | 3′ to 6.5′ |
| Draft | 1.25′ to 2.75′ |

This type of boats are being used in Bat-Dwarka, Salaya and in Jamnagar District. This boat is classified in mechanized boats. In this boat In Board Machine (IBM) are installed between 10 to 50 HP capacity. In the construction of this boat teak wood planks are used while inside knee, hig, shoft log etc., are prepared from "Sajad" wood. Front part (Bow) is pointed while Aft is flat and square prepared with horizontal wooden planks. Mast of this boat is bigger than the OAL. Mast is fitted in the center and very big sail is fitted for additional propulsion, when needed. (Chhaya, 1990).

Fig : 37.I.5 Halar Machwa (Jadia)

## Navlakhi Hoda

Specifications of the boat:

| | |
|---|---|
| OAL | 16′ to 18′ |
| Beam | 4′ to 4.5′ |
| Depth | 3′ to 3.5′ |
| Draft | 1′ to 1.25′ |

This type of boat are being used in Gulf of Kutch, Navlakhi-Kandla area. Hence it is called Navlakhi Hoda. This boat is classified in mechanized boats. In this boat IBM is installed. For part of boat is pointed center part is bulged and broad. While back side flat on which radar is fitted to give proper direction to boat movement. For maintenance every year cocking is to be done.

Fig : 37.I.6 : Navakhi Hoda

## Vahan or Matada

In the northern region of Gujarat, plank-built bots of 10 to 14 m length, 1.8 to 2.3 m depth and a beam of 3 to 4 m were used as "Vahan or Matada". The draught of Vahan was generally 1 to 2 m. Now a days the use of these "Vahan" is reduced.

This type of fishing boats are being operated in South Gujarat at Navsari, Bilimora, Kolak, Umarsadi etc. Similarly it was used in North Gujarat as well as in Gulf of Khambhat

Fig : 37.I.7 : 32' Fiberglass Boat

Fig : 37.I.8 : 10'–0" FRP Dinghy Boat

Fig : 37.I.9 : 20'–0" FRP Fish Transport Boat

Fig : 37.I.10 : 37'–0" FRP Trawler

Fig : 37.I.11 : 35'–0" FRP Fishing Canoe

Fig : 37.I.12 : 49' FRP Trawler

*Source* : M/s Elite Shipyard Pvt. Ltd. Veraval

(Cambay). Fishermen of South Gujarat fishing in Jafrabad, Nawabandar, Rajpara area for Bombay duck fishing with this type of boat during November to March.

With this big boat (OAL 32′ to 45′) small boats known as "Doria" are being used. Fishing is done by "Doria" and fish are stored in this big size "Matada".(Chhaya 1990).

## Fishing Boats of India

### Catamaran

It is keel-less raft formed by rigging together several logs, which are curved and shaped like a canoe. One end of the craft is shaped in to a canoe rising above the water, forming the stem from where the rudder is used. The use of the catamaran is restricted to the east coast from Orissa to Cape Comorin with a short extension northwards on the Kerala coast. Four types of catamarans are known:

(a) ***Orissa and Ganjam type*** It is made up of five logs which are not tied together by rope but are pegged with wood; it is boat shaped.

(b) ***Andhra type*** It is larger than the preceding type (5-7 m. in length) and made of heavy wood. Planks used in fitting the sides often include strong median logs.

(c) ***Coromandal type*** It is made up of 3-5 logs, with many variations in pattern. It is most commonly used around Madras. A specialized type is the seven logged catamaran or *kolamaram* used in the flying fish fishery off Nagapatnam.

(d) ***Boat-catamaran*** It consists of three logs fitted into a regular boat shape and is used on the coast around Manapam and Mukkr areas. Wide variation of this type is found in the Tuticorin, Cape Comorin and Colachel areas.

(e) ***Masula boat*** It is non-rigid boat constructed with planks, sewn together with coir rope, but without frames or ribs so as to withstand the severe knocking of the surfs. Thre are various pattern, viz., *bar* boats in Orissa and Padava or *Padaguon* the Andhra coast. A variant with ribs inside has been developed in the area between Kakinada and Masulipatnam. Masula boats are up to 9 m in length, although generally smaller.

***Dinghi and nauka :*** These are curved boats of Orissa and W.Bengal. Nauka, which are well designed and constructed up to a size of 13 m x 3 m x 2 mare quit spacious and are used for a variety of purposes including fishing operations.

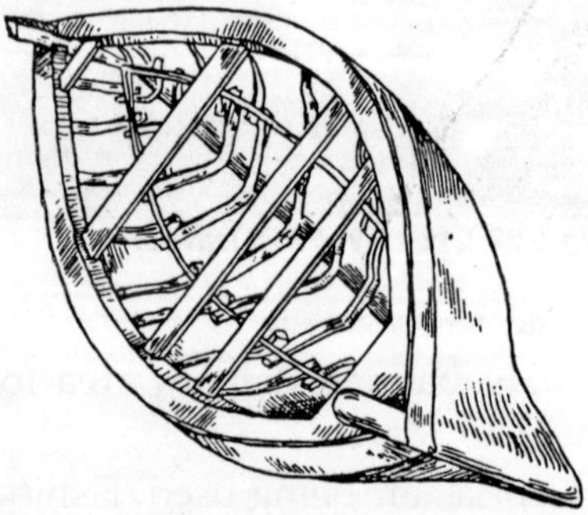

**Fig : 37.I.13 : Dinghi - A Carvel Boat of N. Orissa and W. Bengal**

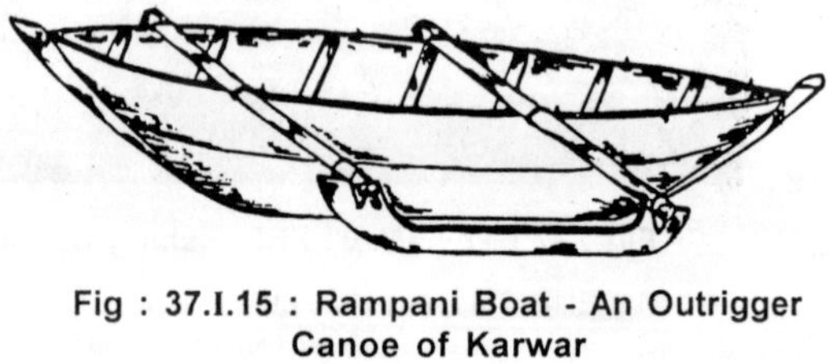

**Fig : 37.I.15 : Rampani Boat - An Outrigger Canoe of Karwar**

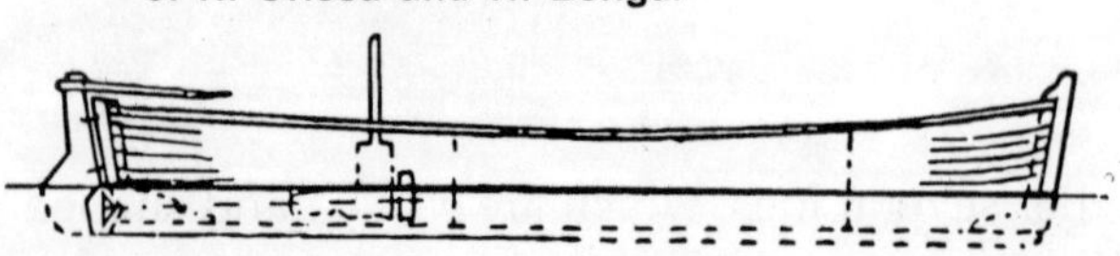

**Fig : 37.I.14 : Tuticorin Type Fishing Boat**

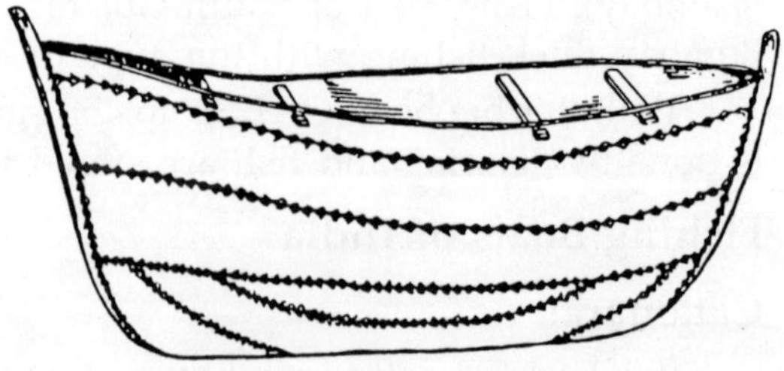

**Fig : 37.I.16 : Padava - A Masula Boat of Andhra Pradesh.**

*Tutticorin boats :* Also called the fishing-luggers, they are carvel boats (11m x 2m x 1m), which are sea-worthy in inshore waters. They are used more as mother ships and cargo boats than directly for fishing. (Anon, 1962).

## Different parameters, which are harmful to boat wood

1. Marine Bacteria
2. Marine Fungus
3. Wood Borers
4. Fouling Organisms

### *Marine Bacteria*

There are many types of bacteria in marine water. They can be digest cellulose. Generally they are found in mud and in upper column of the water. *Siyefiga* group of bacteria is well known for such activity. *Nocdria, Streptomysis* etc., bacteria can also digest cellulose.

#### *Protection*

Use of Ethyl cellulose, Cellulose acetate, Carboxyl Methyl Cellulose and Cellulose Try acetate are good against such bacterial attack.

### *Marine Fungus*

Marine fungus attached to wood of boat. Where they secretes digestive enzymes, which digest cellulose. *Fycomaysis* and *Duteromysetis sp.* are such fungus, which are harmful to wood.

### *Ship worms and woodborers*

Some marine mollusks are known as shipworms.

### *Fouler*

There are certain marine animals, which stick on the surface of boat. Following marine animals are known for such damages. They harm wood of the boat. It is necessary that surface of boat is smooth for easy sailing. When such animal sticks on the surface, surface becomes rough and made hindrance in the smooth movement of boat.

Following marine animals are known for such damages:

i) *Acedia*
ii) *Botrilus*
iii) *Barnecals*
iv) *Mayatilus*

#### *Protection*

1. Foulers cannot stick on metal plates. Hence use either copper or Aluminum sheet on the surface, which remains under water.
2. Apply poisonous material on the surface which will kill such foulers.
3. Apply antifouling colour like Zinc Cromate.
4. Keep boat on deck. In absence of water foulers will die.

## Gear

For fishing different type of fishes, as per their habitat, their feeding habit, their swimming pattern etc., different types of fishing gears are used. A detailed classification of fishing gear is given in Table 37.II.1. Where 13 main fishing gear and their further classification is mentioned.

Similarly depending upon type of gear, method of capture, personal liking of fishermen of particular area etc., fishing gear materials are used. Which includes natural fibers, natural polymer, chemical fibers, mineral source and inorganic material. Table 37.II.2 gives the detail of classification of fishing material.

In fabrication of different gear, different type of knots are used. Fig 37.II.1 to Fig 37.II.17 gives the diagramatic formation of different type of knots. Details of each knots are as under.

In gear, knowledge of different type of Bends, Hitches and Knots are very important. One must know how and where it could be used. In this chapter a bird-eye-view is given for Bends, Hitches and Knots.

I **Bends:** Bends are used for joining the ends of two ropes together. Bends are divided into (i) Single sheet bend (ii) Double sheet bend.

II **Hitches:** Hitches are used for fastening one rope to another rope (not at the ends).

### *Hitches are further divided in to (1) Clow hitch (2) Cow hitch*

III **Knots:** Knots involve the use of only one rope.

There are many types of knots:

*Over hand knot*

This may be used for temporarily whipping a rope or preventing a rope unreeling from a block or for providing foot holds in a lifeline.

*Figure of eight knot*

Use of same as above.

*Reef knot*

Used for tiding two ropes at the end when tied a bundle with a single rope.

*Sheep shank*

Used for temporary shortening a rope or for strengthening the portion of the rope.

*Bow line*

Used in the ship, very useful knots whenever it is desired to make a temporary eye at the end of any rope.

*Bowline on the Bight*

May be used in emergency for lowering a man.

### Different type of Bends are as under

(1) Single sheet bend/fishermen knot:

It is also called trawl knot and English knot, commonly used for fabrication of nets. It can also be used for joining two ropes.

(2) Double sheet bends:

Used for fabrication of nets, mainly to avoid sleepage. Also used for joining two ropes of different diameters.

## Different type of Hitches are as under

*Bucket Hitches*

Used for tying a rope to the handle of bucket. It is also known as fishermen bend.

*Rolling Hitches*

Used for securing a small rope to a large rope under strained.

*Clore Hitches*

Best way to secure a rope to a post after tightening.

*Crown Knots*

Mainly used to prevent unlacing of strands. Also used before doing back spice.

*Wall Knots*

It is used as a stopper knots

There are three major type of fishing gear in operation in Gujarat. (1) Trawl nets (2) Gill nets (3) Dol nets. However, when the catches and catch rate of trawler gradually decline from November to February, fishermen of Gujarat, specially in Veraval area undertake line fishing along with trawlers to supplement catches. Details of different nets and their operational procedure are as under.

## Line Fishing

Line fishing is carried out from small boats in the coastal waters. This type of fishing is very old fishing practice. There are two type of line fishing (a) Hand line (b) Long line.

### *Hand line*

In this type wooden dug out canoes and FRP canoes are used. Fishing lines made of Nylon monofilament (0.3 mm to 2.00 mm) are used by fishermen and are whipped to barbed round bent hooks of different sizes. A lead pieces or an iron ring weighing 20 to 30 g is attached at one end of the main line to keep the hooks closer to the bottom. Mackeral, squids and prawns are commonly used as baits. This type of fishing is locally known as "Lodhwa" fishing.

**Fishing gear material can be classified as under**

1. Natual Fibers
2. Natural Polymers
3. Chemical Fibers
4. Mineral Source
5. Inorganic material.

Detailed classification of each type is given in Table 37.II.1. Using these gear material different types of fishing nets are being fabricated. Similarly fishing is being carried out using different fishing gears. Table 37.II.2 give the details of classification of such fishing gear.

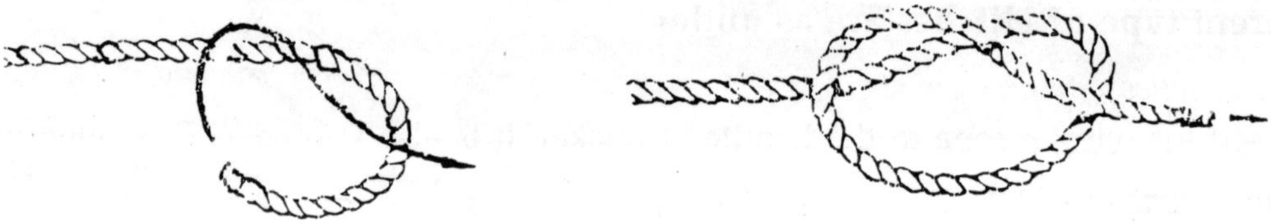

Fig. 37.II.1 : Overhand Knot

Fig. 37.II.2 : Figure of Eight Knot

Fig. 37.II.3 : Reef Knot

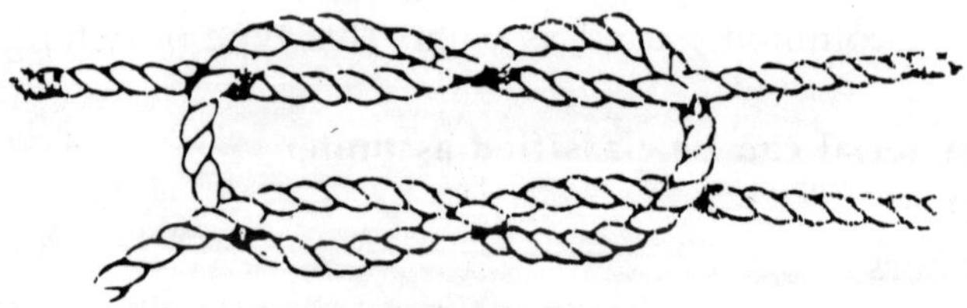

Fig. 37.II.4 : Granny Knot

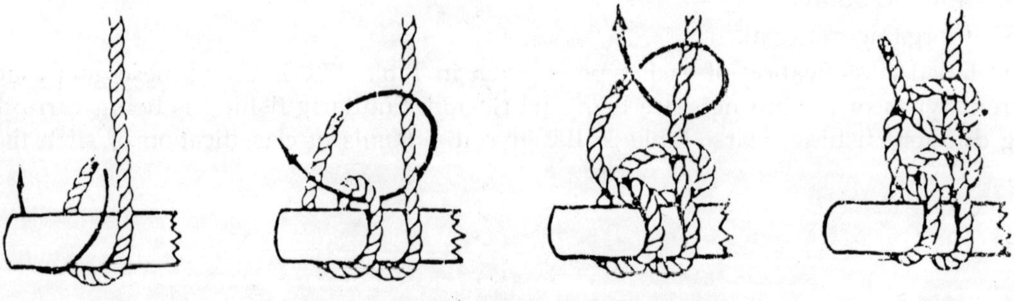

Fig. 37.II.5 : Fisherman's Bend

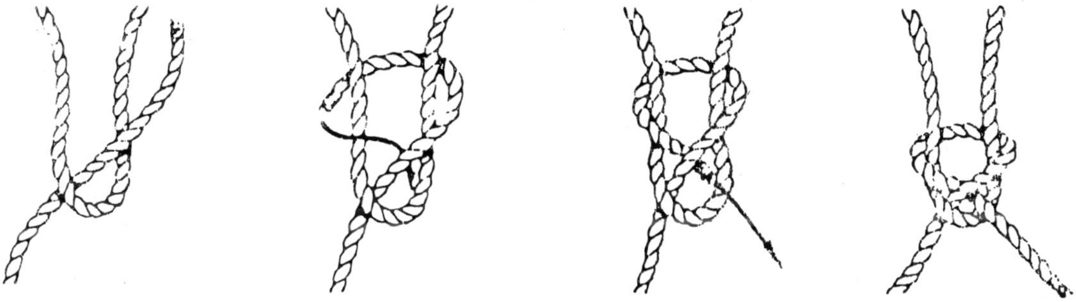

Fig. 37.II.6 : Single Sheet Bend

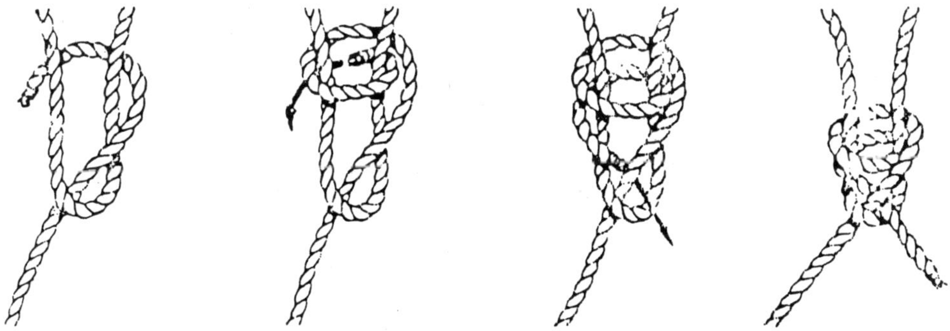

Fig. 37.II.7 : Double Sheet Bend

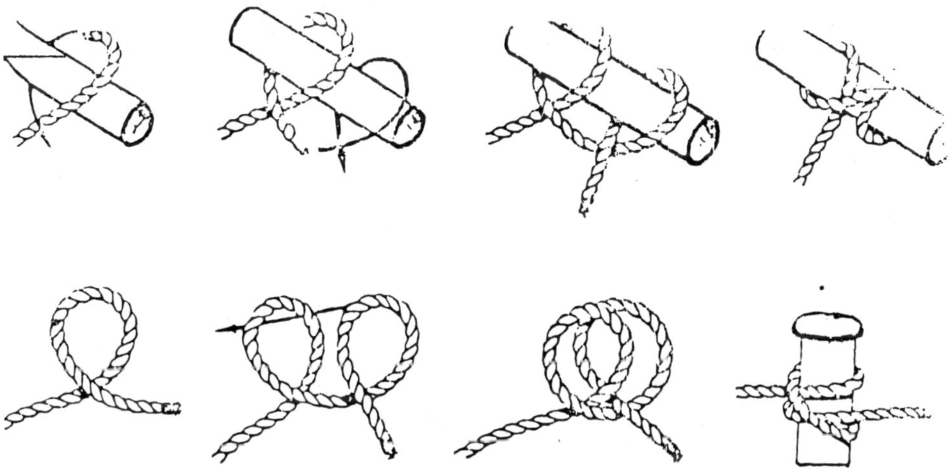

Fig. 37.II.8 : Clove Hitch

Fig. 37.II.9 : Rolling Hitch

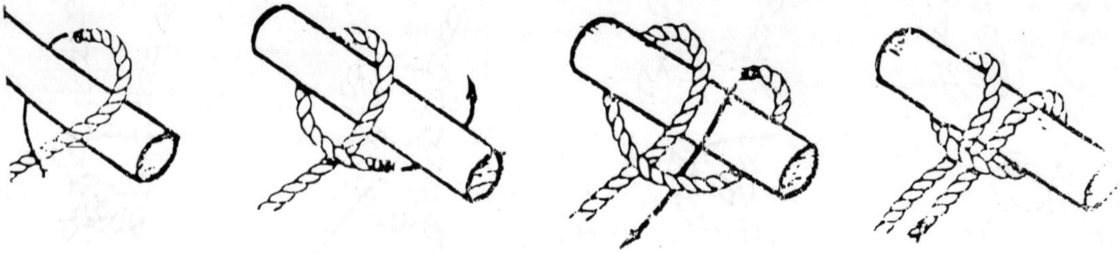

Fig. 37.II.10 : Cow Hitch

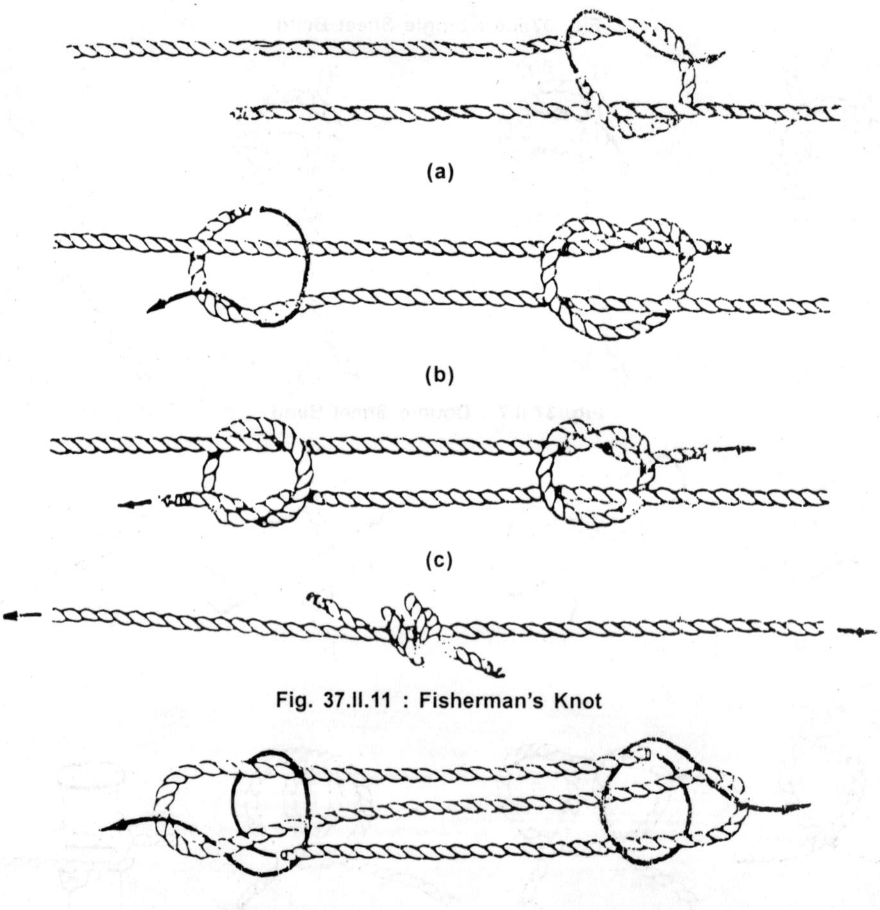

(a)

(b)

(c)

Fig. 37.II.11 : Fisherman's Knot

Fig. 37.II.12 : Sheep shank

Fig. 37.II.13 : Bowline

Fig. 37.II.14 : Heaving Line Knot

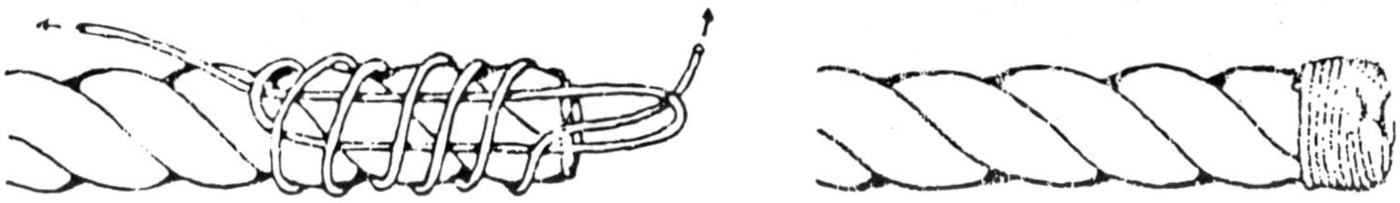

Fig. 37.II.15 : Whipping

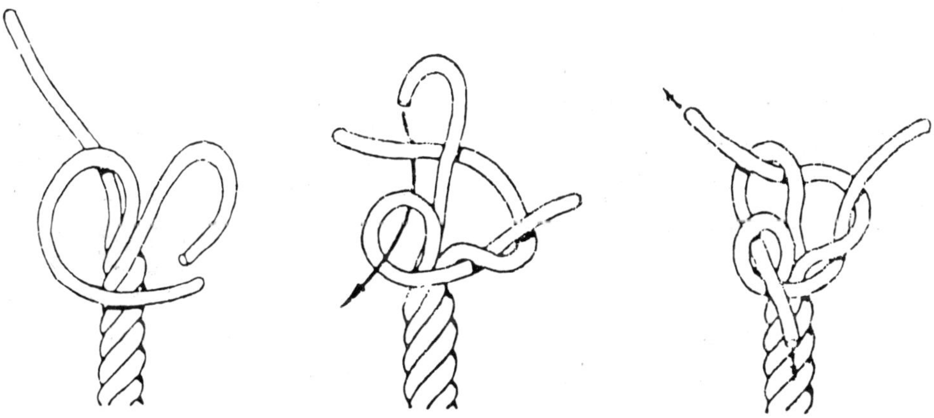

Fig. 37.II.16 : Wall Knot

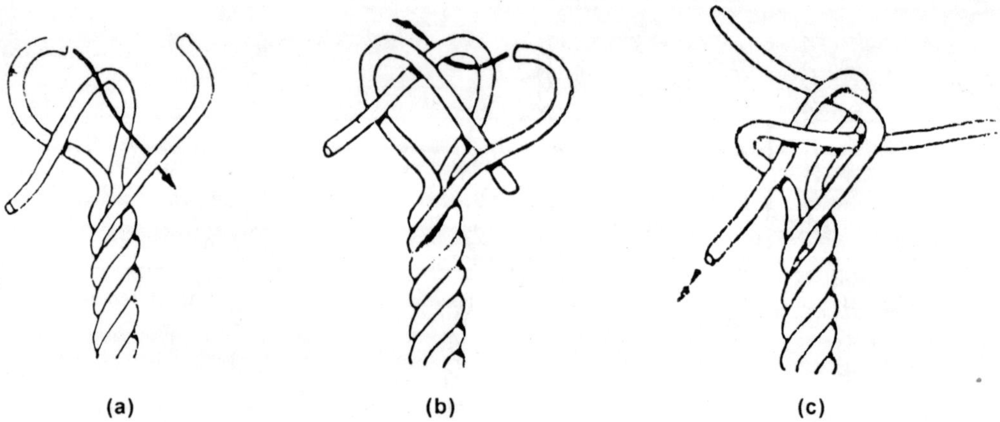


Fig. 37.II.17 : Crown Knot

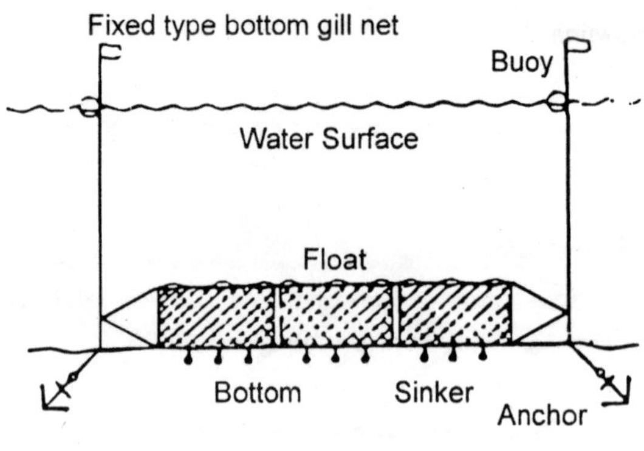


Fig. 37.II.18.

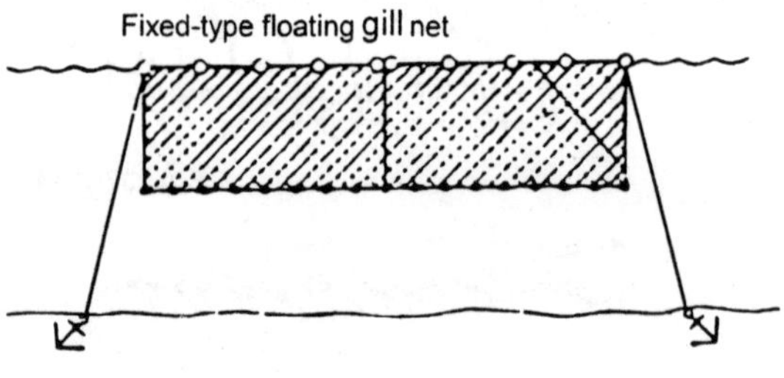


Fig. 37.II.19

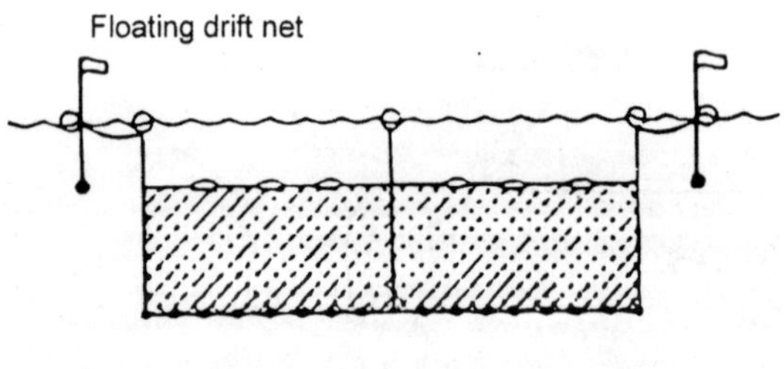


Fig. 37.II.20

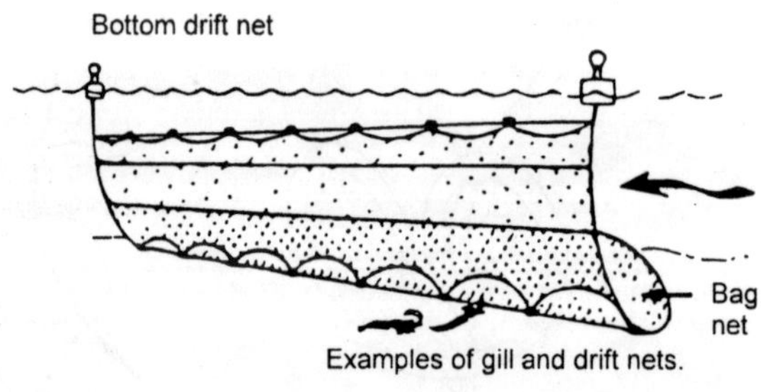


Fig. 37.II.21

This net is being used in creeks and tidal inshore areas of Gulf of Kutch and South Gujarat. In this net cotton, polyethylene or monofilament is used.

Net is without bag size is variable usually 12 to 13 m long, 1.5 to 2.5 m high with mesh size of 1.0 to 2 cm depending upon the season of fish to be caught. Several pieces were fixed on stakes driven at intervals in mud. No floats or sinkers are used.

## Operation of the Net

Net laid pleated and in the inter tidal regions, ends tied to trees or poles fixed for the purpose. When tide is almost full and water has risen up fishermen wade through the water carrying stakes in their arms, fix them at intervals, lift up the net and tie them to the stakes when tide recedes prawns and fishes caught in the net. (Kurian and Sebastian, 1976) This type of net is widely used in Surajbari area near Morbi.

## Bottom Drift Net

### *Details of Net*

This type of net is made from cotton hemp or Nylone. It is made up of 16 to 25 pieces, each 3 to 5 m long 2 to 3 m high and mesh size of 5 to 6 cm. Head rope has floats made from PVC, aluminium, sometimes Thermocol. Foot rope with sinkers at intervals to keep the net in position. In Gujarat bottom set drift net is called "Tangli".

### *Operation of Net*

2 to 3 fishermen in canoe allow the net to drift with the current of the bottom for some time, and lift for the catches. Prawns get entangled along with fishes which are gilled.

## Surface Set Drift Gill Net

### *Details of Nets*

It is drift fill net and used to catch fishes which are on surface. It is made up of 40 to 50 pieces, each 5.8 x 1.8 m. Mesh size of this net is 3.0 cm. Two men in dugout canoe or fiberglass boat can operate the net.

### *Operation of Net*

This net is being used in shallow water near shore.

## Cast Nets

### *Details of Nets*

On account of the highly diverse nature, methods ranging from catching fish indigenously designed nets are adopted for fishing.

Cast nets of both the stringed and stringless varieties are suitable for operation in all inland waters as well as shallow waters near sea shore. These are also known as "Falling Nets" and form one of the simplest type of gear used in fishery. They are more effective on muddy or sandy bottom without any rocky projections. They are made up from cotton, nylon, polyethylene or monofilament.

## Design of Net

It is conical in shape forming a circle when spread out. The foot rope along the circumference of the net has small lead weight attached all around about 15-20 cm intervals. Each weight is cylindrical, 3 cm long and 1 cm diameter, in the center. A string or line passes from the center which is held in the hand of a person for operating net. There are two types of nets stringed and stringless. In the stringed one the central line branches out into several lines and also into sub-branches, finally connected to the free edge of the net are overturned inwardly and fixed by twines which are not connected to the central string. The ordinary one have a radius of about 2-5 m to 3 m, mesh size being 1.0 to 2.0 cm. (Kurian and Sebastian, 1976).

## Operation of Net

These nets can be operated by one man. He operates the net from the shore if the fishes are nearly. The net is taken, part of it in each hand and thrown in the air with a strong swing to spread it and it falls on the water and sinks down covering the fishes, prawns etc., if any, within is circumference. The central string in held in hand, and the net is slowly pulled till the edges close.

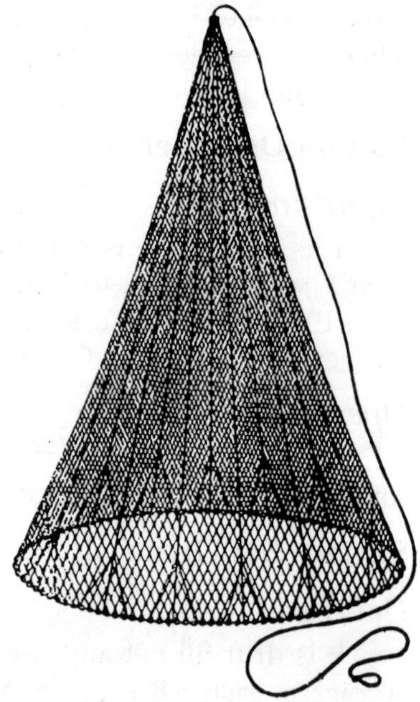

Fig. 37.II.22 : Stringed Cast-Net

### *Long line*

Long lines are locally known as "Wagha". The main line is made up of polyethylene and is about 400 to 600 m long has a diameter of 4 to 6 mm. The branch lines are one m in length and 3 to 4 hooks are attached to each branch line. Ghol, Koth, Dara, Catfishes, Perches are the main varieties of fish landed in this type line. The gear is operated at Okha, Jamnagar, Sikka and other places in northern ports of Gujarat, (Pravin and Ramesan, 1998).

## Hand Line Fishing by Trawlers

The main line of this type is around 100 m long having diameter of 2.0 to 2.5 mm. The stainless steel branch lines are having a length of 5 to 6 cm of 1.5 to 2.0mm diameter. Three branch lines with barbed hooks used for the handline are generally smaller in size. The most common size of hooks used are No. 5 and No. 6. Iron muts or bolts, approximately weighing 50 g are used as sinkers.

On reaching the fishing ground the boat is allowed to drift along with the current and wind. All crew except the "Tandel" sit on one side of the boat with the lines ready for operation. The lines are dropped in the direction of the water current with one end of the line fastened to the boat. Each fishermen operate a single line from the side of the boat after baiting it with whole ribbonfish. Around 50 to 80 m of the main line is released into the water depending upon the depth and the catch. One end of the line is held by the fishermen so as to feel the bait of the fish. As soon as the fish bite is felt the line is given a sharp jerk so that the fish gets hooked properly. The lines then hauled up gradually. The fish is removed from the hook and fishing operation is continued after putting fresh bait. This hand line catch mainly consists of Seer fish of average weight about 10 kg.

## Stationary Gill Net

Local name : Patti.

Gill net is one of the most prominent fishing gear in the artisanal sector in Gujarat contributing 27% of the total catch. Various changes have taken place in the designs of gillnets from time to time. The efficiency of gill nets has increased several times with the introduction of synthetic fibers as a replacement for natural fibers. Polyamide monofilament, polyamide multifilament and polyethylene monofilament (twisted) are mainly used in the fabrication of gill nets. The length of nets, number of units and depth of operation have increased with the advancement of motorisation. Gill nets are generally operated as drift or set nets. Large meshed gill nets for seer, tuna, bill fishes etc., are also popular among the fishermen.

## Fixed or Stationary Nets

### *Stake Net*

Stake net being used in the Gulf of Khambhat is bag type and measures 6 m length having mesh size of 25 mm at the mouth, decreasing to 7 mm at the cod end, made of Nylon yarn. These are mostly rectangular or conical in shape and of a wide variety of sizes. These nets are fixed in the tidal regions of inshore waters during the low tide with stakes, floats or sinkers. The high tide brings the fish in the net and when the tide recedes the fish are trapped. In Gujarat the net is called as Gholu Jal.

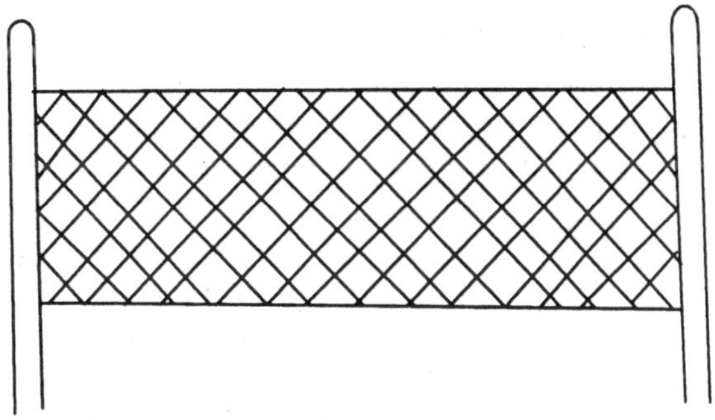

Fig. 37.II.23. Stationary Gill Net

## Operation of the Net

Gholu jal is set against the tide in muddy ground near rocky shores. The net is fixed as Saurashtra dol net with two wooden barrels as floats and two heaps of heavy stones as anchors, but in a simpler way.

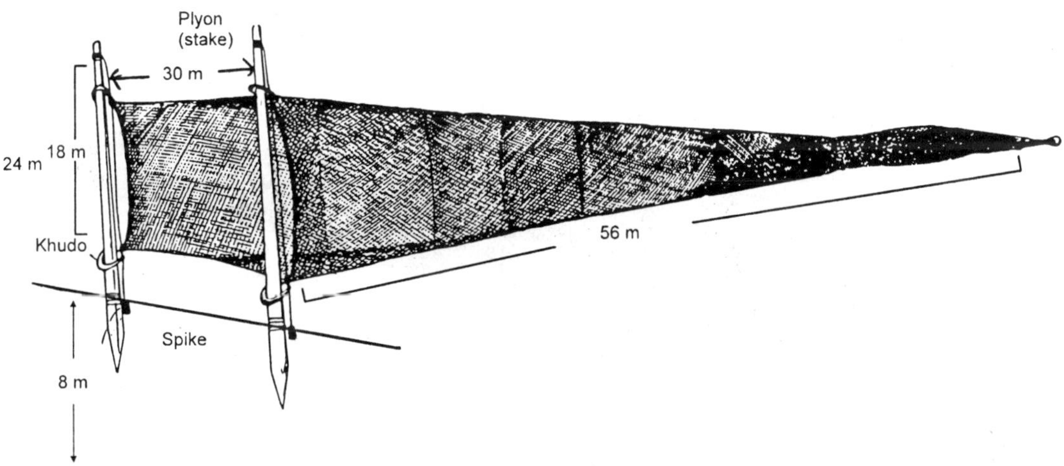


Fig. 37.II.24 : Gunja Net.

## Gunja Net

The gear used in Saurashtra area is a bag net, locally known as "Gunja" having 397cm length, 200 cm height and 114 cm width. The mesh size at mouth is 40 mm, at belly 20 mm and at cod end it is 15 mm. The net has a square mouth and gradually tapers, as a cone, to an opening at the end. Generally the net is made up of one anterior and one posterior piece.

## Operation of net

Two fishermen hold the sticks that are tied to the sides of the net and drag the net along the bottom. This operation is generally conducted at depth less than 0.75 m where there is little tidal force.

## Dol Net

Local name of this net in Gujarat is Dor or Dol Net. "Dol" is conical in shape with wide rectangular opening (mouth) and tapering cod end. The net varies from 45 to 85m in length with circumference of mouth ranging from 45 to 70 m. It is made up of "Polyethylene monofilament plastic twine" Marginal coir rope or monofilament rope is 2.5 cm in diameter. "Dol" resembles trawl net but being stationary it is used to capture column fish. In Saurashtra, eight sections of "Dol" are named as *Bochi (1.2-1.5 m), Mathdu (12.0-13.7m), Malkhadu (2.4-3.6 m). Oar (16.7-20.0 m), Trijo (9.0-12.0 m), Bangu (3.0-4.6 m), Chothi (2.4-4.6 m) and Jalo (3.0-5.5 m).*

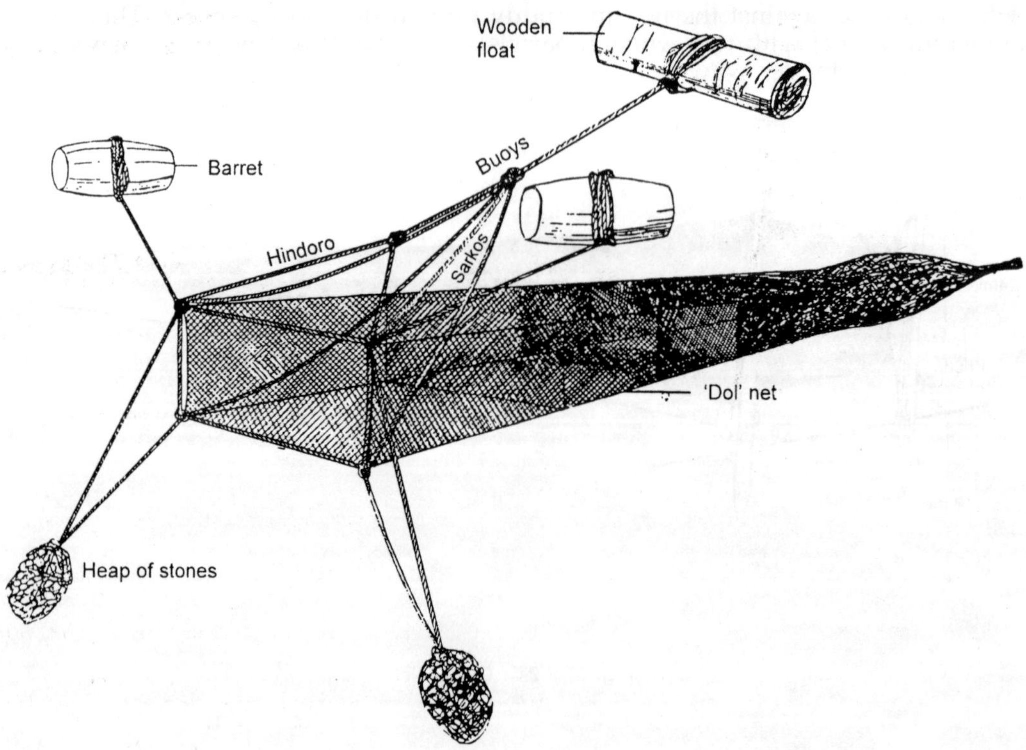


Fig. 37.II.25 : Dol Net.

The mesh size differs in each section and is reduced from mouth to cod end. The mesh size at the mouth is about 9.2 cm decreasing to 0.6 cm at the cod end. The cod end is made up of double layers to give additional strength as entire catch is collected in the cod end with the force of current. The net is set to face the incoming current and when the tide starts recending the position of the net is reversed.

### *Operation of net*

It is fixed in the sea by stakes or buoys. These nets are used in waters where the current is strong and high enough to keep the net in horizontal, expanded position.

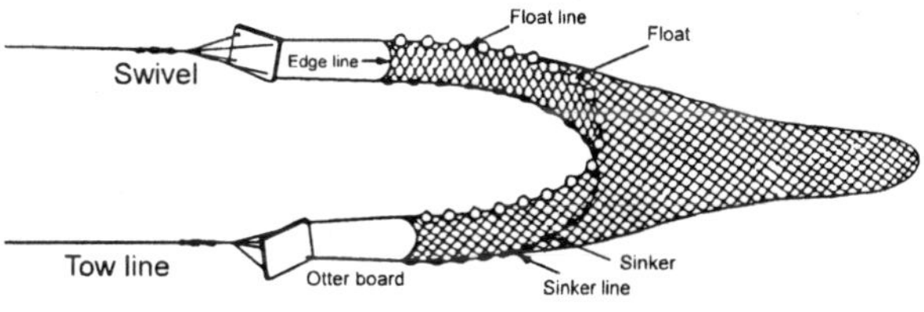


Fig. 37.II.26 : Trawl Net.

## Trawl Net

In recent fishing activity bottom trawl net is considered as most effective and most commonly used net. It is primary a bag net. The initial version of Trawl net had a rectangular frame to which the bag was attached at one side and towing ropes on the other. (Fig. 37.II.26). It was used to catch bottom dwelling fish like flat fish. This net was found not suitable fishing on rocky and uneven grounds.

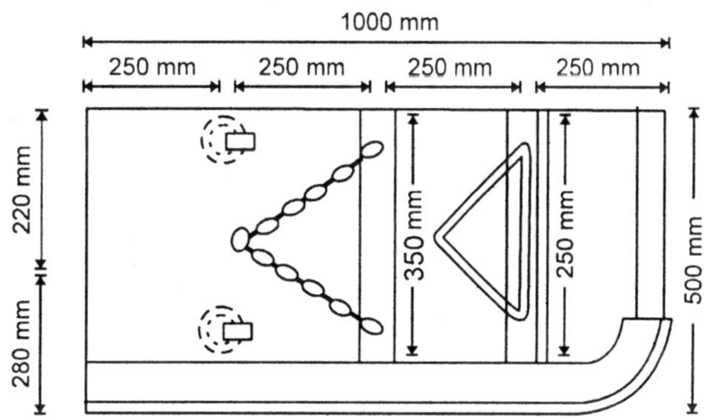


Fig. 37.II.27 : Otter Board

In present trawl net following components are considered most effective:

1. Otter Boards
2. Head rope
3. Bottom rope
4. Cod end.

1. The otter boards are rectangular boards, eithr wooden or metal, to which the rope from net and boat is attached. This otter boards help in horizontal stretching of the net. Secondly at the time of operation otter boards are disturbing fish, crustaceans etc., which are resting on bottom.
2. Head rope is attached at the upper side of the bag net fitted with hollow floats, which will keep net mouth on upward stretches.
3. Bottom rope is attached at the bottom of the bag net (Now trawl net) fitted with iron chain or with heavy lead beads. This will keep net attached to the bottom. When the net is in operation fish, crustaceans and other animals which are disturbed due to otter boards are trying to jump out. They are caught in the net.
4. Cod end is made up with small mesh size and thick twine. This is the place where all fish, crustaceans etc., caught are collected. The end is tied with rope. Catch are released from that end by detaching the said rope.

## Larged Meshed High Opening Fish Trawl Net

High opening trawl is a recent introduction and is found very successful and efficient for exploitation of column and off bottom fishes. The net is found more efficient than the bulged belly trawl in capturing different varieties of column fishes comprising of squids, pomfrets, barracudas, cat fishes, sharks and rays. The net offers 18% lesser resistance which in turn results in utilization of lesser horse power.

Design of 32 m large meshed high opening trawl is given in Fig. 37.II.28.

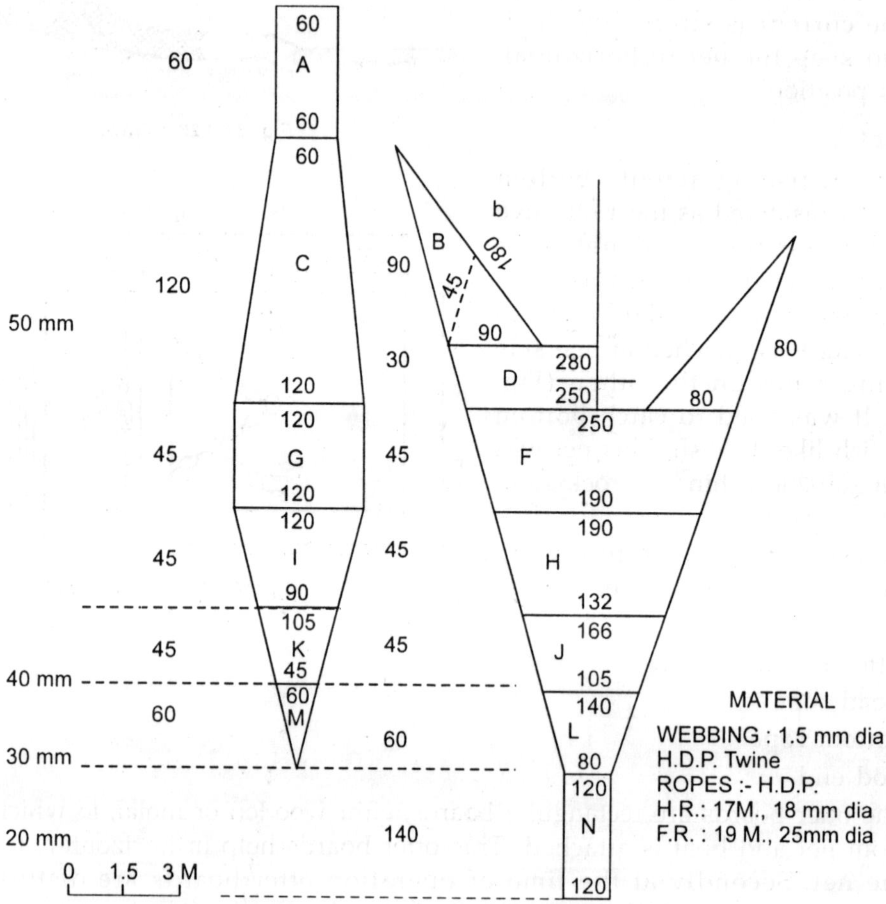


**Fig. 37.II.28 : Large meshed high openning Trawl**

## *References*

Anon. 1962. The wealth of India–Fish and Fisheries. Published by Council of Scientific and Industrial Research, New Delhi. pp. 86-96.

Anon. 1981. All India census of Marine Fishermen, Craft and Gear. *Mar. Fish. Infor. Serv., T & E. Ser.* No 30: 27-30.

Anon. 1988. Squid and Cuttle fish Resources, Distribution, Handling and Processing. Booklet Published by M.P.E.D.A. Cochin.

Chhaya Niranjan D. 1990. Fishing Instruments (I) Fisheries Management and Science published by University Granth Nirman Board, Govt. of Gujarat, Ahmedabad pp. 11-31.

Dholakia, Anshuman D. and Girish D. Vasavda. 1988. Modern fishing boat and fish production. "Matsya Vikas", 109-110. Published by Commissioner of Fisheries, Govt. of Gujarat. Gandhinagar.

Kurian, C.V. and V.O. Sebastian. 1976. Prawn and Prawn Fisheries of India. Published by Hindustan Publishing Corporation (India), New Delhi.

Panikar, P.A., T.M. Sivan, H.N. Mhalathkar, and P. George Mathai. 1978. Selectivity of gill nets for *Hilsa toli and Pampus argenteus. Fish. Technol.*, 15: 61-68.

Pravin, P. and Ramesan, M.P. 1998. An emerging low energy fishing technique at Veraval: Hand Lining. *Fishing chimes*, 18 (5): 30-31.

Ramamurthy, S. and Muthu, M.S. 1969. Prawn Fishing Methods, Bulletin No 14, C.M.F.R.I. Cochin.: 235-257.

Sehara, D.B.S. and J.P.Karbhari. 1987. A study on "Dol" net fishery at selected centres in north-west coast, with special reference to coasts and Reuters. *Mar. Fish. Infor. Serv., T & E. Ser.*, 78 : 1-14.

Sitarama, Rao J., Percy Dawson and Y. Sreekrishnan 1980. Effect of colour on the catch of Gill Nets. *Fish Technol.*, 17: 75-77.

Sudhakar Rao, G.C. Suseelen and S.Lalithia Devi. 1980. Impact of mesh size reduction of Trawl Nets on the Prawn Fishery of Kakinada in A.P. *Mar. Fish. Infor. Serv.*, T & E *Ser.*, 21: 1-6.

Yohannan, T.M.K.K. Balasubramanian, 1989. Drift Net Fishery at Calicut with special reference to *Scombroides. Mar. Fish. Infor. Serv.* T & E *Ser.*, 95: 1-5.

38

# Fisheries Harbours and Landing Centers

## Fisheries harbours

The contribution so far made by the fisheries sector towards creation of employment opportunities, supply of protein food, and earning of foreign exchange is sufficient and bears a promise for further exploitation of the enormous resources. It is well known that the fisheries harbours play an important role in the exploitation of the marine wealth of the country. In view of the importance of the fisheries harbours it has become necessary to develop additional facilities and to take care for the proper maintenance of the fisheries harbours.

In this context different state Governments have constructed and developed fisheries harbours. Details of Fisheries harbours in each maritime state are given hereunder.

Table 38.1 gives the details of type of harbours (Major, or minor), Latitude and longitude, year of starting, facilities, No of MFV etc. Table 38.2 gives the details of fish landing centers of respective states.

**Table 38.1 : Details of Fisheries Harbours of India**

| Name and type of Fisheries Harbour | Year of starting | Cost of construction | Length of Jetty | No of MFV Registered and details |
|---|---|---|---|---|
| GUJARAT STATE (coast line 1600 km, Total 188 major Fish landing centers) | | | | |
| **Major Harbour** | | | | |
| Veraval (20°54' N, 70°22'E) | 1986 | | 1100 M | 3467 Mechanised<br>29 Non-Mech.<br>3496 Total MFV |
| Shivrajpur (Not yet started) | | | | |
| **Medium Harbour** | | | | |
| Porbandar (21° 38 N, 69° 36'E) | | | | 2231 Mech.<br>50 Non-Mech.<br>2281 Total MFV |
| **Minor Harbour** | | | | |
| Mangrol (21° 07' N 70° 06'E) | 1985 | | 375 M | 1492 Mech.<br>17 Non-Mech.<br>1509 Total MFV |

*Contd.......*

*Table 38.1 – Contd...*

| Name and type of Fisheries Harbour | Year of starting | Cost of construction | Length of Jetty | No of MFV Registered and details |
|---|---|---|---|---|
| Okha (22° 29'N, 69° 05'E) | | | | 18 Mech.<br>20 Non-Mech<br>38 Total-MFV |
| Jakhau (23° 14'N, 68° 36'E) | | | | 31 Mech<br>3 Non-Mech<br>34 Total MFV |
| MAHARASHTRA STATE (coast line 720 km, 184 fish landing center) | | | | |
| **Major Harbour** | | | | |
| Sassoon Dock (Mumbai) | | | | 700-800 Trawlers<br>50 Perse Seiners<br>125-159 Dol netter<br>70 Gill netter |
| New Jetty Mumbai | | | | 1000-1100 Trawler<br>25-30 Dol netter<br>25-30 Hooks & lines |
| **Minor Harbour** | | | | |
| Mirkarwada at Ratnagiri | | 344 Lakh | | 400 Trawlers<br>10-12 Gill netters |
| KARNATAKA STATE (UTTER KARNATAKA) Coast line 300 km | | | | |
| **Major Harbour** | | | | |
| Honnavar/Kasaragod | 1982 | 25 Lakh | | 40 MFV |
| Tadri | 1996 | 145 Lakh | | 50 MFV |
| **Minor Harbour** | | | | |
| Karwar | N.A. | N.A. | | 33 MFV |
| SOUTH KARNATAKA | | | | |
| **Major Harbour** | | | | |
| Mangalore<br>Malpe | 1993-94<br>1985 | 1.35 Cr<br>2.065 Cr | | 1871 Trawlers<br>293 Purse seiners |
| Gangolli | 1995 | 8.0 Cr | | 245 Drift gillnetter<br>19 long liners |
| **Minor Harbour** | | | | |
| Hangarcutta | | 72 Lakh | | |
| Bhatkar | | N.A. | | |
| Hejmadi Kodi | | 95 Lakh | | |
| Thengiagundi | | N.A. | | |
| GOA STATE | | | | |
| **Minor Harbour** | | | | |
| Talpona | N.A. | N.A. | 30 M | 30-40 MFV |
| Cortalim | N.A. | N.A. | 34 M | 30-40 MFV |

*Contd......*

*Table 38.1 – Contd...*

| Name and type of Fisheries Harbour | Year of starting | Cost of construction | Length of Jetty | No of MFV Registered and details |
|---|---|---|---|---|
| Cuthona | N.A. | 18.59 Lakh | 144 M | 150-160 MFV |
| Malim | N.A. | 72.28 Lakh | 111 M | 150-200 MFV |
| Chapora | N.A. | 80.03 Lakh | 80 M | 75-100 MFV |
| Gulf of Mannar | | | | |
| Mandapam | | | 150 M | 100 Trawlers |
| **Tamil Nadu State** | | | | |
| Tuticorin | | 2.1 Cr. | 800 M | 450 Trawlers |
| NORTH TAMIL NADU AND ANDHRA PRADESH. | | | | |
| **Major Harbour** | | | | |
| Nizampatnam | N.A. | 315 Lakh | N.A. | 360 Trawlers |
| Madras (Chennai) | 1984 | 1297 Lakh | 495 M | 500 Trawlers<br>25 Motorised<br>200 Non-motorized |
| **Minor Harbour** | | | | |
| Pondicherry | N.A. | 963 Lakh | 332 M | 350 MFV |
| Cuddalore | 1985-86 | 10.74 Lakh | 218 M | 60 MFV |
| Pazhayar | 1986 | 67.32 Lakh | 44 M | 75 MFV |
| NORTH ANDHRA PRADESH ORISSA AND WEST BENGAL | | | | |
| **Major Harbour** | | | | |
| Visakhapatnam (17° 40' N, 83° 18'E) | 1976 renewed 1991 | 4.26 Crores | 503 M | 450 Small MFV<br>71 Trawlers |
| Paradeep (20° 18'N, 86° 52'E) | N.A. | 3.81 Crores | 466 M | 2 Deep Sea Trawler<br>500 Small MFV<br>150 Gill netter<br>50 Trawlers |
| Sankarpur (21° 48'N, 87°52'E) | 1987 | 5.00 Crores | N.A. | 2 Trawlers<br>140 Small MFV<br>150 Gill netters |
| **Minor Harbour** | | | | |
| Diamond Harbour (W.B.) (22° 14'N, 86°50'E) | N.A. | 1.07 Crores | N.A. | Trawlers<br>130 Gill netters |
| Dharma (Orissa) (20° 48' N, 86°50'E) | N.A. | 1.05 Crores | 200 M | 30 Small MFV<br>20 Gill netters |
| Fresurgunj (W.B.) (21° 45'N, 88° 18'E) | N.A. | N.A. | N.A. | 200 Small MFV |
| Lakshadweep Minicoy | N.A. | N.A. | 147 M | N.A. |

*Source* : Mar. Fish. Infor. Serv. T&E. Ser. 153.

**Table 38.2 : State Wise Major Fish Landing Centres of India**

Gujarat : (Total 188 Fish Landing Centers)

| | | |
|---|---|---|
| **Valsad District** | | |
| 1. Magod dubgri | 2. Mehdungri | 3. Survada |
| 4. Bhadeli-Jaga | 5. Bhagal | 6. Kakvadi |
| 7. Nani Danti | 8. Moti Danti | 9. Mohothiya |
| 10. Dani-bhagal | 11. Kolak | 12. Udvada |
| 13. Umersadi | 14. Umaragoan | 15. Dandi- Maroli |
| 16. Dehri | 17. Khatalvad | 18. Nargol |
| 19. Tadgam | 20. Maroli | 21. Fans |
| 22. Kalai | 23. Manghar | 24. Dholayi |
| 25. Vagharech | 26. Bhath | 27. Krishnpur |
| 28. Onjal | 29. Vansi-Borsi | 30. Chormalbhatha |
| 31. Bilimora Kosamba | 32. Navsari | |
| **Surat District** | | |
| 33. Bhimpor | 34. Vannsva | 35. Dumas |
| 36. Suvali | 37. Hajira | 38. Dandi |
| 39. Delasa | 40. Bhagava | 41. Nor |
| 42. Pardi-zakhari | | |
| **Bharuch District** | | |
| 43. Kantiyajal | 44. Luvara | 45. Kavi |
| 46. Sarod | 47. Tankari | 48. Nada |
| 49. Zamdi | 50. Malpur | |
| **Khambhat District** | | |
| 51. Khambhat | 52. Bhavnagar | 53. Ghogha |
| 54. Osara | 55. Mithaivirdi | 56. Sartanpur |
| 57. Gadhula | 58. Katpar | |
| **Amreli District** | | |
| 59. Chanch | 60. Jafrabad | 61. Shiyalbet |
| 62. Madhwad | 63. Kotda | 64. Mul-Dwarka |
| **Junagadh District** | | |
| 65. Veraval | 66. Hirakot | 67. Sutrapada |
| 68. Vadodra Bara | 69. Damlej | 70. Chorwad |
| 71. Mangrol | 72. Shil | 73. Bara |
| 74. Porbandar | 75. Madhupur | 76. Muyani |
| 77. Navibandar | 78. Rajapara | 79. Simar |
| **Jamnagar District** | | |
| 80. Bedi | 81. Sikka | 82. Sachna |
| 83. Sarmat | 84. Bed | 85. Rupen-Dwarka |
| 86. Betbalapur | 87. Salaya | 88. Bharana |
| 89. Vadinar | 90. Jodiya | 91. Balambha |
| 92. Balachadi | 93. Navadhra | 94. Harsad |
| 95. Nana-Ambala | | |
| **Kutch District** | | |
| 95. Mundra | 96. Navinar | 97. Zarapara |
| 98. Shekhadiya | 99. Bhadreshwar | 100. Luni |
| 101. Salaya | 102. Mohava | 103. Tragdi |
| 104. Layaja | 105. Jakhau | 106. Narayan Sarovar |
| 107. Lakhpat | 108. Kandla | 109. Sangadhvira |
| 110. Mithpor | 111. Kukadsar | 112. Takara |
| **Rajkot District** | | |
| 113. Navlakhi | | |

*Source* : Commissioner of Fisheries, Gandhinagar with permission.

*Contd.........*

*Table 38.2 – Contd...*

| | | | | | |
|---|---|---|---|---|---|
| MAHARASTRA STATE (Total 184 Fish landing centers) | | | | | |
| **Major** | | | | | |
| 1. | Agaro | 2. | Sarjekot | 3. | Agardamda |
| 4. | Skirnata | 5. | Satpati | 6. | Rewas |
| 7. | Deogad | 8. | Karanja | 9. | Trombay |
| 10. | Rajapuri | 11. | Ambao | 12. | Buraundi |
| 13. | Pedwe | | | | |
| **KARNATAKA STATE (Total 62 Fish Landing Centers)** | | | | | |
| 1. | Majab | 2. | Karwar | 3. | Binaga |
| 4. | Arga | 5. | Chendia | 6. | Kodar |
| 7. | Harwad | 8. | Keni | 9. | Gangavaali |
| 10. | Tadri | 11. | Kumut | 12. | Honnavar |
| 13. | Talapady | 14. | Someswara | 15. | Ullal |
| 16. | Panambur | 17. | Balkampady | 18. | Kaulal |
| 19. | Hosabettu | 20. | Surathkal | 21. | Mukka |
| 22. | Lachil | 23. | Sashithithla | 24. | Hejmadi |
| 25. | Padabidri | 26. | Thenkayermal | 27. | Badayermal |
| 28. | Badauchila | 29. | Kaupu lighthouse | 30. | Plipu |
| 31. | Kalipunjal | 32. | Muttukoppal | 33. | Padukare |
| 34. | Thottam | 35. | Hoode | 36. | Kodikanya |
| 37. | Manoor | 38. | Gopadi | 39. | Bijadi |
| 40. | Hale Ative | 41. | Kundampura Kodi | 42. | Gangoli |
| 43. | Kanchi Kodi | 44. | Maravanthe | 45. | Navunda |
| 46. | Gangebailu | 47. | Kodiyari | 48. | Paduvari |
| 49. | Kesorakoddi | 50. | Govte | 51. | Belke |
| 52. | Mundali | 53. | Mavinkurva | 54. | Karikal |
| 55. | Jai | 56. | Honnagdde | 57. | Kaikini |
| 58. | Murdeswere | 59. | Byloorbelshe | 60. | Hosahitlu |
| 61. | Alvehitlu | 62. | Manki-Madi | | |
| **GOA STATE** | | | | | |
| 1. | Topona | 2. | Assoina | 3. | Marvel |
| 4. | Caranzalem | 5. | Odoxel | 6. | Sinquert |
| 7. | Oeda Candolim | 8. | Keripernem | 9. | Nerul |
| 10. | Marcaim (Tona) | 11. | Cundal | 12. | Bonddora |
| 13. | Cude (Siolim) | 14. | Bande (Assagao) | 15. | Vagator |
| 16. | Betul | | | | |
| **Gulf of Mannar** | | | | | |
| 1. | Mandapam | | | | |
| **Palk Bay** | | | | | |
| 1. | Thondi | | | | |
| **Southern Tamil Nadu** | | | | | |
| Tuticorin (Major) | | | | | |
| **North Tamil Nadu** | | | | | |
| 1. | Nizampatnam | 2. | Krishnapatnam | 3. | Chennai |
| 4. | Cuddalore | 5. | Pondicherry | 6. | Pazhyar |

*Contd.....*

*Table 38.2 – Contd...*

| | | | |
|---|---|---|---|
| **North Andhra Pradesh, Orissa and West Bengal.** | | | |
| 1. | Visakhapatnam (Major) | 2. | Paradeep Fisheries Harbour (Major) |
| 3. | Sankarpur Fisheries Harbour (Major) | 4. | Roychowk Fisheries Harbour (Minor) |
| 5. | Diamond Fisheries Harbour (Minor) | 6. | Astrarang Fisheries Harbour (Minor) |
| 7. | Fressurgunj Fisheries Harbour (Minor) | | |
| **Lakshadweep** | | | |
| | Minicoy (Major) | | |
| **Karnataka** | | | |
| 1. | Hejmadi Kodi Fisheries Harbour (Minor) | 2. | Hanzarcutta Fisheries Harbour (Minor) |
| 3. | Bhatkal Fisheries Harbour (Minor) | 4. | Manglore Bunder Fisheries Harbour (Major) |
| 5. | Malpe Bunder Fisheries Harbour (Major) | 6. | Gangolli Bunder Fisheries Harbour (Major) |
| 7. | Over and above there are total 58 small fish landing centers. | | |

The Government of India initiated a scheme for the establishment of Fishing Harbours in India in 1955 and undertook survey with the help of F.A.O experts. In 1967, a pre-investment survey of fishing harbour project was established for this purpose under as FAO/UNDP programme.

The Government envisaged and increase in mechanized small boats in India. It was in 1986 that India needs 50 more new fishing harbours and landing centers. (Fish Farming International, 1986, Vol 25, No 5:1)

The survey including the selection of suitable sites, engineering and constructional aspects, economic feasibility on the basis of fishing resources, scope of introduction of suitable fishing crafts, production, handling of fish and the lay out of harbour and supporting facilities. The landing of small fish and low priced (Trash fish) is to be dried cause low hygienic standards and could damage the quality of prime fish catch. To improve this situation construction of Fisheries Harbours is essential (Fishing News 1985).

As part of the World Bank Project, Fisheries Terminal Division to be operated under the auspices of the Commissioner of Fisheries of the Government of Gujarat has been created. The Fisheries Terminal Division will be responsible for the management of zones allocated to the development of two principal capital works components of the project when they are completed. These are. (1) Improvement of the fishing harbour of Mangrol and Veraval to permit more Mechanized Fishing Vessel (MFV) to use these facilities and to provide better protection against adverse weather conditions. (2) Improvement of shore facilities and services at these two ports to provide suitable support for the existing fleet and expansions which are expected to occur as a result of the project.

Mangrol and Veraval which are within 40 km of each other have been selected initially for development table 38.4 and 38.5 gives the details of Fisheries harbours Veraval and Mangrol of Gujarat. (Dholakia & Vasavda, 1988).

Table 38.6 gives the details of other eight fishing harbours covered under World Bank Project.

Fig 38.1 to 38.14 gives the layout plan of different Fisheries Harbours of India.

**Table 38.3 : Latitude and Longitudes of Fish Landing Centres and Ports of Gujarat**

| Name of Fish Landing Centre & Port | Latitude | Longitudes |
|---|---|---|
| Kandla | 22°58' N | 70°13' E |
| Okha | 22°29' N | 69°05' E |
| Bhavnagar | 21°46' N | 72°09' E |
| Bedi | 22°33' N | 70°02' E |
| Porbandar | 21°38' N | 69°36' E |
| Navlakhi | 22°58' N | 70°27' E |
| Veraval | 20°54' N | 70°22' E |
| Mandvi | 22°50' N | 69°21' E |
| Sikka | 22°31' N | 69°49' E |
| Jakhau | 23°14' N | 68°36' E |
| Mundra | 25°45' N | 69°43' E |
| Salaya | 22°19' N | 69°36' E |
| Mangrol | 21°07' N | 70°06' E |
| Nawabandar | 22°44' N | 71°05' E |
| Jafrabad | 20°52' N | 71°23' E |
| Albert Victor | 20°57' N | 71°32' E |
| Bheri Rajula | 20°57' N | 71°31.5' E |
| Broach | 21°42' N | 73°02' E |
| Lakhpat | 23°48' N | 68°52' E |
| Jodia | 22°43' N | 70°17' E |
| Beyt | 22°30' N | 69°20' E |
| Dwarka | 22°15' N | 69°03' E |
| Lamba | 21°53' N | 69°19' E |
| Madhavpur | 21°15' N | 69°57' E |
| Sutrapada | 20°50.5' N | 70°29' E |
| Kodinar | 20°48' N | 70°50' E |
| Kotda | 20°45' N | 70°30' E |
| Madhwad | 20°45' N | 71°15' E |
| Ghogha | 21°41' N | 72°16' E |
| Talaja | 21°18' N | 72°11' E |
| Cambay | 22°19' N | 72°37' E |
| Onjal | 20°45' N | 72°55' E |
| Navsari | 20°50' N | 72°15' E |
| Bilimora | 20°48' N | 72°57' E |
| Umarsadi | 20°45' N | 72°50' E |
| Umaragoan | 20°07' N | 72°44' E |
| Katpar | 21°09' N | 71°46' E |
| Chorwad | 21°04' N | 70°15' E |

*Source* : Dholakia, A.D.

**Table 38.4 : Details of Fisheries Terminal Veraval**

| The shore facilities provided at Veraval are as under | | |
|---|---|---|
| 1. | Breakwater | 814 m |
| | (a) Western | 334 m |
| | (b) Eastern | 480 m |
| 2. | Landing quay and repair quay | 1,100 Running m. |
| 3. | Berthing jetties (5 Nos.) | 500 R.m. |
| 4. | Concrete Breaking slops for cannoes | 520 R.m. |
| 5. | Deepening of basin, up to 3.3 m | 1,20,000 Sq. m. |
| 6. | Boat repair Yard | 449 m. |
| 7. | Market halls 2 Nos. with close booths for fish merchants 100 m x 25 m. | 500 Sq. m. |
| 8. | Open market hall 1 No. 100 m x 25 m. | 250 Sq. m. |
| 9. | Gear Shed | |
| | (a) 120.3 x 7.15 m 1 No. | |
| | (b) 44.3 x 7.15 m 2 Nos. | 1494 Sq. m. |
| 10. | Workshop 26.4 x 17 m | 449 Sq. m. |
| 11. | Canteen Building | 355 Sq. m. |
| 12. | Trash Fish Platform 108 x 24.5 m | 2641 Sq. m. |
| 13. | Water supply | 0.5 m.g.d. |
| 14. | Diesel Supply unit | Two with three outlets each i.e., total six |
| 15. | No of booths in market hall | 46 Nos. |

*Source* : Dholakia & Vasavda. 1988

**Table : 38.5 : Details of Fisheries Terminal Mangrol**

| The shore facilities provided at Mangrol Fishing harbour are as under: | | |
|---|---|---|
| 1. | Break waters | 633 m. |
| (a) | Western | 448 m. |
| (b) | Eastern | 185 m. |
| 2. | Quay and jetties | 375 m. |
| 3. | Deepening of basins | 11900 m$^2$ |
| 4. | Size of basin | 180 m x 70 m |
| 5. | Market Hall | 108.6 m x 18.6 m |
| 6. | Canteen Building | 296 Sq. m |
| 7. | Two gear sheds for net repairing | 40 x 5 m each i.e., 400 Sq. m. |
| 8. | Water Supply | 1.3 lakh lit/day |
| 9. | Diesel Supply unit | One with three outlets. |
| 10. | No of Booths in Market hall | 23 No. |
| 11. | Workshop | 296 Sq. m. |

*Source* : Dholakia & Vasavda. 1988

## Fishing Harbour Mangrol

At Mangrol the first 48 feet long trawler was introduced in 1971. Which was increased to 205 in 1985. While 12 cannoes in 1977 were increased to 175 no in 1985 for the utilising this modern harbour. After ten years of proper utilization of this harbour, total number of trawlers are now increased to 1492 (1998-99).

**Table 38.6 : Details of Eight Fishing Harbours under World Bank Project–GUJARAT.**

| Sl. No. | Name of Harbour | Approach Road | Water supply facilities length of pipeline | Working Shed. | Distance from Veraval N=North S=South |
|---|---|---|---|---|---|
| 1. | Hirakot | 1550 m | 10.6 km | 10x8 m shed | 16 km (N) |
| 2. | Sutrapada | 600 m | | 12x8 m shed | 20 km (N) |
| 3. | Dhamloj | 800 m | 10.0 km | 30x6 m open platform | 40 km (N) |
| 4. | Mul Dwarka | 5.4 km. | 5.4 km | 30x6 m open platform | 45 km (N) |
| 5. | Madhwad | 12.0 km | 18 km | 30x6 m open platform | 65 km (N) |
| 6. | Chorward | 4 km | 12 km | 30x6 m open platform | km (S) |
| 7. | Mangrol Bara | 4 km | | 30x6 m open platform | km (S) |
| 8. | Mangrol | 1 km | | FTD Market hall | km (S) |

*Source* : Dholakia A.D. 1986

## *Facilities*

The market hall has 23 merchants' booths, each 4m x 5m size and each of supplied with fresh water tap, lighting power points, and bins for ice and fish. Available for the market hall users are weighing machines, ice crushers, and platform trolleys, and canteen facilities. A toilet block is also provided. Handling fish in the early morning and in the late evening is now possible as the whole area is served with sodium lighting system, replacing the kerosene lights which were completely inadequate for illuminating the working area.

Available at the quay side are fuel points operated by the Gujarat Fisheries Development Corporation eliminating the traditional task of transporting containers and transferring to the boat by small hoses.

Ice crushing facilities at quay side are provided. Adequate water hoses for potable water are available, replacing the old system of fisher women bringing water from their homes to the boats. (Dholakia and Vasavda, 1987).

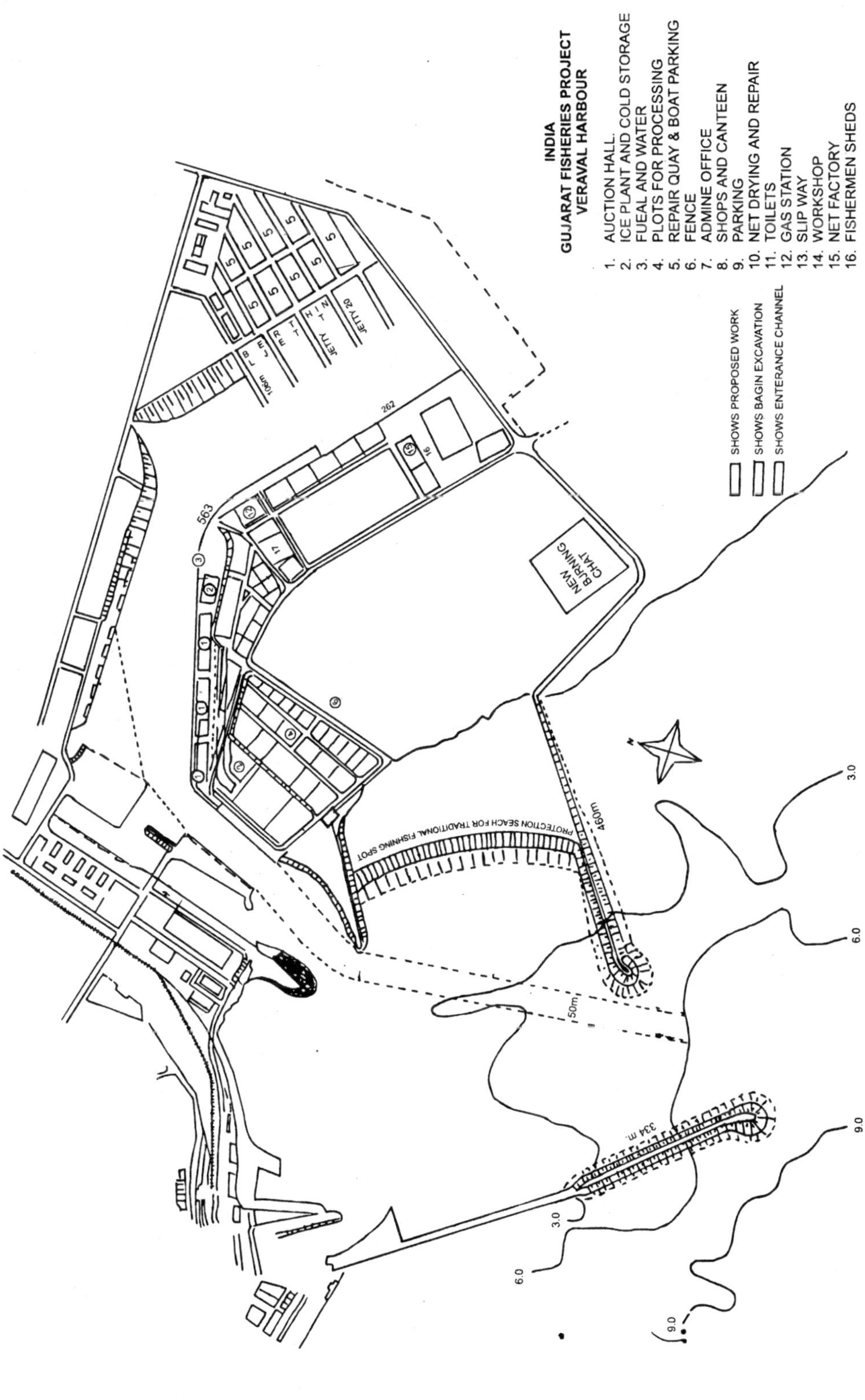


Fig. 38.1 : Lay out of Veraval Fisheries Harbour.

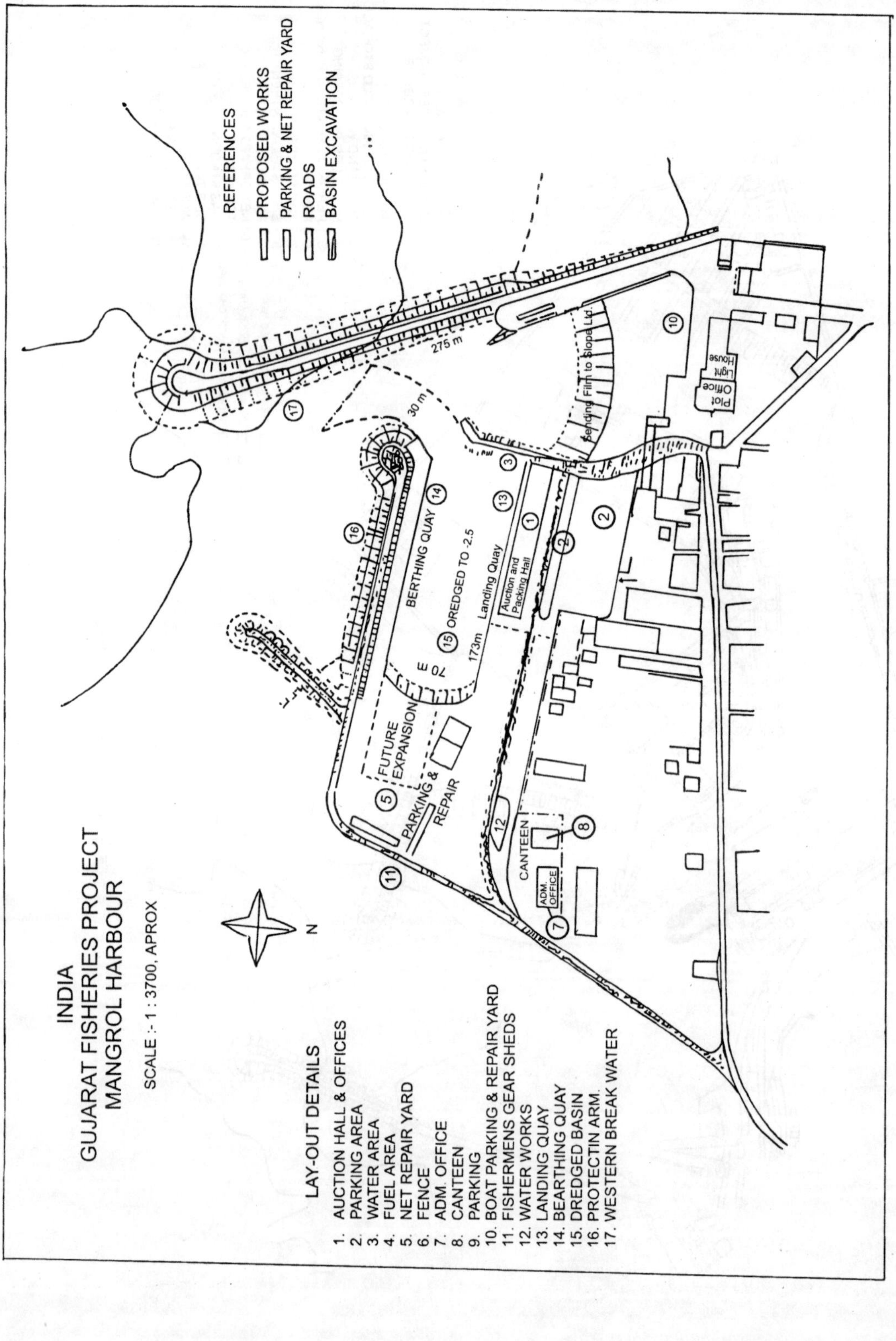


Fig. 38.2 : Lay out of Mangrol Fisheries Harbour.

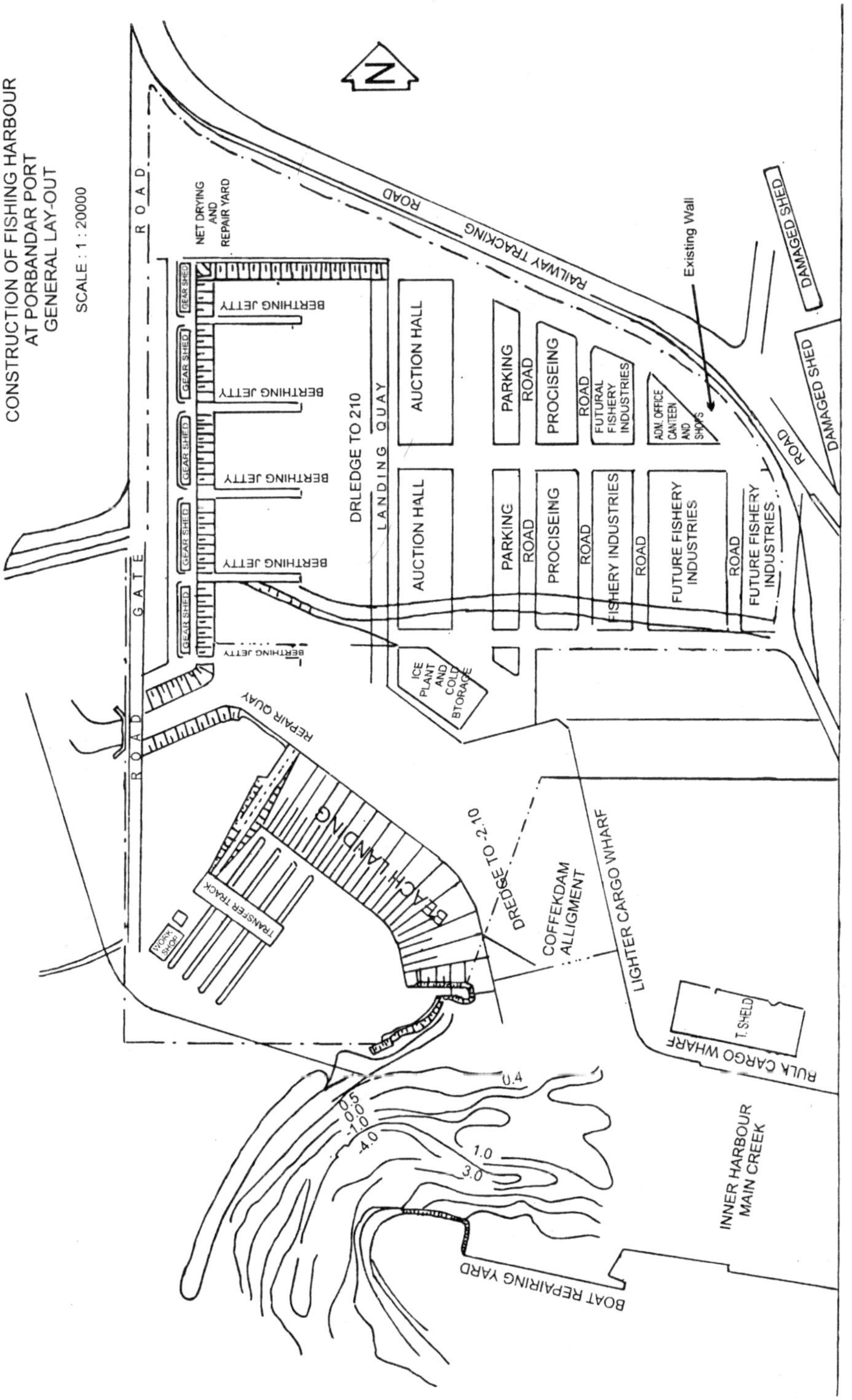


Fig. 38.3 : Lay out of the Porbandar Fisheries Harbour

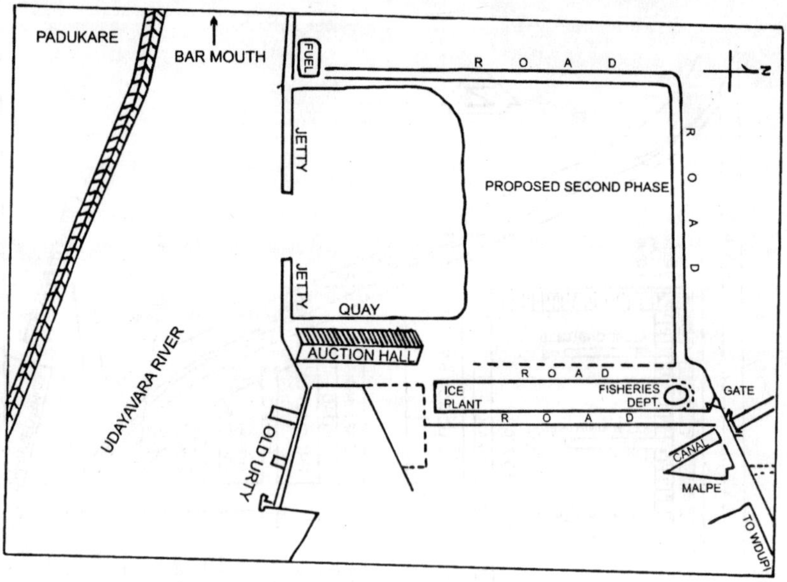


Fig. 38.4 : The lay out of the Malpe Fisheries Harbour (major)

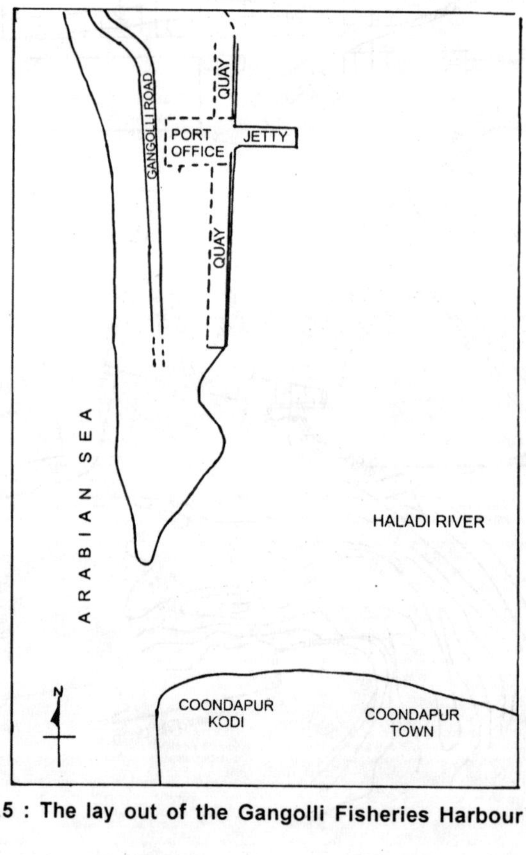


Fig. 38.5 : The lay out of the Gangolli Fisheries Harbour (major)

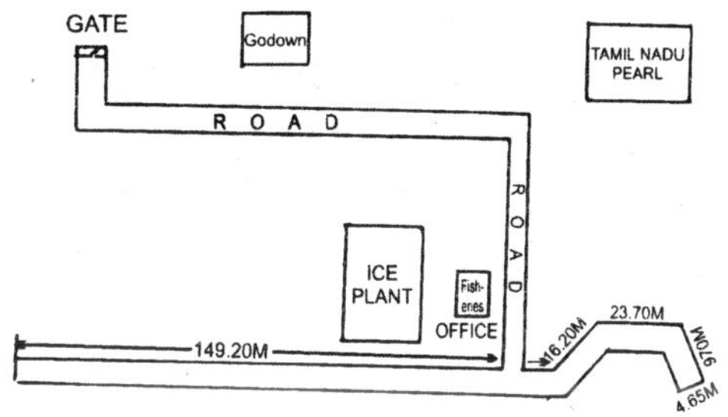


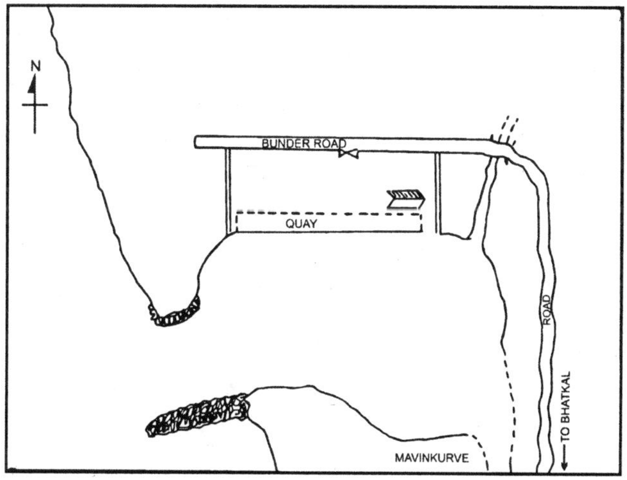


Fig. 38.6 : The lay out of the Bhatkal Fisheries Harbour (major)

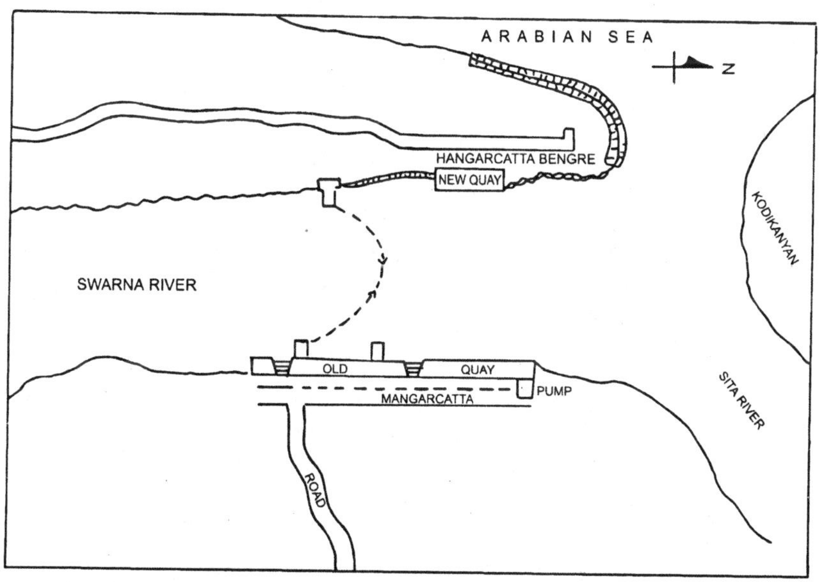


Fig. 38.7 : The lay out of the Hangarcatta Fisheries Harbour (major)

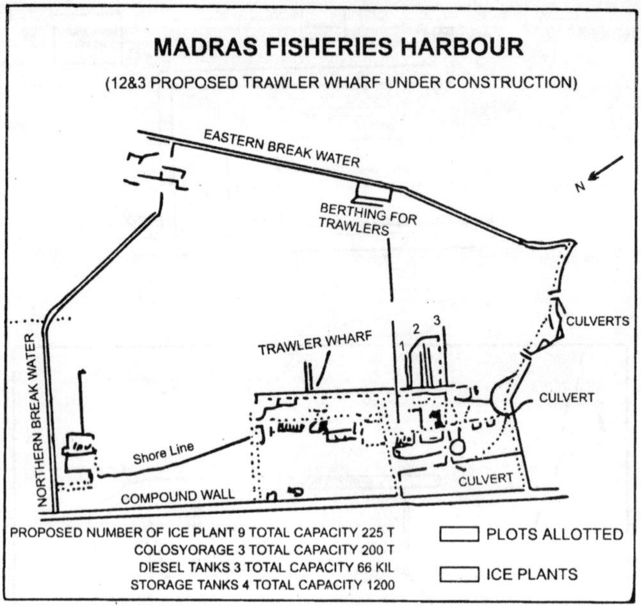


**Fig. 38.8 : The lay out of the Madras Fisheries Harbour**

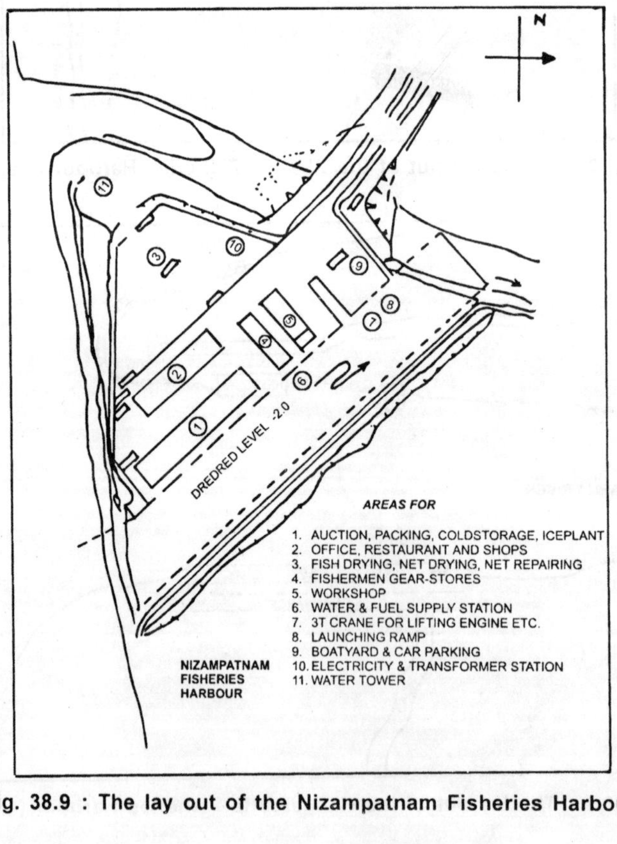


**Fig. 38.9 : The lay out of the Nizampatnam Fisheries Harbour**

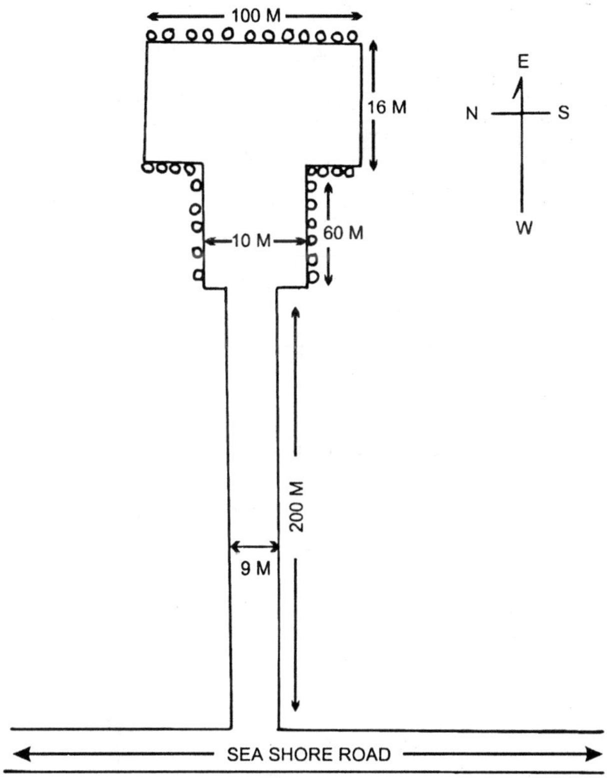


Fig. 38.10 : Lay out of the Thondi Fisheries Harbour

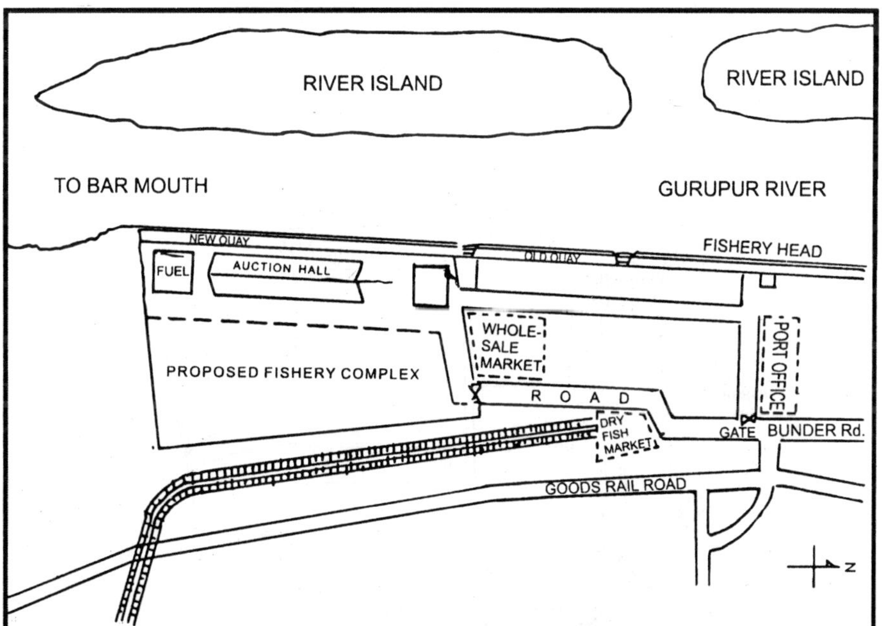


Fig. 38.11 : The layout of the Mangalore Fisheries Harbour (major)

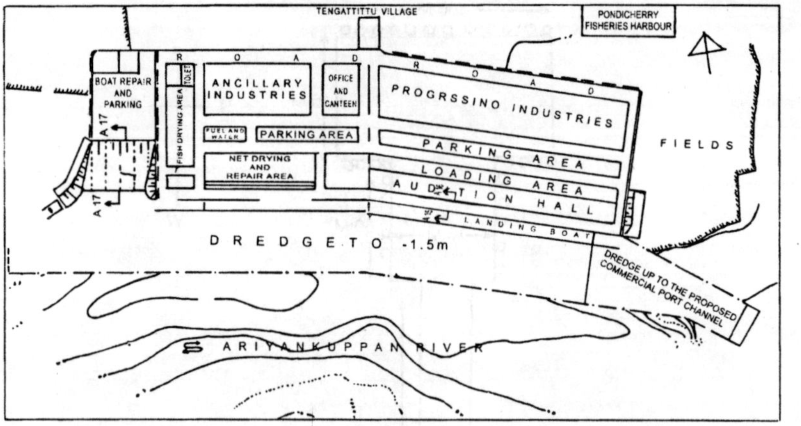


**Fig. 38.12 : The lay out of the Pondicherry Fisheries Harbour**

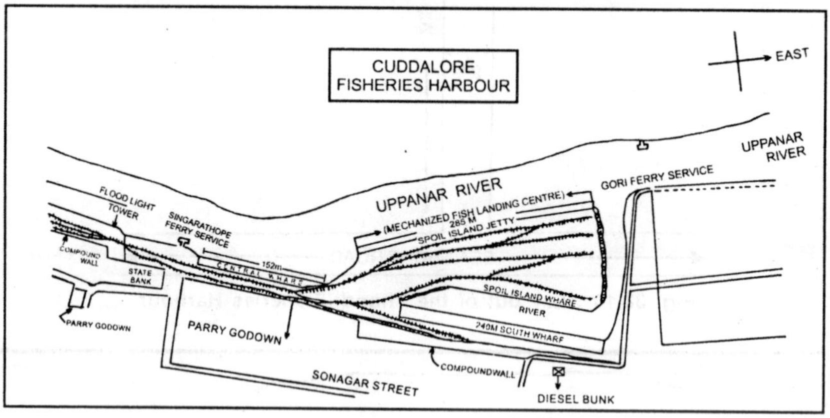


**Fig. 38.13 : The lay out of the Cuddalore Fisheries Harbour**

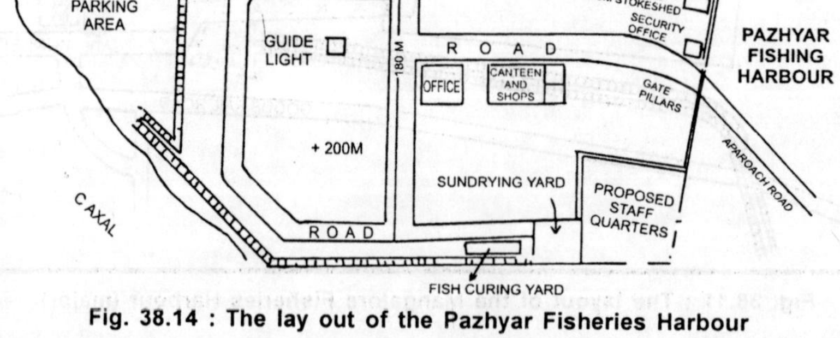


**Fig. 38.14 : The lay out of the Pazhyar Fisheries Harbour**

## *References*

Anon, 1985. Mangrol cleans up fish handling. *Fishing News International,* Published from London Aug. 78-79

Anon, 1886. Requirement for increasing fishing facilities, *Fish Farming International,* Published from London, 25(5): 1

Dholakia, A.D. 1986. Manual of Fisheries Terminal Division of Gujarat. Department of Fisheries, Govt. of Gujarat.

Dholakia, A.D. and G.D. Vasavda 1987. Mangrol Fishing port adopts clear and Hygienic standards. *Fishing Chimes,* Jan: 75-79.

Ganapthy, A. 1998. Fisheries Harbours along the Palk Bay Coast. *Mar. Fish. Tech. Serv. T & E Ser,* 153 : 16-17.

Krishnamurthy, V.S. K.M.S. Ameer Hamsa, M.V. Somaraju, C.V. Seshagiri Rao, M. Chandrasekhar, K. Narayana Rao., R.V.D. Prabhakar, P.V. Krishina, Rao, Sukhdev Bar., Sapan Kumar Ghosh, Pullin Behri Day, Bijoy Krishna and Swapan Kumar Kar. 1998. Fisheries Harbours along the Coasts of North Andhra Pradesh, Orissa and West Bengal. *Mar. Fish. Tech. Serv.* T & E *Ser,* 153: 27-31

Marichamy, R.D.B. James, H.M. Kasim, M. Rajamani, V.S. Rangaswamy, K.M.S. Ameer Hamsa, T.S. Balasubramanian, M. Manickaraja, S. Rajapachim and G. Arumugam. 1998. The Fisheries Harbours and Fishery at Tuticorin. *Mar. Fish. Tech. Serv.* T & E *Ser,* 153: 17-22

Sethuraman, V., 1998. Jetty for Fishing Boats at Mandapam along Gulf of Mannar coast. *Mar. Fish. Tech. Serv.* T & E Ser, 153: 15-16

Singh, V.V. and Kuber Vidyasagar, 1998. Fisheries Harbours in Maharashtra. *Mar. Fish. Tech. Serv.* T & E Ser, 153: 1-6

Sivadas, M., 1998. Fisheries jetty at Minicoy Lakshadweep. *Mar. Fish. Tech. Sev.* T & E *Ser,* 153.

Sunil Kumar, Mohamad K., P.U. Zacharia, Pratchicha Rohit, S. Kemparaju, C.Purandhra and Lingappa. 1998 The Fisheries Harbours and Fisheries in South Karnataka. *Mar. Fish. Tech. Serv.* T & E *Ser,* 153: 8-15.

Vasavda, G.D. and A.D. Dholakia 1988. Development of Veraval Fisheries Terminal, *Fishing Chimes,* 8(3): 15.

Vivekanandan, E., R. Paul Raj, KVS. Seshagiri Rao, G.C. Lakshmiah, L. Chidambaram, A. Srinivasan and M. Radhakrishanan 1998. The Fisheries Harbours and Fishery along the North Tamil Nadu and South Andhra Pradesh Coast. *Mar. Fish. Tech. Serv.* T & E *Ser,* 153: 23-26.

# 39
# FISHERIES MANAGEMENT

Management of natural resources conventionally pertains to the activities of conservation, regulation and allocation. Three aspects of management relate only to the modulation of the rates of flow of a resource from the ecosystem in which it exists. There is fourth dimension of management that is to highlight nurturing of the resources. Nurturing of a natural resource pertains to the concerns for the health of its stock in the live context of the ecosystem in which it exists.

There is also an increasing acceptance as well as a greater appreciation about the close interaction between successful regimes of property rights pertaining to them. Other factors like technology, markets and database can be define in property rights. If property rights are not defined properly lead to breakdown its management.

While the general principles of property right regimes are applicable across contexts, the specific details of the resources as well as the related human context are critical to success in particular applications. Property right regimes co-evolve with human activity, but not necessarily in any set pattern, no single type of property rights regimes can be prescribed as a remedy for problems of natural resources use and management. (John Kurian, 1998)

As per the definition given by World Humanity Action Trust (WHAT) in their report (WHAT-2000)"Fishing for the Future", The systematic framework of social economic, legal and political structures within which humanity chooses and/or accepts to manage its affairs is known as Governance the system.

Governance of the vast and diverse seas and oceans must necessarily be nested-multi-tired and overlapping. The regional, national and local levels attain their own specific significance depending on the ecosystem service, which people intend to use form aquatic terrain. Harvesting fish needs a multi-layered approach from the local to global. (John Kurian, 2001)

Considering all these aspects, different authorities and authors like Ministry of Agriculture, New Delhi (1993), NIO, (1995), Deveraj M. *et. al.*, (1998), Keilekar, (1999), Dholakia and Desai (2003) in India has made suggestion for the management of marine resources of India and Gujarat.

The management in modern economics, promoting the growth of the marine fisheries sector, is basically the responsibility of the Fisheries Department and Fisheries Corporations of the maritime States, with assistance from and through the fisheries co-operatives, fishermen's associations, processed traders', exporters and various other service organizations. The man power required for the development and managerial services is met largely by the Fisheries Colleges, Universities and various Central and State training organizations.

The pattern of marine fish landing in India during the past fifty years clearly reveals that the contribution by the artisanal sector to the total production was significant up to sixties. As a result of popularization and consequent expansion of mechanized fishing during subsequent periods along with the motorization of artisanal craft, the contribution by the artisanal sector declined considerably. The contribution by the mechanized and motorized sector accounts for 87% of the total catch while the artiosanal sector accounts for only 13% (Anon. 1996).

The marine fisheries scenario of the country warrants adequate supports for the following:

1. Effective implementation of the regulatory measures for the optimum exploitation of the resources in the inshore waters.
2. Continental Shelf of Gujarat is very vide, it should be exploited optimally.
3. Effective prevention of third country fishing in our EEZ water.
4. Establishment of artificial roofs for increasing productivity and production in shore waters.
5. Improvements in the system of data generation pertaining to the exploited stocks.
6. Protection and conservation of biodiversity.
7. Improvement of domestic and export marketing through value addition.
8. Man power generation in Research and development. Fishing and Fish processing and Fish Culture.
9. Development of marine fisheries information system.
10. Development of infrastructure with particular emphasis on fishing harbours and berthing facilities.
11. Popularization of sea farming technologies.
12. Establishment of sea farmers development agencies.
13. Improvement of wild stock through sea ranching programs

(*Source* : Deveraj, M. *et. al.*, 1998)

Fisheries does not only provide valuable and healthy food to the humanity, but also provides employment, recreation and business opportunities to the people. Sea food Industry had been witnessing gradual dwindling of resources since couple of years regularly. It is, therefore, necessary to study different aspects of fisheries management, actions to be taken by the Govt., Industries and fishermen.

Some of the critical attributes to today's raw material scarcity which call for immediate action may be recognized as follows:

### *Stock depletion & Reclamation of land*

Most of the crustaceans and fish breed in shallow, low saline waters. Brackish waters lagoons and estuaries with less saline waters and ample food supply in form of planktons along with safer environment serve as best nurseries for fish/larval. These juveniles stay and grow in these waters for a time to reaches a particular size, may be fingerling stage or beyond before they return to sea for further growth and life cycle. Due to uncontrolled and haphazard industrial developments/reclamation of land for construction etc., has destroyed these eco-systems so the breeding grounds and the nurseries. Therefore, most of fish larvae perish as they do not get suitable environment for further growth. Hence, no fresh recruitment, causing stock depletion.

### *Over Exploitation*

Uncontrolled fishing on sizes and seasons, fishing of brooders and juveniles are major attributes to depleting stocks. In fact attention of the fishery scientists and researchers is required to work out breeding patterns for all commercial species and fishing holidays should be enforced.

Fishing of fish fingerlings/juveniles of other soecies should be completely prohibited. Fishing juveniles results not only in weight loss but also loss in spawners. As they are not allowed to grow to reach maturity.

### *Pollution*

Since last decade country has witnessed quantum jump in industrialisation. Industries where on one side have brought in spurt and plenties in consumer market, have considerably increased pollution by either direct by-products, effluents, industrial wastes and discharges or by either associated pollutions. Treated and untreated Industrial pollutants along with pollutions from human resources are willy-nilly discharged to the sea through rivers, rivulets, canals etc. These pollutions destroy natural habitats and habitants of eco-system. It may also render the fish unsuitable for human consumption. Build up in microflora in putrefying organic waste may also lead to various diseases in aquatic flora and fauna.

### *Conservation and Rehabilitation*

To sustain fishery, time has come to think about rehabilitating species under strain. Spawners and the juveniles should be saved. The fishermen should be educated about conservations on their own lives. Govt. should also take appropriate measures to establish hatcheries all along coast to bread strained species and natural stocking of young ones.

## Key point in Management of Marine Protected Areas of India

The IUCN has prepared guidelines for Marine Protected Areas (MPAs) after summarizing the global scenario of management, which are also used in this case. Establishing the MPAs, preparing the plans and managing them effectively in India should be based on principles which are contained in following relevant points (Kelleher, 1999).

1. The MPAs in the country contribute to the maintenance and conservation of marine biological diversity and abundance, which are relevant to sustainable fisheries and maintenance of the coastal processes.
2. It is not feasible in many of the MPAs to divorce the questions of resources use for exclusive preservation, because, marine natural resources and their living space are all sought now by many different users for many different purposes.
3. The tendency in some areas to oppose the traditional use for fisheries in the MPAs seems to be counter productive, inhabiting co-operation between traditional users and Forest Department in creating and managing MPAs.

4. There has been a conflict and lack of co-operation between Forest Department and Fisheries management agencies in almost all areas of the country. This lack of joint action inhibits progress in establishing MPAs and managing them, wherever it is manifest. Management plans should be designed to serve both sustainable use and conservation objectives and relevant agencies should work together in planning and management in transparent manner.
5. Local people and stakeholder must be involved from the earliest possible stage in any MPA that is to succeed. Co-operative action and effective involvement of coastal communities depend on proper understanding among partners. This involvement should extend to them receiving clearly identifiable benefits from the MPA.
6. Socio-economic considerations usually determine the success or failure of MPAs. In addition to biophysical factors these considerations should be addressed from the outset in identifying sites for MPAs, and in selecting and managing them.
7. An MPA must have clearly defined objectives against, which its performance is regularly checked, and a monitoring programme to assess management effectiveness. Management should be adaptive, meaning that it is periodically reviewed and revised as directed by the results of monitoring.
8. There is a need of appropriate legislation, which provides scope of managing an MPA on the basis of the principle and point of management discussed above.
9. The high degree of linkage between land and marine eco-systems, and their connectivity of the sea regime that MPAs be integrated into management regimes that deal with all human activities that effect marine life. Thus, management of MPAs should be integrated with other policies of coastal land use and use of the sea Especially gulfs and bays.
10. The management plan of MPA should operate in conjunction with other policies and laws related to the coastal zone. Management plan of the MPA has to be different from the plans of the protected areas in terrestrial ecosystems, where Forest Department is the single owner of the land. In this case Department of Fisheries, State Maritime Boards, Port Authorities and other stakeholders have equal user ship rights in the area. (H.P. Singh. 2002)

### *Suggestions for exploitation of resources are as follows :*

Some recommendations for further development of marine resources with reference to Gujarat.

1. Molluscan resources like Edible, Pearl Oyster, Chanks, Window Pane Oyster etc., were once in abundance but are in increasingly reducing state of existence. Serious efforts must be made to transplant juveniles and spats in abundance into the Gulf of Kutch and other suitable areas till the time they stabilize can be reproduced in new habitats till the population can be regenerated at sustainable harvestable levels.
2. Much of the area beyond 50 fathoms including EEZ area is open to resources and commercial survey for various species. We know very little of it. Fishery Survey of India (FSI) has mainly concentrated on the west coast up to Bombay and very little beyond Veraval. Even the Indo-Polish Survey is very schay above Veraval. Since we have come up to Maximum Sustainable Yield (MSY) levels up to 50 fathoms levels. We will have to find other areas of commercial exploitation.

3. Except for a few concentrated centers like Veraval, Mangrol, and Porbandar, there is very little infrastructure facilities available along with coastline of Gujarat. Without this there can not be quality control and/or value addition and there can be no proper development without value addition.
4. There is a urgent need for the State and national policy for effective regulatory measures to maintan sub optimal exploitation of marine fisheries resources between 0 to 50 fathom zone. Over exploitation in any form has to be curtailed and stopped. An effective implementing authority to administer these measures has to be created.
5. Sudden and fast industrial growth along the coastline together with urbanization has created serious pollution and environmental hazards. Gujarat has opted for chemical and hydrocarbon Industries in major way and string of thermal power stations are cropping up along the coast. All these have created serious problems for fishery sector. The night marish week fish catch during last two years resulting in a sever economic depression in fisheries and ancillary industries should be eye opener. Standards of pollution control have to be maintained and effective monitoring system has to developed and stringent punitive actives have to be taken by concern agencies or else the coastal marine fisheries of Gujarat will slowly but surely wither out.
6. Mangroves and corals create and effective ec-osystem for fish and other bio-resources reproduction and nurturing future population. Both mangroves and corals are being destroyed. Inspite of coastal regulation zones this has to stop and a big effective mangrove a forest station programme has to be implemented urgently.
7. A big sector like fishery and related subject requires a very strong Human Resource Development Programme in form of Under Graduate, Post Gradaute and Doctorate level studies as well as short term and medium term job oriented vocational training. At present there is no Post Graduate or Doctorate programme in this field professionally. What ever exists at Graduate and vocational level needs to be tightened up and strengthened urgently.
8. Coastal and other types fish and prawn culture programmes if they are stagnant this area needs to be developed and strengthened.
9. Gujarat lacks in proper data base which is prerequisite for any thriving industry. What ever is available is primitive and patchy. Strong informative system on fisheries, marine ecology biodiversity environment etc., has to be developed and made available at push button level.
10. The Department of Fisheries has to be truly technical. At present it mostly deals with files. All district officers and possibly taluka offices should have their own laboratory and field equipment to respond to the local level Technical and Scientific requirements. It should be made mandatory for all technical staff of the Govt. in the Department to be able to respond intelligently when needed. Because all of them have received training of one type or other for short or long period, inside and outside India. The Agricultural University should also fulfill its responsibility similarly. If the officers, academicians and scientist in this field are not able to cater to the requirements of fishermen and the industry there is no purpose in marinating such vast organizations. (Anon. 1995) and (Dholakia and Desai, 2003).

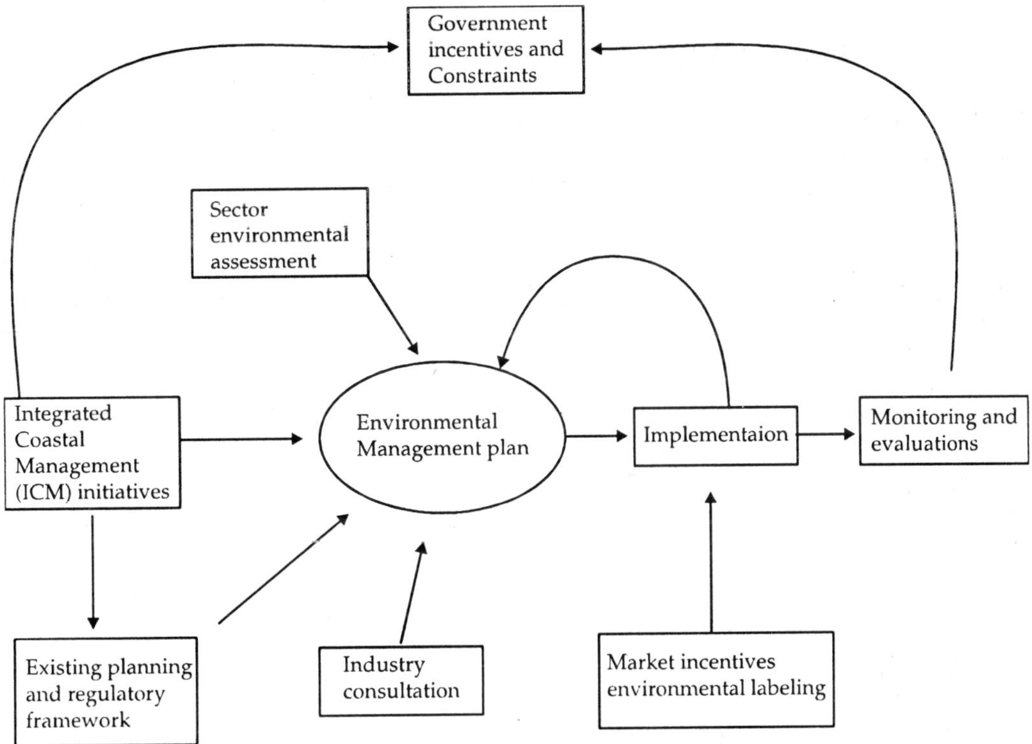


**Fig. 39.1 : Environmental Management Plan**

## *Suggestion for development of High Sea Fisheries*

If following information is collected and data bank is created, it will help in deciding developmetnal steps to be taken for High Sea Fishery.

1. Surface temperature charts on a monthly basis particularly from 20° S. latitude to the continent and island chain on the north.
2. Variation in thermo cline depth with monsoon change particularly within 200 miles of land in the Andaman Sea, Bay of Bengal, Arabian sea.
3. Variation in direction and flow of the mixed layer currents in the Indian Ocean, particularly north of 20° S. latitude and in particular as related to variation in monsoon period and strength.
4. Variation in basic biological productivity rate in the area north of 20° S. latitude and particularly within 200 miles of shore.
5. Relationship of variations in these parameters to variations in the season and yield of the high seas fisheries of the areas. (Anon 1993)

## Environmental Management of Aquaculture Development

The social and environmental impact of coastal aquaculture has been widely reported. It is generally agreed that aquaculture development needs to be better planned and managed if it is to achieve its undoubted potential and develop a sustainable manner. Many countries already attempt to manage the aquaculture sector through some form of regulation like Environment Impact Assessment (EIA), which allows for a more through appraised of social and environmental problems, related to the sitting, design and operation of individual farms (Dholakia, 1994).

In Practice, in most suitable – and especially where aquaculture is developing rapidly– completely voluntary approaches are likely to be inadequate. Key elements for improved environmental management of the Aquaculture sector is given in Fig. 39.1.

Integrated Coastal Management (ICM) has been widely promoted as an appropriate framework for the assessment and management of aquaculture and other activities in the coastal zone. Under ICM, aquaculture is just one of many activities which must be planned and managed with a view to meeting some broader coastal development or environmental objectives. (John Hambrey, 2000).

## Under Sector Environmental assessment following points are covered

Agreed environmental quality standards against which assessment of impact significance should be measured.

Description and comparative evaluation of actual and potential coastal aquaculture systems in the region should be done.

Discussion of the relationship between possible land/water use for coastal aquaculture and existing land/water use policies may be decided.

Description of the conditions and locations in which coastal aquaculture development might take place.

Discussion of environmental capacity for these locations.

Comparative evaluation of aquaculture development alternatives, covering significant adverse and beneficial impact, and migration possibilities for different technical options.

An environmental management plans for the sector, as a whole should be prepared.

A monitoring plan and response procedures should be drawn.

## *References*

Anon, 1979. A compendium on Indian Fisheries and Seafood Export Industry MPEDA on third Indian Sea Food Trade Fair, 1-11 Feb p. 8

Anon, 1993. Techno Economic Survey of Gujarat, Ministry of Agriculture, New Delhi.

Anon, 1995. Marine environment impact assessment of propsoed refinery at Moti Khavdi, N.I.O. Report.

Anon, 1996. CMFRI Special publication No 38; 47.

Anon, 1998. Report of the working group on revalidation of the potential Marine Fisheries Resources of EEZ. Government of India.

Chhaya, N.D. 1997. Minding our Marine Wealth-An Appraisal of Gujarat Coastal Resources Center for Environment Education Ahmedabad.

Chhaya, N.D. 1999. Observation on Environmental impact - Assessment of coastal Marin Eco-system in Kutch District. P. 37, Forum for Planed Industrialization of Kutch.

Devraj, M., S. Srirama Chandra Murty, R. Santhiadhas and K.K. Joshi 1998. *Fishing Chimes*, 18(8) : 18-29.

Dholakia, A.D. 1994. Ph.D. Thesis "Marine Prawn fisheries and its culture in Saurashtra with special reference to *Penaeus merguiensis* (deMan)." Saurashtra University, Rajkot.

Dholakia, A.D. 1998. Annual Report 1998-99. Fisheries Research Station, GAU, Sikka.

Dholakia, A.D. and A.Y. Desai 2003. Marine Resource of Gujarat and its management. Seminar on Fishing and Fish Processing Industries of Gujarat. Present status and Future Needs. Organized By SOFR(T), CIFTand EIC, Govt. of India, on 7-March 2003.

Jhingran, V.G. 1982. Fish and Fisheries of India. P. 563.

John Kurian, 1998. Property rights, Resource Management and Governance. Center for Development Studies South Indian Federation of Fisherman Societies, Thiruvananthapuram. pp. 1-6

John Kurian, 2001. People and the Sea-A "Tropical-majority" SISWO and Center for Maritime Research : p. 23.

John Hambrey 2000. Environmental management of aquaculture development INFOFISH International 5/2000 : 25-29.

Kellekar, 1999. Guidelines for Marine protected Areas. Best Practice protected Area Guidelines series No 3. World Commission on Protected Area (WCPA).

Sebastean Mathew 2000. Gujarat Fisheries Time to move from exploitation to conservation and management regime. Workshop on Current Situation in Fisheries Sector in Gujarat. Gujarat Institute of Development Research Gota, Ahmedabad and International Collective in Support of Fish workers.

Singh, H.S. 1994. Management plan for Marine National Park, Conservator of Forest, Marine National Park. Jamnagar

Singh, H.S. 2002. Marine Protected Education & Research (Geer) Foundation. Indroda National Park, Gandhinagar, pp. 46-47.

Unnikrishnan, N. and Jacob Thomas 1997. Resour, Infor, Ser. Vol 3 (4).

# 40

# Remote Sensing in Marine Fisheries

## Introduction

Marine resources development, specifically, Fisheries development is one of the major areas demanding immediate attention. In this fieldwork carried out in other countries have shown that remote sensing can be successfully used in mapping and monitoring of ocean features like Thermal front, eddies, upwelling, concentration of sediments and biomass. For locating probable areas in the ocean having fish schools such information is very useful (Bhavsar, Praful. D. 1985)

In India, use of remote sensing technique for fishery application were initiated in mid-seventies under FAO/UNDP Pelagic Project. Under the Remote Sensing Applications Mission launched by ISRO (Indian Space Research Organization), it has been operationally demonstrated that satellite remote sensing has a key role to play in exploration of marine resources. The observation on eddies, upwelling, temperature, ocean circulation, salinity etc., provides more accurate forecast. Details of each parameter are as under.

## Eddies

The eddy is an area of closed horizontal streamlines where vertical motion is included or sustained. Fronts and eddies are the two major classes of water motion varying in time and space with the smallest of these beign the largest ones as the mostly vigorous and highly localized within eddies; it is slow and spread over larger area. Therefore, the gradients of temperature, salinity and accumulation of marine organisms are sharper across the fronts than eddies.

Warm/cold core eddies and persistence are to be considered as potential fishing grounds. Eddies are current of waters often seen on the sides of a main current; specially the one moving in a circle. Eddies transport phytoplankton and zooplankton to take place where they would not normally occur. Fig. 40.4 shows clear occurrence of eddy and its persistence in the Gulf of Khambhat.

## Upwelling

Upwelling is an ascending motion of some minimum duration and extent by which water from surface layer and is removed from the area of upwelling by horizontal flow. Upwelling generally occurs along the ocean's eastern boundaries where prevailing winds carry the surface waters away from the coast. Upwelling alters the distribution of physical and chemical properties of waters involved. Upwelling centers are characterized by cool, low phytoplankton, more saline and dense water containing less dissolved oxygen and high nutrients than usually found at a given depth.

It has been generally reported that due to the regular occurrence of upwelling the Arabian Sea is nutrient-rich with low temperature, high salinity, higher primary productivity and consequently yields high fish catch compared to the Bay of Bengal.

It has been recorded that the upwelling occurs during October-November off Bombay. Location of upwelling areas and studying its dynamics in terms of initiation, maturation and stabilization is useful in guiding the fishermen for profitable fishing. For example with the onset of upwelling, favourable winds, cool nutrient rich but phytoplanton poor water would get established in a area representing initation. Favourable grounds will get formed with the sea water movement of upwelled waters coupled with "Biological ageing" of the surface water largely due to the growth of phytoplankton. Based on monitoring the progress of upwelling the steps in guiding the fishermen would be in terms of avoiding areas showing cool waters due to low forage and operate in water showing the maturation and stabilization for getting good catch.

Fig. 40.2 shows persistent upwelling off Saurashtra-Gujarat. Its implication on fisheries was observed. A band of cool waters between Okha and Diu (~230 km) was observed within a 40 mm depth during September-October 1991. Surface temperature of ~22 C in upwelled areas as against 23-26 C in adjacent waters was observed during above study period. Feedback from the Department of Fisheries, Government of Gujarat, indicated a high catch outside upwelled waters (between 40 m - 70 m). The reason could be the seaward movement of upwelled waters, nutrient uptake, increased biomass and higher temperature. Phytoplankton pigment is shown to play a dominant role in the surface heating.

## Thermal Fronts

Thermal fronts are sharp boundaries between two water masses separated by temperature gradient. They are known to be potential zones for the accumulation of phytoplankton.

Thermal characteristics of surface waters can be used as natural tracers of dynamic patterns. Sea surface temperature is the most easily observed environmental parameter and therefore, it has been frequently used in correlation with fish availability, on an operational base, both conventionally as well as through remote sensing. It serves as a useful indicator of prevailing and changing environmental conditions. This is one of the parameter which decides suitable environmental conditions for fish aggregation. Tendency of fish to accumulate in clear water on warmer side of the front in near shore water can be logically explained through their feeding behaviour. It was observed that aggregation of fish may occur in the vicinity of temperature gradient or front in the "Water pockets" close to it. This means all fronts observed in the processed image need not necessarily be indicative of fish availability. They just indicate relatively higher probability of finding fish. Effectiveness of front or temperature gradient for fishing location depends on narrowness of the temperature zones, sharpness or stability of boundary in space and time, availability of food etc.

Fishes are generally not found in abundance in features waters with uniform temperature. These sites should be avoided. The seasonal variation in CPUE and temperature do not indicate any definite relationship.

Thermal fronts are sharp boundaries between two water masses separated by temperature gradient. They are known to be potential zones for the accumulation of phytoplankton because of its richness in nutrient availability, hence a potential fishing ground.

The geographical position of India with peninsular portion extending deep in to the central part of the Indian Ocean gives her a locational advantages in marine fishing activities. At present through India contributes about 40% of the fish landing of the Indian Ocean when viewed against the world production, her share representing less than 2%. It has already been indicated that with an annual net organic production for about two-third area of the Indian Ocean, the relative productivity of this area is in no way less than that of the rest of the world ocean.

Remote sensing of Sea Surface Temperature (SST) from multi-spectral infrared satellite observations offer movement of Tunas and other pelagic fishes which are largely depending on the variations in SST. Based on the surface isotherms, 'thermal equator' can be identified and the movements of pelagic fishes can be studied on tropical area, localized differences in the SST also may help to locate areas of current boundaries, upwelling etc. High surface temperature gradients where the optimum temperature zones are narrow are the places of concentration of albacore and southern blue-fin tuna. It has recently been indicated that the 15°C, 20°C and 23°C isotherms are the lower normal boundary of occurrence of albacore, skipjack and yellow fin tunas respectively. (Sailas *et. al.*, 1985).

## *Location of Potential Fishing Grounds*

Location of potential fishing grounds involves selection of oceanic features from the processed satellite image and subsequently transferring the window enclosing the selected feature on to the respective georefereced map. It is also kept in mind to monitor the water masses in terms of three phase *i.e.*, initiation/enrichment (cool waters), maturation (relatively warm waters) and its stabilization (warm water). Phytoplankton pigment is the most dominant parameter in intercepting the short wavelength radiation in the oceanic waters. In view of this, moderately warm waters were considered as rich in forage and therefore, potential grounds.

Pockets of cold/warm water also considered as potential sites for fish accumulation.

Meanders are a turn or winding of current that may detach from a main stream. Meanders are considered to be one of the best potential grounds for fishing because they cover a large production area.

By and large, warm waters are seen between Dwarka and Diu in ~45 m depth and cool water within 30 m. Cool waters were also seen in the Gulf of Khambhat and areas off Jakhaw. From the CPUE a pelagic resource like ribbon fish were more in warm waters off Dwarka, Mangrol and Diu, and low catch in relatively cool waters off Veraval.

## Role of Remote Sensing in Assessing Pelagic Fishery Resources

Remote sensing of marine fishery resources is comparatively recent development in India. Though the nature and magnitude of some of the pelagic fisheries is partially known, there are some constraints in direct assessment of the resources mainly due to mobility of the resources and greater time involved in collection of resources data. Moreover, the exploitated pelagic fisheries are mostly from traditional/artisnal sector and large scale mechanization and modernization of the fleet for exploitation of resources in

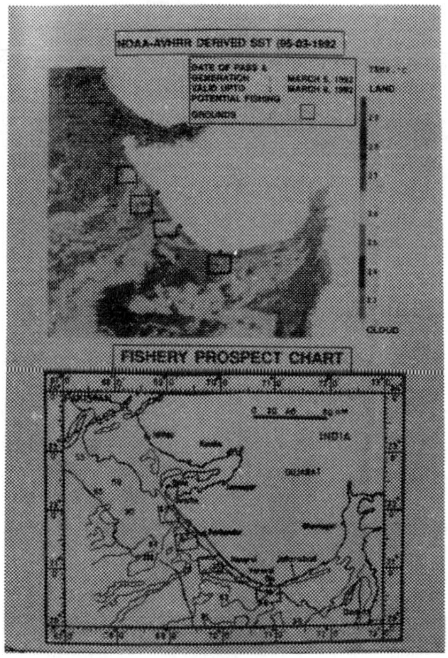


*Source:* Head, Marine Water Resources SAC, Ahmedabad.

**Fig. 40.1 : Fishery prospect chart of Gujarat**

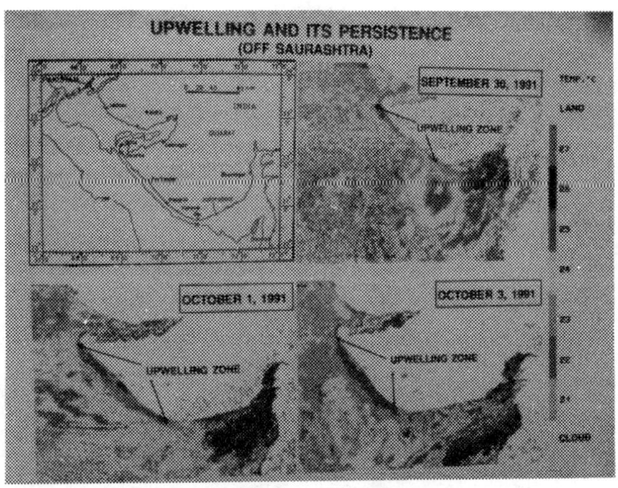


*Source:* Marine Water Resources SAC, Ahemedabad.

**Fig. 40.2 : Upwelling and its persistence off Saurashtra.**

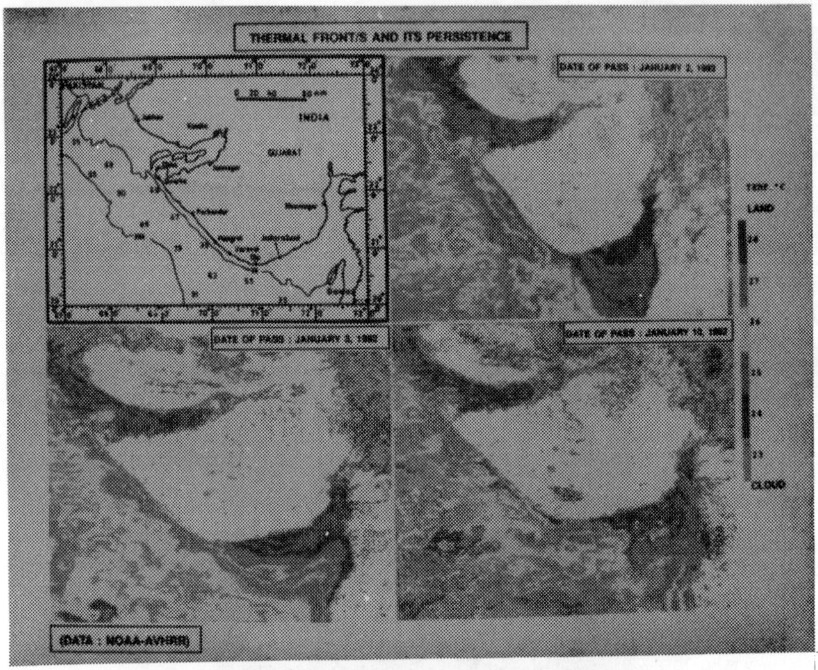


*Source:* Head, Marine Water Resources SAC, Ahmedabad.

**Fig. 40.3 : Thermal feature and its persistence off Saurashtra**

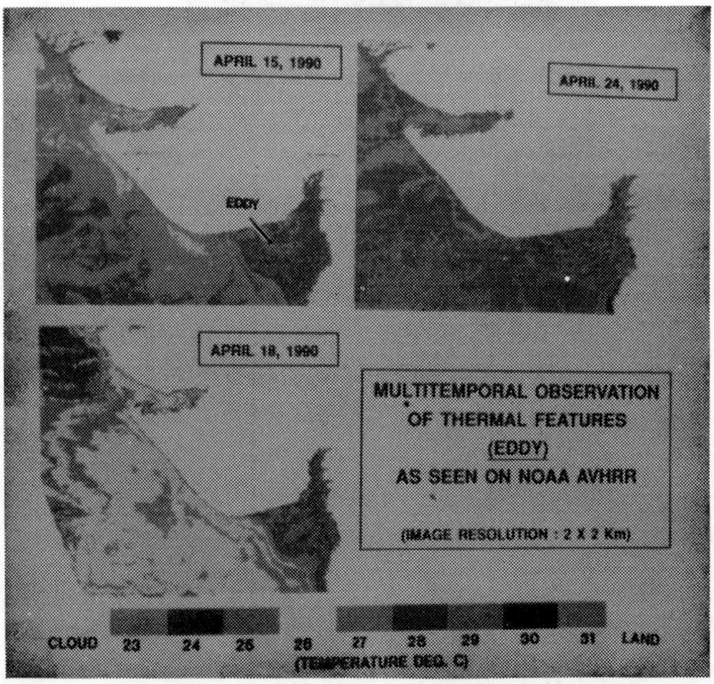


*Source:* Head, Marine Water Resources SAC, Ahmedabad.

**Fig. 40.4 : Eddy observation off Gujarat coast**

*Source:* Head, Marine Water Resources SAC, Ahmedabad.

**Fig. 40.5 : Thermal front of Gujarat and Indian coast**

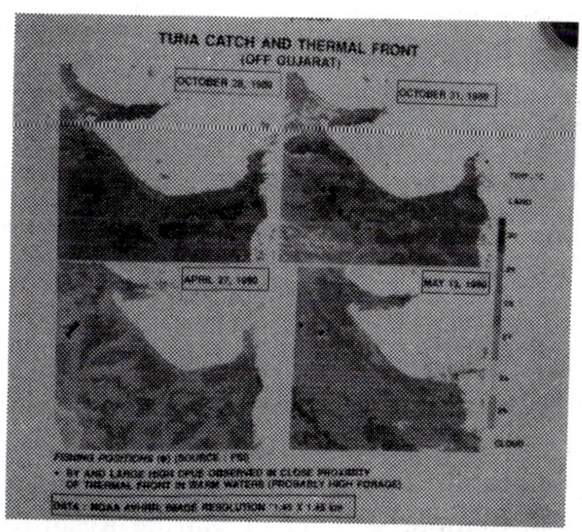


*Source:* Head, Marine Water Resources SAC, Ahmedabad.

**Fig. 40.6 : Tuna catch and thermal front sea surface temperature off Gujarat.**

Exclusive Economic Zone (EEZ) will require long lead time. The remote sensing techniques may therefore, play vital role in management of the marine fish resources (Joseph and Sanvanshi, 1985).

## Ocean Colour Mapping

Phytoplankton pigments in the ocean waters are the prime synthesizers in marine food chain, which in turn terminate as pelagic or benthic nekton. They perform about half of the total global photosynthesis and comprise the upper consumption and production. There are substances that have definite spectral characteristics and thereby govern the ocean colour. Though its correlation with various biological and physiological processes, the quantification of chlorophyll concentration in the ocean waters provide information, which is indirectly relevant to the management of fisheries. With the help of sea truth data, upwelling spectral radiance can be measured at several depths and can be calculated radiance rations at different wavelengths. The resulting values is being used to develope an algorithm for estimating diffuse attenuation co–efficient, K. (Beena Kumari, *et al.*, 1985).

Remote sensing from airborne sensors has been used for studying the well known pelagic schools of oil sardine and mackeral in Arabian sea of the Indian Ocean. A high productivity is reported in the oceanic waters off Chochin. Fish schools could be directly spotted equally on special; films like convential colour (Kodak 2448), Colour infrared (Kodak 2443) and Panchromatic black-and-white (Kodak 2402). An indirect method of mapping spectral distribution of phytoplankton pigment may in turn help in estimating the fish potential through an understanding of its conversion to other tropic states in the ocean food chain. (Narain *et. al.*, 1985).

## Selection of Brackish Water Aquaculture using GIS

Based on parameters were assigned weight ages according to their relative importance and sensitivity within the individual parameters. The engineering parameters of sites viz., topography, elevation with respect to tidal amplitudes, brackish water supply and fresh water supply were assigned maximum weight age as 4, 6, 5, 4 per cent respectively. The ecological parameters such as water salinity, water pH impact of industries and agriculture on water quality, water temperature, water turbidity, soil type, soil pH, seeds availability were assigned maximum weight ages as 4, 4, 5, 3, 3, 4, 4, 4 per cent respectively. Similarly other important parameters such as population (demographic), road net work, availability of palletized feed (infrastructure), annual rainfall (meteorological) etc., were assigned maximum weight ages as 5, 4, 4 and 3 per cent respectively.

Based on weight ages Vansi-Borsi, Onjal, Umargaon in South Gujarat, five sites in West Bengal, five sites in Orissa and four sites in Andhra Pradesh were selected using Geographic Information System (GIS). (Gupta, and Nayak, 1995).

## *References*

Beena Kumari, R.M. Dwivedi, A. Narain, 1985. Sea truth data collection: Estimation of diffuse attenuation co-efficient in ocean colour mapping. Space Application Center, Ahmedabad.

Bhavsar Praful D. 1985. Remote Sensing in Marine Resources. Indian Space Research Organization and Indian Council of Agricultural Research, Ministry of Agriculture, New Delhi.

Gopalan, A.K.S. E.G. Silas 1985. Remote sensing in Marine Resources. Indian Space Research Organization, ICAR, Ahmedabad.

Gupta, M.C. S.R. Nayak, A. Sinha, Anjali Bahuguna 1995. Brackish water Aquaculture Site Selection using Techniques of Geographical Information System (GIS). Space Application Center, ISRO, Ahmedabad.

Joseph, K.M.and V.S. Somvanshi 1985. Marine Fisheries Resources Survey and Role of Satellite Remote Sensing in the assessment of pelagic Fishery Resources of India. Fishery Survey of India, Govt. of India, Bombay.

Narain, A., R.N. Jadhav, K.L. Majumdar, G.P. Sharma, 1985. Remote Sensing of Ocean colour and targeting of fish schools from airborne sensors. Proceeding of Indian Space Organization and ICAR, New Delhi.

Sailas, E.G., P.V.R. Nair, P.P. Pillai, G. Subbaraja, V.K. Pillai, and V.K. Balachandran. 1985. Biological productivity of the Indian Ocean. C.M.F.R.I. Cochin.

# Index

## N



## O



## P

**R**